The Paleobiological Revolution

The Paleobiological Revolution

ESSAYS ON THE GROWTH OF
MODERN PALEONTOLOGY

Edited by
David Sepkoski and Michael Ruse

The University of Chicago Press CHICAGO & LONDON

The University of Chicago Press, Chicago 60637
The University of Chicago Press, Ltd., London
© 2009 by The University of Chicago
All rights reserved. Published 2009.
Paperback edition 2015
Printed in the United States of America

24 23 22 21 20 19 18 17 16 15   2 3 4 5 6

ISBN-13: 978-0-226-74861-0 (cloth)
ISBN-13: 978-0-226-27571-0 (paper)
ISBN-13: 978-0-226-74859-7 (e-book)
10.7208/chicago/9780226748597.001.0001

Library of Congress Cataloging-in-Publication Data
The paleobiological revolution : essays on the growth of modern
paleontology / edited by David Sepkoski and Michael Ruse.
        p.   cm.
Includes index.
ISBN-13: 978-0-226-74861-0 (cloth : alk. paper)
ISBN-10: 0-226-74861-8 (cloth: alk. paper)
1. Evolutionary paleobiology. 2. Paleobiology. 3. Paleontology.
I. Sepkoski, David, 1972– II. Ruse, Michael.
QE721.2.E85P347 2009
560—dc22
2008028286

♾ This paper meets the requirements of
ANSI/NISO z39.48-1992 (Permanence of Paper).

In Memory of

STEPHEN JAY GOULD

J. JOHN SEPKOSKI, JR.

THOMAS J. M. SCHOPF

CONTENTS

ACKNOWLEDGMENTS

Bringing this volume together from concept to completed product has been challenging and rewarding, and we would like to thank a few people without whom this book could not exist. First and most obviously, we are grateful to our many contributors, who have made this (we think!) an excellent and important collection of essays about a major transformation in recent paleontology. Many of the authors in this volume are scientists who contributed directly to the "paleobiological revolution," and we especially thank them for putting their thoughts and experiences on paper as a permanent record of the history they helped shape. We would also like to thank our editor at the University of Chicago Press, Christie Henry, whose unflagging enthusiasm and support for this project has been deeply appreciated. We received helpful guidance from several anonymous reviewers at different stages in the project that have undoubtedly made this a better book. Finally, we thank our families, as always, for their continual love and support.

Paleontology at the High Table

David Sepkoski and Michael Ruse

In 1984, Stephen Jay Gould's Tanner Lectures at Cambridge University presented an overview of the major advancements paleontology had made since he and Niles Eldredge first unveiled their infamous theory of punctuated equilibrium in 1972. Later that year, the English geneticist John Maynard Smith offered his own assessment of the recent contributions of paleontologists to evolutionary theory, which on the whole was quite positive. Beginning by lamenting the relative lack of evolutionary contributions from paleontologists from the 1940s onward, he explained that the tendency among even the more theoretically minded paleontologists had been merely "to show that the facts of palaeontology were consistent with the mechanisms of natural selection and geographical speciation . . . rather than to propose novel mechanisms." In response, "the attitude of population geneticists to any palaeontologist rash enough to offer a contribution to evolutionary theory has been to tell him to go away and find another fossil, and not to bother the grownups."[1] However, over the last ten years, he reports, this attitude has changed, thanks in large part to the work being done by theoretical paleontologists like Gould. His now-famous closing sentence (at least among paleontologists) reads, "the palaeontologists have too long been missing from the high table. Welcome back."[2]

Generally speaking, Maynard Smith's assessment of the contributions of paleontology to evolutionary theory between the formation of the modern evolutionary synthesis and the late 1970s is an accurate reflection of, at the very least, the general view of evolutionary biologists towards their fossil-collecting cousins. One might indeed go back much further, to the very establishment of the theory of evolution by Charles Darwin, to find the source of biologists' uncharitable attitude toward paleontology. Despite the fact that paleontological evidence played a vital role in demonstrating evolutionary succession, in what appears in retrospect to be a profound irony, even as Darwin elevated the significance of the evidentiary contribution of paleontologists, he also had a major hand in condemning paleontology to second-class disciplinary status. One of his greatest anxieties was that the incompleteness of the fossil record would be used to criticize his theory: that the apparent gaps in fossil succession could be cited as, at the very least, negative evidence for the gradual and insensibly graded evolution he proposed. (He also fretted at the complete absence of Precambrian fossils—a topic addressed by more than one of the chapters in this volume). At worst, he worried that the record's imperfections would be used to argue for the kind of spontaneous special creation of organic forms promoted by theologically oriented naturalists whose theories he hoped to obviate. His strategy in the *Origin*, then, was to scrupulously examine every possible vulnerability in his theory, and as a result he spent a great deal of space apologizing for the sorry state of the fossil record.

The metaphor Darwin chose in his apology for the fossil evidence was that of a great series of books from which individual pages had been lost and were likely unrecoverable. "I look at the natural geological record" he continued, "as a history of the world imperfectly kept, and written in a changing dialect; of this history we possess the last volume alone, relating only to two or three countries. Of this volume, only here and there a short chapter has been preserved; and of each page, only here and there a few lines."[3] Darwin's theory revolutionized paleontology, since the fossil record became the only potential source of evidence that evolution had occurred and for interpreting the history of organic change. Darwin's dilemma, however, was that he both needed paleontology and was embarrassed by it. Even though he celebrated the contributions of paleontologists, he simultaneously undercut any claims their emerging discipline might have for autonomy within evolutionary theory, and this attitude was internalized both by paleontologists and biologists over much of the subsequent history of evolutionary theory.

This view is in evidence as early as the late nineteenth century: in an 1889 review essay in *Nature*, an anonymous author (known only as "E. R. C.") put the attitude of many nonpaleontological observers succinctly: "The palaeontologist has been defined as a variety of naturalist who poses among geologists as one learned in zoology, and among zoologists as one learned in geology, whilst in reality his skill in both sciences is diminutive."[4] Aside from his rather negative opinion of paleontology (he later calls the study of fossils "a definite hobby"), this short essay is significant because it marks the first instance in the literature of the use of the term *neontology* to refer to the study of living organisms. Overall, though, "E. R. C." expressed a view that would become common among biologists, geologists, and even some paleontologists themselves: that subordinating paleontology to geology would provide "a better chance for the cultivation of true geology, which now, to some extent, has its professional positions, its museums, and its publications invaded by these specialists [i.e., by paleontologists]."[5] Clearly, there were more than intellectual issues at stake here. With limited resources available in academic and museum institutions for geology, it was natural for scientists in established departments to want to preserve what they viewed as the traditional core of their disciplines, which for geologists was the study of sediments, minerals, and stratigraphy.

A survey of statements regarding paleontology and its potential contributions by important biologists between 1900 and 1945 reveals the extent to which the discipline had sunk in the eyes of the larger evolutionary community. For example, T. H. Morgan, whose study of the genetics in populations of fruit flies was a landmark in twentieth-century biology, offered the following sneering evaluation of paleontology in 1916:

> Paleontologists have sometimes gone beyond this descriptive phase of the subject and have attempted to formulate the 'causes,' 'laws' and 'principles' that have led to the development of their series. . . . The geneticist says to the paleontologist, since you do not know, and from the nature of your case you can never know, whether your differences are due to one change or to a thousand, you can not with certainty tell us anything about hereditary units which have made the process of evolution possible. And without this knowledge there can be no understanding of the causes of evolution.[6]

A much more widely read echo of this attitude is found in Julian Huxley's *Evolution: The Modern Synthesis* (1942), which served as a kind of manifesto for the synthetic movement. Huxley opined that many paleontologists had been misled toward orthogenesis and Lamarckism because "the paleontologist, confronted with his continuous and long-range trends, is prone to misunderstand the implications of a discontinuous theory of change."[7] Paleontological data is inherently suspect because the fossil record is unreliable: it is incompletely preserved, and the material that *is* preserved is insufficient to inform theoretical conclusions. Or, as Huxley bluntly puts it, "paleontology is of such a nature that its data by themselves cannot throw any light on genetics or selection. . . . All that paleontology can do . . . is to assert that, as regards the type of organisms which it studies, the evolutionary methods suggested by geneticists and evolutionists shall not contradict its data."[8] In other words, paleontologists should be content with the role assigned them ever since Darwin—to document and verify historical confirmation of the processes biologists proposed guided evolution, but not to leap to independent conclusions.

As Maynard Smith's essay notes, even the great George Gaylord Simpson, the American paleontologist who, along with Ernst Mayr and Theodosius Dobzhansky, helped establish the modern synthesis in the 1940s, shied away from demanding complete theoretical equality for paleontologists. One of the broadest and most lasting contributions of his masterpiece, *Tempo and Mode in Evolution* (1944), was its suggestion that paleontology and the fossil record have something unique to say about macroevolution. However, when Simpson sat down to revise *Tempo and Mode* in the early 1950s, a major shift in his thinking appears to have taken place, which led him to significantly downplay his earlier assertion of the theoretical autonomy of paleontology. Exactly why he did this is the subject of some debate, but Gould and others have suggested that Simpson capitulated to pressure from biologists and geneticists to endorse, in Gould's words, "a more rigid selectionism" that favored the selection of small genetic changes—which take place on a scale nearly invisible in the fossil record—as the primary mechanisms of evolution.[9]

From one perspective, it was a legitimate triumph for paleontologists that their discipline was recognized so prominently in the institutionalization of the synthetic theory. Without question, this was largely due to Simpson's efforts, which were, undeniably, heroic. It would be a mistake, however, to conclude that paleontology was, in 1944 or afterward, a fully equal and respected partner in the community of neo-Darwinian evolutionary biology. Paleontologists would certainly benefit from greater participation in the evolutionary biology community—more secure institutional positions, greater respect for their data, better access to mainstream publications and conferences, and a larger stake in theoretical discussions all followed over the next few decades. But there was a cost as well: the synthetic party line discouraged paleontologists from approaching macroevolutionary analysis of the fossil record with confidence that paleontology had unique access to patterns and processes of evolution undetectable by genetics or systematics. Even Maynard Smith proved to be a fickle friend to paleontology: writing in the *New York Review of Books* in 1995, he infamously said of Gould "the evolutionary biologists with whom I have discussed his work tend to see him as a man whose ideas are so confused as to be hardly worth bothering with," adding "he is giving non-biologists a largely false picture of the state of evolutionary theory."[10]

THE PURPOSE OF THIS BOOK

What this book is about is the major transformation in the approach of paleontologists toward questions of a theoretical, evolutionary nature, most of which have taken place over the last forty-odd years, or roughly between 1970 and the present. Given the history of the discipline, it seems all the more remarkable that, as Maynard Smith put it in 1984, "in the last ten years . . . this situation has been changed by the work of a group of paleontologists, of whom Gould has been a leading figure."[11] The 1970s saw a host of exciting new ideas in paleontology, not least of which were areas Maynard Smith himself pointed to in his essay: the theory of punctuated equilibrium, the hierarchical model of selection, species sorting, and mass extinction theory. What Maynard Smith appears to be describing is a bona fide revolution in science—in Thomas Kuhn's terms, the establishment of a new paradigm to overthrow the old, stodgy order. Unknown to Maynard Smith (or at least unremarked), however, was a tradition in paleontology that goes back to at least the 1950s, and features the work of pioneers in the field such as Norman Newell in the United States and Otto Schindewolf in Germany, each of whom practiced a theoretical paleontology, and each of whom trained important figures of the 1970s revolution. Indeed, there are as many elements of continuity as there are of revolution in the history of modern paleontology, and this book will take no official position on the significance, revolutionary or otherwise, of that innovative burst of ideas. Nonetheless it is widely acknowledged that *something* important happened to paleontology in the last few decades, and the chapters that make up this volume will tell versions of that story from a variety of perspectives.

The core of this book are the chapters—more than a dozen in all—written by the paleontologists themselves, all of whom were on hand to experience the exciting events of the 1970s and 1980s first hand. Many of our contributors had major roles in shaping the character of paleontology during this period, and it is the position of the editors that the most valuable contribution of this volume is that it records, in many cases for the first time, these first-hand accounts, recollections, and retrospective evaluations by the scientists who have been most active in the field. Because the contributors have been asked to write for an educated general audience, this book also provides an opportunity for readers who wish to become more intimately acquainted with paleontology's technical contributions to evolutionary theory to learn directly from leaders in those areas of the field. Several historians and philosophers have also been invited to participate, since we hope this volume will also help stimulate interest in further study of the history and conceptual development of modern paleontology.

This book is divided into three sections. Part I, Major Innovations in Paleobiology, examines the growth and development of some of the central ideas in modern paleobiology. A major theme is the relationship between the study of broad patterns of evolution, biodiversity, and extinction, which often rely on statistical examination of large computer databases, and the continued empirical study and collection of fossil data. Most of the chapters in Part I discuss ongoing debates and issues within and surrounding paleobiological research into evolutionary trends and processes, though historical background is often provided to give the reader context to appreciate the significance of current thinking. David Sepkoski begins by providing a historical overview of paleobiology, from the origin of the term itself through the emergence of a distinct set of paleobiological methods and questions in the 1950s and 1960s. He emphasizes that while paleobiology experienced an accelerated period of activity during the 1970s and 1980s, its roots were firmly established by the work of the previous generation of paleontologists, particularly by George Gaylord Simpson and Norman Newell. Next, Michael Benton examines modern approaches to perhaps the oldest problem in paleontology: the perceived incompleteness of the fossil record. Benton examines recent approaches in molecular phylogenetics and statistical tests of paleontological sampling error, concluding that while accurately interpreting the fossil record still presents many challenges, future prospects in paleobiology are quite good. Richard Fortey then turns to the question of the distribution and evolution of marine life in the early Paleozoic, and examines the lessons of recent studies of biogeography and biodiversity for paleobiology. Fortey stresses that both continued study of broad patterns and detailed empirical taxonomic work are necessary to solve problems in biogeography and evolution.

The next chapters in Part I focus on paleobiological studies of specific groups of organisms, and on the lessons the histories of these particular investigations have for paleobiology generally. Richard Aldridge and Derek Briggs' chapter discusses the study and interpretation of the conodonts, a group of extinct animals whose identification and anatomical reconstruction was a mystery and a challenge throughout much of the twentieth century. Their chapter highlights the difficulties posed to paleontologists by fossils whose soft-tissue structure is unclear, and the particular case of the conodonts is presented as a fascinating paleontological detective story in which the authors were major participants. J. William Schopf turns the discussion to the early history of life by describing the solution to one of the great mysteries in the history of science—the absence of Precambrian fossils. Schopf, himself one of the pioneers in Precambrian paleontology, narrates the story of the gradual expansion of the fossil record over the past several decades, including the discovery of microfossils dating some 3.5 billion years old. Next, Jack Horner's contribution examines dinosaur paleobiology, and in particular looks at interpretations of dinosaur evolution. Horner argues that dinosaur paleontologists were among the first to reach the evolutionary high table, and his chapter provides a summary of the many exciting discoveries that he and other researchers have made in recent decades. Tim White's survey of hominid paleobiology rounds out this group of chapters, which describes the history of modern interpretations of human evolution. White examines institutional, intellectual, and popular factors in evolutionary interpretation of hominids, including the influence of larger debates such as cladistics and punctuated equilibrium and the disciplinary identity of paleoanthropology.

The last two chapters in Part I address major theoretical questions in contemporary paleobiology. Patricia Princehouse's provocative essay investigates whether Niles Eldredge and Stephen Jay Gould's theory of punctuated equilibrium challenges the central foundations of Darwinism. Princehouse argues that, ultimately, punctuated equilibrium sparked important work on hierarchy in natural selection that was instrumental in gaining recognition and respect for paleobiology among evolutionary biologists. Finally, Francisco J. Ayala discusses the relationship between evidence for evolution provided by molecular biology and the paleontological evidence in the fossil record. Ayala asks whether molecular data establishes a more reliable evolutionary clock than the fossil record, and his chapter underscores the continued tension between paleontological and biological approaches to evolutionary theory.

Part II, The Historical and Conceptual Significance of Recent Paleontology, moves from a consideration of specific and ongoing problems in current paleobiology to a historical and philosophical examination of themes and issues central to the last forty years of paleobiological thought. The first three chapters of this section discuss primarily philosophical topics: Derek Turner asks whether paleontology, as a historical science, can perform experiments, and whether paleobiological experiments are like experiments in other disciplines. Turner distinguishes between important methodological strategies in empirical testing used by paleobiologists, focusing specifically on the problem of apparent trends in the evolution of body size. Todd Grantham examines the emergence of taxic paleobiology during the early 1980s, which was advocated by its proponents as a solution to the problem of independent levels of selection within the evolutionary process. Despite the prominence of these claims, Grantham argues that taxic paleobiology did not produce consensus between paleontologists and other evolutionary biologists, and may have even widened disagreement between those two groups about fundamental macroevolutionary mechanisms. Finally, David Fastovsky shifts attention to questions about the popular appeal and relevance of paleontology by examining the social and political meaning of dinosaurs in popular culture. Focusing on three case studies, Fastovsky argues that important discoveries about the biology, behavior, and extinction of dinosaurs were influenced not just by empirical developments, but also by the social climate of the times in which they were produced.

The next chapters in Part II address historical foundations of paleobiology outside the United States. Susan Turner and David Oldroyd describe the discovery, by the Australian paleontologist Reginald Sprigg in 1946, of the famous Ediacaran fauna, which included some of the earliest fossils then known. Sprigg's discovery, they argue, helped pave the way for Precambrian paleobiology, and furthermore illustrates the sometimes tortuous path of ideas to scientific acceptance: Sprigg's findings, made by a young and relatively unknown scientist, were not accepted until their later appropriation by Charles Walcott, of Burgess Shale fame. Next, Manfred Laubichler and Karl Niklas examine the important morphological tradition in German paleontology, which in many ways developed its own paleobiology independently of both the Anglo-American tradition and the Modern Evolutionary Synthesis. By investigating the careers of three mid-twentieth-century German scientists—Otto Jaekel, Walter Zimmermann, and Otto Schindewolf—Laubichler and Niklas identify a pluralistic, biologically oriented German paleontology that both predated and anticipated many of the concerns of the paleobiology movement in the United States.

The last four chapters in Part II are historical studies of major theoretical shifts in paleobiological thinking. David Sepkoski investigates the origin and early history of the theory of punctuated equilibrium, from its first articulation in 1971 through subsequent revision and reaction into the early 1980s. Sepkoski addresses the central claim by many observers—that Gould's version of the theory was intended as a challenge to orthodox Darwinism—by examining immediate and later reactions in both published and unpublished sources, and focuses especially on the relationship between Gould and his friend Thomas J. M. Schopf. The subject of John Huss' contribution is another theory in which Gould and Schopf figure prominently: the so-called "MBL" model of simulated clade shape, which was first proposed shortly after punctuated equilibrium was announced. Huss discusses the factors that led the model's authors (who included David M. Raup and Daniel Simberloff, in addition to Gould and Schopf) to propose their idea, and examines both the significance of and objections to the model, concluding that the MBL model offers important lessons about theory testing and simulation that resonate beyond paleobiology. Rounding out a trio of chapters examining the contribution of Gould and his colleagues is Joe Cain's examination of the sometimes testy relationship between members of the new generation in paleobiology and the older guard. Specifically, Cain asks why Gould attacked legendary paleontologist G. G. Simpson, arguing that this "ritual patricide" was central to Gould's efforts at establishing a new disciplinary identity for his favored brand of macroevolutionary paleobiology. Lastly, Arnold Miller presents a historical analysis of the publication and reception of the famous "Consensus Paper" (1981), in which five competing interpretations of global marine diversity were reconciled. Miller uses this paper and the work of its authors (Jack Sepkoski, Richard Bambach, David Raup, and Jim Valentine) as a way of examining disagreements over trends in biodiversity, consensus-building in science, and the shaping of the paleobiological agenda in the early 1980s.

The final section, Part III, offers personal reflections on careers in paleobiology by many of the scientists who shaped the paleobiological revolution. These chapters provide valuable insight into many of the ideas, questions, and themes discussed in the earlier sections, but will also be valuable for scholars and students as an original contribution to the historical record of the growth of modern paleobiology. James W. Valentine, one of the pioneers in the field of paleoecology during the 1960s and 1970s, describes his personal experiences as a scientist committed to integrating paleontological and biological study. In particular, Valentine discusses early attempts to relate faunal associations and fossil distributions, and also the current and future significance of molecular biology for paleontology. Richard Bambach, another leader in the establishment of paleobiology during the early 1970s, presents his personal journey as a paleoecologist and evolutionary paleobiologist from the 1950s through the 1990s. Bambach's account sheds light on the important connections between paleoecology and paleobiology, and his personal experiences track many of the central developments of the paleobiological revolution. Rebecca German offers the fascinating insights of someone who was a student of paleobiology during the 1970s, and her personal account of mentors such as David Raup, Tom Schopf, and Stephen Jay Gould sheds important light on the role of pedagogy during the paleobiological revolution. German describes how formal and informal instruction shaped the next generation of paleobiologists, and provides a glimpse of the field not normally accessible through published research.

Breaking with the largely American perspective of the other personal accounts, Anthony Hallam offers the perspective of a paleobiologist trained in the United Kingdom, and describes the development of paleobiological interests among British paleontologists from the 1950s to the 1980s. Hallam focuses particularly on punctuated equilibrium and the associated theory of species selection, which played an important role in his own research, concluding with an assessment of the significance of the idea of punctuational speciation for paleobiology generally. The next chapter, by Arthur Boucot, also examines punctuated equilibrium, which he compares to the idea of community evolution that emerged from ecological study of the fossil record. A longtime critic of punctuated equilibrium, Boucot identifies what he considers several weaknesses in that theory, and argues that community evolution is based on a more reliable empirical foundation. The personal reflections conclude with the transcript of an interview with David M. Raup, one of the most important theoretical paleontologists of his generation and a major architect of the paleobiological revolution. In this interview Raup discusses his involvement in major debates over macroevolution and extinction throughout the 1970s and 1980s, and also discusses his pioneering work in computer-based simulation and analysis of fossil data.

The final two chapters in this volume offer a general assessment of both the past and future of paleobiology. David Jablonski, one of the leaders of the current generation of paleobiologists, presents his view of the current state of the discipline, and identifies six major areas for investigation that will define paleobiology's future. Jablonski argues for the necessity of continued efforts to unite paleobiology with the wider community of evolutionary biology, and provides an important manifesto for students and practitioners of paleobiology. In the conclusion to the volume, Michael Ruse explicitly addresses the title of this book, asking whether the development of paleobiology over the past several decades constitutes a genuine scientific revolution. Using philosopher of science Thomas Kuhn's definition as a starting point, Ruse evaluates a number of criteria by which this era of paleobiology might be judged to have been revolutionary, and compares the emergence of paleobiology with other major transformations in the modern natural sciences. His conclusion, which would likely be endorsed by all contributors to this book, is that whatever label one uses to describe the growth of paleobiology, it was an event of significant importance in the history of recent science and a subject worthy of continued and serious historical and philosophical study.

NOTES

1. John Maynard Smith, "Palaeontology at the High Table," 401.
2. Ibid., 402.
3. Charles Darwin, *On the Origin of Species*, 310–11.
4. "E. R. C.", "Review of Die Stamme Des Thierreiches, by Von M. Neumayer," 364.
5. Ibid.
6. Thomas Hunt Morgan, *A Critique of the Theory of Evolution*, 25–27.
7. Julian Huxley, *Evolution, the Modern Synthesis*, 31.
8. Ibid., 38.
9. Stephen Jay Gould, "G. G. Simpson, Paleontology, and the Modern Synthesis," in *The Evolutionary Synthesis; Perspectives on the Unification of Biology*, 161.
10. John Maynard Smith, "Genes, Memes, & Minds," *New York Review of Books* 42, no. 19 (1995).
11. Maynard Smith, 1984, 401.

REFERENCES

Darwin, Charles. 1964. *On the origin of species*. Cambridge, MA: Harvard University Press.
"E. R. C." 1889. Review of die Stamme des Thierreiches by Von M. Neumayer. *Nature* 39: 364–65.
Gould, Stephen Jay. 1980. G. G. Simpson, paleontology, and the modern synthesis. In *The evolutionary synthesis: Perspectives on the unification of biology*, ed. Ernst Mayr and W. B. Provine, 153–72. Cambridge, MA: Harvard University Press.
Huxley, Julian. 1942. *Evolution, the modern synthesis*. New York: Harper.
Maynard Smith, John. 1995. Genes, memes, & minds. *New York Review of Books* 42 (19): 46–48.
———. 1984. Palaeontology at the high table. *Nature* 309: 401–02.
Morgan, Thomas Hunt. 1916. *A critique of the theory of evolution*. Princeton, NJ: Princeton University Press.

* I *

Major Innovations in
Paleobiology

CHAPTER ONE

The Emergence of Paleobiology

David Sepkoski

In a sense, paleobiology has been around since the beginning of the modern discipline of paleontology. If it is defined simply as the "study of the biology of extinct organisms," as it is in the *Dictionary of Ecology, Evolution and Systematics*, then paleobiology describes what many, if not most, paleontologists have done since at least the early nineteenth century.[1] After all, paleontology is usually defined as the study of fossils—and what are fossils, other than the physical remains of past life? But this simple equation hardly captures what was distinctive, exciting, or different about the research that was promoted over the past forty years by members of the paleobiological movement, which has radically changed the profession of paleontology. The nature of that change—which, broadly, involved theoretical, quantitative reinterpretations of patterns of evolution and extinction—is documented in the chapters that follow, many of which are written by the scientists themselves, who were actively involved in effecting this transformation. This introductory essay puts these fairly recent developments in broader historical context and attempts to identify some of the defining features of paleobiology.

In the modern, disciplinary sense, paleobiologists address their research toward biological questions about fossils and the fossil record, as opposed to investigating geological questions such as the deposit of fossils and their stratigraphic sequence. In practice, this means particularly a focus on the evolution, adaptation, ecology, function, and behavior of extinct organisms. Paleobiologists study both vertebrate and invertebrate fossils, but since the middle of the twentieth century, a major focus has been on invertebrates, which are more richly documented in the fossil record by some orders of magnitude than vertebrate remains. An important reason for this orientation has to do with the methodology of modern paleobiology: since the computer revolution of the 1960s and 1970s, methods of quantitative modeling, analysis, and tabulation have been at the center of the paleobiological approach. The great number of invertebrate fossils collected over the past century has provided an extensive resource for paleobiologists especially interested in studying the large-scale patterns and processes in the history of life, and sophisticated statistical tools developed over the past several decades have made quantitative, computer-assisted research an essential component of paleobiology.

Another, related characteristic of recent paleobiology has been its closer professional relationship with biology and its greater integration into the mainstream of evolutionary studies than was the case fifty years ago. The great American vertebrate paleontologist George Gaylord Simpson was a major architect of the Modern Evolutionary Synthesis during the 1940s and 1950s, but his adamant belief that paleontology and biology are sister disciplines was not widely shared by colleagues in either discipline. One hallmark of paleobiology since that time has been a much greater integration of the study of biology into pedagogy and practice and a concerted effort to both borrow from biological disciplines (especially from genetics and ecology), as well as to bring the fruits of paleontological research to a wider biological audience. This has, of necessity, prompted institutional reorganization among paleontologists—not always an uncontested or uncontroversial endeavor—that has even resulted in the establishment, in a few cases, of autonomous departments or programs in paleobiology at universities and museums. One of the most significant developments on this front was the establishment, in 1975, of the journal *Paleobiology*, which has ever since been the leading outlet for specifically paleobiological research and a major tool for the establishment and promotion of the paleobiological agenda.[2]

That said, there is no single definition that would be agreed upon by all paleobiologists, and even among the various authors in this volume there is substantial difference of opinion about what the central methods and assumptions of paleobiology should be. This essay, then, takes a historical approach to understanding the character of paleobiology—from the emergence of the term to the fairly recent past—and will let paleontologists speak for themselves in the following chapters about its current meanings. One conclusion that all observers would agree on, though, is that the discipline has changed remarkably over the past several decades; this essay locates the roots of that change even further back, to the beginning of the twentieth century, and examines some of the themes and debates that have characterized paleobiology's emergence from the shadow of geology and biology.

PALEONTOLOGY AFTER DARWIN

When Charles Darwin was developing his theory of evolution, biology and paleontology had not yet become firmly established as independent disciplines, and as a naturalist, Darwin simply marshaled and interpreted the available evidence from all fields as they best supported his argument. These arguments drew freely on paleontology, biology, geology, and related subjects, and it would be fair to say that *Origin of Species* presented evidence from the fossil record that might be considered paleobiological. But the aftermath of the publication of *Origin* was a period that saw significant disciplinary reorganizations, and one result was that scientists became increasingly aware of distinct disciplinary identities. A number of historians have written about the emergence of the experimental tradition in biology during the second half of the nineteenth century, which contributed greatly to the trajectory evolutionary study took after 1859.[3] In mimicking some of the laboratory practices and methods of established disciplines like physics and chemistry, biologists greatly enhanced the prestige and autonomy of their field. The emphasis in post-*Origin* biology was on uncovering the mechanisms of heredity, which, although not fully accomplished until the turn of the twentieth century and the rediscovery of Gregor Mendel's work, nonetheless made great strides in identifying cell structures responsible for heredity (e.g., chromosomes) and studying the physiological processes of biological development (such as patterns in ontogeny).[4]

This turn toward biology as the central evolutionary discipline indirectly contributed to the formation of a disciplinary identity for paleontology. Darwin had stated, more or less, that paleontology had already provided everything it was likely to contribute to understanding evolution, so for supporters of Darwin there was no great urgency to scrutinize the fossil record. In fact, Darwin's supporters were more likely to want to push paleontology into the background: as William Coleman argues, "to the biologist that [fossil] record posed more problems than it resolved . . . the incompleteness of the recovered fossil record, in which a relatively full historical record for any major group was still lacking, was the very curse of the transmutationist."[5] As a result, there were really only three alternatives available to paleontologists with regard to evolutionary theory: (1) to ignore any special theoretical relevance of paleontological data and focus purely on descriptive studies of morphology and stratigraphy; (2) to accept the Darwinian line, but to nonetheless try to improve the quality of the record of isolated fossil lineages to support Darwin's theory; or (3) to reject Darwinian evolution and seek some other theoretical explanation of evolution in which fossil evidence could be brought more directly to bear.

Over the next hundred years (and perhaps even longer), the majority of working paleontologists tended toward a position that was essentially agnostic toward evolutionary theory. This did not mean rejecting Darwin or evolution—it simply meant their *work* did not attempt to make any comment or contribution to the theory. In the early twentieth century this attitude became even more prevalent, as the burgeoning petroleum industry's demand for paleontological expertise swelled the ranks of paleontologists with scientists whose interest in the field was economic.[6] Those nineteenth- and early twentieth-century paleontologists who did pursue larger interpretive questions about the fossil record tended to subscribe to non-Darwinian, directional evolutionary models like Lamarckism and orthogenesis, which had the effect of further marginalizing paleontology from biology.[7]

By the early twentieth century, paleontology was fairly isolated from biology and other evolutionary disciplines: the most spectacular advances in the field had been in the collection of large vertebrate fossils, and broad, empirical studies of evolutionary pattern and process were not actively pursued. Rightly or wrongly it was also perceived that paleontologists had abandoned Darwinism and natural selection, which alienated those evolutionary biologists who were still committed to Darwinian orthodoxy, and for which paleontology would pay heavily when Darwinism emerged triumphant in the mid-twentieth century. Finally, from an institutional perspective, paleontology was in danger of losing all contact with biology: isolated in geology and museum collections departments, paleontologists had little regular interaction with experimental biologists. This led to mutual mistrust and incomprehension between the two fields, which was only exacerbated after the genetic turn in biology following the rediscovery of Mendel. Darwin may have considered paleontology, geology, and biology to be equal partners in the enterprise of evolutionary natural history, but as the twentieth century began, they were separated by a fairly wide gulf.

Nor, as it turned out, was the methodology adopted in the nineteenth century by most vertebrate paleontologists adequate to meet the demands of biologists' emerging conception of rigorous quantitative science: vertebrate paleontology was descriptive first and foremost, and quantitative paleontological analysis was limited to the most cursory kinds of anatomical measurements and tabulations. Biology, on the other hand, underwent a quantitative revolution in the first several decades of the twentieth century, where "the attachment of numbers to 'nature'—and the growing measurability and testability of natural selection within a populational framework" helped produce "a mechanistic and materialistic science of evolution that could rival Newtonian physics."[8] The impetus for this transformation was the discovery of quantitative laws of heredity, such as the Hardy-Weinberg principle of stable genetic equilibrium, which established a mathematical basis for confirming the expectations of natural selection in populations.[9] Paleontologists simply had no way of translating their data into terms that population biologists and geneticists could make use of, and, until G. G. Simpson stepped to the fore in the 1940s, remained mostly invisible to evolutionary biologists.

Despite paleontology's rather lowly status among biologists, it was in fact during the early decades of the twentieth century that the term *paleobiology* first began to be used. The earliest record of the word comes from an 1893 paper in the *Quarterly Journal of the Geological Society of London*, by S. S. Buckman, who commented on the usefulness of a term that "I may call 'palæobiology.'"[10] However, the much more likely source for the eventual widespread usage of the word is the Austrian vertebrate paleontologist Othenio Abel, who began using the term *päleobiologie* to describe biologically informed paleontology as early as 1912. In that year, he published *Grundzüge der Paläobiologie der Wirbeltiere (Fundamentals of Vertebrate Paleobiology)*, followed by *Paläobiologie der Cephalopoden* in 1916; his most widely read work among English-speaking paleontologists was *Paläobiologie und Stammesgeschichte (Paleobiology and Phylogeny)*, published (but never translated) in 1929. Abel was a distinguished professor of paleontology at the universities of Vienna and Göttingen, where he had a significant influence on German paleontology before the war, and in Vienna was also responsible for founding the journal *Palaeobiologica* in 1928 as the official organ of the Viennese *Paläeobiologischen Gesellschaft*.[11]

Abel is an interesting case. Theoretically, he was sympathetic to the idealist tradition of directional evolution, and supported a version of orthogenesis, but as Peter Bowler notes, he "made at least a pretense of conforming to a mechanistic language" in describing his theory.[12] For example, in *Paläobiologie und Stammesgeschichte* Abel wrote "we need assume neither a supernatural principle of perfection, nor a principle of progression, nor a vital principle"; nonetheless "the phenomenon of orthogenesis, which has often been disputed but now can no longer be denied, is transmitted by the mechanical law of inertia into the organic world."[13] However, he was also a strong proponent of the biological basis for paleontological theory, and his orthogenetic beliefs were not cultured in isolation from biology (as were the views of many American paleontologists, including Cope and Osborn), nor did he reject all of the adaptationist tenets of Darwinism. In fact, he argued that "research on adaptation had originally cultured the nucleus of paleobiology" in Darwin's day,[14] and he lamented the subsequent exclusion of paleontology from biology:

> One should think that through the appearance of this work [*Origin of Species*], which produced such an enormous revolution in biology, paleontological research would all at once be steered onto a new path, and that the basis for these sorts of [paleobiological] investigations were prepared here. It is so much the more astounding that paleontology held itself in the background for so long, and can scarcely take its place in that eternally lively discussion, in any case not to the extent as the depth of the available knowledge of fossils allowed at that time.[15]

In other words, Abel argued that paleontology had been prevented from taking its place at the evolutionary high table in part by its subordination to geology, a complaint that would become more and more common among paleontologists over the next several decades. Abel concluded, however, that paleobiology had a decisive role to play in evolutionary theory: "Among all phylogenetic research disciplines paleobiology stands alone in being able to demonstrate historical documentation, and to make readable and to draw conclusions from these facts."[16]

Abel's work was read reasonably widely by American paleontologists, and he is cited repeatedly in Simpson's groundbreaking *Tempo and Mode in Evolution* (1944). Simpson did not approve of Abel's reliance on orthogenesis to explain the evolution of horses (at one point calling Abel's belief "naïve"), but it is certain that Abel's general message about the ambitions of paleontology were received more warmly.[17] For example, in 1926 Simpson published a paper on the evolution of Mesozoic mammals, which he described as "a study in paleobiology, an attempt to consider a very ancient and long extinct group of mammals not as bits of broken bone but as flesh and blood beings."[18] This was Simpson's first use of the term *paleobiology*, and the paper prominently cites Abel's *Grundzüge der Paläobiologie der Wirbeltiere*. He also recalled many years later that "while still in graduate school I found Othenio Abel's books particularly interesting and useful."[19] Simpson was a committed reader of German scientific literature (he later reviewed German paleontological publications in the journal *Evolution* for his linguistically challenged colleagues), and at the very least his (and others') familiarity with Abel's work probably accounts for the origin of the term *paleobiology* in its modern context.[20]

Nonetheless, paleontologists in the first several decades of the twentieth century generally reflected the attitudes projected onto their discipline by biologists. Ronald Rainger concludes that despite the biological interests of people like Abel, "interest in such biological questions did not transform the discipline of paleontology. In the 1920s, just as in the 1880s, many students of the fossil record remained preoccupied with descriptive, taxonomic questions, and vertebrate paleontology was still primarily a museum science."[21] According to Rainger, this state of affairs persisted until Simpson offered his radical reevaluation of paleontological goals and methods. This assessment is probably accurate for the bulk of paleontological practice in the first part of the twentieth century, but it is important not to diminish the continuity between Simpson and his predecessors, or to overstate the discontinuity between Simpson's approach and prior paleontological theory. Paleontologists up to and during the synthesis elaborated a theoretical agenda for their discipline, and Simpson's voice was perhaps just the loudest and most persuasive among many of his contemporaries'.

PALEONTOLOGY AND THE MODERN
EVOLUTIONARY SYNTHESIS

The "Modern Synthesis" of evolutionary biology has been fairly consistently defined by historians as the sum total of theoretical development, roughly between 1937 and 1950, whereby the genetic principles of Mendelian heredity were accommodated to Darwin's theory of natural selection.[22] In other words, biologists applied the knowledge of heredity accumulated by geneticists in the first decades of the twentieth century to the principles of gene flow as determined by ecologists and biologists studying adaptation and selection in populations. The resulting synthesis defined evolutionary biology as a study of the movement (via inheritance and mutation) of *genes* within *populations*. One of the most important aspects of the synthetic approach was the development of a quantitative understanding of gene flow in populations, which allowed biologists to confirm that Darwin's qualitative assessment of the sufficiency of natural selection to produce evolution agreed with the modern understanding of genetics.[23] The doctrine produced by the end of the synthesis period became commonly known as "neo-Darwinism."

G. G. Simpson, who came into close collaboration with biological colleagues like Ernst Mayr and Theodosius Dobzhansky at Columbia University and the American Museum of Natural History, was a major figure in establishing the Modern Synthesis, and he undeniably brought new attention and respectability to paleontology within evolutionary biology. It would be a mistake, however, to conclude that paleontology was, in the 1940s or afterward, a fully equal and respected partner in the neo-Darwinian community. Paleontologists would certainly benefit from greater participation in the evolutionary biology community—more secure institutional positions, greater respect for their data, better access to mainstream publications and conferences, and a larger stake in theoretical discussions all followed over the next few decades. But there was a cost as well: as Patricia Princehouse argues, "in large part the modern synthesis served to sideline major research traditions in paleontology."[24] One of those traditions involved approaching macroevolutionary analysis of the fossil record with confidence that paleontology had unique access to patterns and processes of evolution undetectable by genetics or systematics.

Simpson's role in this aspect of the story is complex. He was perhaps the most influential paleontologist of the twentieth century, and his masterpiece, *Tempo and Mode*, has been read by generations of paleontologists and biologists. He was also, however, in many respects an iconoclast, and even as he appealed to the biological understanding of evolution promoted by the other major architects of the synthesis, his vision was significantly at odds with many of his paleontologist colleagues. Simpson appears to have been aware of how radical his views were—both for biologists *and* paleontologists. Echoing the sentiments of many contemporary biologists, he later recalled that "at the time when I began to consider this subject [of evolution] I believe that the majority of paleontologists were opposed to Darwinism and neo-Darwinism, and most were still opposed in the early years of the synthetic theory."[25]

Even before publishing *Tempo and Mode*, Simpson had already announced his intention to revitalize paleontology by increasing the discipline's analytical rigor. With his wife, the psychologist Anne Roe, Simpson began preparing a book on biological statistics in the mid-1930s. The product of this collaboration was a textbook titled *Quantitative Zoology* (1939), which was a primer in mathematical and statistical analysis for zoologists and paleontologists. In the preface, Simpson and Roe note that while it is "proper" for zoologists to avoid relying on an a priori mathematical framework, nonetheless the behaviors and characteristics of actual organisms can be profitably translated into a symbolic language.[26] The central problem the authors hoped to address was that "whether from inertia, from ignorance, or from natural mistrust . . . most zoologists and paleontologists distrusted the overt use of any but the very simplest and most obvious numerical methods."[27]

Simpson began writing *Tempo and Mode* while he and Roe were still finishing *Quantitative Zoology*, but publication was delayed until 1944, after Simpson returned to the United States from active military duty. In the preface, he notes that "the final revision was made under conditions of stress," and that because of the circumstances several "important studies" relevant to his subject were omitted.[28] Simpson is obliquely referring here to Huxley's *Evolution: The Modern Synthesis* and Mayr's *Systematics and the Origin of Species*, both of which were published in 1942, after he had begun his service. He had, however, read Dobzhansky's *Genetics and the Origin of Species*, and that work had a profound influence on his vision of evolutionary paleontology. In later years, Simpson stressed the importance of this encounter: "The book profoundly changed my whole outlook and started me thinking more definitively along the lines of an explanatory (causal) synthesis and less exclusively along lines more nearly traditional in paleontology."[29] It also "opened a whole new vista to me of really explaining the things that one could see going on in the fossil record and also by study of recent animals," and allowed him to relate his own paleontological research to the exciting new work in genetics.[30]

Probably the single most important influence Dobzhansky had on Simpson was to push the latter to think about the history of life (and the evidence of the fossil record) in terms of the genetics of once-living populations. The major argument of *Tempo and Mode* is that what happens on the Darwinian population level explains transformations in the fossil record, and that those transformations can be explained using models of population genetics. Paleontology, Simpson stressed, could be useful for uncovering the mechanisms that drive evolution, and not just for documenting the physical historical record. As he wrote in the introduction, "like the geneticist, the paleontologist is learning to think in terms of populations rather than of individuals and is beginning to work on the meaning of changes in populations."[31] Simpson's great insight was that paleontology could be modeled after population biology with the additional dimension of *time*—he described *Tempo and Mode* as a work in "four-dimensional" biology, where the distribution and transformation of organisms could be tracked in a temporal geography analogous to physical geography. He emphasized that the temporal (or historical) element of paleontology offered a unique and critical perspective to evolutionary theory, and the importance of this message cannot be overstated: after *Tempo and Mode*, temporal biogeography became central to paleontological evolutionary theory.

One of *Tempo and Mode*'s broadest and most lasting contributions to evolutionary theory was its suggestion that paleontology and the fossil record have something unique to say about macroevolution. The synthetic view of macroevolution—endorsed by Dobzhansky, Mayr, Huxley, and many others—held that major evolutionary patterns at the higher taxonomic levels are simply extrapolated effects of microevolution. As a paleontologist, however, Simpson had a keener eye for the apparent discontinuities in the fossil record than his colleagues Mayr and Dobzhansky, and his approach to macroevolution reflected this. In *Tempo and Mode* he argued that the ubiquity of extrapolation from microevolution was not a settled matter. Simpson's theory broke evolution into three tiers: the first, microevolution, basically followed the Synthetic account. Macroevolution, the second tier, accounted for broader patterns, but it was a third process, "mega-evolution," which brought about major taxonomic changes. In order to account for the seemingly abrupt transitions in the fossil record, Simpson introduced the idea of *quantum evolution*, which described accelerated evolutionary change among small populations that due to geographic isolation had come into disequilibrium.[32] Simpson suggested that quantum evolution probably utilized basic microevolutionary mechanisms of random mutation and natural selection (and not saltations), but he emphasized that such accelerated change might constitute an independent process. While his theory was not necessarily opposed to the broader Synthetic view, it certainly raised eyebrows with the provocative statement that "if the two [macro- and microevolution] proved to be basically different, the innumerable studies of micro-evolution would become relatively unimportant and would have minor value in the study of evolution as a whole."[33]

It is sufficient to note here that, while not explicitly in conflict with the synthetic theory of evolution, Simpson's approach to macroevolution and the fossil record was somewhat idiosyncratic. Stephen Jay Gould notes that while Simpson's approach to the fossil record was "consistent with genetic models devised by neontologists" in that it held "adaptation as the primary cause and result of evolutionary change," and maintained that "continuous transformation of populations" explains directional patterns in evolutionary history, nonetheless Simpson left the door open for other explanations.[34] In particular, he argued that sequential discontinuities in the fossil record (especially across taxonomic categories) might not always be artifacts of imperfect preservation: "the development of discontinuities between species and genera, and sometimes between still higher categories, so regularly follows one sort of pattern that it is only reasonable to infer that this is normal and that sequences missing from the record would tend to follow much the same pattern."[35] In fact, Simpson continued, "the face of the fossil record really does suggest normal discontinuity at all levels," an observation whose significance for evolutionary theory is unclear.[36] While he notes that many observed gaps in the fossil record are likely "a taxonomic artifact," this does not adequately explain the systematic occurrence of the gaps between larger units.[37]

A final, important feature of *Tempo and Mode* is its commitment to a quantitative, analytical method. As suggested earlier, many biologists and even some paleontologists had dismissed all hope of paleontology ever reaching the quantitative sophistication of most other scientific disciplines, including biology. This pessimism contributed greatly to paleontology's theoretical subordination as a discipline, and quite likely even discouraged many bright, analytical students from pursuing the profession. Simpson, however, was determined to apply the methodology he promoted in *Quantitative Zoology* to paleontological data, and in this regard his effort was genuinely revolutionary. Gould calls Simpson's "use of quantitative information . . . [his] second greatest departure from traditional paleontological practices," which he characterizes as "a novel style . . . [of] drawing models (often by analogy) from demography and population genetics and applying them to large-scale patterns of diversity in the history of life."[38]

Perhaps Simpson's most significant use of quantification is his treatment of taxonomic survivorship, or the measure of the longevity of a particular taxon or group. Simpson's approach was to gather taxonomic data from fossil catalogues like K. A. von Zittel's *Grundzüge der Palaeontologie*, from paleontological monographs, or from other systematics literature, and then to tabulate the longevity of the group based on first and last appearances in the record. Next, following the method Raymond Pearl established for statistical demography, he plotted curves representing survivorship over time as a percentage of the initial population.[39] By modifying this data with a number of straightforward statistical devices, Simpson was able to draw out several very interesting conclusions: general patterns of survivorship appear, on the whole, to follow the same, diminishing parabolic curves, although different groups (he compares pelecypod mollusks to carnivorous placental mammals) have widely differing rates.[40] These curves can also be correlated with extant fauna, comparisons can be made within groups over different periods of time, and generalizations about major fauna can be made, all of which, Simpson noted, can shed important light on evolutionary patterns. In succeeding chapters, we will again and again confront examples of "Simpsonian" analytical techniques applied to paleobiology; Simpson's presentation of paleontology's amenability to theoretical modeling and statistical techniques had the greatest possible influence on later generations of paleontologists.

All in all, Simpson's work substantially helped to bring paleontology in line with mainstream attitudes in biology and genetics, and also worked to carve out an independent place within evolutionary biology. As such, Simpson made a major contribution to the early stages of the development of paleobiology. What was necessary following the publication of *Tempo and Mode* (and its successor, *Major Features of Evolution*, a completely revised version, published in 1953) was an active community of paleontologists dedicated to pursuing paleobiological questions. Simpson helped provide the motivation, but others would need to take up the challenge he posed.

THE DEVELOPMENT OF PALEOBIOLOGY

During the 1950s and 1960s a major transformation was quietly taking place in paleontological approaches to evolutionary theory and the fossil record, one which Simpson certainly played a role in starting. This shift involved several distinct but related aspects: first, paleontologists began to actively assess the institutional status of their discipline—asking whether it belonged, for example, with geology, with biology, or rather constituted an independent discipline on its own. Second, paleontologists began more and more to explicitly connect their work with the agenda of the Modern Synthesis, and to publish in outlets (such as the journal *Evolution*) that were read by biologists and geneticists. Even papers in paleontology-specific publications like *Journal of Paleontology* took on a more theoretical cast during this period. Third, and perhaps most significantly, paleontology became quantitative. This is not to say that quantitative methods (measurements and statistical analysis) had been absent from the work of paleontology in the past, but the period between 1950 and 1969 saw a burgeoning interest in addressing broad, synthetic questions about the fossil record (e.g., biodiversity, evolution, extinction) with quantitative rigor and sophistication not previously seen in paleontological literature.

Between 1940 and the later 1960s, a number of paleontologists began to publicly question paleontology's longtime association with geology, and to argue that paleontology—as the study of ancient *life*—belonged more properly among the biological disciplines. For example, in his 1946 presidential address to the Paleontological Society, J. Brookes Knight made a forceful call to arms for paleontologists to throw off the restrictive shackles chaining them institutionally to geology departments. "Because paleontology is not truly

a branch of geology," he wrote, "it does not best serve geology when culti-
vated and applied by geologists," concluding that "because paleontology is
the study of the life of the past it is a biological science."[41] These comments
touched off a minor controversy in the paleontological community. The first
to respond was J. Marvin Weller of the Walker Museum at the University of
Chicago, who rejected Knight's call entirely. Arguing that paleontological
stratigraphy is "the heart of geology" and its "single great unifying agency,"
Weller urged paleontologists to stick close to their geological roots.[42] "Invertebrate paleontology is much more closely related to geology than biology"
he reasoned, and the two fields are mutually interdependent, whereas biology
and paleontology can each "get along" without the other. He had little time for
vertebrate paleontologists, whom he considered hardly even geologists, and
even less interest in the kind of paleontological-biological synergy preached
by his many of his peers: "any student of fossils who does not have a strong,
abiding, and well-founded interest in geology . . . is not a paleontologist. He
is simply a paleobiologist."[43]

It is especially interesting that Weller used the term *paleobiologist* as an
epithet rather than a compliment. However, there were other paleontologists
at the time who regarded paleobiology as an approach to be actively pursued,
rather than avoided, and none had a greater influence than the invertebrate
specialist Norman Newell. It may fairly be argued that nobody did more to
promote the agenda of paleobiology in the 1950s and 1960s than Newell, and
his influence, measured directly through his work, and indirectly through his
mentoring of students and younger paleontologists, was profound. Newell's
hand touched nearly every major aspect of paleobiology during his career,
and he can be said to have been directly responsible for, in no particular
order, the investigation of broad patterns in the fossil record, the development of quantitative approaches to fossil databases, the study of the evolutionary significance of mass extinctions, and the creation of the subdiscipline
of paleoecology. Throughout his career, Newell also tirelessly promoted the
institutional agenda of paleobiology, and he trained many of the leaders of the
movement's next generations.

According to Gould, a student of Newell's who would become one of the
most active of that later generation of paleobiologists,

> When virtually all paleontologists were trained as geologists and had
> no biological knowledge beyond the basics of invertebrate morphology,
> Norman Newell saw, virtually alone, that the most exciting future direc-
> tion in paleontology lay in its relationship to evolutionary theory and to

biological thought in general. I think that only a few very old-fashioned paleontologists would deny today that this prediction, has been fulfilled and that American invertebrate paleontology is now in its most exciting phase since the era immediately following Darwin's *Origin of Species*. With his early monographs, and his persistent encouragement of biological thinking, Norman Newell was the godfather of this movement.[44]

This is strong praise indeed, but leaving aside Gould's somewhat hyperbolic assessment of the status of paleobiology, it is probably an accurate characterization of Newell's contributions. Preston Cloud, with whom Newell pioneered the study of important fossil beds of the Permian of Texas and elsewhere, has remarked that "by his imaginative researches, Newell has been instrumental in a rejuvenation of biological invertebrate paleontology. One of America's foremost invertebrate paleontologists, he is outstanding for his interpretation of fossil invertebrates in the light of the ecology and life histories of living species."[45] And Ernst Mayr agrees, reflecting that "Norman has served as an important bridge between specialized paleontology and evolutionary biology as a whole . . . [and was] quite instrumental in introducing the evolutionary synthesis into invertebrate paleontology."[46]

An example of Newell's vision for paleontology can be seen in an essay that he and Columbia University colleague Edwin Colbert coauthored in response to Marvin Weller's criticism of paleobiology. While the authors noted that "it is not likely that many universities could be persuaded to erect separate paleontology departments," they respectfully offered that "Professor Weller's point of view admirably expresses the traditional (and 'narrow') attitude of the geologist toward paleontology," which "is being modified only too gradually." Paleontology is only considered a branch of geology, Newell and Colbert reasoned, "because paleontologists, through lack of adequate training in biology, have made it so."[47] They proposed a division of paleontology into two categories—stratigraphic and "paleobiology"—and emphasized that even this dichotomy obscured significant areas of overlap between the two approaches. Many of the goals of paleontology transcend stratigraphy, they stressed, such as phylogeny reconstruction and the restoration of the fossil record, but are also beyond the ken of biologists who lack paleontological training. And turning the tables on Weller, Newell and Colbert argued that it is its close traditional association with geology that has, "as much as anything . . . [caused] the lack of mature growth of this branch of [invertebrate] paleontology." In their conclusion, Newell and Colbert centered the issue on paleontology's engagement with evolution: "the invertebrate paleontologist in North America has suffered because of his lack of an *evolutionary* viewpoint, the result of a lack of training in biology."[48]

In addition to pursuing his agenda publicly in *Journal of Paleontology*, Newell also worked to change the mentality at his home institution. In 1948 or 1949, Newell sent a memo to his colleagues in the geology department at Columbia titled "Instruction in Paleobiology," which he described in a hand-written note to Simpson as "part of an unavoidable campaign of missionary work." In it, he outlined his programmatic agenda for revising the way pale-ontology was taught, and ultimately practiced. "The period between the two world wars," he wrote,

> was characterized by development in invertebrate paleontology chiefly along utilitarian lines, seemingly at the expense of fundamental progress in the science. . . . Because of the traditional union between invertebrate paleontology and geology it has come to be forgotten that the roots of paleontology are in biology, just as geophysics rests on physics. It is a tragedy that paleontology has at last become a 'handmaiden to geology.' Yet the techniques and mass of data of paleontology are now so distinct from geology and biology that the majority of biologists and geologists do not even know what constitutes urgent problems in paleontology. Although it is seldom accorded the status of a separate science, paleon-tology is just that.

Newell drew particular attention to the problems in the current pedagogical climate: with "the majority of teachers of paleontology" being "stratigraphers or petroleum geologists, concerned entirely with the application of paleontol-ogy to geology. . . . Little progress is being made toward an understanding and interpretation of fossils and their life environment." However, Newell saw an opportunity to change this at Columbia, drawing on the rich resources at the American Museum of Natural History (AMNH), to develop "a program of instruction in invertebrate paleontology, or paleobiology, at a professional level, adequate for the development of research specialists."[49]

Newell also promoted paleobiology through the example of his research, which, from the 1950s forward, became more and more concerned with an-swering broad questions about evolution and extinction using quantitative analysis of the invertebrate fossil record. In a 1959 symposium sponsored by the Paleontological Society celebrating "Fifty Years of Paleontology," Newell gave an overview of the growth of paleobiology that expresses important ele-ments of his agenda for the field. He begins by noting that "from the very beginnings of our science there have been two schools, those who study fossils in order to understand stratigraphy, and those who study fossils in order to learn about past life," and he applauds others who have called for greater biological orientation in paleontology.[50] Newell was pleased to report that "the fossil record is much richer than we formerly supposed," but cautioned that paleontology needed to produce more biologically sensitive workers to meet the demands of the changing profession. He also cited five "truly revolutionary developments of the past three decades": (1) improved collection and preparation of fossils; (2) "recognition of the special importance of populations in taxonomy and evolution"; (3) more attention to ecological context; (4) "the application of statistical methods . . . [to] all sorts of paleontological problems"; and (5) greater understanding of the geochemistry of fossils.[51] Newell contrasted the "gradual increase in appreciation of the positive merits of the fossil record" with Darwin's earlier "preoccupation with the deficiencies in the record," and while he noted a continued "lively debate" over interpretations of the record, he cited a "general agreement . . . that many striking patterns of fossil distributions have been confirmed hundreds of times."[52] In terms of the sheer quantity of paleontological data, Newell pointed to the dramatic improvement of knowledge of the record: whereas Charles Schuchert estimated, in 1910, some 100,000 extant fossil species, Curt Teichert's calculation in 1956 raised that number to ten million.[53] Overall, Newell predicted "the future prospects for paleontology are, indeed, very bright."[54]

Two of Newell's most important contributions to the growth of paleobiology were his study of trends in the sequential succession of invertebrate evolution and his analysis of the role of mass extinction in the history of life. In the first instance, Newell drew attention to the unique set of problems paleontology faces in applying taxonomic divisions to fossil populations. Here his major concern is preservational bias: while "the fossil record is in fact astonishingly rich and meaningful," the 'time dimension' in paleontology complicates matters, since "the selection of species limits in a vertical series might be arbitrary."[55] In other words, the added dimension of time is both a boon and a hindrance to paleontology: within a given horizontal sample (i.e., a group of organisms taken from the exact same stratum or moment in geological time) it might certainly be possible to distinguish taxa, including species and perhaps even subspecies or varieties. But paleontology also has a vertical dimension, and as the taxa identified from horizontal samples continue forward in time it is extremely difficult to discern where taxonomic limits or divisions should be placed. This situation is further complicated by the fact that vertical sequences are almost always interrupted, and the paleontologist is not guaranteed to fill in these gaps by further collection. Finally, as Newell notes, horizontal and vertical perspectives must be combined to get an accurate picture of the influence of geography on phyletic evolution.[56] Nonetheless, in the face of such apparently insoluble difficulty, Newell remains confident that "properly conceived and diagnosed, palaeontological species and subspecies can be consistently recognized and studied by the same methods as those employed in neontology."[57] How does he imagine this might be possible?

The answer, Newell determines, is to apply quantitative analysis to the confusing array of fossil data—to let statistics do what the paleontologist is unable to accomplish using traditional, descriptive techniques. In the past, paleontologists had relied on a typological basis for identifying species and higher taxa, but ecological and evolutionary study requires paleontology to reorient itself to the neontological population understanding; according to Newell, the "crude procedure" of typology "does not measure up to modern requirements in studies of stratigraphic and evolutionary palaeontology."[58] This is mainly because the typological species concept ignores population variability, which should in each instance follow a normal population curve. A type specimen is normally chosen (i.e., sampled) arbitrarily, and the paleontologist has no guarantee that it "represent[s] the most frequent condition of populations" (i.e., that it would fall in the middle of a normal variability curve). Instead, the procedure should be to select, ideally as randomly as possible, a group of examples from a population and to estimate, using "biometrical analysis," the range of variation for that population. The trick, according to Newell, "is to summarise in a reasonably accurate way the characteristics of a vast assemblage of individuals, perhaps numbering billions, by means of data provided by a few specimens."[59]

The only way such a drastic extrapolation is justified is if we can have confidence that the few specimens chosen give a reasonable indication of the limits of variability in their parent population. Surprisingly, Newell argues, most populations *can* be estimated in such a way, and individual samples are in fact reliable indicators of average variability, provided that they are sampled *randomly*. The mistaken belief that only large and well-documented collections can be analyzed this way has meant "very little headway has been made toward the establishment of uniform practice in quantitative palaeontology"; what we are seeing in Newell's proposal is the solidification of a major argument that statistical analysis can correct for the inadequacies of fossil preservation. This would be perhaps the single most important future direction in paleobiology, but it ultimately depended on a serendipitous convergence of paleontological thinking and technology. As Newell noted several years later, "the recent application of electronic IBM computers in the solution of paleontologic problems" is "more than just another statistical technique"; rather, as Newell went on to predict, the advent of inexpensive, readily available digital computing meant that "in the near future, we may have at our disposal the means for more or less routine quantitative solutions of all sorts of paleontological problems involving complex interrelationiships of many variables."[60] In other words, evolutionary paleontology was about to become a quantitative discipline.

In Newell's second major area of contribution—the study of mass extinctions—he set out to examine patterns in the invertebrate fossil record with a particular eye for relationships between organic and physical histories, and his work directly influenced some of the most important paleobiological theories of the next generation. In the first of three important papers, "Catastrophism and the Fossil Record," published in *Evolution* in 1956, Newell addressed German paleontologist Otto Schindewolf's arguments about "the enigmatic, apparently world-wide, major interruptions in the fossil record which mark the boundaries of the eras."[61] In granting that "abrupt paleontological changes at these stratigraphic levels are real, [and] apparently synchronous," Newell helped to legitimize the study of mass extinctions as a significant evolutionary process, and provided important groundwork for David Raup and Jack Sepkoski's later analysis of mass extinction patterns. This legitimization was important: up until the time of Newell's essays, the prevailing attitude in the paleontological community was that to seriously discuss the possibility of cyclical mass extinctions was to invoke the specter of catastrophism, which was associated either with old, discredited ideas, or with the lunatic fringe. But in even being willing to discuss catastrophism publicly Newell was taking a brave stand, and his series of papers may have helped erase some of the taint that surrounded discussion of mass extinctions.

A fairly definitive statement of Newell's understanding of the role of mass extinctions in evolutionary history can be seen in two of his later essays on the subject: a paper on "Revolutions and the History of Life" delivered at a special Geological Society of America symposium in 1963, and a more popular piece published in *Scientific American* the same year.[62] Newell's symposium paper opens with the bold claim that "the purpose of this essay is to demonstrate that the history of life . . . has been episodic rather than uniform, and to show that modern paleontology must incorporate certain aspects of both catastrophism and uniformitarianism while rejecting others."[63] Noting that most geologists think "change" is "uniform and predictable rather than variable and stochastic," he calls for greater openness toward discontinuity and unpredictability, and opines that "catastrophism rightly emphasized the episodic character of geologic history, the rapidity of some changes, and the difficulty of drawing exact analogies between past and present."[64] This statement is a fairly remarkable repudiation of the ubiquity of uniformitarianism, a pillar of both Darwin's theory and the neo-Darwinian interpretation of the Modern Synthesis, and Newell makes it clear that he intends it as such. A major assumption of uniformitarianism is that gaps in the fossil record are the result of biases in deposition, preservation, or collection, but here Newell endorses Schindewolf's argument that when such "abrupt changes occur in relatively complete sequences over a large part of the earth, they indicate episodes of greatly increased rate of extinction and evolution."[65] He also points to other factors, such as the stratigraphic correlation of extinctions of totally unrelated groups, and the tendency for episodes of apparent extinction to be followed by evidence of "episodes of exceptional radiation." This latter point is especially important, since it contributes to a model of how extinction and evolution function hand in glove: Newell proposes that major extinction events clear the adaptive landscape and open new niches for surviving organisms to exploit, leading to massive and relatively sudden migrations and the production of new forms.[66]

This paper also includes a lengthy consideration of causal factors in mass extinctions, and here Newell presses more urgently the need to develop explanations for the regular extinction of unrelated groups. After first dismissing proposed causes such as cosmic radiation, oxygen fluctuations, changes in ocean salinity, and saltation, he presents a tentative hypothesis of selective elimination via environmental change as the major cause of mass extinction. According to Newell, "this hypothesis postulates widespread, approximately synchronous, environmental disturbances and greatly increased selection pressure," for which he suggests three possible causes.[67] The first of these is migrations "involving better adapted immigrants and less adapted natives," which might become more frequent during times of environmental stress. This he poses as a direct challenge to Darwin's assertion that migrations are "selective and continuous," although he notes this is probably the least likely source of very sharp discontinuities in the fossil record. The second factor is "severe climate changes," such as global ice ages, but while Newell observes this has been the most popular explanation for major extinctions (e.g., the dinosaurs) he discounts its importance since (a) evidence of major climate shifts does not correspond with extinction events, and (b) plants (which we would expect to be especially responsive to climate fluctuations) are not affected during major mass extinctions of animals.[68] Finally, he addresses paleogeographic factors such as changes in sea level, which unsurprisingly emerge as the most likely culprit. According to Newell "it seems clear that rapid emergence of the continents would result in catastrophic changes in both terrestrial and marine habitats and such changes might well trigger mass extinctions among the most fragile species."[69]

Overall, Newell's contributions to the study of mass extinction are significant primarily because of their legitimizing factor within the discipline. While sea level is no longer considered a major factor in the most dramatic extinction events in the history of life, Newell, as a past president of the Paleontological Society and a widely respected figure in the field, lent considerable respectability to this area of study. By challenging some of the tenets of uniformitarianism, he also opened the door to more radical critiques of neo-Darwinism presented by paleobiologists over the next two decades. Indeed, two of the more active proponents of such revisions were directly influenced by Newell: Gould was Newell's doctoral student at Columbia between 1963 and 1967, and Niles Eldredge studied with Newell throughout the 1960s as both an undergraduate and a graduate student. As Eldredge recalls, it was not lost on either Gould or himself that "Newell was the only person in twentieth century paleontology who was talking about the importance of extinction," a fact that led directly to Eldredge's own interest in patterns of evolution and extinction.[70] Perhaps even more importantly, however, Newell stressed that characterizing evolution and extinction as episodic, discontinuous, and stochastic did not mean abandoning a quest for general regularities, nor did it necessitate abandoning a systematic, quantitative study of the fossil record. As he put it, "yet the record of past revolutions in the animal kingdom is understandable by application of basic principles of modern science. In this sense, the present is the key to the past."[71] As many of the chapters in this volume will explore, one of the central themes in the modern paleobiological movement would be the explainability—and even predictability—of complex, dynamic phenomena such as evolution and extinction. And in this regard most of the paleontologists at the forefront of this research over the next two decades were, either directly or indirectly, Newell's students.

THE PALEOBIOLOGICAL REVOLUTION

Between roughly 1970 and 1985, paleobiology went through what might properly be called a revolution, which saw the goals and methods of theoretically minded, biologically oriented paleontology promoted on a wider stage than ever before. As this period of paleobiology's history is the central subject of this book, I will let the following chapters speak for themselves. However, broadly speaking, beginning in about 1970, paleobiology entered its more active proselytizing phase, and paleobiologists self-consciously worked to raise the status of their discipline, both by promoting the theoretical products of quantitative, theoretical paleontology, and by establishing new institutional and disciplinary footholds, including pedagogical reform and the establishment of new outlets for publication. From an intellectual perspective, the most spectacular example was Gould and Eldredge's "Punctuated Equilibria: An Alternative to Phyletic Gradualism," which appeared in 1972 in a collection of essays entitled *Models in Paleobiology*, edited by Thomas J. M. Schopf. This work was joined by studies by Raup, Steven Stanley, Sepkoski, and others, of species diversity, taxonomic survivorship, and rates of evolution and extinction using stochastic (random) modeling and multivariate analysis to fundamentally reorient many of the questions paleontologists were asking about the nature of evolutionary change. These studies were based on techniques that were not part of typical graduate education in paleontology, and required their authors to cross disciplinary boundaries to import new methodologies.

In particular, paleobiologists drew heavily on statistical techniques developed during the previous few decades in population biology, which had undergone a kind of quantitative revolution of its own in the 1950s and early 1960s. A transitional moment for paleobiologists also came in 1975, when a new journal—titled simply *Paleobiology*—was launched by the Paleontological Society, under the guidance of Schopf, who served as editor until 1980. The explicit intention behind this journal was to promote new paleontological methods and questions, and from its inception it served as the primary organ for quantitative studies in macroevolution and extinction. Another, equally important role the journal played, however, was as a mouthpiece for manifestos promoting the new agenda. Gould, in particular, published a number of essays of a general, theoretical nature touting the significance of his approach to evolutionary modeling in the first ten years of the journal's existence.

The early 1980s saw the establishment of paleobiology as a mainstay in many university and museum departments, and the contributions of paleontologists to evolutionary theory became standard literature in evolutionary biology. However, its status was not uncontested, and this volume offers perspectives on several key debates within paleobiology. One locus for controversy was Gould's promotion of a purportedly non-Darwinian, antiadaptationist program, which drew fire from biologists and paleontologists alike. Even as it drove innovative studies of the patterns and processes involved in macroevolution, the work of Gould, Eldredge, Stanley, and others provoked controversy in many quarters. While this debate took place in a variety of forums (including journal publications and correspondence), a central event that is examined is the notorious macroevolution conference that took place in 1980 at the Field Museum of Natural History in Chicago, where paleontologists and biologists clashed over the interpretation of punctuated equilibrium and other macroevolutionary hypotheses advanced by paleontologists. A second major topic of controversy during the 1980s were studies of mass extinction, authored by Raup and Sepkoski, that used statistical analysis to propose a twenty-six-million-year cycle of periodic mass extinctions in the fossil record. This work was closely tied to the discovery, by Louis and Walter Alvarez, of physical evidence of the impact event that may have killed off the dinosaurs, and to the wider (and more controversial) Nemesis or death star hypothesis. Macroevolution and mass extinction became the signature themes of recent paleobiology, and they were also the topics of greatest controversy.

But as this introductory essay has argued, the rapid development of paleobiology over the past several decades was preceded by a less visible, but vitally important, period when paleobiology began to emerge from traditional descriptive paleontology. I have focused particularly on the work of G. G. Simpson and Norman Newell, two of the most active early promoters of paleobiology, but I might have just as easily focused on other, equally important developments, including the growth of paleoecology during the 1960s, the role of non-English language paleontological theory (such as the German "morphological tradition" discussed by Laubischler and Niklas in this volume), the advent of mathematical models and computing technology, the discovery of Precambrian fossils, the authorship of textbooks, the proliferation of journals, or any of a variety of additional topics. The chapters that follow offer additional perspectives on the history and philosophy of paleobiology, including first-hand accounts by several of the leading figures of the paleobiological revolution, which shed light on many of these issues. The point is that modern paleobiology has important antecedents in earlier lines of intellectual and institutional development, all of which are necessary to understand why paleobiology exists in the form it does today, and, as a continually evolving scientific discipline, where it may lead in the future.

NOTES

1. Roger J. Lincoln, Geoffrey Allan Boxshall, and P. F. Clark, *A Dictionary of Ecology, Evolution, and Systematics,* 179.

2. See David Sepkoski, "The 'Delayed Synthesis': Paleobiology in the 1970s," in *Descended from Darwin: Insights into American Evolutionary Studies, 1925–1950*, ed. Joseph Cain and Michael Ruse. Forthcoming.

3. See, e.g., Elizabeth B. Gasking, *The Rise of Experimental Biology;* Garland E. Allen, *Life Science in the Twentieth Century;* Philip J. Pauly, *Controlling Life: Jacques Loeb and the Engineering Ideal in Biology.*

4. See particularly Peter J. Bowler, *The Mendelian Revolution: The Emergence of Hereditarian Concepts in Modern Science and Society.*

5. William Coleman, *Biology in the Nineteenth Century: Problems of Form, Function, and Transformation*, 66.

6. See Ronald Rainger, "Subtle Agents for Change: The Journal of Paleontology, J. Marvin Weller, and Shifting Emphases in Invertebrate Paleontology, 1930–1965."

7. See Peter J. Bowler, *Life's Splendid Drama: Evolutionary Biology and the Reconstruction of Life's Ancestry, 1860–1940.*

8. Vassiliki Betty Smocovitis, *Unifying Biology: The Evolutionary Synthesis and Evolutionary Biology*, 122 and 127.

9. For the history of population genetics, see William B. Provine, *The Origins of Theoretical Population Genetics.*

10. S. S. Buckman, *Quarterly Journal of the Geological Society* 49 (1893), 127. The *Oxford English Dictionary* records this as the first appearance of the word. It is possible that the term has an earlier, independent origin, but no earlier usage has been established.

11. W. E. Reif, "The Search for a Macroevolutionary Theory in German Paleontology," *Journal of the History of Biology* 19 (1986); W. E. Reif, "Deutschsprachige Paläontologie Im Spannungsfeld Zwischen Makroevolutionstheorie Und Neo-Darwinismus (1920–1950)," in *Die Entstehung Der Synthetischen Theorie. Beitruage Zur Geschichte Der Evolutionsbiologie in Deutschland 1930–1950*, ed. T. Junker and E.-M. Engels.

12. Bowler, *Life's Splendid Drama*, 359.

13. Othenio Abel, *Palaeobiologie Und Stammesgeschichte*, 399. All translations are mine unless otherwise noted.

14. Ibid., v.

15. Ibid., 5.

16. Ibid., vi.

17. George Gaylord Simpson, *Tempo and Mode in Evolution*, 149.

18. George Gaylord Simpson, "Mesozoic Mammalia, IV; the Multituberculates as Living Animals," *American Journal of Science* 11 (1926) 228.

19. Simpson, quoted in Ernst Mayr and William B. Provine, *The Evolutionary Synthesis: Perspectives on the Unification of Biology*, (Cambridge, MA: Harvard University Press, 1980), 456. See Othenio Abel, *Grundzèuge der Palaeobiologie der Wirbeltiere*, (Stuttgart,: E. Schweizerbart, 1912).

20. U. Kutschera, "Palaeobiology: The Origin and Evolution of a Scientific Discipline," *Trends in Ecology and Evolution* 22, no. 4 (2007).

21. Ronald Rainger, "Vertebrate Paleontology as Biology: Henry Fairfield Osborn and the American Museum of Natural History," in *The American Development of Biology*, ed. Ronald Rainger, 1988, 244.

22. Prominent histories of the synthesis include *Provine, The Origins of Theoretical Population Genetics; Smocovitis, Unifying Biology; Mayr and Provine, The Evolutionary Synthesis: Perspectives on the Unification of Biology; Ernst Mayr, The Growth of Biological Thought: Diversity, Evolution, and Inheritance;* Joseph A. Cain, "Common Problems and Cooperative Solutions: Organizational Activity in Evolutionary Studies, 1936–1947," *Isis* 84 (1993); Joseph A. Cain, "Epistemic and Community Transition in American Evolutionary Studies: The 'Committee on Common Problems of Genetics, Paleontology, and Systematics' (1942–1949)," *Studies in History and Philosophy of Biological and Biomedical Sciences* 33 (2002); and Allen, *Life Science in the Twentieth Century.*

23. Some historians, including Provine, view this as *the* major accomplishment of the synthesis. Provine, *The Origins of Theoretical Population Genetics,*

24. Patricia M. Princehouse, "Mutant Phoenix: Macroevolution in Twentieth-Century Debates over Synthesis and Punctuated Evolution" (PhD diss., Harvard University, 2003), 21.

25. Simpson, quoted in Ernst Mayr, "G. G. Simpson," in *The Evolutionary Synthesis: Perspectives on the Unification of Biology*, ed. Ernst Mayr and William B. Provine, 455.

26. George Gaylord Simpson and Anne Roe, *Quantitative Zoology; Numerical Concepts and Methods in the Study of Recent and Fossil Animals*, 1st ed., vii.

27. Simpson and Roe, *Quantitative Zoology,* viii.

28. Simpson, *Tempo and Mode in Evolution,* vi.

29. Simpson, quoted in Mayr, "G. G. Simpson," 456.

30. Simpson, quoted in Leo F. Laporte, *George Gaylord Simpson: Paleontologist and Evolutionist,* 25.

31. Simpson, *Tempo and Mode in Evolution,* xvi.

32. Ibid., 206.

33. Ibid., 97.

34. Stephen Jay Gould, "G. G. Simpson, Paleontology, and the Modern Synthesis," in *The Evolutionary Synthesis; Perspectives on the Unification of Biology*, ed. Ernst Mayr and W. B. Provine, 161.

35. Simpson, *Tempo and Mode in Evolution,* 98.

36. Ibid., 99.

37. Ibid., 107.

38. Gould, "G. G. Simpson, Paleontology, and the Modern Synthesis," 158–59.

39. See Raymond Pearl and Lowell J. Reed, "On the Rate of Growth of the Population of the United States since 1790 and Its Mathematical Representation," *Proceedings of the National Academy of Sciences of the United States of America* 6, no. 6 (1920).

40. Simpson, *Tempo and Mode in Evolution*, 24–26.

41. J. Brookes Knight, "Paleontologist or Geologist," *Bulletin of the Geological Sociey of America* 58 (1947) 282–83.

42. J. Marvin Weller, "Relations of the Invertebrate Paleontologist to Geology," *Journal of Paleontology* 21, no. 6 (1947) 570. See also Rainger, "Subtle Agents for Change."

43. Weller, "Relations of the Invertebrate Paleontologist to Geology," 572.

44. Stephen Jay Gould to Niles Eldredge, March 9, 1978. American Museum of Natural History (AMNH) Invertebrates Department Archive.

45. Preston Cloud to Roger Batten, February 22, 1978. AMNH Invertebrates Archive.

46. Ernst Mayr to Niles Eldredge, March 1, 1978. AMNH Invertebrates Archive.

47. Norman Dennis Newell and Edwin Harris Colbert, "Paleontologist; Biologist or Geologist," *Journal of Paleontology* 22, no. 2 (1948) 265.

48. Newell and Colbert, "Paleontologist; Biologist or Geologist," 267.

49. Norman Newell, "Instruction in Paleobiology," American Museum of Natural History Department of Vertebrate Paleontology Archive (n.d.), Box 67, Folder 21.

50. Norman Dennis Newell, "Adequacy of the Fossil Record," *Journal of Paleontology* 33, no. 3 (1959) 489.

51. Ibid., 490.

52. Ibid., 490–91.

53. Newell, "Adequacy of the Fossil Record," 492; Charles Schuchert, "Biologic Principles of Paleogeography," *Popular Science* (1910) 591–92; Curt Teichert, "How Many Fossil Species?" *Journal of Paleontology* 30, no. 4 (1956).

54. Newell, "Adequacy of the Fossil Record," 499.

55. Norman D. Newell, "Fossil Populations," in *The Species Concept in Palaeontology: A Symposium*, ed. P. C. Sylvester-Bradley, 67.

56. Ibid., 70.

57. Ibid., 70–71.

58. Ibid., 71.

59. Ibid., 74.

60. Newell, "Adequacy of the Fossil Record," 490.

61. Norman Dennis Newell, "Catastrophism and the Fossil Record," *Evolution* 10, no. 1 (1956a) 97.

62. Norman D Newell, "Revolutions in the History of Life," in *Uniformity and Simplicity. Special Paper–Geological Society of America;* Norman D Newell, "Crises in the History of Life," *Scientific American* 208, no. 2 (1963). Because the two pieces were composed at the same time and cover substantially similar topics, reference here will be made only to the more scholarly presentation from the GSA symposium.

63. Ibid., 64.

64. Ibid., 65.

65. Ibid., 74.

66. Ibid., 82.
67. Ibid., 84.
68. Ibid., 85.
69. Ibid., 88.
70. Interview with Niles Eldredge, conducted by David Sepkoski, 1/19/06. Transcript in author's possession.
71. Newell, "Revolutions in the History of Life," 89.

REFERENCES

Abel, Othenio. 1912. *Grundzèuge der Palaeobiologie der Wirbeltiere.* Stuttgart: E. Schweizerbart.
———. 1980. *Palaeobiologie und Stammesgeschichte (The history of paleontology).* New York: Arno.
Allen, Garland E. 1975. *Life science in the twentieth century.* New York: Wiley.
Bowler, Peter J. 1989. *The Mendelian revolution: The emergence of hereditarian concepts in modern science and society.* Baltimore: Johns Hopkins University Press.
———. 1996. *Life's splendid drama: Evolutionary biology and the reconstruction of life's ancestry, 1860–1940.* Chicago: University of Chicago Press.
Buckman, S. S. 1893. *Quarterly Journal of the Geological Society* 49.
Cain, Joseph A. 1993. Common problems and cooperative solutions: Organizational activity in evolutionary studies, 1936–1947. *Isis* 84:1–25.
———. 2002. Epistemic and community transition in American evolutionary studies: The 'Committee on Common Problems of Genetics, Paleontology, and Systematics' (1942–1949). *Studies in History and Philosophy of Biological and Biomedical Sciences* 33:283–313.
Coleman, William. 1971. *Biology in the nineteenth century: Problems of form, function, and transformation.* New York: Wiley.
Gasking, Elizabeth B. 1970. *The rise of experimental biology.* New York: Random House.
Gould, Stephen Jay. 1980. «G. G. Simpson, paleontology, and the Modern Synthesis. In *The evolutionary synthesis; Perspectives on the unification of biology*, ed. Ernst Mayr and W. B. Provine, 153–72. Cambridge, MA: Harvard University Press.
Knight, J. Brookes. 1947. Paleontologist or geologist. *Bulletin of the Geological Society of America* 58:281–86.
Kutschera, U. 2007. Palaeobiology: The origin and evolution of a scientific discipline. *Trends in Ecology and Evolution* 22 (4): 172–73.
Laporte, Leo F. 2000. *George Gaylord Simpson: Paleontologist and evolutionist.* New York: Columbia University Press.
Lincoln, Roger J., Geoffrey Allan Boxshall, and P. F. Clark. 1982. *A dictionary of ecology, evolution, and systematics.* Cambridge: Cambridge University Press.

Mayr, Ernst. 1980. G. G. Simpson. In *The evolutionary synthesis: Perspectives on the unification of biology*, ed. Ernst Mayr and William B. Provine, 452–63. Cambridge, MA: Harvard University Press.

———. 1982. *The growth of biological thought : Diversity, evolution, and inheritance.* Cambridge, MA: Belknap Press of Harvard University Press.

Mayr, Ernst, and William B. Provine. 1980. *The evolutionary synthesis : Perspectives on the unification of biology.* Cambridge, MA: Harvard University Press.

Newell, Norman Dennis. 1956a. Catastrophism and the fossil record. *Evolution* 10(1): 97–101.

———. 1956b. Fossil populations. In *The species concept in palaeontology: A symposium*, ed. P. C. Sylvester-Bradley, 63–82. London: The Systematics Association.

———. 1959. Adequacy of the fossil record. *Journal of Paleontology* 33 (3): 488–99.

———. 1963. Crises in the history of life. *Scientific American* 208 (2): 76–92.

———. 1967. Revolutions in the history of life. In *Uniformity and simplicity*, 63–91. Boulder, CO: Geological Society of America (GSA).

Newell, Norman Dennis, and Edwin Harris Colbert. 1948. Paleontologist: Biologist or geologist. *Journal of Paleontology* 22 (2): 264–67.

Pauly, Philip J. 1987. *Controlling life : Jacques Loeb and the engineering ideal in biology.* Oxford: Oxford University Press.

Pearl, Raymond, and Lowell J. Reed. 1920. On the rate of growth of the population of the United States since 1790 and its mathematical representation. *Proceedings of the National Academy of Sciences of the United States of America* 6 (6): 275–88.

Princehouse, Patricia M. 2003. Mutant phoenix: Macroevolution in twentieth-century debates over synthesis and punctuated evolution. PhD diss., Harvard University.

Provine, William B. 1971. *The origins of theoretical population genetics.* Chicago: University of Chicago Press.

Rainger, Ronald. 1988. Vertebrate paleontology as biology: Henry Fairfield Osborn and the American Museum of Natural History. In *The American development of biology*, ed. Ronald Rainger, 219–56. Philadelphia: University of Pennsylvania Press.

——— 2001. Subtle agents for change: *The Journal of Paleontology*, J. Marvin Weller, and shifting emphases in invertebrate paleontology, 1930–1965. *Journal of Paleontology* 75 (6): 1058–64.

Reif, W. E. 1986. The search for a macroevolutionary theory in German paleontology. *Journal of the History of Biology* 19:79–130.

———. 1999. Deutschsprachige Paläontologie im Spannungsfeld Zwischen Makroevolutionstheorie und Neo-Darwinismus (1920–1950). In *Die Entstehung der Synthetischen Theorie. Beitruage Zur Geschichte der Evolutionsbiologie in Deutschland 1930–1950*, ed. T. Junker and E.-M. Engels, 151–88. Berlin: Verlag für Wissenschaft und Bildung.

Schuchert, Charles. 1910. Biologic principles of paleogeography. *Popular Science* 76:591–600.

Sepkoski, David. Forthcoming. The 'Delayed Synthesis': Paleobiology in the 1970s. In *Descended from Darwin: Insights into American evolutionary studies, 1925–1950*, ed. Joseph Cain and Michael Ruse. Philadelphia: American Philosophical Society Press.

Simpson, George Gaylord. 1926. Mesozoic mammalia, IV: The multituberculates as living animals. *American Journal of Science* 11: 228–50.

———. 1944. *Tempo and mode in evolution*. Columbia Biological Series No. 15. New York: Columbia University Press.

Simpson, George Gaylord, and Anne Roe. 1939. *Quantitative zoology: Numerical concepts and methods in the study of recent and fossil animals*, 1st ed. New York: McGraw-Hill.

Smocovitis, Vassiliki Betty. 1996. *Unifying iology: The evolutionary synthesis and evolutionary biology*. Princeton, NJ: Princeton University Press.

Teichert, Curt. 1956. How many fossil species? *Journal of Paleontology* 30 (4): 967–69.

Weller, J. Marvin. 1947. Relations of the invertebrate paleontologist to geology. *Journal of Paleontology* 21 (6): 570–75.

CHAPTER TWO

The Fossil Record: Biological or Geological Signal?

Michael J. Benton

New species have appeared very slowly, one after another, both on the land and in the waters. Lyell has shown that it is hardly possible to resist the evidence on this head in the case of the several tertiary stages; and every year tends to fill up the blanks between them, and to make the percentage system of lost and new forms more gradual. —Charles Darwin, *On the Origin of Species* (1859).

Darwin is referring here to Charles Lyell's famous nomenclature for the Tertiary System, in which the proportion of extinct to modern forms increased as one went back in time. He notes how new finds are plugging the gaps in the record. Charles Darwin famously devoted two chapters in *On the Origin of Species* (Darwin 1859) to the fossil record, and one of these was entitled "On the Imperfection of the Geological Record." Here he outlined the sequence of fossils in the rocks that showed how life changed from simple to more complex organisms up through the stratigraphic succession. He also highlighted the gaps in the fossil record, and reasons why every organism, and every species, would not necessarily be preserved. His main aim in covering these topics was to explain why the fossil record did not demonstrate a complete picture of the evolution of life, preserving all the intermediate forms demanded by his theory. And yet, Darwin ended with the hope that fossils would paint the pattern of the history of life as paleontologists continued their collecting efforts.

The next hundred years saw a to-and-fro in confidence about whether fossils could actually tell us much about the history of life. In reviewing this theme, Stephen Gould (1983) traced how Darwin's high expectations for the fossil record were quickly dashed: paleontologists failed to identify many successions of fossils that told the story of evolution. Indeed, by 1909, most evolutionists saw the future in the new science of genetics, and could see little use for fossils. After the modern synthesis of the 1930s and 1940s, George Gaylord Simpson and others had shown that paleontology at least confirmed all aspects of Darwinian evolution, but perhaps could show nothing more. Gould characterizes the change through these years as a move from "irrelevance" in 1909 to "submission" in 1959. By 1982, Gould argued, in typical gung-ho style, that paleontology was truly in "partnership" with biology in delivering major insights on larger-scale patterns and processes that could never be predicted from a study of living organisms alone. At the time when Gould (1983) wrote, the "consensus paper" by Sepkoski and colleagues (1981) had just been published (see chapter 18 of this volume), and leading paleontologists seemed to agree that they could expect to find the large patterns of evolution from numerical studies of fossil databases. But where are we now?

Since 1981, there was perhaps a decade of relatively unchallenged development of statistical and empirical approaches to macroevolution and the fossil record, but concerns have since been expressed from two main directions. First, the growth of molecular phylogenetics cast doubt on patterns of relationships and dates from the fossils. Second, paleontologists themselves began to highlight the disturbing fact that the shape of the fossil record seems to map directly onto geological signals, such as the sea level curve or the volume of rock deposited. Could it be that paleontologists had been living in a fools' paradise, doggedly plotting patterns of fossils through time that told us nothing about evolution, but a great deal about sampling?

Note that, in this chapter, I concentrate on the larger-scale patterns that may be gleaned from the fossil record—global diversification and mass extinction, for example—and not the medium-scale, lineage-specific aspects of evolution, where quality and sampling concerns are rather different.

MOLECULES AND FOSSILS

The Molecular Clock

Fossils held the hegemony of deep time in evolution until 1962. In that year, in a classic paper, Emil Zuckerkandl and Linus Pauling made a modest proposal—that perhaps proteins and other molecules changed at predictable rates through long timescales. They later called this concept the *molecular clock*. It was known then, for example, that all vertebrates, and various other organisms, possess the protein hemoglobin, which transports oxygen and makes our blood red. Human beta-hemoglobin is chemically identical to the hemoglobin of a chimp, but it differs in twenty-five of the 146 amino acid positions from the hemoglobin of a cow, in forty-five positions from the chicken, and in ninety-six positions from the shark. So, surely what this showed was that the amount of molecular difference was proportional to the time since any pair of species diverged from their last common ancestor. Humans and chimps diverged only a few million years ago, and so their hemoglobin has had little time to accumulate any changes, whereas humans and sharks last shared a common ancestor perhaps 500 million years ago, and so there has been 1,000 million years of independent evolution between the two.

The molecular clock was soon put to use as a tool in drawing phylogenetic trees, even though the labor in acquiring protein sequences in the 1960s was Herculean; the first such tree, with a set of proposed dates, was published by Vincent Sarich and Allan Wilson in 1967. These authors compared the hemoglobins of the great apes, showed that chimps were more closely related to humans than to gorillas or orangs, and as if that were not enough, suggested that the human and chimp lineages (evolutionary lines) separated a mere five million years ago. This predictably caused outrage on two fronts, but the date is our main concern. In 1967, paleoanthropologists were pretty clear that *Proconsul* from the early and mid-Miocene was on the human line, and that meant the split must have been fifteen to twenty million years ago. In the end, of course, the paleontologists reexamined their fossils and discovered that *Proconsul* was neither an ape nor a human, but an outgroup (relative of the ancestor) of both, and so it said nothing at all about the date of the chimp-human split. Even after four decades, the Sarich and Wilson (1967) date for the chimp-human split is pretty much correct, if perhaps slightly too young: the oldest human fossils are now *Sahelanthropus* and *Orrorin*, reasonably securely dated as six to seven million years old.

Since 1970, and with a few such high-profile debates, most molecular studies have tended to confirm patterns of relationship established using fossils and morphological characters, and the order of branching in these trees has tended to match the order of fossils in the rocks. An uneasy truce existed, where molecular phylogeneticists and paleontologists might snipe at each other from time to time, but they agreed about most things. This perhaps uneasy coexistence came to an end rather dramatically in the mid-1990s, partly as a result of the increasing ease in obtaining molecular sequences, as well as an important renewal of interest in the computing algorithms used to draw trees, but also perhaps reflecting some mutual misunderstandings between both camps.

Molecular Age Doubling

A couple of papers published in 1996 perhaps exemplify the renewal of debate, and a sudden renewal of qualms about the quality of the fossil record.
Gregory Wray, Jeffrey Levinton, and Leo Shapiro argued in a paper that attracted a great deal of interest that modern animal groups, ranging from
sponges and corals at one end to vertebrates and echinoderms at the other,
had originated some 1,200 million years ago, rather than 500–600, as the fossils indicated. This new date threatened the whole story of the rise of multicellular life through the late Precambrian, and especially the so-called Cambrian
Explosion, the time when marine animals with skeletons suddenly appear in
the fossil record at the beginning of the Cambrian Period, 542 million years
ago. The headline message of the new paper was clear: animals had diversified
600 million years before the paleontologists thought, and so half the history
of all those groups was simply missing. The new evidence understandably
caused paleontologists to doubt their evidence: could it be that a crucial half
of the record was missing?

The second paper was by Blair Hedges and colleagues, also published in
1996, and it looked at the timing of the origin of modern orders of birds and
mammals. Here again the fossil record told a story of sudden diversification,
this time after the demise of the dinosaurs some 65 million years ago. Birds
and mammals have a rich fossil record in the Mesozoic, and both groups had
long been known to have existed side by side with the dinosaurs. But these
were primitive orders—toothed birds like *Archaeopteryx*, and an array of mammals of modest dimensions that were classified outside the modern orders.
Hedges and colleagues presented their molecular evidence that modern birds
and mammal orders had originated perhaps 120–130 million years ago, well
before the first fossils. What was going on here? Paleontologists noted that
the new molecular dates were roughly twice the fossil dates in both cases.
Similar challenges followed through the 1990s, and to many commentators,
the paleontologists were giving way. No longer could they claim that the fossil
record told us the history of life—it told us only bits and pieces, and we didn't
know which bits, so perhaps it was time for the hoary old fossil hound to
hang up his hammer and leave the field to the molecular sequencers and their
sparkling new labs.

Paleontologists responded in three ways at the time: either as ostriches,
lapdogs, or mules. I hasten to classify myself as a mule in this. The ostriches,
perhaps the majority of paleontologists, ignored the molecular challenge and
rather hoped it would go away. The lapdogs accepted the new dates without

question and tried to find ways to accommodate their data: perhaps the molecular dates were the true dates and they recorded the moment of genealogical separation between major groups, but there had been a long cryptic history when the organisms were soft-bodied, rare, or living in restricted areas, and they then burst onto the scene much later, which marks their appearance in the fossil record. This is no explanation, of course, just a statement of ignorance, since it's unlikely a major group of organisms could sustain itself in obscurity for tens or hundreds of millions of years. The mules clung doggedly, or perhaps mulishly, to the view that the new molecular dates must be wrong, or at least that there must be a single story and both data sets—the fossils and the molecules—have to agree somehow. Were the ostriches, lapdogs, or mules right?

The Counter-Debate

Wray, Levinton, and Shapiro had drawn a new tree of relationships of the major animal groups based on particular genes they all shared, and they had calibrated the tree against spot dates from the fossil record. It was the calibration method that mattered. I remember reading their paper in 1996, and hearing the concerns of paleontological colleagues. I was inclined to doubt the new evidence, and oddly enough it was the easiest part of the exercise that gave rise to doubts. Sequencing genes and running the sequences through alignment and tree-building algorithms is a complex process, and methods are constantly debated and revised. Fixing the dates of the branching points has hitherto been a rather simpler procedure. Imagine a triangular elastic branching structure—fix it at the forked end, to represent the present-day groups, and then you can stretch it back as far as you like in time, using generally a single fixed point to date one of the nodes. Fix this calibration point at ten million years, and the tree stretches back to accommodate that. Fix the same point at fifty million years, and the whole tree stretches back five times as far.

Many people were concerned about the choice of calibration dates and, some years later, Kevin Peterson and colleagues showed a good reason for caution. Wray had used fixed dates from vertebrates to date the root of the animal tree. But vertebrate molecules evolve more slowly than those of nearly all other animal groups. So, this meant that Wray was using a slow molecular rate to project his estimates back in time, and the calculated date was much more ancient than it should have been. On recalculating, Peterson and colleagues (2004) found that the molecular evidence suggests a basal divergence of animal groups at 650 to 700 million years ago, much closer to the fossil estimates. There is still a time gap of 100 million years or so, and indeed this might be reduced on further recalculation of the molecular estimates or fossils may eventually be found that fill some of the gap.

What of the bird and mammal dates? Were they really twice as old as the oldest fossils, and was that first half of the fossil record of modern birds and mammals missing? Michael Foote and colleagues (1999) showed that the probability that fossils of modern mammalian groups would be entirely missing from huge spans of the Cretaceous was most unlikely, based on their preservation probabilities, and that is a reasonable statement of paleontological concern. Their view was countered by critics who argued that preservation probabilities for modern mammals in the Cretaceous were perhaps lower than those after sixty-five million years ago because either there were far fewer species and they were rare, or that the rocks were wrong (all marine), or that they were living in unsampled parts of the globe. But, in support of Foote, is the "missing mastodon" argument that I expressed in the same year (Benton 1999): we can go on explaining an apparent gap for so long, but eventually we have to accept that the missing taxon just isn't there, after searching hard and long. In the 1700s, some scientists explained the fossil bones of mastodon from Ohio as relics of living elephants that were yet to be found in some remote region in the American West. Eventually, of course, when all such remote areas had been explored, and no living mastodons were spotted, the bones had to be accepted as evidence of extinction. Paleontologists have identified a number (admittedly not many) of Late Cretaceous mammal localities, some yielding superbly preserved complete skeletons of placental and other mammals, but so far not a whisker of a member of a modern order.

Is there any sign of rapprochement between molecular and paleontological estimates for the branching of modern birds and mammals? There is still a substantial difference in the molecular and fossil dates for divergence of modern bird orders, although some unequivocal ducklike birds are now known from the latest Cretaceous, whereas before most such records were doubtful. The time gap is, however, still some twenty to thirty million years. A year or two ago it seemed that the mismatch in molecular and fossil ages for the modern placental mammals had been resolved, but perhaps not.

At one level the mammal paleontologists have not changed their position: there are indeed few, if any, fossils of mammals in the Cretaceous that can be assigned with confidence to a modern order. So, reputed records of basal monkeys, hedgehogs, and rodents living side by side with *Tyrannosaurus rex* and other dinosaurs are as dubious now as they always were. However, there are many Cretaceous records of placental mammals that lie outside the modern orders. Such fossils have been known for a long time, and they have been much enhanced recently by reports of complete specimens such as *Eomaia* from the 125-million-year-old Liaoning deposits of China. But these early forms cannot enlighten us about the molecules-versus-fossils debate because the fossils lie outside the tree of living groups. But there were some Cretaceous fossils that apparently lie within the tree of modern placental mammals but that do not belong to any modern mammalian order. These are the zhelestids, and possibly also the zalambdalestids, from Uzbekistan, described by Archibald, Averianov, and Ekdale (2001). The zhelestids were identified as basal relatives of the hoofed mammals and carnivores, while the zalambdalestids were said to be relatives of primates or rodents. These Uzbek fossils are dated as ninety-five to one hundred million years old, well down in the Cretaceous.

The eighteen or so modern orders of mammals are divided into four super-orders: the Xenarthra from South America, the Afrotheria from Africa (of course), and the Laurasiatheria and Euarchontoglires from the northern hemisphere. Clearly, these four great superorders branched first in the early history of the placental mammals, perhaps from ninety to one hundred million years ago, and then, after the extinction of the dinosaurs, the modern orders flourished from sixty-five million years onward. So, the paleontologists had failed to sort out the deep relationships of the modern orders of placental mammals, and they had apparently failed to assign the Uzbek fossils, and some others, to their correct positions in the tree. Equally, the molecular biologists had perhaps been too quick to accept the first dates they calculated, and some had not fully grasped the difference between orders and superorders.

Sadly, this apparent consensus, or rapprochement, has been shattered by a full-scale cladistic analysis of the zalambdalestids and zhelestids by John Wible and colleagues (2007). They argue unequivocally that both groups of fossils from Uzbekistan lie low in the tree, outside the four great superorders of modern placental mammals, and that they tell us nothing about the timing of radiation of the modern groups. Further, they argue that none of the placental mammal fossils from the Cretaceous belong to modern orders or superorders, so the twenty- to forty-million-year gap between the molecular dates and the first fossils remains unbridged.

Partnership

Paleontologists and molecular biologists share two major enterprises: drawing the tree of life and dating it. There is only a single tree, and each branching point must have a single date, in my view. Are we doomed to continuing spats like those just described? Practitioners in both fields need to respect each other and exercise caution. First results are not always correct results, especially if they contradict everything that has been put together by generations of other scientists. The morphologists and paleontologists tussled with the tree of placental mammals for years, and they failed miserably to resolve deep relationships. Then, molecular phylogeneticists discovered the Afrotheria in 1997, and the remainder of the tree fell into place rapidly. Numerous independent molecular investigations confirm the tree, and paleontologists are still puzzling over the morphological evidence for the new superorders. What is it that elephants, dugongs, hyraxes, tenrecs and golden moles share—apart from having originated in Africa? The schnozzle? Maybe. They all seem to share the character of testicondy, retention of their testes in the abdomen, and not exposed in a scrotal sac—but this might be the ancestral character of all mammals. Maybe no one will ever identify the morphological character that unites Afrotheria—perhaps they just evolved so fast in the Cretaceous that they never acquired a unique character, or such a character or characters has been lost, overwritten by their very different patterns of evolution since. Even the most hard-bitten paleontologist has to admit his or her data are not up to the task here.

In dating the tree, it seems to me there is a clear partnership. The to-and-fro debate about dates is daft. The irony, on the one hand, has been that molecular phylogeneticists have promoted their own calculated dates against the perceived weakness of existing paleontological dates—or should that be *most* existing paleontological dates, because at least one paleontological calibration date is needed to date the tree, and such calibration dates have sometimes been treated as some kind of holy grail, immune to challenge or criticism.

The partnership in dating the tree of life is simple: paleontologists supply as many dates as they can, and these are vetted for consistency and then used in multiples to determine otherwise unknown dates. Multiple dates and vetting are important. Paleontologists have a sense of which dates they can determine with some confidence, and which might be rather weak. But it's possible also to take a set of best-estimate dates and assess them for consistency, or congruence, on an established tree: if any should appear to be wildly out of line with the others, by being either too young or too old, they can be rejected. Then, the set of congruent dates can be used to determine missing dates. The weakness of this approach is, of course, that all the paleontological dates might be congruent because they are all too old or too young by the same amount. Equally though, such errors might emerge on subsequent checking with other sets of dates.

What of the dates produced by paleontologists? Up to now, most, but certainly not all, molecular phylogeneticists have tended to use point dates, without error bars. Many now try to encompass an error term in their calculations, say plus-or-minus 5% of the age. I have been studying this issue with Phil Donoghue, also at the University of Bristol, and we have realized that the distribution of uncertainty around a fossil date is not symmetrical, a point made previously by Robert Reisz, Johannes Müller, Marcel van Tuinen, Elizabeth Hadly, and others (Reisz and Müller 2004; van Tuinen and Hadly 2004). In fact, it is possible to give a rather accurate minimum constraint on the date based on the oldest known fossil in a clade (that is, the oldest definite fossil, ignoring scrappy and uncertain remains that might be older). Then, the distribution of probability on the date follows a logistic, or S-shaped, curve (fig. 2.1), mapping the rough shape of diversification of the group back to a point where there are no more fossils, even dubious ones, and one or more fossil beds can be identified that ought to contain fossils of the clade in question, but do not. Molecular phylogeneticists have methods to accommodate distributions of dates like this, with a so-called hard minimum constraint and a soft maximum constraint. We hope these developments (Benton and Donoghue 2007) mark a new era of collaboration across the systematics community, where molecular biologists and paleontologists each do what they are best at, and by combined efforts perhaps come closer to a reasonable estimate of the truth than has been possible up to now.

PLOTTING THE FOSSIL RECORD

A Biological Signal?

As noted at the start of this chapter, Charles Darwin assumed that the fossil record, plagued with gaps as it was, would show the shape of the history of life. And that is broadly the popular perception, too. Reports of new fossils and studies of mass extinctions in the press generally accept the dating of those fossils or events, and any statements about the shape of an evolutionary pattern inferred from the new discovery.

This common perception is based on qualitative evidence, and that qualitative evidence satisfies most paleontologists, and indeed most nonpaleontologists in the scientific community. The key observation is that new fossil finds rarely rewrite the textbooks, although we all conspire with our institutional publicity offices when they feed that line to the press. In fact, Darwin knew about early Paleozoic trilobites and brachiopods, Silurian and Devonian fishes,

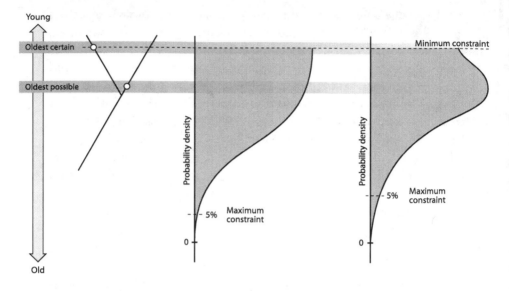

Figure 2.1 Dating the origin of a clade from fossils. The minimum date estimate may be known rather precisely, based on the oldest fossil that is known confidently from the group—this is a so-called hard estimate. The lower age estimate is, however, "soft," meaning that it is based on an assumption about the shape of the origin of the clade, here shown as an S-shaped, or logistic, curve, and an estimate of the 95% confidence interval on the lower date. This maximum estimate may be based on bracketing (known dates of the oldest fossils in outgroups of the clade in question), bounding (absence of fossils from a lower fossil deposit of the kind that ought to include such fossils), or a model of preservation likelihood and speciation rates. It is 95% probable that the true date of origin of the clade lies above the 95% maximum estimate, but there is a 5% chance it lies on the long tail below.

Carboniferous trees, giant insects and tetrapods, dinosaurs, Mesozoic mammals, *Archaeopteryx* (the oldest bird, then and now), Eocene horses, and so on. The two big areas of wholly new discoveries since 1860 are Precambrian life and human fossils. But, were the fossil record a patchy *and misleading* record, one might expect some real surprises, like a Carboniferous dinosaur or a Triassic trilobite. So, I would argue, the record may be patchy, but it is not necessarily misleading.

This was borne out by quantitative studies through the 1990s. A number of paleontologists realized that they had the remarkable good fortune to have at their disposal two, or arguably more than two, independent sources of data on the history of life. Not only are there the fossils in the rocks, there are also phylogenetic trees. A phylogenetic tree may contain fossil species, or it may be based entirely on living forms. The tree is drawn up by quantitative analysis of character data, either morphological or molecular. So, the age of a fossil has no influence on the placement of a species in the tree—its placement is based entirely on rationalizing the character distributions until they fit a single tree solution best. The disjunction between fossils in the rocks and phylogenetic trees is especially clear in the case of molecular trees that are produced entirely from protein or nucleic acid sequences of living organisms. It would be a wonder if molecular trees gave the same sequence of events as the order of fossils in the rocks—and yet they normally do!

This was first shown by Mark Norell and Mike Novacek from the American Museum of Natural History in 1992, who found that 75% of trees of mammals agreed significantly with the order of fossils in the rocks. Now, this is not meant to be a test of the fossil record, assuming the trees are correct, nor is it a test of the quality of the trees, assuming the fossil record is correct. It's an assessment of congruence between two data sets. If they disagree, then there is no way to tell whether the tree or the fossils, or both, are wrong. But if the trees and the fossils agree, then it's hardly credible to claim that one or other is at fault. Congruence in such a case indicates that both methods, based on different kinds of data, are telling the same, true, story.

After this analysis, several groups began pursuing this approach, and in the end, my colleagues Matt Wills and Becky Hitchin and I accumulated a set of 1,000 trees and fossil records spanning all groups of organisms, plants, animals and microbes, through the last 600 million years of geological time. We looked at congruence between trees and fossils using many different approaches and found remarkable levels of agreement, and the agreement seemed comparable for different groups of organisms and through geological time (fig. 2.2). In a 2000 paper, we argued that this confirmed that the fossil record gave us the correct broad pattern of the history of life, and that the quality of the record did not diminish markedly back in time.

This last finding was a surprise to us and everyone else. It is a commonplace observation that fossils are abundant and easy to collect in the Cenozoic, but obscure and rare in the Cambrian. Fossils are rarer in older rocks because large volumes of those older rocks have been lost to burial beneath younger rocks, subduction, folding, pressure, or erosion, and these processes often damage or destroy any fossils that can still be collected. So how can the trees versus fossil ages congruence measures be the same for truly ancient and less ancient fossil records? The answer has to do with the scale of observation. At the local scale, working in a quarry and collecting individual specimens, there is no doubt the fossil record gets worse with increasing age. However, at the global scale, and the levels of stratigraphic stages (mean duration seven

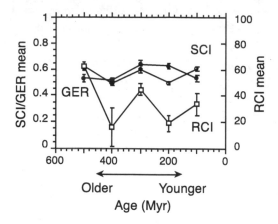

Figure 2.2 Congruence of the fossil record and phylogenetic trees. Many examples show good matching between the order of appearance of fossils in the rocks and the order of branching in a phylogenetic tree. Comparisons of a large set of such trees with known fossil records suggest that levels of congruence (or matching) remained constant through the last 500 million years. Here, a sample of 1,000 trees, divided into five time bins (each of about 200 trees) shows remarkably constant mean congruence values for the SCI (stratigraphic consistency index) and GER (gap excess ratio) measures through time. The RCI (relative completeness index) declines toward the present, but this is because the numerator and denominator include absolute time (so the older the base of a tree, the higher the value). In all cases, the higher the value of the metric, the better the congruence. (Based on information in Norell and Novacek [1992] and Benton, Wills, and Hitchin [2000].)

million years) and genera or families (not species or specimens), the quality of the fossil record seems constant—maybe constantly good or constantly bad, but constant nonetheless.

Critics have pointed out that all we were measuring was the proportion of the fossil record that is known. The contrast to be grasped is between the diversity of life as it actually was in the past, and the sample represented by fossils (namely, the fossil record). So, it could be that the fossil record documents 80% of the life of the Miocene, but only 5% of the life of the Cambrian—and this is quite likely. Our clade-versus-age measures might show that we have uniform coverage of the fossil record through time, but that would not address the issue of the overall very weak documentation of the Cambrian. Our response is that clade-versus-age measures are applied to cladograms, and those cladograms can include fossilizable and unfossilizable organisms. So, a tree based on modern organisms can extend back in time as far as the Cambrian, for example, and indicate that jellyfish or worms were there; their absence as fossils would then translate into a major gap that would lower the values of the clade-versus-age metrics. Nonetheless, the use of clade-versus-age metrics does not provide a strong link between reality and the fossil record, because it presumably works less and less well the further back in time one goes.

If this is the case, it confirms the assumptions made in the consensus paper by Sepkoski and colleagues (1981) that the fossil record retains enough of a biological signal that we can use it to study some aspects of evolution. But that consensus has been challenged by some paleontologists in the past ten years.

A Geological Signal?

First, John Alroy and colleagues (2001) have suggested that sampling might diminish markedly back in time. Here they are resurrecting the view expressed by David Raup in his classic 1972 paper where he outlined all the reasons, some of them just indicated, why the decline in marine diversity back into the Paleozoic is not a real signal, but actually documents the ever-worsening quality of sampling. John Alroy and colleagues have presented numerous statistical studies based on sectors of the fossil record that appear to support this view. If they are correct, the fossil record presents more a geological signal than a biological signal. Is there independent evidence for such a claim?

Shanan Peters and Mike Foote argued in 2002 that the fossil record largely mimics the distribution of rocks. They plotted extinction and origination curves against the number of named stratigraphic formations in the United States. At times when diversity and origination were apparently high, so too were the numbers of formations; at times of extinction, the number of named formations plummeted. They interpreted this to mean that our record of the diversity of fossils through time is based largely on the volume of available rock, or at least the volume of studied rock. The more paleontologists work on rocks of a particular age, the more fossils they find and the more formations they name. If fossil volume mirrors rock volume, what might control that?

Sea level might be the driver. Through time, global sea levels have risen and fallen, and at present we are in a time of relatively low sea levels because huge volumes of water are locked up in polar ice. In the Cretaceous, for example, when there was no, or very little, polar ice, sea levels were up to 200 meters higher than today. When sea levels are high, more marine rocks are deposited, especially on the continental shelf, where life is most abundant. So, when sea level rises, the volume of marine rocks deposited worldwide increases, and so too does the apparent diversity of life in the sea. The corollary might be that when sea levels are high and marine fossils are being buried in abundance, there should be low levels of deposition and fossil preservation on land, a view suggested by Peters and Foote (2001; [Equally, high sea levels might mean the continents become more fragmented, and so endemism increases, and global diversity on land could also increase]).

The sea level idea has been tested by Andrew Smith (2001, 2007) and Emmanuel Fara (2002). Smith found that the marine fossil record of much of the Mesozoic largely mirrored the sea level curve, and there is no doubt that particular rises and falls in apparent diversity must relate to particular up and down movements of the sea. But Smith found that the global marine diversity curve diverged from the sea level curve over the past one hundred million years: as sea levels have steadily fallen through that time, marine diversity has rocketed. Further, Fara found no inverse correlation between the sea level curve and the apparent diversity of life on land. The prediction that rises in sea level might be matched by falls in terrestrial diversity, and vice versa, just did not happen.

So, if the rock and fossil records are linked to some extent, but not completely, is one driving the other, or are both driven by a third cause? Following publication of his paper in 2002, I commented to Mike Foote at a conference that the linked patterns might not indicate that rock volume was driving fossil diversity, but perhaps the opposite. It might well be that geologists name lots of formations when fossils are abundant, and name fewer when fossils are rare. That was perhaps a cheeky volte face of the rocks-drive-fossils argument, but Shanan Peters (2005) has now resurrected an older idea first suggested by Jack Sepkoski and others in the 1970s (Sepkoski 1976; Flessa and Sepkoski 1978), and even by Norman Newell (1967) with regard to extinctions, that the linked formations/fossil richness curves might be driven by a third common cause—maybe sea level change, or maybe sea level and continental plate movement. The point is then that the matching curves are giving *both* geological and biological signals, not merely a geological signal that masks any biological signal. The *common-cause* hypothesis of Sepkoski, Peters, and others is that times of high sea level and voluminous sediment deposition were also times of rich life in the seas, and so both rocks and fossils are abundant. There is an analogy here with the long-established species-area effect in ecology: large islands support many species, while small islands support few.

PERSPECTIVE

I have never doubted that the fossil record tells the story of the history of life, and neither did Jack Sepkoski. The 1981 consensus paper may have been naive in some ways—after all, the five matching patterns just proved that the five samples of the fossil record matched, not that this was the true pattern. All five databases, whether sampling fossil species, genera or families, or trace fossil diversity, were all sampling from the same rocks, although the sampling methods differed. So the sampling was congruent, but was the record in the rocks anything like the true biological pattern? What about those episodes in Earth history when no rock has been preserved and so we have a complete blank? What about those groups of soft-bodied, or microscopic, organisms that we know nothing about? This is impossible to assess because what is required is a comparison of the known (i.e., the fossil record) with the unknown. There are perhaps two responses to this point that I believe to be correct, though neither is decisive.

The commonest response would be to say that modern oceans and landscapes contain a mixture of readily fossilizable, and essentially unfossilizable, organisms. We know the soft-bodied and microscopic groups today, and so these can be restored for past times. So, we know the Ordovician ocean was populated by brachiopods, trilobites, corals, sea lilies, and bryozoans, and it can also be assumed to have harbored jellyfish and worms. And indeed, sites of exceptional fossil preservation, such as the Burgess Shale or the Solnhofen Limestone, provide some kind of test: we know the jellyfish and worms from those times because they are occasionally preserved. We just have to join the dots a little more in reconstructing the evolution of such soft-bodied groups. The sites of exceptional fossil preservation suggest that the proportions of skeletonized to nonskeletonized organisms were pretty much the same in the past as they are today, so perhaps this conservative assumption can be applied to fill out the known fossil record.

The second answer comes from our work on comparing trees and fossil records. Trees contain a mix of readily fossilizable and unfossilizable organisms. Indeed, a molecular tree can be constructed for worms or jellyfish just as readily as for clams or fishes. A tree of entirely soft-bodied organisms clearly cannot be compared with the fossil record because there is little or no fossil record. But a tree of entirely skeletonized organisms can be compared with one that mixes skeletonized and nonskeletonized organisms, such as a tree of all animals. Both kinds of trees match the fossil records equally well. There is no evidence (yet) from molecular trees that massive sectors of biodiversity are being missed or that such soft-bodied organisms, if we knew them, would entirely turn our picture of ancient ecosystems on its head.

I see a golden future for paleontology in drawing and dating the tree of life, in seeking to understand major diversifications and extinctions, trends, large-scale ecosystem shifts, and the dance of faunas and floras as continents and oceans moved stately around the globe. The challenges to fossil record quality are good for the vigor of the subject, and sampling standardization is crucial. However, it would be wrong to overstate the problems: paleontologists can confidently follow Jack Sepkoski in celebrating and promoting their astonishing database on the history of life.

Acknowledgments

I am very grateful to David Sepkoski and Michael Ruse for the invitation to write for this volume, and to Mike Foote and Arnie Miller for their incisive (and kindly expressed) criticisms of an earlier draft.

REFERENCES

Alroy, J., C. R. Marshall, R. K. Bambach, K. Bezusko, M. Foot, F. T. Fürsich, and T. A. Hansen. 2001. Effects of sampling standardization on estimates of Phanerozoic marine diversification. *Proceedings of the National Academy of Sciences, USA* 98:6261–66.

Archibald, J. D., A. O. Averianov, and E. G. Ekdale, 2001. Late Cretaceous relatives of rabbits, rodents, and other extant eutherian mammals. *Nature* 414:62–65.

Benton, M. J. 1999. Early origins of modern birds and mammals: Molecules vs. morphology. *BioEssays* 21:1043–51.

Benton, M. J., and P. C. J. Donoghue. 2007. Paleontological evidence to date the tree of life. *Molecular Biology and Evolution* 24:26–53.

Benton, M. J., M. Wills, and R. Hitchin. 2000. Quality of the fossil record through time. *Nature* 403:534–538.

Darwin, C. 1859. *On the origin of species*. London: Murray.

Fara, E. 2002. Sea level variations and the quality of the continental fossil record. *Journal of the Geological Society* 159: 489–91.

Flessa, K. W., and J. J. Sepkoski, Jr. 1978. On the relationship between Phanerozoic diversity and changes in habitable area. *Paleobiology* 4:359–66.

Foote, M., J. P. Hunter, C. M. Janis, and J. J. Sepkoski, Jr. 1999. Evolutionary and preservational constraints on origins of biologic groups: Divergence times of eutherian mammals. *Science* 283:1310–14.

Gould, S. J. 1983. Irrelevance, submission, and partnership: The changing role of paleontology in Darwin's three centennials, and a modest proposal for macroevolution. In *Evolution from molecules to men*, ed. D. S. Bendall, 347–66. Cambridge: Cambridge University Press.

Hedges, S. B., P. H. Parker, C. G. Sibley, and S. Kumar. 1996. Continental breakup and the ordinal diversification of birds and mammals. *Nature* 381:226–29.

Newell, N. D. 1967. Revolutions in the history of life. *Special Paper of the Geological Society of America* 89:63–91.

Norell, M. A., and M. J. Novacek. 1992. The fossil record and evolution: Comparing cladistic and paleontological evidence for vertebrate history. *Science* 255: 1690–93.

Peters, S. E. 2005. Geologic constraints on the macroevolutionary history of marine animals. *Proceedings of the National Academy of Sciences, USA* 102:12326–31.

Peters, S. E., and M. Foote. 2001. Biodiversity in the Phanerozoic: A reinterpretation. *Paleobiology* 27:583–601.

———. 2002. Determinants of extinction in the fossil record. *Nature* 416:420–24.

Peterson, K. J., J. B. Lyons, K. S. Nowak, C. M. Takacs, M. T. Wargo, and M. A. McPeek. 2004. Estimating metazoan divergence times with a molecular clock. *Proceedings of the National Academy of Sciences, USA* 101:6536–541.

Raup, D. M. 1972. Taxonomic diversity during the Phanerozoic. *Science* 177:1065–71.

Reisz, R. R. , and J. Müller. 2004. The fossil record and molecular timescales: A paleontological perspective. *Trends in Genetics* 20:237–41.

Sarich, V. M., and A. C. Wilson. 1967. Immunological time scale for hominid evolution. *Science* 158:1200–03.

Sepkoski, J. J., Jr. 1976. Species diversity in the Phanerozoic: Species-area effects. *Paleobiology* 2:298–303.

Sepkoski, J. J., Jr., R. K. Bambach, D. M. Raup, and J. W. Valentine. 1981. Phanerozoic marine diversity and the fossil record. *Nature* 293:435–37.

Smith, A. B. 2001. Large-scale heterogeneity of the fossil record, implications for Phanerozoic biodiversity studies. *Philosophical Transactions of the Royal Society, Series B* 356:1–17.

———. 2007. Marine diversity through the Phanerozoic: Problems and prospects. *Journal of the Geological Society* 164:731–45.

van Tuinen, M., and E. A. Hadly. 2004, Error in estimation of rate and time inferred from the early amniote fossil record and avian molecular clocks. *Journal of Molecular Evolution* 59:267–76.

Wible, J. W., G. W. Rougier, M. J. Novacek, and R. J. Asher. 2007. Cretaceous eutherians and Laurasian origin for placental mammals near K/T boundary. *Nature* 447:1003–06.

Wray, G. A., J. S. Levinton, and L. H. Shapiro. 1996. Molecular evidence for deep Precambrian divergences among metazoan phyla. *Science* 274:568–73.

Zuckerkandl, E., and L. Pauling. 1962. Molecular disease, evolution and genetic heterogeneity. In *Horizons in Biochemistry*, ed. M. Kasha and P. Pullman, 189–225. New York: Academic Press.

CHAPTER THREE

Biogeography and Evolution in the Early Paleozoic

Richard A. Fortey

The inventory of past marine life is much more than a list of extinct species. Each of those species had its own biography: its habitat preferences, its mode of life, and reasons for its eventual demise. The present-day biota has a complex history, and it is likely that the geological past was every bit as interesting. Even before evolution came to be a respectable intellectual option, the geographical distribution of organisms had been a subject of scrutiny. Some eighteenth-century scientists, for example, considered that animals claimed as extinct may actually have lurked in then-unexplored parts of the world. Even today, myths of exciting survivors have currency among so-called cryptozoologists. Founding the science of biogeography, Darwin's contemporary, Alfred Russel Wallace, famously recognised the differences in geographical and taxonomic distribution of organisms across "Wallace's Line." It soon became clear that the evolutionary history of the biological world could not readily be disentangled from that of the landmasses of which the earth is composed. Much of the biological richness of the planet derived from the fact that organisms were quite different in separate locations. How this came to be—the history of animal distributions—soon brought paleontologists into the discussions. What was seen today could be understood by looking to the past—and to fossils.

In the first studies, animal and plant distributions were plotted out on existing geography. Patterns soon began to emerge, such as the concentration of marsupial animals in Australia. Curious anomalies—such as the presence of marsupials also in southern America—soon demanded to be explained. The acceptance of continental drift, later to be formalized as plate tectonics, placed the historical geographical distributions of animals and plants in a different context. From the first, fossils were to have an important part to play in scientific advances into ancient geography. Maybe, the reasoning went, fossil distributions could be used to map out the shape and extent of former continents, rather like postage stamps issued from a lost world. Sets of fossils of limited distribution could be used to plot the limits of former geological and geographical entities—ancient continents, supercontinents, or the smaller tracts of continental or oceanic crust known as *terranes*.

However, during the earlier half of the twentieth century there were many other paleontologists who claimed that distributions might rather be explained by migrations or invasions from centers of origin. The arguments were somewhat analogous to those raging among the contemporary anthropologists: there were those who claimed that similarities between the ancient civilizations of, say, Egypt and South America—such as the construction of pyramids— were the result of parallel cultural evolution in separate origins of civilization. Others thought the similarities so persuasive that they reasoned there had to have been migration of people between the two areas: there must have been Drakes or Van Deimens of prehistory. Expeditions were mounted (those of Thor Heyerdahl come to mind) to demonstrate that one journey or another was possible. This made for great adventures and best-selling books. Questions of human migration were not finally settled until genetic markers could be sequenced—the biomolecular equivalent of those postage stamps. South America and the Middle East were firmly separated at last. Nonetheless the same science also proved that mankind could and did travel far.

Paleontological evidence was to the fore in the debates about the former extent of the Gondwana continent, and the truth or otherwise of continental drift. Proponents of the connections between Africa, India, and South America used fossil distributions, most famously that of the large-leafed late Paleozoic tree *Glossopteris*, as proof that there had once been a single continent in the southern hemisphere. How else could one account for the presence of identical species on what were now widely separated continents? The similarities of rocks and fossils over this area were already well known to Eduard Suess at the end of the nineteenth century, and Alfred Wegener and Alex du Toit were quick to exploit the information in support of mobilistic theories two or three decades later. Dispersalists provided the opposition, citing evidence of land bridges, and accounts from sea travellers of tortoises adrift on floating logs: for a while they held the day, when geophysicists were of like mind. Only when the evidence from palaeomagnetism supporting mobile continents had been introduced did the opposition subside, and Gondwana and the greater supercontinent of Pangaea at the end of the Paleozoic become intellectually acceptable; even so, I can still recall hearing a lecture from an antimobilist as late as the eighties (I mean the 1980s, of course).

It is now accepted that much of the evolutionary history of life in the Phanerozoic has been intimately related to continental configuration. Times of wide continental dispersal accompanied by well-marked climatic zonation have stimulated vicariant speciation, and hence a diversity of species. This is the situation at the present day, when we enjoy a rich assemblage of different faunas on different continents running from poles to equator: a time of plenty in terms of biodiversity (at least, until mankind's actions took their toll). By contrast, the Pangaea phase of continental unification also coincided with the end-Paleozoic diversity low: the giant supercontinent was not conducive to great biodiversity. Climate and geography conspired to make this the major dip in the diversity curve of the history of life that Jack Sepkoski drew, a time when extinction also took its most dramatic toll.

By the late 1960s it was realized that Pangaea itself had been assembled from still earlier paleocontinents, which had converged to weld into a single body. The crucial paper here was that by J. T. Wilson (1966), published in *Nature*. Moving back in time before the Permian to greater than 290 My became a major geological and paleontological challenge. By contrast to that connected with the break-up of Pangaea, the "palaeomag" was still very primitive at the time, and fossil evidence was crucial in reconstructing the form of those long-vanished continents that were destined to fuse into Pangaea. Wilson himself had used the contrast in Ordovician trilobite faunas to either side of the Appalachian-Caledonian mountain chain as the crucial piece of evidence for the existence of what was called a former "proto-Atlantic" ocean. The idea was that this Ordovician ocean ran between the European and North American (Laurentian) continents, following broadly, but not exactly, the line of the present Atlantic, which, of course, had only opened again as Pangaea came apart. This neatly accounted for some observations that had first been made a hundred years earlier, noting how similar the rocks of Northwest Scotland were to those of the "Beekmantown" of Quebec province and parts of New York State. Of course they were similar—they were part of the same ancient continent! Closure of this "proto-Atlantic" ocean resulted in the major tectonic signature of a linear mountain belt now comprising the Appalachians, the Caledonides of Ireland, Wales, and Scotland, and the grand coastal range of Norway—these were the Himalayas of the Devonian.

A frisson of excitement passed through both the geological and paleontological community after the publication of Wilson's paper. Here was a new paradigm with which to interpret any number of geological phenomena: an ancient mountain chain beckoned. A century of fieldwork had supplied masses of information, but only now was it possible to conceive of a synthesis. I was a research student at the time, and John Dewey lived in the next office to mine in Cambridge (UK). I use the work "live" advisedly, because during the phase of the late 1960s and early 70s Dewey did indeed spend most of his time in the office, devising elegant summaries to show how much of the geology of the eastern seaboard of North America and the western seaboard of northern Europe might be accounted for by recognizing features appropriate to the closure of a major ocean basin (Dewey 1969). He worked for some of this time with Jack Bird, and the sounds of their guffaws and quipping formed a kind of leitmotif to my own research at this time. For a profoundly new vision of this great tract of land was evolving. Who could resist seeing in the mind's eye the rising of the great Caledonian chain as Europe and America came together to form a seam in the center of Pangaea? And as the mountains rose it was easy now to envisage the initiation of the fresh water river systems and basins that became the sites for Old Red Sandstone (Devonian) deposition, which were also the places where fishes became transformed into land animals, and plants greened the landscape for the first time. Life and earth and geography were linked together in a new explanation that seemed an illumination rather than a prosaic piece of history.

I might be indulged to introduce my own research at this point. While Dewey remade the world, I was sitting in the next room digging out Ordovician trilobites collected from the Arctic island of Spitsbergen. This apparently esoteric study turned out to have direct connections with the plate tectonic revolution. The Ordovician was a time when continents were almost as dispersed as they are at the present day. It was also when Wilson's proto-Atlantic ocean—by now renamed Iapetus—was wide, but closing. In my research I soon realized that there were different natural assemblages of bottom-dwelling trilobites related to depth across the Ordovician continental shelf of Laurentia (of which Spitsbergen was a part; Fortey 1975). The deeper-water communities could be used to recognize the proximity of former continent edges, and they had special biogeographical properties, too, with the deeper trilobites including more widespread taxa. Recognition of former continents was most reliable if one looked at shallower organisms, most specifically adapted to the ambient conditions. In these kind of facies the faunas on opposite edges of Iapetus were most strikingly different. This work was significant because up to that time (Whittington 1966) it had been claimed that there *were* no natural trilobite communities related to ancient environments. Now we could relate them to vanished oceans. In 1977, I was lucky enough to jump across the present Atlantic Ocean to go to Memorial University of Newfoundland, where Hank Williams was producing his magisterial summary maps of Appalachian geology. It was clear that the broad picture of faunas and geography applied there, too. The island of Newfoundland spanned both sides of Iapetus with the "mobile belt" between. Within this belt there were islands of various kinds whose positions within Iapetus were uncertain: some of their faunas had decidedly mixed properties, and one might have concluded that they were volcanic islands stranded within the ancient ocean. These examples led to the development of the concept of the terrane (Williams and Hatcher 1982) that in turn led to a more realistic modeling of the history of the ancient oceans.

If the recognition of Iapetus set the scene, that still left the rest of the world in the Lower Paleozoic to consider. Early models had been made using pencil and paper: John Dewey in particular was a masterly creative draughtsman in three dimensions. It would not be too much of an exaggeration to say that early models of the early Paleozoic globe were made by pushing around cardboard cut-outs. Computers were obviously destined to make more sophisticated (and sometimes more accurate) syntheses from data relevant to ancient geography. Then paleontologists could see how their data mapped out on these base maps. This endeavor began seriously with a symposium held in Cambridge under the title *Organisms and Continents through Time* (Smith, Briden, and Drewry 1973), which became the first in the Special Papers series of the Paleontological Assocation and remains the best-selling number in the series. The base maps for different slices of time produced by Smith, Briden, and Drewry were on the basis of paleomag evidence. Meanwhile, on the other side of the Atlantic, Fred Zeigler and Chris Scotese at the University of Chicago were developing a more subtle system that maximized continental configurations according to a swathe of different data derived from sedimentology, stratigraphy, and paleomagnetism. Fred had realized that information on past paleogeography would be useful to industry in resource prediction, and obtaining industrial sponsorship allowed for a much more powerful computing base than had hitherto been possible. Chris Scotese continued this work for many years—and indeed still does (www .scotese.earth.com), steadily refining and improving the continental reconstructions, and animating them in a time series. For a party piece he allows the continents to keep moving from the present day for the next hundred million years or so. Probably the most cited Paleozoic continental compilations were those made by Chris Scotese in conjunction with the Oxford paleontologist Stuart McKerrow (Scotese and McKerrow 1990); their early Ordovician map is reproduced here. One might say that this work was the paleogeographical equivalent of Jack Sepkoski's databasing efforts in other directions. There were other groups at work with computer reconstructions at the same time—for example, Clive Burrett in Australia, using paleontological evidence in particular. The problems with remagnetization that had bedeviled some of the earlier paleomagnetic work were also gradually being resolved. With several teams at work it was perhaps not altogether surprising that there were sometimes vigorous differences of opinion as to the position of a given continent at certain time slices in the Ordovician.

The North Atlantic region was a case in point. There had been a long tradition of mapping out the distribution of Ordovician faunas around this region, since the fossils were better known here than anywhere else in the world. Just prior to the plate tectonic revolution, the distinguished brachiopod worker Alwyn Williams had plotted the distributions of his animals (fig. 3.1), showing how ocean currents around a narrow mid-Pangaean sea might account for the patterns he observed. The recognition of Iapetus completely restructured these data within a year or two. On J. T. Wilson's original sketch the eastern side of Iapetus comprised a single, large European continent stretching from the Norwegian Arctic to Africa. However, paleontologists soon recognized that there were major differences in the earlier Ordovician in fossil faunas, such as brachiopods and trilobites, between the Baltic regions and more southerly areas such as Wales and Brittany. Faunas from the latter strongly resembled those of Gondwana, the western part of which lay near the pole at the time. The conclusion could be drawn that the southern part of Europe (including Avalonia—i.e., Wales and eastern Newfoundland) was close to Gondwana and at high paleolatitudes—and that the Baltic continent was separate and at temperate paleolatitudes. Sedimentologic evidence pointed to the same conclusion. Cocks and Fortey (1982) named an Ordovician seaway—Tornquist's—that separated a Baltica continent from Gondwana proper. This was in contradiction to a paleomagnetic "fix" current at the time that had placed Avalonia at low paleolatitudes. How dare mere fossils counter an answer that had come out of the latest technology? I recall a vigorous, not to say rude discussion at an international meeting shortly after the 1982 publication where the well-known "paleomagician" Jim Briden stated that the fossils simply had to be wrong. Fortunately, time heals all such controversies,

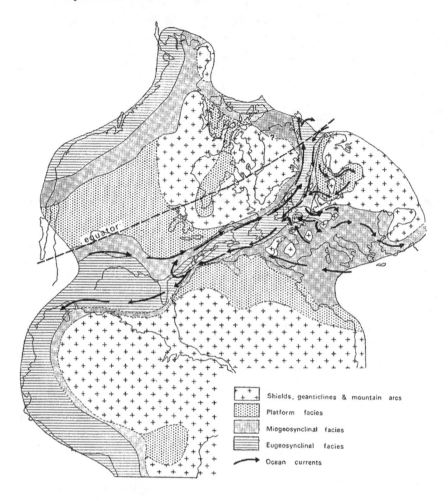

Figure 3.1 Faunal distributions explained by invoking oceanic circulation patterns, prior to the "plate tectonic explanation." Ordovician brachiopods from Williams (1969, fig. 29).

and better Ordovician pole positions were subsequently obtained that supported the high-latitude position of Avalonia in the early Ordovician. The fossil people and the paleomagnetic people have converged in the end. Furthermore, as the paleomagnetic data have got better, it has suggested configurations that probably could not be detected by the faunal evidence alone. For example, Cocks and Torsvik (2002) have suggested that the Baltica continent rotated during the Ordovician. The "three continent" version of the North

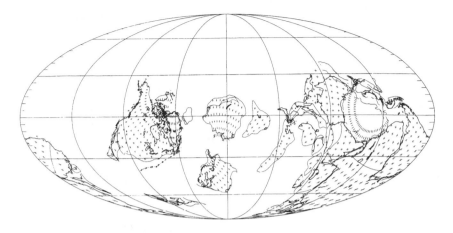

Figure 3.2 Early Ordovician (Floian) continental distributions, which explain faunal distributions more elegantly. From Scotese and McKerrow (1990, fig. 7).

Atlantic area is now accepted by most workers, but one can still find dissenters today (e.g., Carls 2003).

On a global scale, there are still differences in the placement of continental masses between different schools, but the major features of the Scotese and McKerrow maps (e.g., fig. 3.2) would probably be accepted by the majority of workers. Smaller continental masses and terranes are a different matter: research is still actively ongoing in these areas. The complex territories in central Asia are a good example. In early world maps, Kazakhstan is shown as a kind of triangular mass, separate from the Siberian plate. It is actually a whole series of concatenated terranes that were strung out originally as separate entities across oceans during the Ordovician. Central Asia is a perplexing jigsaw puzzle, and working out what was where, and when, is going to take a while. When I reviewed faunal evidence for these areas in conjunction with Robin Cocks in 2003, the evidence from trilobites and brachiopods seemed to point to most of Kazakhstan lying off the Chinese plate. This was in contradistinction to a model that had spread out the terranes in a long "Kipchak Arc" between former Baltica and Siberia (Sengor and Natal'in 1996). But not everyone agrees with this revised model, and I would not like to predict the outcome. The world cannot be remade too quickly. One outcome is certain: paleontologists will have an important role in future discussions. There will be attempts to use faunal data more objectively to determine an optimal position for a given terrane (for example Lees et al. 2002). In a decade or so there may be sufficient agreement for ancient geography to have passed from the realm of science into the province of history.

*   *   *

With continental distributions better understood, the next step is to look at their effect on global marine biodiversity. Jack Sepkoski's diversity curves had shown that the Ordovician was the time when variety at most taxonomic levels increased rapidly, and when many of the major groups of marine organisms either originated (such as bryozoans, many brachiopod groups) or rapidly diversified (molluscs in general, cephalopods in particular). It was a time when ecosystem structures, such as reefs and within-sediment burrowing, achieved something of their modern design. These different aspects were famously encapsulated in Jack Sepkoski's (1981) factor analysis of the fossil record and the recognition of the so-called "evolutionary faunas," the Ordovician being when the Cambrian evolutionary fauna declined as the Paleozoic evolutionary fauna increased—and some elements of the Modern evolutionary fauna appeared—the result being an overall increase in marine biodiversity during the period.

If the Cambrian was the time when many of the lifestyles of marine organisms first appeared, the Ordovician was as crucial as a time when many ecological aspects of the marine biosphere anticipated those pertaining at the present day. This "great Ordovician biodiversification event" has been recently summarised by Webby et al. (2004), who provided a compendium of the differing histories of all the taxonomic groups. It would be impossible to summarize this information here. Some studies (Adrain et al. 1998) indicate that the main division between Cambrian-style faunas and those typical of the rest of the Paleozoic was at the Lower-Middle Ordovician boundary rather than at the base of the system. The biogeographic component of Ordovician biodiversification will be emphasized in this essay. It is likely that the history of continental movements in the Ordovician was as important as any other factor in shaping the future design of the marine biosphere.

We have seen that the Ordovician was a time when the continents were widely dispersed—but also a period of continuous change between them, when oceans were both destroyed by subduction and initiated by continental fragmentation. Ordovician paleogeography is best understood in the context of a few large paleocontinents—Laurentia, Baltica, Siberia, and Gondwana. Between and around them were a number of island arcs, microcontinents, and oceanic islands (including terranes). Because the Ordovician was typified by shallow epeiric seas that spread widely over continental platforms, the fossil record from the main continental areas is, in general, excellent. That from the island arcs and microcontinents is altogether more patchy. In the case of island arcs the record is preserved in what are now eroded mountain belts, such as the Appalachian/Caledonian orogen or the Urals. Island arcs are preserved in areas of soft collision like Newfoundland, and are often destroyed in areas of former continent-continent collision. In a few sites, continent margin deposits have been obducted onto the adjacent craton, the best known of which is the Cow Head Group of western Newfoundland (James and Stevens 1986). Overall, our knowledge of marginal and oceanic faunas is deficient, and of oceanic areas—now subducted away—we know very little indeed. This should be borne in mind in the context of the generalizations that follow.

The major continental areas have positions that are now well constrained from paleomagnetic evidence. It is clear that Laurentia straddled the paleo-equator for the duration of the Ordovician, so it was largely a tropical area with abundant limy substrates. The western part of Gondwana, by contrast, surrounded the pole, and noncalcareous, clastic deposition naturally dominates the whole Ordovician in this region. The Gondwana continent was sufficiently vast, however, to span all paleolatitudes from the pole to the tropics—indeed, the rocks of the tropical side in central Australia could pass for those in New Mexico. Baltica was at temperate latitudes for the earlier part of the Ordovician; Siberia was also tropical although independent from Laurentia. Such strong climatic zonation and geographical separation was conducive to high biodiversity and to the development of what used to be termed "faunal provinces." There is some evidence that some groups of organisms radiated separately in the different paleocontinents; for example, clams were abundant, and soon varied, in some parts of Gondwana in the early Ordovician, but decidedly rare in Laurentia. Conversely, many early corals were Laurentian. The history of widespread groups like trilobites was different on separate paleocontinents. As the continents moved during the course of the Ordovician, so the trilobite faunas changed in response (Lees, Fortey, and Cocks 2002). Faunas only merged as continents approached one another. It was fortunate that some groups of extinct organisms, graptolites and conodonts in particular, included sufficiently widespread species to permit international correlation of strata.

This is only a partial caricature of the faunal picture. There were other important controls on biodiversity besides the disposition of paleocontinents. Recall the natural depth-related fossil assemblages on which I had done early work in the Ordovician of Spitsbergen. Faunas from one habitat to another could be as distinct in their way as those from one continent to another. As so often before, Jack Sepkoski (1988) neatly summarized the different components of biodiversity: that within a given fauna or community, between different communities, and between geographically isolated areas, typically paleocontinents. Art Boucot had done sterling work on depth-related communities (which he termed *benthic assemblages*—BAs) and I always regretted that Art and Jack seemed to be coming from different cultures—no doubt the reason for their intellectual estrangement was the fact that Jack had never described a fossil (to my knowledge), something that Art regards as a ticket of admission to the profession of paleontologist. Nonetheless, the recognition of different benthic assemblages showed interesting new patterns. Deeper faunas were often comparatively widespread, even extending from one continent to another, in contrast to shallower faunas. Jack and David Jablonski investigated patterns of origination—there seemed to be a correlation between first appearances of major taxa and shallow-water habitats. Only later would faunas spread out down the continental shelves. The fish (and therefore, ultimately, ourselves) may provide an example: this is important stuff.

We are starting to have a picture of the Ordovician biological world and its history, and it cannot be separated from the geography of the time. The game of "interrogate the database" which Jack Sepkoski started continues unabated. Arnie Miller has been focusing particularly on the geographic aspects of the appearance of taxa in the Ordovician, and has recorded paleoenvironmental information alongside the usual taxonomic and stratigraphic data in his huge database. He believes (Miller and Mao 1995) that the Ordovician radiation may be associated with orogenic activity—and the Ordovician was indeed an active time—since faunas in associated sites are more prolific than they are on stable epicontinental areas. This might hark back to some suggestions made by Bob Neuman years previously, that Ordovician volcanic islands might be the places where brachiopods evolved novelties. There is a better understanding of climate during the Ordovician, and a diversity peak seen late in the Ordovician in many groups (Webby et al. 2004) has been suggested as a time of global warming preceding the famous extinction accompanying the ice age at the end of the period. The list goes on. What is clear is that fossil data is pivotal in answering many interesting questions about the great radiation of the Ordovician.

I might finish on a personal note. During the time period dealt with in this essay, data gathered by earlier generations of scientists were mined for generalizations, and successfully so. There arose, I think, a perception of "synthesists versus taxonomists." A few scientists tried to do both activities, as I did myself, however inadequately. But some specialists began to feel like the little people, as Mrs Helmsley memorably described them. There is no doubt that the broad-brush approach to the fossil record, and the kind of numerical analyses that Jack Sepkoski did so well, widened the appeal of paleontology, tackled important questions, and got the trade invited to that high table. But Jack himself knew very well that he depended on a school of dedicated specialists to provide data that were sound. There need to be people who know *in detail* about ammonites, or trilobites, and so on. I know from my own experience that jobs on the shop floor of paleontology are getting rare and hard to find. Twenty years ago there were probably twice as many people who worked on trilobites as there are now, and ammonite specialists are almost a threatened species. Even the great national museums do not employ specialists in many important fields. It takes a curious kind of dedication to spend a lifetime working on fossils, and it would be a retrograde step, in my view, if there were nobody to check and amplify databases, describe new fossil faunas, and continue to make the public at large realize that fossils are intrinsically exciting.

REFERENCES

Adrain, J. M., R. A. Fortey, and S. R. Westrop. 1988. Post Cambrian trilobite diversity and evolutionary faunas. *Science* 280:1922–25.

Carls, P. 2003. Tornquists' Sea and the Rheic Ocean are illusive. *Courier Forschung-institut Senckenberg* 242:89–109.

Cocks, L. R. M., and R. A. Fortey. 1982. Faunal evidence for oceanic separations in the Palaeozoic of Britain. *Journal of the Geological Society, London* 139:465–78.

Cocks, L. R. M., and Torsvik, T. 2002. Earth geography from 500 to 400 million years ago: A faunal and palaeomagnetic review. *Journal of the Geological Society, London* 159:631–44.

Dewey, J. F. 1969. Evolution of the Appalachian/Caledonian orogen. *Nature* (London) 222:124–29.

Fortey, R. A. 1975. Early Ordovician trilobite communities. *Fossils and Strata* 4:331–52.

Fortey, R. A., and L. R. M. Cocks. 2003. Palaeontological evidence bearing on global Ordovician-Silurian continental reconstructions. *Earth Science Reviews* 61:245–307.

James, N. P., and R. K. Stevens. 1986. Stratigraphy and correlation of the Cambro-Ordovician Cow Head group, western Newfoundland. *Bulletin of the Geological Survey of Canada* 366:1–143.

Lees, D. C., R. A. Fortey, and L. R. M. Cocks. 2002. Quantifying paleogeography using biogeography: A test case for the Ordovician and Silurian of Avalnia based on brachiopods and trilobites. *Paleobiology* 28:343–63.

Miller, A. I., and S. G. Mao. 1995. Association of orogenic activity with the Ordovician radiation of marine life. *Geology* 23:305–8.

Scotese, C. R., and W. S. McKerrow. 1990. Revised world maps and introduction. *Memoir of the Geological Society of London* 12:1-21.

Sengor, A. M. C., and B. A. Natal'in. 1996. Palaeotectonics of Asia: Fragments of a synthesis. In *The tectonic evolution of Asia*, ed. A. Yin and T. M. Harrison, 486–640. Cambridge: Cambridge University Press.

Sepkoski, J. J., Jr. 1981. A factor analytic description of the fossil record. *Paleobiology* 7:36–53.

———. 1988. Alpha beta or gamma: Where does all the diversity go? *Paleobiology* 14:221–34.

Smith, A. B., J. C. Briden, and G. E. Drewry. 1973. Phanerozoic world maps. *Special Papers in Palaeontology* 12:1–42.

Webby, B. D. E., F. Paris, M. L. Droser, and I. G. Percival (eds.) 2004. *The great Ordovician biodiversification event*. New York: Columbia University Press.

Whittington, H. B. 1966. Presidential address. Phylogeny and distribution of Ordovician trilobites. *Journal of Paleontology* 40:696–737.

Williams, A. 1969. Ordovician faunal provinces with reference to brachiopod distribution. In *The Precambrian and Lower Palaeozoic rocks of Wales*, ed. A. Wood, 117–54. Cardiff, UK: University of Wales Press.

Williams, H., and R. D. Hatcher. 1982. Suspect terranes and the evolutionary history of the Appalachian orogen. *Geology* 10:530–36.

Wilson, J. T. 1966. Did the Atlantic close and then re-open? *Nature* (London) 211:676–81.

CHAPTER FOUR

The Discovery of Conodont Anatomy and Its Importance for Understanding the Early History of Vertebrates

Richard J. Aldridge and Derek E. G. Briggs

After a meeting of the Council of the Palaeontological Association at the Natural History Museum in London on March 17th, 1982, Derek Briggs (then a lecturer at Goldsmiths' College in London) asked Dick Aldridge if he had to head home right away. Normally such an inquiry would be followed by a suggestion that we repair to a local pub (it being St. Patrick's Day), but on this occasion Derek had a different purpose. We instead went to Goldsmiths', where Euan Clarkson was on an extended visit from Edinburgh to collaborate with Derek on remarkably preserved fossil shrimps from the Carboniferous of Scotland. The reason for getting the three of us together lay in a fossil—not a shrimp, but an even more intriguing specimen from the same rocks. This fossil was a discovery of potentially exceptional significance, but Briggs and Clarkson were seeking corroboration of their identification before making an announcement to the world. What they had found, of course, was the first conodont with preserved soft tissues (Briggs, Clarkson, and Aldridge 1983), which was to become a significant player in international debates on the nature, origin, and early evolutionary history of the vertebrates.

BACKGROUND: WHAT ARE CONODONTS?

Conodonts are extinct animals, represented in the fossil record by tiny tooth-like elements made of calcium phosphate. These "teeth" generally became disassociated and scattered in the sediment when the animal died and decomposed. Conodont elements were first discovered by Christian Pander, who reported them from several horizons, spanning the Ordovician to Carboniferous periods, in 1856. The composition of the conodont elements was the same as that of the remains of associated fishes, so Pander interpreted them as the teeth of an otherwise unknown fish group, comparing them to cyclostomes and squalids. This may have been prescient, but the true nature of conodonts was destined to become the subject of considerable speculation for more than 130 years.

Conodont elements are common in many Paleozoic sedimentary rocks, but the group became extinct at the end of the Triassic Period. The elements of the earliest true conodonts are found in rocks of Upper Cambrian age. These are almost universally coniform in shape, with a hyaline, translucent crown in-filled by a more opaque basal body. In time, more complicated morphologies evolved, bearing sharp ridges and, occasionally, serrated margins, and early in the Ordovician there was a major radiation producing complex and highly varied forms with ramifying denticulate processes and ornate platforms.

Conodont elements normally range from a fraction of a millimeter to two mm in size, but rare examples reach lengths of up to twenty-five mm. Although they were usually dispersed in the sediment, complete skeletons are occasionally preserved on bedding planes. Such natural assemblages were first discovered in the 1930s in North America and Germany, and about thirty localities that yield intact conodont skeletons are now known from rocks of different ages around the world. These finds demonstrate that each individual conodont animal possessed a skeletal apparatus comprising fifteen or more separate elements, which show strong morphological differentiation in a bi-laterally symmetrical arrangement.

Although natural assemblages reveal something about the structure of the conodont skeleton, they provide little clue to the nature of the organism itself. In the *Treatise on Invertebrate Paleontology* (1981), Klaus Müller wrote "the origin of conodonts is considered by many paleontologists to be one of the most fundamental unanswered questions in systematic paleontology." Conodont elements have been compared with structures in a remarkable variety of animals, including the teeth or rasping radulae of various worms, molluscs, and fishes, the spines of arthropods, the skeletal walls of conulariids, and the copulatory spicules of nematodes; some authors even postulated affinities with plants or algae. Some of these ideas are more plausible than others, but in the absence of consensus, some authorities regarded conodonts as representative of a distinct body plan, that is, as a separate phylum.

The nature of conodonts has two aspects: the function of conodont elements, and the identity of the organism that bore them. In the 1970s some paleontologists tried to illuminate the second question by tackling the first.

The structure of conodont apparatuses suggests that they were involved in catching and processing food, but two rival theories emerged to explain the mechanics of feeding (Bengtson 1980). One school of thought argued that the toothlike appearance of many elements indicates that the conodont apparatus served to capture and process relatively large prey in a similar manner to jawed vertebrates. The other school considered that the elements provided a rigid base for a soft structure that filtered particulate matter from the water. It soon became apparent, however, that the only way to unravel the biological nature of conodonts was to find a fossil that preserves not only the elements but also evidence of the soft tissues that are normally lost to decay.

Soft tissues are rarely preserved in the fossil record. A combination of exceptional conditions is necessary, including rapid burial, low oxygen levels, and the precipitation of minerals on decaying organic material. In the late 1960s, well-preserved soft-bodied fossils containing conodont elements were discovered in the Carboniferous Bear Gulch Limestone of Montana. The specimens in question, now known as *Typhloesus*, are cigar-shaped, with a long, flat, ovoid body and a finlike structure at one end. The name *conodontochordate* was coined for them. The first examples caused a sensation when they were unveiled by Harold Scott at the 1969 North American Paleontological Convention. It was appropriate that Scott should be involved in these discoveries, as he had diligently sought evidence for the nature of conodonts throughout his career, but as more specimens were uncovered, problems began to arise with their interpretation as a conodont animal. The conodont elements are not arranged as integrated skeletons, but are scattered in the gut. Some specimens of *Typhloesus* do not contain any conodont elements, while others incorporate the remains of several apparatuses and sometimes even structures from other organisms, including worm jaws and arthropod mandibles. Simon Conway Morris (1990) confirmed that these specimens, remarkable though they are, are not the bodies of conodonts but of conodont-eating animals.

Conway Morris himself put forward a second contender in 1976 while he was at Cambridge working on the Middle Cambrian Burgess Shale. While examining Walcott's collections in the Smithsonian, Conway Morris came across a single specimen of a flat, ovoid animal with an annulated trunk and impressions of conelike structures, broadly similar to some Upper Cambrian conodont elements, in the head. He named this animal *Odontogriphus*, meaning "toothed riddle," and suggested that it might be an example of a conodont animal in which the elements formed a rigid support for tentacles making up a soft filtering apparatus. The 'elements' are preserved only as molds, and any original biomineral has been dissolved away. It was, therefore, not possible to determine whether the 'elements' of *Odontogriphus* were composed of calcium phosphate like conodont elements, nor whether they possessed the characteristic internal structures of conodont hard tissues. *Odontogriphus* predates the earliest true conodonts, and has recently been interpreted as a mollusc on the basis of nearly 200 new specimens from the Burgess Shale (Caron et al. 2006). So the nature of the conodont animal remained unresolved.

THE CHANCE DISCOVERY OF THE FIRST CONODONT SOFT TISSUES

Paleontological research advances in several ways, with the discovery of new fossils still of major importance. Progress in our understanding of conodonts and their place in chordate phylogeny is a direct result of a serendipitous discovery, but it also reflects the impact of a number of new ideas and approaches to paleontological research over the last twenty years, particularly in taphonomy and systematics. Briggs and Clarkson investigated many localities in their research on Carboniferous shrimps, but only one, along the Granton-Muirhouse shore within the city of Edinburgh, yielded a conodont body. The rocks at Granton are dominantly shales, which were deposited in a brackish lagoon that was occasionally flooded by the sea. The Granton Shrimp Bed, a thin laminated impure limestone within the shales, preserves multitudes of shrimps, and smaller numbers of worms, nautiloids, hydroids, fishes, and gastropods. These animals died in the quiet waters and lay undisturbed in the oxygen-depleted bottom sediments, where their soft tissues were quickly replicated by the deposition of calcium phosphate.

As part of the investigation of this deposit Clarkson, in a classic example of the importance of specimen repositories to paleontology, worked through collections in the British Geological Survey at Murchison House in Edinburgh that had lain unexamined for some sixty years. There he found the part and counterpart of an unusual segmented worm-like fossil about four centimeters long and two millimeters wide (fig. 4.1). One end showed a tail with fins and the other, a bilobed feature in front of a set of minute toothlike structures. It immediately occurred to Clarkson that this specimen was a fossil cyclostome, related to the modern hagfishes or lampreys. This made it a highly significant find, as fossil hagfishes were then unknown and the fossil record of lampreys was extremely limited. The possibility that the teeth were conodont elements also crossed his mind.

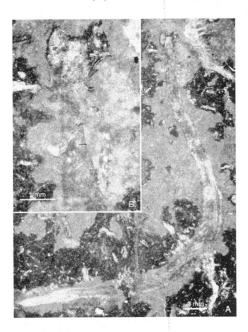

Figure 4.1 The first specimen of a nearly complete conodont found from Granton, Edinburgh, Scotland (A), and (B), the anterior portion of the conodont showing the two lobes in the head and the conodont apparatus.

Clarkson brought the specimen to London to share his discovery with Briggs. During their deliberations they compared the arrangement of the toothlike structures with published drawings of conodont bedding plane assemblages, and Briggs spent some time teasing grains of sediment away from the partly concealed posteriormost pair of elements with the help of weak acid in an attempt to confirm their nature. But they needed an expert opinion, and so the scene was set for our meeting at Goldsmiths' in March 1982. Aldridge had no forewarning of the discovery, but on seeing the specimen had no doubt that the teeth comprised a complete conodont skeleton. This left three possibilities: that the specimen represented two organisms, either a conodont skeleton and a worm-like animal fortuitously superimposed, or a soft-bodied animal that died in the act of ingesting a conodont; or that this was indeed the first specimen of a conodont body. It was clear that the apparatus lay within the animal, so the association between the two was not a result of chance superposition. It was more difficult, on the basis of a single specimen, to rule out the possibility that the fossil preserved one animal in the act of eating another. But there was no sign of other parts of a prey animal extending beyond the head, and the apparatus lay in exactly the expected place for a feeding structure, behind the mouth opening. Furthermore, the arrangement of the elements closely matched several known assemblages and represented a natural death configuration. So, as Briggs and Clarkson suspected, they had discovered the remains of an entire conodont nearly 130 years after the elements first became known.

The initial plan was to submit an announcement of the discovery to *Nature*, so Briggs and Clarkson prepared a draft, entitled "The Conodont Animal Is a Chordate," which they sent to colleagues for comment. Stefan Bengtson observed that the evidence for conodont affinity remained equivocal and offered the journal *Lethaia* as an alternative vehicle, promising very rapid turnaround. Briggs and Clarkson invited Aldridge to join them in authoring this longer manuscript and the paper, now entitled simply "The Conodont Animal," appeared early in 1983. In it we tried to analyze the evidence for affinities as objectively as possible. The apparent bilateral symmetry of the specimen, the serial structures along the trunk, and the possession of a tail, appeared to eliminate all but two groups of living animals: the chordates and the chaetognaths. The evidence for one or the other, however, was equivocal, and we opted to follow the then status quo, retaining a separate phylum for the conodonts.

Stephen Jay Gould (1983) applauded our decision, doffing his hat to conodonts for providing additional evidence of widespread experimentation in body plans among early animals. Here he was heralding the significance he would subsequently attach to the "weird wonders" of the Cambrian Burgess Shale in his 1989 book *Wonderful Life*. Other authors, however, favored either a chaetognath or chordate affinity. Two French scientists, Simon Tillier and Jean-Pierre Cuif, demurred, and used the features of the animal to advance a quite different hypothesis—that conodonts belonged to the aplacophoran molluscs—but resemblances between the two organisms were shown subsequently to be superficial. It was clear, however, that we needed additional specimens to improve our understanding of conodont anatomy and to determine conodont relationships. A grant from the Natural Environment Research Council enabled Aldridge to assemble a research team, with Paul Smith as a postdoctoral conodont specialist. So a new focus for conodont research, on the paleobiology of the animal, was initiated with a systematic and careful search for more material in the Granton Shrimp Bed.

MORE CONODONT CARCASSES

Additional Examples from Granton

Equipped with sledgehammers, crowbars, and high hopes, we made several visits to the foreshore at Granton. Initially, our efforts were frustrated. We found large numbers of beautifully preserved shrimps and other fossils, but no more conodont remains. Our reward finally came in the summer of 1984, when Neil Clark, collecting under Clarkson's direction, found three specimens in a single afternoon! Careful preparation using a customized dental drill and fine dissecting needles, a technique developed for Burgess Shale fossils, exposed the complete length of each specimen, revealing more details of the anatomy, including the feeding apparatuses of two of them. None of the specimens is as complete as the first, but they all added new information. Subsequently, six more specimens were found, bringing the total known from Granton to ten.

The Granton animals range from twenty-one to fifty-five mm in preserved length. The soft tissues of all the specimens are similar, showing V-shaped trunk segments with their apices pointing forward. These structures are myomeres, equivalent to the zig-zag muscle blocks along the body of fishes. Muscles of this type are known only among chordates. The additional specimens also show a pair of axial lines along the length of the trunk. We interpret these as the margins of the notochord, the unmineralized precursor of the backbone. This conclusion was informed by observations on the decomposition of amphioxus (the lancelet *Branchiostoma*) carried out by Briggs and Amanda Kear (1994) in Bristol as part of a program of laboratory experiments to investigate soft-tissue preservation. The gut trace in amphioxus degrades rapidly and disappears, whereas the notochord decays to two parallel lines, equivalent to those preserved in the conodonts. The notochord in the conodont animals extends to the tip of the tail and reaches as far forward as the ramiform elements of the feeding apparatus.

The tail is preserved on only two of the specimens. It consists of a ray-supported fin, apparently more extensively developed along one side than the other. The head likewise is evident on only two specimens, and even those preserve only part of the soft tissues. The absence of traces of gill slits behind the feeding apparatus may be the result of decay. The only structure that is clearly preserved in the head is a pair of lobes, lying anterior to the feeding apparatus and in line with the trunk. It was possible to interpret these lobes as a hood that covered the feeding apparatus, or as paired structures such as the semicircular canals, otic capsules, or most likely, the eyes. Given this last interpretation, a pair of small subcircular dark patches behind the eyes in the original specimen may be the otic capsules, and faint transverse traces along the axis between and behind these may represent a set of gill pouches.

The additional morphological information provided by the new specimens allowed us to reassess the affinities of the conodonts. They clearly were not chaetognaths, and all of the soft tissue features are consistent with an assignment to the phylum Chordata. It was no longer possible to argue separate phylum status for the conodonts unless structures comparable with the chevron-shaped myomeres, the notochord, the asymmetrical ray-supported caudal fin and the paired eyes had all evolved convergently (Aldridge et al. 1986, 1993). The problem now was to determine where within the chordates the conodonts belonged.

The Waukesha Animal

While the conodont specimens were being accumulated at Granton, Briggs found a second conodont with body traces in a collection from Early Silurian dolomites from Waukesha County, Wisconsin. This specimen is not well preserved, but is important in representing a completely different conodont group from those in Scotland (Smith, Briggs, and Aldridge 1987). The feeding apparatus is partly exposed and consists of coniform elements, the curved cusps pointing inward and backward. A broad halo behind the apparatus exposes patches of the trunk that preserve segments, but the boundaries between them appear to be straight, in contrast to the V-shaped myomeres of the Granton specimens. Unfortunately, little more of the soft-tissue morphology can be distinguished, and we await the discovery of more specimens.

The Soom Shale Specimens

At about the same time as the Waukesha specimen was being studied a chain of events was unfolding in South Africa that would lead to additional vital evidence. During the mid-1980s, officers of the Geological Survey of South Africa working in the Cedarberg Mountains north of Cape Town chanced upon a new roadside cutting through the Ordovician Soom Shale. Here they found some curious spiky fossils, which were sent to the paleontological staff at the survey headquarters near Pretoria for identification. The initial interpretation of these structures compared them to early vascular plants known from the Devonian. Another suggestion was that they were graptolites, so they were sent to graptolite expert Barrie Rickards at the University of Cambridge, who thought they might be jaws or even conodont elements. Following this hunch, Rickards contacted Aldridge, who traveled to Cambridge and confirmed that the specimens were conodont apparatuses. They are very unusual conodonts, an order of magnitude larger than normal elements and preserved as moulds, so it is not surprising that their identity was missed by nonspecialists.

This discovery opened up a treasure trove of fossils with preserved soft tissues in the Soom Shale, but, most importantly, it led to the discovery of more than a hundred conodont apparatuses with associated head structures, including one specimen with part of the trunk. The head structures are dark, paired lobes consistent in shape, size, and position with their interpretation as the remains of eyes. The most complete specimen was found by Sarah Gabbott; only the anterior part is present, but it includes the eyes, the feeding apparatus, and about ten centimeters of the trunk (Gabbott, Aldridge, and Theron 1995). If the proportions were the same as those of the Granton specimens, the Soom Shale animal must have been some 40 centimeters long. The trunk is subdivided into simple V-shaped myomeres, and a gap in preservation along the trunk axis marks the line of the notochord, which was apparently lost to decay. At the anterior end a pair of white, ovoid patches represents the eyes, behind which lies the complete feeding apparatus, ventral to the trunk trace. A dark amorphous area in the trunk, below the notochord, may represent the liver. The microscopic structure of the muscle blocks shows the fibers and the component fibrils, a unique example of high-fidelity three-dimensional replication in clay minerals (Gabbott, Aldridge, and Theron 1995). More important was the recognition of muscle fibres in the oval patches that represent the eyes; it is not the eye cups that are preserved here, but the extrinsic eye muscles. Although the Soom Shale specimens come from a different conodont order than those from Granton, the anatomy is strikingly similar, and together they allow a reconstruction that is probably applicable to all conodonts with complex apparatuses (fig. 4.2).

CLUES FROM THE BIOMINERALIZED SKELETON

These new discoveries allowed the relationship of conodonts to other chordates to be reassessed. Several characters provide compelling evidence that the conodonts were at least as advanced as the hagfishes: the ray-supported

Figure 4.2 A reconstruction of the living conodont based on the fossils from Scotland and South Africa (artist: David Baines).

caudal fins, the toothlike feeding apparatus, the paired eyes, and the extrinsic eye muscles. The next step was to consider the evidence provided by the hard phosphatic tissues of the conodont elements.

Among chordates only the armored agnathans and the gnathostomes secrete biomineralized, phosphatic skeletal elements. Forms that are more primitive, including the protochordates, the hagfishes, and the lampreys, lack biomineralized skeletons. Conodont element histology might not match exactly that of other vertebrate tissues but, if conodonts were vertebrates, their skeletal microstructure should show evidence of the same developmental processes.

The earliest observations of the internal structure of conodont elements were made by Pander, who noticed that the translucent crowns of conodont elements are made up of fine layers or lamellae. Pander also observed that these hyaline layers are interrupted by opaque patches full of small cavities. These patches, which commonly fill the core of conodont denticles, are white in reflected light, and have long been known simply as "white matter." Pander concluded that conodont elements grew by the addition of successive layers of calcium phosphate, one *inside* another. In 1938, some eighty years later, it was shown that successive lamellae were added to the outside and not the inside of the crown. Discontinuities were observed in the lamellae that mark levels of damage and repair of the crown tissue during growth and could only be retained by adding new layers to the outer surface.

For many years, opinions about the nature of conodont hard tissues varied, with some workers discounting any resemblance between the histology of conodont elements and the structure of vertebrate hard tissues. New observations, like Klaus Müller and Yasuo Nogami's demonstration in the 1970s that the crown and basal body of conodont elements grew synchronously, marked major steps forward in our understanding. It was with the discovery of the conodont animals, however, that new interest was generated in the histology of conodont elements. New techniques of optical microscopy now allowed much better resolution of microstructure, and among the first scientists to restudy element histology were the Polish paleontologist Jerzy Dzik and a young German, Dietmar Andres. They concluded independently that the tissue that forms the basal body of conodont elements is homologous with dentine and that the lamellar tissue of the crown is homologous with enamel, but not everybody agreed. The American biologist Dick Krejsa, for example, considered that the hyaline crown of conodont elements is homologous with the functional tooth of the hagfish, and that the conodont basal body represents a developing replacement tooth. These interpretations are refuted by the formation of the crown and basal body of conodonts synchronously, not sequentially, and by the addition of lamellae on the outer surface. Furthermore, there is normally little similarity between the upper surface of the basal body and the outer, functional surface of the conodont crown.

The challenge to test whether the histology of conodont elements could be reconciled with those of vertebrates was taken up by a number of British paleontologists, primarily Ivan Sansom, working with his research supervisors, Howard Armstrong and Paul Smith, and with an expert in fish tooth development, Moya Smith. They examined acid-etched sections of conodont elements using advanced optical microscopy and showed that the structure of the lamellar crown tissue is very similar to that of vertebrate enamel (Sansom et al. 1992); the crystallites of the crown lamellae vary from less than a micron to more than thirty microns in length, and their orientation from perpendicular to the element surface to subparallel to its long axis. The incremental lines of the crown lamellae also pass almost imperceptibly into white matter, clear evidence that the two tissues were secreted in continuity. Further studies suggested the presence of dentine tubules in the basal body of some specimens.

But several workers still argued that the apparent similarities of lamellar tissue to enamel were superficial, pointing out that the variable orientations of the tiny crystallites that make up the lamellae do not often correspond to the pattern in living vertebrates. Other researchers, particularly Anne Kemp, an Australian biologist, and Bob Nicoll, an American conodont specialist, used chemical tests to investigate the organic molecules within the conodont apatite. Collagen occurs in the bone, cartilage, and dentine of living vertebrates, but is absent from enamel, which is secreted purely by the epidermis. Kemp and Nicoll found that picrosirius red, a stain for collagen in modern tissues, stained the lamellar crown tissue of conodont elements, but not the white matter or basal tissue. They concluded that the lamellae could not be composed of enamel and that the white matter and basal body could not be bone or dentine. However, these tests were based on the assumption that chemically active collagen could survive in conodont tissues for hundreds of millions of years, which is considered unlikely by organic geochemists.

Phil Donoghue, another young British paleontologist, studied growth patterns in conodont elements and showed that in some coniform elements the entire crown is composed of a fan of crystallites that build up a single homogeneous prism. In more complex elements, individual denticles are constructed of one of these prisms. The variation in the attitude of the crystallites can be attributed partly to this pattern of growth, but also reflects functional requirements. Most significantly, Donoghue demonstrated that a similar range of enamel structures occurs in other vertebrates. Sansom and Donoghue further showed that the basal tissue of conodont elements, which is extremely variable, can be compared to different types of dentine in unequivocal vertebrates, which may be tubular or atubular, lamellar or alamellar, and may include calcispheres. The tissues that make up conodont elements are, therefore, at least comparable developmentally to dentine and enamel.

CLADISTIC STUDIES OF CONODONT RELATIONSHIPS

The combined evidence of conodont soft anatomy and skeletal histology provided a more secure basis for an assessment of the position of the conodonts within the vertebrate clade. The first computer-based parsimony analysis to assess conodont relationships was undertaken by Philippe Janvier, who placed them as a sister group to the lampreys in a 1996 cladogram. This preliminary assessment was built upon by Donoghue, Peter Forey, and Aldridge, who presented a more comprehensive analysis in 2000. This study was based on a character matrix of 103 attributes in seventeen different chordate taxa, including tunicates and cephalochordates (amphioxus) as outgroups. The results consistently showed conodonts to be vertebrates, more derived than hagfishes and lampreys, but more primitive than all other living and fossil forms with phosphatized skeletons (fig. 4.3). This position was maintained even in analyses where the most controversial characters (enamel, dentine, extrinsic

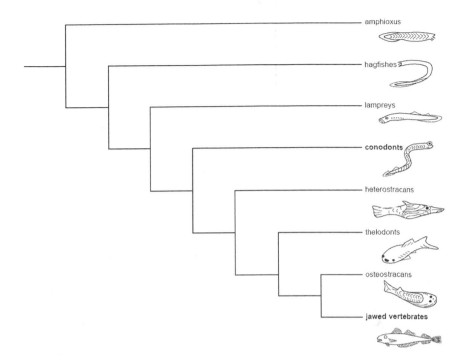

Figure 4.3 Cladogram of relationships between some representative chordate groups, showing conodonts as more derived than the hagfishes and the lampreys, and more closely related to the jawed vertebrates (gnathostomes). After Donoghue, Forey, and Aldridge (2000).

eye muscles) were coded as questionable or absent. Hence, on the basis of all currently available information, the conodonts resolve as stem-group gnathostomes (albeit without backbone or jaw), the first animals along the line to the jawed vertebrates that possessed a biomineralized skeleton. This analysis predicted the existence of early Cambrian agnathans without a biomineralized skeleton, a prediction fulfilled by the discovery of Lower Cambrian vertebrates from Chengjiang in south China as the analyses were being completed. Some authorities continue to posit a protochordate affinity for conodonts, but this hypothesis emphasizes shared primitive characters and is not supported by cladistic analyses. The resolution of conodonts as stem-group gnathostomes confirms that they are not a separate phylum and concords with the assignment in recent years of several enigmatic fossils to the stem lineages of well-known extant phyla.

THE FUNCTION OF THE CONODONT APPARATUS

With the new understanding of conodonts, the functional morphology of conodont elements and apparatuses could now be assessed in an anatomical context. First it was necessary to understand the three-dimensional arrangement of the elements in the living animal. Evidence for this came from two sources: complete skeletons on bedding planes and the fossils preserving soft-tissue anatomy.

Our first attempts to model apparatus architecture began in the mid-1980s, in collaboration with Paul Smith and Rod Norby. Norby (Illinois Geological Survey) had investigated many natural bedding plane assemblages as a graduate student, and we started by building on this work. At first sight, natural assemblages show a rather bewildering variety of patterns. Some are clearly symmetrical or nearly so, whereas in others different parts of the apparatus lie at varying angles to each other. Such variability had led some workers to suggest that the different patterns are a result of rigor mortis. Using models and photography, in a method developed to interpret the flattened fossils of the Burgess Shale, we were able to explain all the patterns by simple collapse of a three-dimensional structure as the animal decayed lying in different orientations to the bedding (Aldridge et al. 1987).

Following this success the approach was applied to the giant apparatuses from the Soom Shale, and was further refined by Mark Purnell and Donoghue, who produced the most accurate models to date of the arrangement of elements in an apparatus (Purnell and Donoghue 1997). These three-dimensional models allowed hypotheses of how elements functioned to be generated and tested. If the elements were tentacle supports in a filter-feeding array, there would be no systematic wear on the surfaces. However, if the elements occluded like teeth, then there should be wear where the surfaces came into repeated contact. Examination of the model indicated the likely contact points, and Purnell used scanning electron microscopy to reveal evidence of wear in the predicted places on the robust posterior elements. It is clear that these elements not only look like teeth, but that they functioned as teeth. They processed prey that was probably grasped by the array of ramiform elements at the front of the apparatus. Examination of platform elements dissected from natural assemblages revealed matching ridges and grooves that allowed the elements to occlude. Purnell was even able to demonstrate similarities between the pitting and scratching on the wear facets and those on the teeth of modern carnivorous and scavenging mammals.

The evidence shows that conodonts equipped with complex apparatuses were not filter-feeders but active carnivores and/or scavengers that fed on macroscopic prey. There is less direct evidence for the function of apparatuses composed entirely of coniform elements, but preliminary models show that they, too, could have served to grasp prey.

CONCLUSION

Alfred Romer, doyen of vertebrate paleontogists in the last century, wrote in *Man and the Vertebrates* (1933) that the earliest vertebrates were jawless mud grubbers in freshwater ponds and streams, using a skeleton for protection. The new understanding of conodonts, together with other recent finds of early fishes in the Cambrian and Ordovician, reveals a different scenario. Vertebrates originated in the sea, and the mineralized vertebrate skeleton first appeared in the conodonts as a raptorial device that consisted of tissues comparable to dentine and enamel. External armor appeared later. The earliest vertebrates with skeletons were not grubbers in the mud, but active, agile predators. Although conodont origins and intrarelationships are poorly known and much remains to be done to understand their biology and evolution, the discovery of a small specimen in a metaphorically dusty museum drawer has led us to unanticipated insights into our deepest ancestry.

REFERENCES

Aldridge, R. J., D. E. G. Briggs, E. N. K. Clarkson, and M. P. Smith. 1986. The affinities of conodonts—new evidence from the Carboniferous of Edinburgh, Scotland. *Lethaia* 19:279–91.

Aldridge, R. J., D. E. G. Briggs, M. P. Smith, E. N. K. Clarkson, and N. D. L. Clark. 1993. The anatomy of conodonts. *Philosophical Transactions of the Royal Society, London B* 340:405–21.

Aldridge, R. J., M. P. Smith, R. D. Norby, and D. E. G. Briggs. 1987. The architecture and function of Carboniferous polygnathacean conodont apparatuses. In *Palaeobiology of conodonts*, ed. R. J. Aldridge, 63–75. Chichester, UK: Ellis Horwood.

Bengtson, S. 1980 Conodonts: The need for a functional model. *Lethaia* 13:320.

Briggs, D. E. G., E. N. K. Clarkson, and R. J. Aldridge. 1983. The conodont animal. *Lethaia* 16:1–14.

Briggs, D. E. G., and A. J. Kear. 1994. Decay of the lancelet *Branchiostoma lanceolatum* (Cephalochordata): Implications for the interpretation of soft-tissue preservation in conodonts and other primitive chordates. *Lethaia* 26:275–87.

Caron, J.-B., A. Scheltema, C. Schander, and D. Rudkin. 2006. A soft-bodied mollusc with radula from the Middle Cambrian Burgess Shale. *Nature* 442:159–63.

Conway Morris, S. 1976. A new Cambrian lophophorate from the Burgess Shale of British Columbia. *Palaeontology* 19:199–222.

———. 1990 *Typhloesus wellsi* (Melton and Scott 1973), a bizarre metazoan from the Carboniferous of Montana, U.S.A. *Philosophical Transactions of the Royal Society of London B* 327:595–624.

Donoghue, P. C. J. 1998. Growth and patterning in the conodont skeleton. *Philosophical Transactions of the Royal Society of London B* 353:633–66.

Donoghue, P.C.J., P. L. Forey, and R. J. Aldridge. 2000. Conodont affinity and chordate phylogeny. *Biological Reviews* 75:191–251.

Gabbott, S. E., R. J. Aldridge, and J. N. Theron. 1995. A giant conodont with preserved muscle tissue from the Upper Ordovician of South Africa. *Nature* 374:800–803.

Gould, S. J. 1983. Nature's great era of experiments. *Natural History* 83(7):12–21.

Janvier, P. 1996. The dawn of the vertebrates: Characters versus common ascent in the rise of current vertebrate phylogenies. *Palaeontology* 39:259–87.

Müller, K. J. 1981. Zoological affinities of conodonts. In *Treatise on invertebrate paleontology, Part W, miscellanea, supplement 2, Conodonta*, ed. R. A. Robison, 78–82. Geological Society of America, Boulder, CO. Lawrence: University of Kansas Press.

Purnell, M. A. 1995. Microwear on conodont elements and macrophagy in the first vertebrates. *Nature* 374:798–800.

Purnell, M. A., and P. C. J. Donoghue. 1997. Skeletal architecture and functional morphology of ozarkodinid conodonts. *Philosophical Transactions of the Royal Society of London B* 352:1545–64.

Romer, A. 1933. Man and the vertebrates. Chicago: University of Chicago Press.

Sansom, I. J., M. P. Smith, H. A. Armstrong, and M. M. Smith. 1992. Presence of vertebrate hard tissues in conodonts. *Science* 256:1308–11.

Smith, M. P., D. E. G. Briggs, and R. J. Aldridge. 1987. A conodont animal from the Lower Silurian of Wisconsin, U.S.A., and the apparatus architecture of panderodontid conodonts. In *Palaeobiology of conodonts*, ed. R. J. Aldridge, 91–104. Chichester, UK: Ellis Horwood.

CHAPTER FIVE

Emergence of Precambrian Paleobiology: A New Field of Science

J. William Schopf

INTRODUCTION

In the early 1960s, when paleobiology first took root, the known history of life—the familiar progression from spore-producing to seed-producing to flowering plants, from marine invertebrates to fish, amphibians, then reptiles, birds, and mammals—extended only to the beginning of Cambrian Period of the Phanerozoic Eon, some 545 million years ago. Now, after a half-century of discoveries, life's history looks strikingly different—an immense early fossil record, unknown and assumed unknowable, has been unearthed to reveal an evolutionary progression, dominated by microbes, that stretches seven times further into the geologic past than was previously known. Yet despite its apparent newness, the discovery of this Precambrian missing record of life has antecedents that extend to the days of Darwin. Such is the way of science, with each generation building on and learning from the successes, as well as the failures, of its predecessors.

This essay is an abbreviated history of how the assumptions accepted by early workers influenced the development of this science, of how and by whom breakthrough advances were brought to the fore in the mid-1960s, and of lessons learned in the still-ongoing search for records of ancient life. At the request of the editors of this volume, this essay is in part a personal account; my recollection of the events recounted may not be as precise as I or scholarly historians of science might prefer, but it is as accurate as I can muster.

Darwin's Dilemma

Like so many aspects of natural science, the beginnings of the search for life's earliest history date from the mid-1800s and the writings of Charles Darwin (1809–1882) who, in *On the Origin of Species*, highlighted the missing Precambrian fossil record and the problem it posed to his theory of evolution:

> There is another . . . difficulty, which is much more serious. I allude to the manner in which species belonging to several of the main divisions of the animal kingdom suddenly appear in the lowest known [Cambrian-age] fossiliferous rocks. . . . If the theory be true, it is indisputable that before the lowest Cambrian stratum was deposited, long periods elapsed . . . and that during these vast periods, the world swarmed with living creatures. . . . [But] to the question why we do not find rich fossiliferous deposits belonging to these assumed earliest periods prior to the Cambrian system, I can give no satisfactory answer. The case at present must remain inexplicable; and may be truly urged as a valid argument against the views here entertained (Darwin 1859, Chapter X).

Darwin's dilemma begged for solution. Though this problem was to remain unsolved—the case "inexplicable"—for more than 100 years, the intervening century was not without bold pronouncements, dashed dreams, and more than a little acid acrimony.

PIONEERING PATHFINDERS

J. W. Dawson and the "Dawn Animal of Canada"

Among the first to take up the challenge of Darwin's theory and the problem of the missing early fossil record was John William Dawson (1820–1899; fig. 5.1), principal of McGill University and a giant in the history of North American geology. The son of strict Scottish Presbyterians, Dawson was a staunch Calvinist and devout antievolutionist. In 1858, a year before publication of Darwin's opus, specimens of distinctively green- and white-layered limestone collected along the Ottawa River near Montreal were brought to the attention of William E. Logan, director of the Geological Survey of Canada. Because the samples were known to be ancient (from Laurentian strata, now dated at about 1,100 My) and had layering that Logan supposed too regular to be purely inorganic (fig. 5.2), he displayed them as possible pre-Cambrian

Figure 5.1 The official portrait of Sir John William Dawson upon his appointment (1853) as principal of McGill University, Canada.

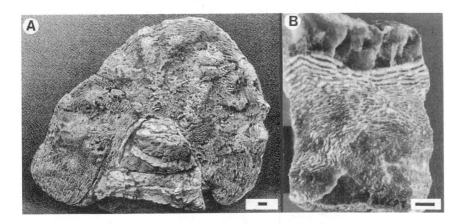

Figure 5.2 *Eozoon Canadense*, the "dawn animal of Canada," (A) as illustrated in Dawson's *The Dawn of Life* (1875), and (B) shown by the holotype specimen archived in the U.S. National Museum of Natural History, Washington, D.C. (Bars for scale represent 1 cm.)

fossils at various scientific conferences, where they elicited spirited discussion but gained little acceptance as remnants of early life.

In 1864, however, Logan brought specimens to Dawson, who not only confirmed their biologic origin but identified them as fossilized remnants of giant foraminiferans, huge oversized versions of the shells of tiny protozoans. So convinced was Dawson of their biologic origin that a year later, in 1865, he formally named the putative fossils *Eozoon Canadense*, the "dawn animal of Canada." Dawson's claim was questioned almost immediately, the beginning of a fierce debate that raged on until 1894, when specimens indistinguishable from *Eozoon* were found near Mount Vesuvius in southern Italy and shown to be geologically young blocks of limestone, their fossil-like appearance the result of inorganic veining by the green metamorphic mineral serpentine (O'Brien 1970).

Despite such evidence, for the rest of his life Dawson continued to press his case, spurred by his belief that the discovery of this "dawn animal" had exposed a gap in the fossil record so enormous that it unmasked evolution's claimed continuity, leaving Biblical special creation as the only answer:

> There is no link whatever in geological fact to connect Eozoon with the Mollusks, Radiates, or Crustaceans of the succeeding [fossil record] . . . these stand before us as distinct creations. [A] gap . . . yawns in our imperfect geological record. Of actual facts [with which to fill this gap], therefore, we have none; and those evolutionists who have regarded the dawn-animal as an evidence in their favour, have been obliged to have recourse to supposition and assumption (Dawson 1875, 227).

Dawson's complaint is understandable—in the fourth and all later editions of *The Origin*, Darwin cited the great age and primitive protozoal relations of *Eozoon* as consistent with evolution, just the sort of "supposition and assumption" that Dawson found so distressing.

C. D. Walcott: Founder of Precambrian Paleobiology

Dawson's debacle would ultimately prove to be little more than a distracting detour on the path to progress that was soon to be charted by the American paleontologist Charles Doolittle Walcott (1850–1927; fig. 5.3).

Like Dawson before him, Walcott was enormously energetic and highly influential (Yochelson 1997). He spent most of his adult life in Washington, D.C., where he headed powerful scientific organizations—first, as director of

Figure 5.3 Charles Doolittle Walcott, probably in 1894, at the time of his appointment as director of the U.S. Geological Survey. (Courtesy of E. L. Yochelson, Smithsonian Institution.)

the U.S. Geological Survey (1894–1907), then as secretary of the Smithsonian Institution (1907–1927) and president of the National Academy of Sciences (1917–1923). Surprisingly, however, Walcott had little formal education. He received only ten years of schooling, first in public schools and, later, at Utica Academy (from which he did not graduate), and he never attended college and had no formally earned advanced degrees (a deficiency more than made up for in later life when he was awarded honorary doctorates by more than a dozen academic institutions).

In 1878, as a twenty-eight-year-old apprentice to James Hall, chief geologist of the state of New York and acknowledged dean of American paleontology, Walcott was introduced to stromatolites—wavy, layered mound-shaped rock masses laid down by ancient communities of mat-building microbes—Cambrian-age structures near the town of Saratoga in eastern New York State. Named *Cryptozoon* (meaning *hidden life*), these cabbagelike structures (fig. 5.4) would in later years form the basis of Walcott's side of a nasty argument known as the "*Cryptozoon* controversy." A year later, in July 1879, Walcott was appointed to the newly formed U.S. Geological Survey. Over several field seasons, he and his comrades mapped the geology of sizable segments

Figure 5.4 *Cryptozoon* reefs near Saratoga, New York. (Photo by E. S. Barghoorn, November, 1964.)

of Arizona, Utah, and Nevada, including unexplored parts of the Grand Canyon, where in 1883 he first reported discovery of Precambrian specimens of *Cryptozoon*. Other finds soon followed, with the most notable in 1899—small, millimetric black coaly discs that Walcott named *Chuaria* and interpreted to be "the remains of . . . compressed conical shell[s]," possibly of primitive brachiopods (Walcott 1899). Although *Chuaria* is now known to be a large single-celled planktonic alga, rather than a shelly animal, Walcott's specimens were authentic fossils, the first true cellularly preserved Precambrian organisms ever recorded.

After the turn of the century, Walcott moved his fieldwork northward along the spine of the Rocky Mountains, focusing first on the Lewis Range of northwestern Montana, from which he reported diverse stromatolite-like structures and, later, chains of microscopic cell-like bodies he identified as petrified fossil bacteria. His studies in the Canadian Rockies, from 1907 to 1925, were even more rewarding, resulting in the discovery of an amazingly well-preserved assemblage of Cambrian algae and marine invertebrates—the famous Burgess Shale Fauna that to this day remains among the finest and most complete samples of Cambrian life known to science (Gould 1989).

Walcott's contributions are legendary—he was the first discoverer in Precambrian rocks of *Cryptozoon* stromatolites, of cellularly preserved algal plankton (*Chuaria*), and of possible fossil bacteria, all capped by his pioneering investigations of the benchmark Burgess Shale fossils. The founder of Precambrian paleobiology, Walcott was first to show, more a century ago and contrary to accepted wisdom, that a substantial fossil record of Precambrian life actually exists.

A. C. Seward and the Cryptozoon Controversy

The rising tide brought on by Walcott's discoveries was not yet ready to give way to a flood. Precambrian fossils continued to be regarded with incredulity, a view bolstered by Dawson's *Eozoon* debacle but justified almost as easily by the scrappy nature of the available evidence. Foremost among the critics was Albert Charles Seward (1863–1941; fig. 5.5), professor of botany at the University of Cambridge and the most widely known and influential paleobotanist of his generation. Because virtually all claimed Precambrian fossils fell within the purview of paleobotany—whether supposed to be algal, like *Cryptozoon* stromatolites, or even bacterial—Seward's opinion had special impact.

In 1931, in *Plant Life Through the Ages*, a paleobotanical text used worldwide, Seward wrote that the

> general belief among American geologists and several European authors in the organic origin of Cryptozoon is . . . not justified by the facts . . . primitive algae may have flourished in Pre-Cambrian seas and inland lakes; but to regard these hypothetical plants as the creators of reefs of Cryptozoon and allied structures is to make a demand upon imagination inconsistent with Wordsworth's definition of that quality as 'reason in its most exalted mood' (Seward 1931, 86, 87).

He was even more categorical in his rejection of Walcott's report of fossil bacteria:

> It is claimed that sections of a Pre-Cambrian limestone from Montana show minute bodies similar in form and size to cells and cell-chains of

Figure 5.5 Sir Albert Charles Seward on the occasion of his appointment as vice-chancellor of the University of Cambridge in 1924. (Courtesy of Cedric Shute, The Natural History Museum, London.)

existing [bacteria]. . . . These and similar contributions . . . are by no means convincing. . . . We can hardly expect to find in Pre-Cambrian rocks any actual proof of the existence of bacteria" (Seward 1931, 92).

Seward's 1931 assessment of the science was largely on the mark. Mistakes had been made. Mineralic, purely inorganic objects had been misinterpreted as fossils. More and better evidence was much needed. But his dismissive rejection of *Cryptozoon* and his bold assertion that "we can hardly expect to find in Pre-Cambrian rocks any actual proof of the existence of bacteria" turned out to be misguided.

Given his expertise, Seward might have come to a different conclusion. He rejected *Cryptozoon* chiefly because of its similarity to inorganic concretions, and he cast aside Walcott's report of fossilized bacteria on the grounds that such fragile microorganisms were unpreservable (Seward 1931). Of these mistakes, Seward's dismissal of *Cryptozoon* is the more understandable. He was well aware that putatively biogenic stromatolite-like "sea biscuits" had been reported from South Australia, but he also knew there was much debate as to whether these and similar structures were biological or purely inorganic.

His rejection of Walcott's "fossil bacteria" is more difficult to fathom. In his 1931 volume, Seward discusses at length the famous fossil plant beds of Scotland's Devonian Rhynie Chert, a unit from which some ten years earlier Kidston and Lang had reported petrified microscopic cyanobacteria. A master of paleobotany, Seward might have made the connection: microbes, including cyanobacteria, have robust cell walls; cell-walled organisms—land plants and bacteria both—can be preserved by petrifaction; reports of petrified microbes might not be so remarkable after all. As it turned out, no such connection was made. Seward's influence was pervasive, and his dismissal of the bits and pieces of the then-known Precambrian fossil record was a serious blow to progress. It took another thirty years, and a bit of serendipity, to put the field back on track.

EMERGENCE OF A NEW FIELD OF SCIENCE

In the mid-1960s—a full century after Darwin broached the problem of the missing early fossil record—the hunt for early life began to stir, and in the following decades the floodgates would finally swing wide open. But this surge, too, had harbingers, now dating from the 1950s.

A Benchmark Discovery by an Unsung Hero

The worker who above all others sparked the beginnings of modern studies of ancient life was Stanley A. Tyler (1906–1963; fig. 5.6) of the University of Wisconsin, the geologist-mineralogist who in 1953 discovered the now-famous Precambrian (1,900-million-year-old) microbial assemblage petrified in fine-grained coaly cherts of the Gunflint Formation of Ontario, Canada. A year later, together with Harvard paleobotanist Elso S. Barghoorn (1915–1984; fig. 5.7), Tyler published a short note announcing the discovery (Tyler and Barghoorn 1954), a sketchy study that showed fossils to be present in the deposit but that failed to note either the exact location of the find or that the fossils were present within, and were actually the microbial builders of, large *Cryptozoon*-like stromatolites (an association that, once recognized, would prove key to future advances in the field). Substantive reports would come later—though not until after Tyler's untimely death, an event that deprived him from receiving the credit he deserved—but this initial 1954 report of "the oldest structurally preserved organisms that clearly exhibit cellular differentiation and original carbon complexes which have yet been discovered in pre-Cambrian sediments" was a benchmark, a monumental "first."

Figure 5.6 Stanley A. Tyler in 1959, at Van Hise Rock in the Baraboo Hills of Wisconsin. (Courtesy of John Valley and Robert Dott, University of Wisconsin, Madison.)

Figure 5.7 Elso S. Barghoorn in 1964 at a cottage near Schreiber, Ontario, the base camp of his geologic studies of the Gunflint chert.

Figure 5.8 Boris Vasil'evich Timofeev in 1969, during a geologic field trip to collect Precambrian siltstones in the western Ukraine. (Courtesy of Mikhail A. Fedonkin, Paleontological Institute, Russian Academy of Science, Moscow.)

Contributions of Soviet Science

At about the same time, in the mid-1950s, a series of articles by Boris Vasil'evich Timofeev (1916–1982; fig. 5.8) and his colleagues at the Institute of Precambrian Geochronology in Leningrad (St. Petersburg) reported the discovery of the fossilized spores of land plants and similar microscopic cells in Precambrian siltstones of the Soviet Union. In thin sections, like those studied by Tyler and Barghoorn, fossils are detected within a rock, enclosed in the mineral matrix, so the possibility of laboratory contamination can be ruled out. But preparation of thin sections requires special equipment and their microscopic study is tedious and time consuming. A more rapid technique, pioneered for Precambrian studies in Timofeev's lab and now used routinely in the petroleum industry, is to dissolve a rock in strong acid and concentrate the organic-walled microfossils in the resulting sludgelike residue. This maceration technique, however, is notoriously subject to error-causing contamination—and because during these early years of Precambrian paleobiology there was as yet no established ancient fossil record with which to compare new finds, mistakes were easy to make. Though Timofeev's lab was not immune to such mistakes and his claims of ancient land plant spores turned out to be in error, much of his work has since proved sound, and the technique he pioneered to ferret out microfossils in Precambrian shaley rocks is now in worldwide use.

Famous Figures Enter the Field

Early in the 1960s, the fledgling field was joined by two geologic heavyweights, an American, Preston E. Cloud (1912–1991; fig. 5.9), and an Australian, Martin F. Glaessner (1906–1989; fig. 5.10), both attracted by the unexplained abrupt appearance and explosive evolution of shelly invertebrate animals that marks the beginning of the Phanerozoic Eon.

A leader in the development of Precambrian paleobiology, Cloud was full of ideas, opinions, and good hard work. His reading of life's early history was first evident in the late 1940s when he argued in print that though the known Early Cambrian fossil record is woefully incomplete, it is nevertheless the court of last resort—in his view, the only court that mattered (Cloud 1948). By the 1960s, he had become active in the field, authoring an article that to many certified the authenticity of the Tyler-Barghoorn Gunflint microfossils (Cloud 1965) and, later, numerous papers reporting new finds of Precambrian fossil microbes. Above all, he was a gifted synthesist, showing his mettle in a masterful article of 1972 that set the stage for modern understanding of the interrelated atmospheric-geologic-biologic history of the Precambrian planet (Cloud 1972).

In the early 1960s, a second prime player entered this now fast-emerging field—Martin Glaessner, of the University of Adelaide in South Australia. A scholarly, courtly, old-school gentleman, Glaessner was among the first to make major inroads toward understanding the (very latest) Precambrian record of multicelled animal life (Radhakrishna 1991).

In 1947, three years before Glaessner joined the faculty at Adelaide, Reginald C. Sprigg announced the discovery of fossils of primitive soft-bodied animals, chiefly imprints of saucer-sized jellyfish, at Ediacara, South Australia. Though Sprigg thought the fossil-bearing beds were Cambrian in age, Glaessner showed them to be Precambrian (albeit marginally so), making the Ediacaran fossils the oldest animals then known. Together with his colleague, Mary Wade, Glaessner spent much of the rest of his life working on this benchmark fauna, bringing it to international attention in the early 1960s and, later, in a splendid monograph (Glaessner 1984).

Figure 5.9 Preston E. Cloud, about 1975, at a scientific meeting at Snowbird, Utah.

Figure 5.10 Martin F. Glaessner, about 1988. (Courtesy of Brian McGowran, University of Adelaide, Australia.)

With Glaessner in the fold, the stage was set. Like a small jazz combo—Tyler and Barghoorn trumpeting microfossils in cherts, Timofeev beating on fossils in siltstones, Cloud strumming the early environment, Glaessner the earliest animals—great music was about to be played. At long last, the curtain was to rise on the missing record of Precambrian life.

BREAKTHROUGH TO THE PRESENT

My own involvement dates from 1960, when, as a sophomore in college, I became enamored of this problem, an interest that was to become firmly rooted during the following few years and has continued ever since to be the focus of my life.

In the fall of 1960, when I first became fascinated with this area of science, I knew nothing about life's early history. I was young, an eighteen-year-old sophomore at Oberlin College in northeastern Ohio. That semester I was enrolled in my second geology course, "History of the Earth," and I listened intently as my favorite professor, Larry DeMott, lectured about the missing Precambrian fossil record and the dilemma it posed to Darwin and to the acceptance of evolution. To this day I remain puzzled as to why this particular problem kindled in me such immediate, and now long-lasting interest, but I am sure it stems from ideas set in place many years earlier. I had been brought up in a family of scientists—my mom was schooled in botany; my dad, a professor at Ohio State, was a paleobotanist; and my older brother went on to become professor of paleontology at the University of Chicago. I had no doubt that Darwin was right. Evolution was a fact. So, though DeMott told us that this century-old unsolved problem was one of the great mysteries in all of natural science, I was sure that there simply had to be a solution. The absence of a Precambrian fossil record may have been "inexplicable" to Darwin, but I became determined that it would not be so to me. This, I thought, was a problem I could tackle.

That evening, I returned to my dorm room and thumbed through my paperback edition of Darwin's *Origin*, finding at least a dozen entries where he had raised the problem. DeMott was right; the missing early fossil record posed a huge dilemma. At that time, Oberlin had one of the best small college libraries in the world. I read everything I could find about Precambrian life. The more I read, the more intrigued I became. I was particularly taken by the writings of the American paleontologist George Gaylord Simpson, who argued that because the evolutionary distance between humans and trilobites was more or less the same as that between a trilobite and an amoeboid proto-zoan (thought by Simpson and his contemporaries to be among the earliest forms of life), and because the oldest trilobites were about 500 My in age, then the first amoebas—and thus the origin of life itself—should date from about 1,000 My ago. From this it followed that the emergence of life would have required an enormously long period—billions of years—but to Simpson this made good sense, because he thought that there must have been a vast evolutionary gap between nonlife and the earliest living systems.

Clearly, a lot of this was guesswork. But if Simpson's notion was even close to the mark it told me that all I had to do was to trace back the fossil record to about 1,000 My ago, where I might then expect to find actual evidence of life's beginnings. What a great find that would be! I was young, naive, and full of enthusiasm. This was heady stuff.

A year later, in the fall of 1961, by then even more committed to tackling this daunting problem, I screwed up my courage and wrote to the only two Americans in the field I had managed to identify: Elso Barghoorn, at Harvard, and Preston Cloud, then the newly appointed head of the department of geology at the University of Minnesota. My hope was that one or the other would find me suitable as a prospective graduate student.

Both treated me with kindness, and Barghoorn even gave me chunks of the Gunflint chert, which I sectioned and used as the subject of my Oberlin honors thesis. As it turned out, permission for me to do this honors thesis did not come easily. My hope was to be admitted to Harvard, to work with Barghoorn, with such admittance, according to the Harvard catalogue, being granted only to "Honors Students." My search through the Oberlin catalogue revealed that a geology department honors program had been put in place in 1908. To my chagrin, however, I soon discovered that there had never been a Geology "Honors Student" in Oberlin's history. I explained my plight to the geology faculty and was told to submit a petition. The three professors of the department faculty met to decide my fate: one (the long-term head of the department) voted no, arguing that undergraduates "had no business do-ing research"; a second voted yes (largely because he was a recent PhD from MIT, where he had heard of Barghoorn's work on ancient fossils); and the third, newly appointed and as yet unfamiliar with the workings of Oberlin, abstained. Luckily for me, the "yes" vote held sway and I was permitted to do the first geology honors thesis in the history of the college. (At the time, this episode was terrifically worrisome to me, but for the geology department it may have been something of a boon. Years later, when I was periodically in town as a member of the Oberlin board of trustees, I made a point of keeping track of the department's activities, and I was privately pleased to note that six to eight honors students now graduate each year.)

Bringing the Gunflint Fossils to Light

In the summer of 1963, a freshly minted college graduate, I entered Harvard as Barghoorn's first graduate student to focus on early life. I have earlier recounted in some detail my recollections of those heady days (Schopf 1999) and need not repeat the story here. Suffice it to note that virtually nothing had been published on the now-famous Gunflint fossils in the nearly ten years that had passed between the Tyler-Barghoorn 1954 announcement of the find and my entry into graduate school in June 1963. In the interim, however, Tyler and his students had done yeoman work, finding and photographing an enormous array of beautifully preserved fossils, some quite bizarre and all new to science (fig. 5.11). Then, unexpectedly, in October of that year, Stanley Tyler passed away at the age of 57, never to see the fruits of his years-long labor reach the published page. Within a year thereafter, a series of events that would shape the field began to unfold, set off first by a squabble between Barghoorn and Cloud as to who would scoop whom in a battle for credit over the Gunflint fossils. By late 1964 this spat had been settled, with Cloud opting to hold off publication of his work on the Gunflint fossils until Barghoorn had completed his part of a manuscript detailing the find with the by-then deceased Tyler. The two articles appeared in *Science* in 1965, first Barghoorn and Tyler's "Microorganisms from the Gunflint chert" (Barghoorn and Tyler 1965), followed a few weeks later by Cloud's contribution, "Significance of the Gunflint (Precambrian) microflora" (Cloud 1965). Landmark papers they were!

Unlike the largely unnoticed 1954 Tyler-Barghoorn announcement of discovery of the Gunflint fossils, the Barghoorn-Tyler 1965 article—backed by Cloud's affirmation of its significance—generated enormous interest. Yet it soon became apparent that acceptance of these ancient life forms would come only grudgingly. The well had been poisoned by Dawson's debacle, the *Cryptozoon* controversy, Seward's negativism, and other similar misadventures—object lessons handed down from professor to student, generation to generation, that by the mid-1960s had become entrenched in paleontologic lore. Moreover, it was all too obvious that the Gunflint organisms stood alone. Marooned in the remote Precambrian, they were isolated by nearly a billion and a half years from all other fossils known to science, leaving a gap in the fossil record nearly three times longer than the entire previously documented

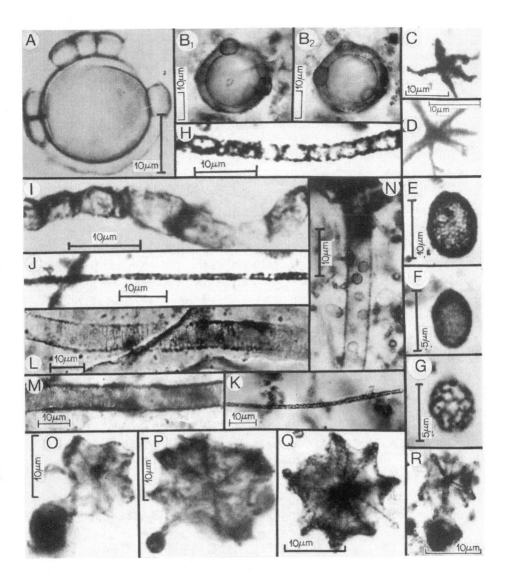

Figure 5.11 Microfossils of the Precambrian (~1,900-million-year-old) Gunflint chert of southern Canada. (A and B) *Eosphaera*, in (B) shown in two views of the same specimen; (C and D) *Eoastrion*; (E, F, and G) *Huroniospora*; (H through K) *Gunflintia*; (L and M) *Animikiea*; (N) *Entosphaeroides*; (O through R) *Kakabekia*.

history of life. Skepticism abounded. Couldn't this whole business be some sort of fluke, some hugely embarrassing, awful mistake?

As luck would have it, such doubts soon could be laid to rest. During the spring of 1964, Barghoorn and his wife-to-be, Dorothy Osgood, had traveled to Australia on a geologic collecting trip. While in Adelaide, he had a chance encounter with German oil geologist Helmut Wopfner, who had recently returned from the central Australian outback where, near the Ross River Tourist Camp east of Alice Springs, he had seen Precambrian *Cryptozoon*-like stromatolites interbedded with jet-black cherts—a pairing just like that of the Gunflint and exactly what Barghoorn was looking for. Thus primed, Barghoorn journeyed to Ross River, where he collected half a dozen hand-sized specimens of these promising fine-grained cherts, by then identified as belonging to the Late Precambrian Bitter Springs Formation. Once the Gunflint paper had been completed, I was assigned to work on these samples, which I soon discovered contained a remarkable cache of new microscopic fossils (fig. 5.12), many indistinguishable from living cyanobacteria and virtually all better preserved than the Gunflint microbes. Though the age of the Bitter Springs cherts was known only approximately (and has since been dated to be ~850 million years), at the time it seemed likely to be about 1,000 million years old, roughly half as old as the Gunflint deposit.

Barghoorn and I soon sent a short report to *Science* (Barghoorn and Schopf 1965), the publication of which, close on the heels of the articles on the Gunflint fossils, not only dispelled doubt about whether such finds were some sort of fluke, but seemed to suggest that the early fossil record might be richer and surprisingly easier to unearth than anyone had dared imagine. Indeed, it is at least arguable that the only odd thing about the Gunflint and Bitter Springs finds is that such discoveries had not been made earlier. Walcott's report of "fossil bacteria" was on the right track, only to be derailed by Seward's skepticism; and given his knowledge of petrified fossils, Seward himself might well have puzzled through the problem—but he did not. Prior to the discoveries of the mid-1960s, workers worldwide had followed conventional wisdom by assuming that the tried and true techniques so successful in the hunt for large fossils in the Phanerozoic would prove equally rewarding in the Precambrian. We now know that this was a mistake.

LESSONS FROM THE HUNT

The Gunflint and Bitter Springs articles of 1965 set a new course, showing for the first time that the previously missing Precambrian fossil record

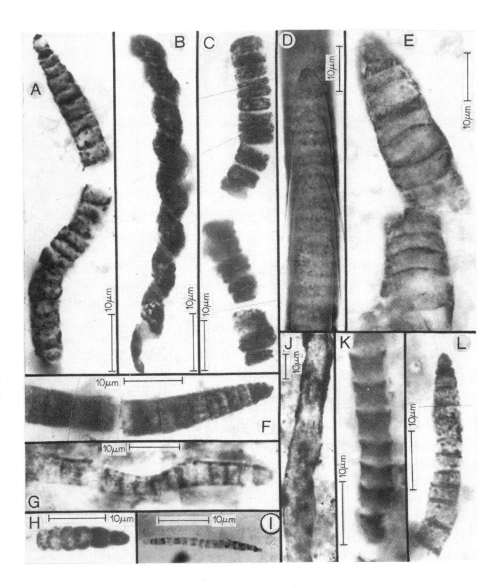

Figure 5.12 Filamentous microfossils of the Precambrian (~850-million-year-old) Bitter Springs chert of central Australia. Because the petrified microbes are three dimensional and sinuous, composite photos have been used to show the specimens in (A through G), (I), (K), and (L). (A, F, I, and L) *Cephalophytarion*; (B) *Helioconema*; (C and G) *Oscillatoriopsis*; (D) unnamed oscillatoriacean cyanobacterium; (E) *Obconicophycus*; (H) *Filiconstrictosus*; (J) *Siphonophycus*; (K) *Halythrix*.

could be unearthed by focusing the search on microscopic forms of life. From this beginning, hundreds of such fossil-bearing Precambrian deposits have now been discovered, the documented fossil record has been traced back to ~3,500 million years ago (Schopf 2006), and astounding new techniques have been devised for studies of the three-dimensional morphology and coaly carbon-chemistry of minute Precambrian fossils embedded in solid rock (Schopf, Tripathi, and Kudryavtsev 2006).

The four keys of the strategy that has led to this progress, as valid today as they were when they were set in place in the mid-1960s, are to search for (1) microscopic fossils in (2) black cherts that are (3) fine-grained and (4) associated with *Cryptozoon*-like structures. Each part plays a role.

1. Megascopic plants and animals, the large organisms of the Phanerozoic, are now known not to appear in the fossil record until near the close of the Precambrian—except in immediately sub-Cambrian strata, the hunt for large body fossils in Precambrian rocks was doomed from the start.
2. The blackness of a chert commonly gives a good indication of its organic carbon content—like fossil-bearing coal deposits, cherts rich in petrified organic-walled microfossils are usually a deep jet-black color.
3. The fineness of the quartz grains making up a chert provides another hint of its fossil-bearing potential—cherts subjected to geologic pressure-cooking are usually composed of recrystallized large quartz grains that give them a sugary appearance, whereas the grains in cherts that have escaped such fossil-destroying processes are microscopic, giving the rocks a waxy, glass-like luster.
4. *Cryptozoon*-like structures (stromatolites) are now known to have been produced by flourishing microbial menageries, layer upon layer of microscopic organisms that make up localized biologic communities. Stromatolites petrified by fine-grained chert before microbial decay and cellular disintegration set in are promising hunting grounds for the fossilized remnants of the microorganisms that built them.

Now, some four decades after the first major discoveries, Precambrian paleobiology is thriving—the vast majority of scientists who have ever investigated the early fossil record are working today, reporting new finds at an ever-quickening clip. The Barghoorn-Tyler studies of stromatolite-forming microbes have been augmented by studies of living microbial communities and of biochemical markers (the bases of rRNA phylogenies) that place such microorganisms on early branches of the universal tree of life. Timofeev's finds of fossil phytoplankton in Precambrian siltstones have been extended to reveal the first major extinctions in the history of life and the early history of biologic diversity. Cloud's environmental syntheses, supported by new understanding of comparative planetology, atmospheric evolution, and geochemical evidence of the history of photosynthesis, have been sharpened and refined to reveal an increasingly focused picture of the developing early Earth. And Glaessner's studies of Precambrian animal fossils have grown into a global blizzard of activity that has provided new insight into the abrupt rise of shelled multicelled animals that marks the close of the Precambrian.

The advances we see today were set in motion by the bold workers who blazed this trail in the 1960s, just as their course was influenced by Dawson, Walcott, and Seward, the pioneering pathfinders of the field. And at its root, all of this progress dates to Darwin and the dilemma of the missing Precambrian fossil record he first posed. After more than a century of trial and error, of search and final discovery, we can be thankful that what was once "inexplicable" to Darwin is no longer so to us.

REFERENCES

Barghoorn, E. S., and J. W. Schopf. 1965. Microorganisms from the Late Precambrian of Central Australia. *Science* 150:337–39.
Barghoorn, E. S., and S. A. Tyler. 1965. Microorganisms of the Gunflint chert. *Science* 147:563–77.
Cloud, P. 1948. Some problems and patterns of evolution exemplified by fossil invertebrates. *Evolution* 2:322–50.
———. 1965. Significance of the Gunflint (Precambrian) microflora. *Science* 148:27–45.
———. 1972. A working model of the primitive Earth. *American Journal of Science* 272:537–48.
Darwin, C. R. 1902/1964. *On the origin of species by means of natural selection, or the preservation of favoured races in the struggle for life.* London: John Murray/Cambridge, MA: Harvard University Press.
Dawson, J. W. 1875. *The dawn of life.* London: Hodder and Stoughton.
Glaessner, M. F. 1984. *The dawn of animal life, a biohistorical study.* Cambridge: Cambridge University Press.
Gould, S. J. 1989. *Wonderful life.* New York: Norton.
O'Brien, C. F. 1970. *Eozoon Canadense,* 'the dawn animal of Canada.' *Isis* 61:206–23.
Radhakrishna, B. P., (ed). 1991. *The world of Martin F. Glaessner, memoir no. 20.* Bangalore, India: Geological Society of India.
Schopf, J. W. 2006. Fossil evidence of Archaean life. *Philosophical Transactions of the Royal Society B* 361:869–85.

―――. 1999. *Cradle of life: The discovery of Earth's earliest fossils.* Princeton, NJ: Princeton University Press.

Schopf, J. W., A. B. Tripathi, and A. B. Kudryavtsev. 2006. Three-dimensional confocal optical imagery of Precambrian microscopic organisms. *Astrobiology* 6:1–16.

Seward, A. C. 1931. *Plant life through the ages.* Cambridge: Cambridge University Press.

Tyler, S. A., and E. S. Barghoorn. 1954. Occurrence of structurally preserved plants in Pre-Cambrian rocks of the Canadian Shield. *Science* 119:606–08.

Walcott, C. D. 1899. Pre-Cambrian fossiliferous formations. *Geological Society of America Bulletin* 10:199–214.

Yochelson, E. L. 1997. *Charles Doolittle Walcott, paleontologist.* Kent, OH: Kent State University Press.

CHAPTER SIX

Dinosaurs at the Table

John R. Horner

In this day and age, dinosaurs are media superstars, arguably the most familiar of extinct organisms. They are icons of popular science, and many of their collectors and examiners have been publicly elevated to among the most well known of scientists. Dinosaurs are so trendy that some of the most tedious of descriptive papers make the news and generate wild Internet discussion among both professional paleontologists and the general public, young and old. Some scientists, including some paleontologists, would probably argue that dinosaur paleontology as a discipline has never quite made it to the high table, on account of its excessive popularity with the public. Unfortunately, science has held this view for more than a hundred years, ever since the feuding days of E. D. Cope and O. C. Marsh. Since as far back as 1877, these two men have loomed as symbols of this idiosyncratic pseudoscience, pervasive in a P. T. Barnum style of amusement. But Cope and Marsh were simply the first in a series of eccentric personalities in dinosaur paleontology. For those ivory tower scientists who judge the validity of rigorous science based on pedantic methods, this historical mixture of eccentric character and extreme public entertainment is said to detract from notable scientific accomplishment.

All in all, I personally think dinosaur paleontology reached the high table long ago, and that its public facade has overshadowed the genuinely interesting and important contributions made by noteworthy individuals. I certainly think it can at least be said that if paleontology as a holistic field has reached the high table, it did so thanks to the rigorous editorial choices of the journal *Paleobiology*. From its inception in 1975 there have been approximately thirty-five papers directly related to dinosaur biology published in the journal, and nearly half of those have been published since 2000. Although some of the proposed hypotheses of the earlier papers have since been falsified, the paleontologists and their articles are revered as notable by scientists in many fields outside paleontology. *Paleobiology* is recognized as the primary conduit for the highest quality of analytical study in paleontology. The fact that there are so many analytical papers concerning dinosaurs attests to the view that dinosaur paleontology sits at the same table as any other branch of paleobiology.

It is certainly true, however, that the early dinosaur paleontologists were primarily collectors, much more interested in descriptive studies or popular displays than in systematic analyses. It is also true that most of these old-timers were not shy to public attention, and ensconced themselves in both scientific and public debate. However, a few of the early dinosaur paleontologists, besides being unconventional, were interested in understanding dinosaurs as biological inhabitants of interesting ecosystems. Some of these paleontologists provided new and provocative hypotheses, a few of which could even be considered methodically analytic. Foremost, I think, was Baron Franz von Nopcsa, a Hungarian nobleman whom writer Adrian Desmond described as "that colourful Transylvanian spy, dinosaur expert, and self-styled heir designate to the Albanian throne" who, although being a "brilliant, if arrogant, misfit, was basically an ideas man." In the early twentieth century Baron Nopcsa proposed a variety of hypotheses, from the function of hadrosaurian crests to the ecology of *Archaeopteryx*, and he used new methods such as bone histology and ideas such as continental drift to support his hypotheses. Dinosaurs were not seen as very interesting animals by most other scientists at the time, however, on account of the perception that by Linnaean standard classification they were reptilian; so for all practical purposes, that meant they were cold-blooded, sluggish, and monotonous. With the exception of Nopcsa's work in the 1920s, and some thoughts by G. R. Wieland in 1942 about the possibility that dinosaurs were warm-blooded, dinosaurs would remain generally uninteresting to science throughout the first half of the twentieth century. Popular books from the early part of the second half, such as *All About Dinosaurs*, by Roy Chapman Andrews, and a series of adolescent-level books by Edwin Colbert maintained high public interest for children and young adults in America. I was one of those children, and when I graduated from high school and went on to college expecting to study dinosaur paleontology, I was told that dinosaurs were of no consequence, and that fossil mammals were the only reasonable, *au courant* pursuit. Fortunately for me, it was 1964, and a scientific revolution was on the horizon.

The leading edge of the transformation in thinking, or what would be described by one of the revolutionists as the "dinosaur renaissance," would be swift; it needed only the two advisor-student generations of Edwin Colbert to John Ostrom, and Ostrom to Robert Bakker. Ostrom was the initiator, and Bakker the insistent prod. Ostrom realized, after a comparative study of *Archaeopteryx* and *Deinonychus*, that dinosaurs were much more closely allied with birds, as T. H. Huxley had proposed a hundred years earlier. Bakker, along with Peter Galton, would take the bird-dinosaur union a step further and propose a new class, the Dinosauria, that included the two reptilian orders Saurischia and Ornithischia plus all of Aves. At the same time, in the late 1960s and early 1970s, the French paleobiologist Armand de Ricqlès, studying bone histology in Paris, would expand on earlier osteohistological studies by Don Enlow and John Currey, observing that the microstructure of dinosaur bone was much more similar to mammalian and bird bone than to typical reptile bone, and would propose that birds and extinct dinosaurs shared similar physiologies. Ricqlès, Ostrom, and Bakker each championed the idea of some sort of dinosaur endothermy. The convergence of these ideas from Ostrom, Bakker, Galton, and de Ricqlés quickly propelled dinosaurs from primarily amusing to scientifically compelling.

Other participants who added significant perspectives to the dinosaur rebirth in the 1970s included Peter Dodson of the University of Pennsylvania, Jim Hopson of the University of Chicago, and Jim Farlow of Indiana State University. Dodson published two important, quantitative studies of dinosaur variation, in which he was able to hypothesize sexual dimorphism and ontogeny based on morphometric analyses. One of these studies, concerning hadrosaurian (duck-billed) dinosaurs revealed for the first time that these dinosaurs went through radical changes during their cranial ontogenies, and that many of the smaller taxa that had originally been named were actually juveniles of other known species. Much of my work on dinosaur behavior, initiated in the late 1970s, would rely on Dodson's hypotheses. Another important study, published by Jim Hopson in 1980, concerned the evaluation of dinosaur brains. Using the encephalization quotient (EQ) developed by Harry Jerison, Hopson determined that most dinosaurs had EQs similar to, or slightly above, those of crocodylians. The smartest of dinosaurs were the dromaeosaurs and troodontids, the closest relatives to birds. In addition to these works, Jim Farlow initiated a number of studies concerning dinosaur ecology. Ostrom, de Ricqlès, Bakker, Dodson, Hopson, Farlow, and a Canadian, Dale Russell, who was studying dinosaur extinction, were the primary architects of the dinosaur renaissance. Some also place myself into this milieu.

I was fortunate to enter paleontology in the late 1970s, during the early part of this conversion period, with discoveries of eggs, embryos, and juveniles, and behavioral theories that supported the new emerging view of dinosaurs as active, warm-blooded ancestors of birds. Evidence that at least some dinosaur species nested in colonies, cared for their nest-bound young, and traveled in gigantic aggregations was controversial at first, but later discoveries by many of our colleagues around the world have tended to support this view.

These new paleontological discoveries added to the seminal works of the renaissance architects, and the arrival of cladistics as a method for placing organismal characteristics in an evolutionary framework gave the dinosaur rebirth both relevance and influence. Dinosaur paleontologists began to address paleobiological questions in the context of both ontogeny and phylogeny, using modern biological methods. If it weren't for one small glitch, exacerbated by the fact that dinosaur paleontology is constantly in public view, I think that by the mid-1980s the field would have reached a position equivalent in reputation to other scientific disciplines.

The glitch was, and continues to be, in the form of dissenters, a small group of paleo-ornithologists, led by Storrs Olson of the Smithsonian and Alan Feduccia of the University of North Carolina, who disagree with the assessment of birds as the evolutionary descendants of dinosaurs. The dissenters freely admit that they have no evidence to support an alternative hypothesis, but that their argument has more to do with the differences they see between birds and dinosaurs, and a perceived time gap between derived dinosaurs and early birds. Unfortunately, the dissenters seem to have had the ear of some editors of the journal *Science*, and a number of early papers supporting the bird-dinosaur link were apparently turned away, while the objections of the dissenters were published, giving a false sense of importance to a nonscientific debate. Phylogenetic hypotheses, as we all know, are proposed and tested on the basis of similarities rather than differences, and time has nothing to do with it. As John Ostrom, Jacques Gauthier, Kevin Padian, and many other dinosaur paleontologists have demonstrated, birds share numerous physical characteristics with dinosaurs, and more are discovered and added to the list each year. The hypothesis meets the requirements of the scientific method in that it is testable, and as yet, unfalsified. But, in the popular press, the debate rages on, denying the apparent stability of a coherent, rigorous discipline and the acceptance of methods that are used by systematists working on everything from molecules to blue whales.

So, where are we now? Ignoring the so-called debate, which in my mind is of no scientific consequence—other than to misinform the public about how science is done—dinosaur paleontology has been advancing exponentially in the last three decades.

From my point of view, the most interesting avenue of research over the past thirty years or so, in which a fair number of people have been engaged, is the question of dinosaurian physiology. Although apparently first suggested by Wieland, it was Ostrom, Bakker, and de Ricqlès who more formally proposed a series of hypotheses that dinosaurs were more advanced physiologically than nondinosaurian reptiles. In 1980 the renaissance architects, together with a few other paleobiologists, published their ideas in the 1980 American Association for the Advancement of Science (AAAS) symposium volume entitled "A Cold Look at the Warm-Blooded Dinosaurs," edited by R. D. K. Thomas and E. C. Olson. This important volume provided definitions, and therefore boundaries, for both sides of a physiological debate that continues to this day. Various authors, including myself, have weighed in on this discussion, and over the years there have been as many physiological hypotheses to explain the dinosaurs as there are combinations to be made. The questions are as open now as they were in 1980. Most of the arguments against dinosaurs having been tachymetabolic endotherms, or endothermic heterotherms, seem to come from the camp supporting the non-dinosaur-bird connection, but some disagreement is also clearly present within the group who think phylogenetically. More data and new questions are obviously required before there is to be a consensus. But then, this *is* the quest of scientific endeavor.

Evaluating the scientific importance of particular discoveries that yield new information about biological aspects such as ecology, behavior, or physiology is very difficult in dinosaur paleontology because so little seems clearly understood. Every new discovery has the potential to provide startling, often contradictory, new information to our currently incomplete record. Whereas the discoveries of the nesting grounds in Montana and Mongolia, and more recently Argentina, have provided important insights into the behavior of dinosaurs, discoveries of new kinds of dinosaurs around the world are arguably as important to enable us to reconstruct an historical record of dinosaur evolution and paleobiogeography. Whereas new sites that reveal behavioral information are rare and unpredictable, new taxa, according to Peter Dodson, are found on an average of about one every seven weeks or so. In addition, numerous Mesozoic geological formations and sediment facies around the world remain unexplored, but they provide the opportunity to yield abundant new taxa and potentially exciting biological interpretations. The point is that field exploration and collection remain a vital part of dinosaur paleontology, probably more so than nearly all other aspects of paleontology, because of historically poor data retrieval, new collecting methods, and the potential of extraordinary new localities. An example of this is the recently discovered quarries in the Yixian Formation near Liaoning, China, where feathered dinosaurs and birds have been found, many preserved with impressions of feathers covering their bodies. Reported at first primarily by paleontologists Ji Qiang and Ji Shu'an, this single site has added enormously to our current knowledge of dinosaur relationships, growth, and physiology. It provides concrete evidence that various nonavian dinosaurs possessed nearly every characteristic that we once ascribed only to birds.

Just mentioning the simplest of these characteristics, which include an open acetabulum (socket to receive the femoral head), the three-toed pes, (hind foot), reduced fifth and fourth digits of the manus (hand), and three or more sacral vertebrae, would seem to convince anyone that it was at least a good, testable hypothesis. A few added characteristics shared by the dromaeosaurids and birds include the furcula (wish-bone), feathers, and oblong eggs. One of the more remarkable synapomorphies, from my point of view, is the elongated calcified cartilage zones of the epiphyseal growth plates seen in the long bones. These extensive epiphyses, plus highly vascularized bone tissues, are what allowed both dinosaurs and early birds the capability of extremely rapid growth rates. This unique feature has been further adapted by living birds, and is the reason why they all grow to full adult size in less than one year.

Do any of these new perspectives help raise the judge's perception of dinosaur paleontology as a rigorous science? I presume that decision will be made years from now, in retrospect. For now, here is my perspective on the results of the dinosaur renaissance, and why I think dinosaur paleontology is not only at the high table but above it, at the cutting edge of science.

In the 1980s, after Jacques Gauthier had established cladistically that dinosaurs were indeed the ancestors of birds, dinosaur research branched off in many directions. I continued my pursuit of new specimens to test ideas about behavior, growth, and evolution, and began histology projects with Armand de Ricqlès, and later also with Kevin Padian. Robin Reed, a researcher from Belfast, also added a few histological insights relevant to the physiology question. Other American paleontologists, such as Jim Farlow and Peter Dodson, delved further into ecology and physiology. David Weishampel of Johns Hopkins University produced his now-famous paper on the acoustics of hadrosaurs, demonstrating that the hollow crests of adult crested hadrosaurs could have resonated when air was blown through their nasal chambers. Weishampel would move from there on to feeding mechanics in hadrosaurs, and together with David Norman (U.K.) looked into functional morphology and feeding in a variety of dinosaurian taxa. Paleontologists seeking new dinosaurian species throughout the world included people like Mark Norell and Paul Sereno (U.S.), Phil Currie (Canada), Dong Zhiming (China), Rinchen Barsbold (Mongolia), Ashok Sahni (India), Philippe Taquet (France), and Jose Bonaparte (Argentina), among others.

By the 1990s dinosaur paleontology had evolved into a significant faction, and new technologies like computerized tomography were providing innovative methods of study. Following the release of the movie *Jurassic Park*, student enrollments in dinosaur paleontology courses rose sharply. In 1995 I was chairing the graduate committees of eighteen vertebrate paleontology students at Montana State University. Several of them would go on to produce important dinosaur research, much of it in the recent news.

My former students Kristi Curry (Minnesota Science Museum in St. Paul), Greg Erickson (Florida State University in Tallahassee), and David Varricchio (Montana State University in Bozeman), together with other paleohistologists such as Anusuya Chinsamy-Turin of the University of Cape Town, South Africa, and German paleontologist Martin Sander, each made considerable strides in studies of bone histology, hypothesizing rates of linear or mass accumulation for a number of dinosaur taxa, and in some cases even predicting the longevity of a few species. What I think is one of the most remarkable discoveries in modern dinosaur paleontology, however, was made by my former student Mary Schweitzer, now at North Carolina State University in Raleigh. The discovery was made while she and her assistant Jennifer Wittmeyer were examining bone tissues that my staff had mined out of a femur of a *Tyrannosaurus rex* specimen that we called B-rex.

When my students and staff excavate dinosaur remains they are always cognizant of the fact that important data can be gleaned from virtually every aspect of a skeleton, as well as the rocks in which they are encased. The skeletal element from which the bone tissues of B-rex were mined for Mary was broken during an attempt to divide a large plaster jacket for helicopter recovery. The bone tissues crumbled from the break, and were immediately collected, wrapped in foil, and saved, to be sent off to Mary for analysis. Mary's interest in bone tissues has always been simple—she's looking for anything out of the ordinary.

For her doctoral dissertation in the early 1990s, Mary had examined cortical tissues from another *T. rex* skeleton, MOR 555. During her analysis she discovered vascular canals that contained round, reddish structures that resembled red blood cells, and test after test failed to falsify that hypothesis. But publishing her results became an unforgiving task, as the peer reviewers for both *Science* and *Nature* simply ignored the possibility of her conclusions. For several years, publication was stifled in virtually every peer-reviewed journal to which she submitted, but it was finally saved by Armand de Ricqlès. The results of the paper, published in *Annales de Paléontologie* in 1999, were immediately dismissed by experts as having any possible biological origin, and relegated to structures of geologic anomaly.

Four years later Mary and Jennifer would identify similar reddish, spherical structures in the femoral tissues of the B-rex specimen. But this time they would be discovered in transparent, flexible tubes, acid etched from the permineralized bone. Upon immersion in water the cell-like structures poured out of the flexible tubes. Scanning electron microscope (SEM) analysis of the tubes revealed morphological structures identical to those of modern, bone entrapped blood vessels. Further etching and analysis revealed intact osteocytes with internal contents and flexible filipodia. As with her *T. rex* analyses of the early 1990s, Mary initiated a series of tests in an attempt to falsify the hypotheses, but to no avail. The paper was written up, submitted to *Nature*, and immediately rejected, with many of the same arguments and dismissals that had been leveled against her initial paper. This time, however, the paper would pass muster and be accepted for publication in *Science*. The announcement made global news, and was almost immediately either sensationalized as evidence supporting a young earth, or once again scoffed at by various paleobiologists and molecular biologists as improbable. Herein lies the discord within the philosophical viewpoint of the ivory tower, and the reason there is anything to discuss about whether a particular field of science has reached the high table.

Keeping in mind that all questions are philosophical—meaning that there are no scientific questions, only scientifically based answers—Mary's original question—"What are these little red things?"—was as good a question as anyone could ask. The structures were red, spherical, and entrapped within the vessel spaces of the bones. Mary hypothesized that the structures were remnants of blood, and assayed them to discover that they were composed of the biological form of iron known as *heme*, the iron component of blood. Further analyses of the bone tissues and the surrounding sediment failed to reveal any evidence of geological iron.

Although there was considerable opposition to Mary's results, no one stepped forward with evidence to falsify any of her testable hypotheses. The naysayers were content to naysay, and the editors of the scientific journals were content to trust the naysayers' opinions, even though they had little or no scientific basis. Regardless of whether Mary's hypotheses turn out to be correct, she has followed the strict procedures of scientific inquiry and proposed hypotheses that are both testable and repeatable. In my opinion, Mary Schweitzer is currently at the head of the high table in dinosaur paleontology. As for her detractors, I doubt the table is in view.

In my opinion, the other young dinosaur paleontologists who currently sit at the table with Mary include John Hutchinson, presently at the Royal Veterinary College near London, Emily Rayfield, University of Bristol, Matt Carrano of the Smithsonian, Larry Witmer of Ohio University in Athens, Hans Larsson of McGill University in Montreal, and Karen Chin at the University of Colorado. For me, what sets these people off with Mary is both their multidisciplinary application and their creative endeavors. John Hutchinson and Matt Carrano both study dinosaur locomotion. John has primarily looked at *T. rex*, while Matt is doing analytical studies involving evolutionary aspects of locomotion between nonavian and avian dinosaurs. Emily uses 3-D-imaged skulls to hypothesize feeding biomechanics, and Larry Witmer, and his students, are interested in reconstruction of cranial soft-tissue features such as noses, ears, and brains. Both use methods in comparative anatomy and computerized tomography. Hans Larsson is working in what might be called the new field of paleo-evo-devo, looking into the possibilities of using bird genetics and embryology to study or possibly even replicate dinosaurian characters. Karen Chin, a paleobotanist, analyzes dinosaur coprolites to evaluate the ecology of certain dinosaur species.

But what of the future? Will dinosaur studies continue to be primarily individualistic endeavors, or will the field make adjustments in order to answer some of the more interesting, bigger-picture questions?

In my mind, the future of dinosaur paleontology is multidisciplinary synthesis, with teams of senior researchers and students tackling problems using a wide variety of methods, but beginning in the field with the collection of new specimens. An example from some of the work I'm involved with concerns a project to attempt to reveal the ecology of the Hell Creek Formation, one of the last dinosaur-dominated ecosystems before the extinction. Called the Hell Creek Project, the participants include a number of paleontologists who specialize in virtually all of the taxa found, plus geologists and statisticians. The data collection portion of the project has been underway for seven years, and there is funding for another four. Collectors gather up everything, making note of both stratigraphic and geographic location and sediment type. Paleontologists who study particular groups formulate hypotheses in the dark from one another, and each year we test our own hypotheses with new data. At the end of the exercise, all of the data will be combined to create testable, synthetic hypotheses about various facets of dinosaur ecology and evolution. Comprehensive collection and data-processing projects like this obviously take considerable time and funding, but have the potential to yield huge (for vertebrate studies) data sets for future analyses.

At the 2006 AAAS annual meeting in St. Louis, Mary Schwietzer and I organized a symposium to discuss "new approaches to paleontological investigation." What was learned was that dinosaur collection and investigation incorporates the latest technologies available to any field of science.

I have no doubt that the future of dinosaur paleontology will continue to take countless different directions, and that new and startling discoveries will continue, probably even increasing in number as new people are added to the field. As I look around at programs around the world I'm encouraged by the quality of up-and-coming students and their seemingly strong focus on quantitative analyses. I'm certain that the near future of dinosaur research will provide significant perspective, and with the novel potential of paleo-evo-devo, will it even be possible to conceive of a nonavian dinosaur, retroengineered from a modern bird?

REFERENCES

Bakker, R. T., and P. M. Galton. 1974. Dinosaur monophyly and a new class of vertebrates. *Nature* 248:168–72.

Currie, P. J., and K. Padian, eds. 1997. *Encyclopedia of dinosaurs*. San Diego, CA: Academic Press.

de Ricqlès, A. 1974. Evolution of endothermy: Histological evidence. *Evolutionary Theory* 1:51–80.

Desmond, A. J. 1976. *The hot-blooded dinosaurs: A revolution in palaeontology*. New York: The Dial Press/James Wade.

Dodson, P. 1975. Taxonomic implications of relative growth in lambeosaurine hadrosaurs. *Systematic Zoology* 24:37–54.

Farlow, J. O. 1976. A consideration of the trophic dynamics of a Late Cretaceous large-dinosaur community (Oldman Formation). *Ecology* 57:841–57.

Hopson, J. A. 1977. Relative brain size and behavior in archosaurian reptiles. *Annual Review of Ecological Systematics* 8:429–88.

Horner, J. R., and R. Makela. 1979. Nest of juveniles provides evidence of family structure among dinosaurs. *Nature* 282:296–98.

Ostrom, J. H. 1973. The ancestry of birds. *Nature* 242:136.

Roger, D. K. T., and C. O. Everett. eds. 1980. *A cold look at the warm-blooded dinosaurs.* AAAS Selected Symposium 28. New York: Westview.

Schweitzer, M. H., J. L. Wittmeyer, J. R. Horner, and J. B. Toporski. 2005. Soft-tissue vessels and cellular preservation in *Tyrannosaurus rex. Science* 307:1952–55.

Weishampel, D. B. 1981. Acoustic analyses of potential vocalization in lambeosaurine dinosaurs (Reptilia: Ornithischia). *Paleobiology* 7:252–61.

CHAPTER SEVEN

Ladders, Bushes, Punctuations, and Clades: Hominid Paleobiology in the Late Twentieth Century

Tim D. White

INTRODUCTION

Humans are but one of millions of terminal twigs on life's tree, but our evolution has understandably commanded a disproportionate amount of scholarly attention. Darwin's 1859 treatise deliberately sidestepped human evolution, but he devoted a book to the subject a little over a decade later. Indeed, from Darwin to Dover, evolutionary biologists have repeatedly turned to hominid evolution for their arguments and examples. In Darwin's day, the hominid fossil record was largely unknown beyond Neanderthals (Hominidae bounds genera in the human clade after the last common ancestor we shared with chimpanzees). Compared to general vertebrate paleontology, recovery of hominid fossils began slowly, but literally thousands of hominid fossils have now been prised from sediments spanning six million years.

Richard Delisle's *Debating Humankind's Place in Nature 1860–2000* (2006) is the most comprehensive attempt at a history of human evolutionary studies. He asks: "In what way then does paleoanthropology's nature differ from the other specialities of paleontology?" He concludes: "It differs in no way since they are all bound by epistemological imperatives which are fundamentally identical. However, their specificities arise from the different portions of the tree of life they each try to reconstruct." (369–70). Furthermore, as Eldredge and Tattersall observed: "Those deficiencies occasionally identified with the study of fossil man are in fact liberally shared with all branches of paleontology." (1975, 219).

Evolution provides the theoretical foundation for hominid paleobiology.

Delisle describes the history of hominid paleobiology as a "complex interplay of the comparative method and of the fossil record." (2006, 369). He concludes: "the relationship between specific evolutionary theories and paleoanthropology . . . since 1860 has been more or less superficial" (368). That seems appropriate for a target clade (hominids) containing a relatively few bizarre lineages descended from a late Miocene ape. Hominoid primates are short terminal twigs of life's tree. Accordingly, they are not a particularly representative group for developing or testing hypotheses about general evolutionary pattern and process.

Ever since Darwin, leading evolutionists have accommodated hominids to their arguments. Several architects of the Modern Synthesis (notably Dobzhansky, Simpson, and Mayr) explicitly addressed hominid evolution. This attention to hominids continued during the rise of contemporary paleontology. Indeed, many of the most important subsequent evolutionary theorists (notably Eldredge, Gould, Stanley, Vrba, Gingerich, and others) continued to refer explicitly and frequently to the data and practice of hominid paleontology.

The architects of the Synthesis chastised human evolutionists for being out of touch with evolutionary biology, but by the mid-1970s, hominid paleontologists and their colleagues were asserting that "Paleoanthropology indeed, far from being, as sometimes claimed by those occupied with other groups, a special and inferior case of paleontological practice, actually epitomizes many of the more established concepts and methods of paleontology in general." (Eldredge and Tattersall, 1975). Indeed, Gould (2002) called paleoanthropology a "small, contentious and vital field," and repeatedly drew upon it to illustrate evolutionary phenomena.

A small but respectable history of hominid paleobiology has developed due to the influence of Spencer, Hull, Bowler, Proctor, Delisle, and others. It is beyond this essay's scope to review that work. Rather, I focus here on the most recent history of human paleobiology, the second half of the twentieth century. This period witnessed the accumulation of a wealth of new hominid fossils during a turbulent period of methodological and theoretical advances in paleobiology.

This essay is divided into three parts. Part I outlines hominid paleobiology's place in science. Part II situates hominid paleobiology within post-Synthetic evolutionary biology. Part III examines the "co-evolutionary" interplay between methodological and theoretical developments in evolutionary biology and the accumulating hominid fossil record during the past thirty-five years, focusing on the contemporary debate over species diversity.

Anthropologists, detectives, and jurists are all acutely aware that living informants have a great capacity for historical revision in nearly real time. Historians of science may therefore justifiably suffer qualms about this volume's format. Our editors have strongly encouraged authors to recount tales of their own roles in late twentieth-century paleontology's alleged move to the "high table" of evolutionary theory. Combined with editorial limitations on the number of allowed citations, these facts offer boundless opportunities for historical revisionism. Under these conditions, the "histories" narrated within this volume's pages are patently idiosyncratic. In general, all autobiography is rendered suspect by the lack of adequate temporal perspective. Those (limited) parts of what follows should not be read as exceptions to that rule.

PART I. THE ROOTS OF CONTEMPORARY HOMINID PALEOBIOLOGY

In 1950 Mayr wrote: "never more than one species of man existed on the earth at any one time." (112). Today, fifty-nine years later, synthetic treatments recognize as many as twenty-six hominid species (half of them named since 1994) and some see seven "adaptive radiations" within the clade (often now demoted to tribal status, "Hominini"). Some see this classificatory exuberance as a breakthrough: "As our science absorbs the lessons of evolutionary complexity that other branches of paleontology have already learned, we can look forward to a new perspective on our origins." (Tattersall, 1997b, 340). Contemporary hominoid classifications reflect a wide diversity of opinion, from the generic to the familial levels. Extremes range from Goodman's inclusion of chimpanzees and gorillas in genus *Homo* to the more conventional Simpsonian familial distinction for the hominid clade. Some conclude that contemporary hominid systematics is presently in its "worst shape" since the 1930s (Marks 2005, 49).

What happened between the simplicity of the Synthesis and contemporary speciose interpretations of hominid evolution? Hominid paleobiology was structured by a complex interplay between empirical findings and intellectual and technological developments. The subject's popular appeal, its peculiar academic settings, and developments in allied sciences all played roles in shaping contemporary manifestations of the discipline.

The Academic Setting of Hominid Paleobiology

Hominid paleobiology has long stood at the crossroads of the earth, life, and social sciences. Its major roots lie in "anthropocentric" disciplines (human anatomy, anthropology, prehistoric archaeology). This anthropological rooting (also suffered by much of primatology) is a historical contingency traceable to Biblical times. It places human evolutionists in a curious position, approaching an enduring identity crisis.

The vertebrate paleontological community (itself usually contingently nested in earth sciences) has often shunned contact with paleoanthropologists, and vice versa. Historical reasons for this are both logistical (revolving around departmental placements in major research universities) and practical (few paleontologists specialize on so narrow a clade; none of their organisms leave an archaeological record). These attitudes may also persist because anthropologists have been historically late to assimilate to the Modern Synthesis and were roundly chastised by its architects.

The unprecedented growth of U.S. research universities and funding agencies after World War II firmly emplaced biological (or "physical") anthropology within expanding departments of anthropology. From Boston to Berkeley, these departments emphasized a "four-field approach" to anthropology that integrated linguistics, archaeology, ethnography, and physical anthropology. Studies initiated during the 1960s in such anthropology departments played major roles in developing paleoanthropology. Behavioral and ecological studies of living primates, especially the great apes, provided useful perspectives. Taphonomic analyses and ethnoarchaeological approaches provided additional data that complemented primatological studies to illuminate early hominid behaviors.

Beginning in the 1970s and gaining momentum in the 1980s, postmodernist and economic forces began to disassemble these holistic anthropology departments. By historical legacy, they became largely controlled by social anthropologists prone to reject science as the best route to knowledge. Consequently, by century's end, many hominid paleobiologists were nested in medical school departments of anatomy, or in biology. This collapse of American anthropological science is still underway. The negative consequences for contemporary paleoanthropologists are counterbalanced by their participation in paleontology's integration into organismal biology.

Developments in Allied Disciplines

Like any other science, hominid paleobiology was thoroughly influenced during the last fifty years by general developments in technology, especially electronic computing. These technological achievements are often taken for granted. However, they form the infrastructure of much of modern science and have transformed the way that science is done and communicated. Computers allow micro–computed tomography (micro-CT) to peer within fossils, microscopes to reveal fine structure, differential global positioning systems (GPSs) to navigate, mass spectrometers to measure isotopes, digitizers to capture morphology, and sequencers to reveal genomes. They also allow us to communicate globally, sharing vast data sets in real time. And they allow international multidisciplinary networks to form and tackle previously unassailable laboratory and field research problems.

The emergence of plate tectonics in the 1960s transformed the earth sciences. Hominid paleobiology was a beneficiary of this transformation. Structural geology and lithology, revealed by satellite platforms, became important tools for identifying fossilferous sedimentary packages, dramatically expanding the fossil record. The rise of radioisotopic chronometry and paleomagnetic studies after World War II led to the accurate appreciation of earth history. The application of potassium-argon dating to the fossils from Olduvai Gorge was a breakthrough in understanding the temporal dimensions of hominid evolution. By century's end, developments in single-crystal laser fusion argon-argon dating provided crucial and comprehensive calibration for the African Cenozoic.

The calibration of African fossiliferous sequences opened windows on global climatic change via oceanic and polar drilling programs. These efforts played important roles in the development of late twentieth-century hominid paleobiology. Vrba, DeMenocal, and many others have probed these records in search of correlations, posited causal relationships, and even hypothesized that punctuated equilibria, heterochrony, and physical environmental changes might be linked. Besides their assistance in reconstructing past environments inhabited by early hominids, allied earth sciences studies have generated evolutionary data, hypotheses, and insights extending far beyond the hominid twig. By their very nature these efforts constitute multidisciplinary and international attempts to decipher climatic and tectonic changes at the global and local scales. Stimulated by efforts to understand hominid evolution, they are currently providing important context for considerations of anthropogenically induced global climatic change at the decadal, millennial, and longer scales.

Bowler (1986) and Delisle (2006) consider the difficulties that nineteenth- and twentieth-century hominid biologists experienced in their attempts to phylogenetically place living humans among the extant primates. After 1950, the rapid rise of molecular phylogenetics helped to resolve these recurrent controversies. The anthropological incorporation of such studies brought a new dimension to hominid paleobiology. A progression of molecular techniques, beginning with serology and moving progressively through immunology, protein sequencing, DNA hybridization, and genome sequencing has increasingly and definitively revealed the phylogenetic relationships among living hominoid primates. Relative branching order of extant hominoid taxa is now resolved, with humans firmly established as the sister clade of extant African apes. Efforts to date the splits among the great apes and humans continue to be compromised by limitations involving required calibration from a still-inadequate fossil record.

In the biological sciences, in addition to developments in evolutionary theory discussed in detail in the following, recent and rapid progress has been made in understanding the genetic and developmental substrates underpinning anatomical and behavioral differences among humans and other primates. Unraveling the polygenic, regulatory basis of morphological change has begun, with an integration of quantitative genetics, gene expression studies, and comparative genomics. Evolutionary developmental biology has already made great strides in understanding model organisms, and its impact on future studies of human evolution seems assured.

The Popular Appeal of Hominid Paleobiology

The target organism of human evolutionary studies guarantees an abiding public interest. From Darwin to Scopes, this meant a prominent public profile for hominid paleobiology. This was further enhanced by two episodes of creationist resurgence during the last fifty years. But even those facts insufficiently explain the subject's current profile, a visibility enhanced by the rise of electronic media and communication.

Contemporary science has profoundly penetrated domains extending from the genome to the cosmic background. Many subjects of science have become increasingly difficult for nonexperts to grasp. In contrast, the raw and frequently tangible data of vertebrate paleontology are largely recovered and appreciated at the human scale. The tooth of a *T. rex* (or the femur of an early hominid) is easier for the public audience to visualize than a string of base pairs or fermionic superpartners of neutral gauge bosons. This is "scaling to the familiar," a phenomenon that today inflates paleontology's already-prominent media niche. High-impact science journals with an eye toward popular consumption (Lawrence's "fashion" journals) frequently achieve interdisciplinary "balance" via paleobiological papers, including some about hominids. This creates niches for paleopundits, ghostwriters of popular and semipopular books, and countless television documentaries—all too often long on personality, but short on science.

Coming decades will provide needed perspective on what is now happening in hominid paleobiology, but the peculiar relationships among its contemporary practitioners, their institutions, and the modern global media are already worth noting. I have written elsewhere about developments in hominid paleobiology during the last two decades, raising long-term concerns about disciplinary health (White 2000).

PART II: FORWARD FROM THE SYNTHESIS:
PALEONTOLOGY, HOMINID PALEOBIOLOGY,
AND SYSTEMATICS

Hominids and the Synthesis

The architects of the Modern Synthesis created a coherent theory of evolution that integrated genetics, paleontology, and systematics. Its implications were explicitly assimilated into hominid paleobiology (although Tattersall would later characterize this integration as an "invasion of the halls of paleoanthropology"). Writing in 1944, Dobzhansky wrote that no more than a single hominid species existed at any one time during human evolution. In a footnote he accused anthropologists of improper taxonomic practice. Six years later, in 1950, he and Washburn organized a Cold Spring Harbor Symposium. Mayr would later state: "It was on that occasion that the study of fossil man was integrated into the evolutionary snythesis (sic)." (1982, 231). He noted the attendant shift from typological to populational thinking—and the subsequent widespread adoption of a polytypic, evolutionary species concept.

In his published contribution for the 1950 conference, Mayr collapsed the entire hominid fossil record into a single evolving lineage of genus *Homo*, basing his interpretation, at least in part, on ecological principles. This classification was therapeutic: by mid-century nearly thirty generic and over one hundred specific names had already been applied to hominid fossils. Extant chimpanzees, by Schultz's count, had been bestowed twenty-one different generic and seventy-three specific names. Thirteen years later, when Washburn convened the next conference on hominid systematics, Mayr would admit a second hominid lineage, based on the work of Broom and Robinson in South Africa and on the new findings of the Leakeys at Olduvai Gorge. But that was as far as he was willing to go: "When one reads the older anthropological literature with its rich proliferation of generic names, one has the impression of large numbers of species of fossil man and other hominids coexisting with each other. When these finds are properly placed into a multi-dimensional framework of space and time, the extreme rarity of the coexistence of two hominids became at once apparent." (1963, 339). Only a few workers (Brace and Wolpoff at Michigan) continued to argue for a single hominid lineage, carrying this argument into the 1970s, when the weight of accumulated fossil evidence ultimately forced its abandonment.

During the 1960s and early 1970s the evolutionary species concept was widely employed within and beyond anthropology. Schultz's studies established high levels of within-species variation based on geographic, idiosyncratic, sexual, and ontogenetic factors in extant higher-primate species. And from the primate fossil record, Le Gros Clark noted the "fallacies" that had led paleontologists astray in their systematic efforts. These and other workers provided a powerful interpretive framework for hominid paleobiology, firmly based on the foundation of the Modern Synthesis. That foundation would be rattled during the 1970s by the rise of parsimony cladistics and punctuationism.

Hominids and Cladistics

In 1955 the anatomist Le Gros Clark presaged phylogenetic systematics in his book, *The Fossil Evidence for Human Evolution:* "The importance of making a general distinction between primitive and specialized characters depends on the fact that the latter may be taken to indicate divergent trends of evolution" (41). Against a widespread Simpsonian-era appreciation that evolutionary novelty was the key to unraveling primate phylogeny, it is not surprising that the basic concepts of Hennigian phylogenetics, or cladistics, were rapidly assimilated into 1970s hominid systematics.

Eldredge and Tattersall's 1975 paper "Evolutionary Models, Phylogenetic Reconstruction, and Another Look at Hominid Phylogeny" was the first semi-formal application of Hennigian phylogenetics to the hominid fossil record. Tattersall later described it as: "a pretty naïve effort . . . given that it was the work of a trilobite specialist and one who had up to then been interested mainly in the lower primates." (1997a, 168). It was followed with a more formal 1977 paper coauthored with Delson. For the first time, this put to print Nelson's concepts of cladograms, trees, and scenarios (Hull, 1988). In this paper, great optimism was expressed regarding the potential of cladistic methods for hominid systematics: "proper methodology would enable us shortly to frame an hypothesis of hominid phylogeny . . . which should be acceptable as the least unlikely scheme to all workers embracing the methodology."(264). The authors concluded that there was "ample justification for our view that a cladistic approach will lead to far more stability in hominid phylogeny reconstruction" (265).

Cladistic analysis and cladistic classification were predictably adopted, early and often, by hominid paleobiologists—albeit with varying degrees of rigor, formality, and success. Dozens of papers appeared beyond the American Museum of Natural History. Clarke's 1977 explicitly cladistic PhD dissertation on South African hominids was exemplary. Abundant new fossils recovered during the 1970s provided ample opportunity for cladistic practice.

A host of paleoanthropologists subjected available samples of hominid fossils to Hennigian parsimony analysis during the 1980s and 1990s. This required that the fossils first be allocated to operational taxonomic units, usually historically recognized "species" whose morphological attributes were then parsed into characters whose distributions were subjected to parsimony analysis. The resulting cladograms (some based on hundreds of characters) gave expected results at the genus level and above—but this hardly constituted a breakthrough. Resolution at the species level (Hennig's lowest operational unit) remained poor, particularly for species whose very existence and definition were rooted in pre-Synthesis typology and historical contingency. As paleoanthropologists parsed their data sets to maximize the number of available "taxa," hominid taxonomic inflation soon returned to pre-Synthesis levels, in what Jolly termed a "schizophilic taxonomic climate" (Jolly 2001, 183).

Tattersall, McHenry, Skelton, Wood, Chamberlin, Begun, Groves, and a variety of others all espoused the virtues of a cladistic approach during the 1980s and beyond, following practitioners throughout paleontology. Compared to many mammalian clades, hominids are relatively well sampled across the past four million years. However, hominid lineages are usually poorly represented at any one time interval, and are often sparsely sampled across their relatively wide geographic ranges. Worse, as Schultz, Le Gros Clark, and Simpson had all recognized from their work on well-established neontological analog species, high levels of within-species variation, parallelism, and convergence were the norm among higher primates. Finally, the possibility of frequent phyletic evolution advocated by the Synthesis, if real among hominids, had the potential to compromise the entire cladistic approach.

Critiques of cladistics were offered by Mayr, Simpson, Bock, Trinkaus, Cartmill, Brace, Habgood, Tobias, Harrison, Cunroe, Kennedy, Walker, Szalay, and others. Objectionists raised criticisms related to ancestors, graded states, homoplasy, character definition, and coding, and other aspects of hominid cladistics. Their committed cladist opponents considered these objections trivial or worse. Perhaps because of their relatively large and young record, many hominid paleobiologists maintained deep reservations about the notion that ancestor recognition in the fossil record was, in the words of one paleontological cladist, an "impossibility."

By the 1990s, cladistics as a basic method of phylogeny reconstruction had become ubiquitous in paleoanthropology. But contrary to initial prognostications, the application of the method had destabilized nomenclature and created even more diversity of interpretation. Part of this turmoil was owed to the accumulation of more fossils. But hominid nomenclature had destabilized as fundamentalist cladistic classifiers chased the elusive dream of paraphyly-free classification. The failure of cladistics to resolve a basic cladogram for hominid "species" became disconcerting, even to its zealous advocates. Wood, an early adopter and heavy user of cladistics, began to question its ability to generate accurate phylogenies based on craniodental remains of extant primates. He and other like-minded workers rediscovered what Simpson, Le Gros Clark, and Schultz had all appreciated—higher primates are highly variable and exhibit a high degree of homoplasy. Hope has recently emerged that developments in evolutionary developmental biology will render character definition less arbitrary.

By 2000, the diversity of phylogenetic trees and consequent classifications held by paleoanthropologists was impressive. Delisle (2001) reviews the historical background. A 1999 book by Gee, a prominent cladist senior editor at *Nature*, set out to show how cladistics represented a paleontological revolution. With human evolution as a central theme, the book's stance was described by one progressive syntheticist as "Nonsense rhetoric." And so it goes.

Hominids and Punctuated Equilibria

Whereas Tattersall and Eldredge are widely credited for the first formal application of the cladistic method to hominid paleobiology, Delson's role in the post-Synthetic reorientation of hominid phylogenetics and classification was also pivotal. As a graduate student working in and around New York's American Museum of Natural History, Delson had early exposure to cladistics and more. His Columbia PhD dissertation on cercopithecine monkey evolution, completed in 1973, was specifically uncladistic. Nevertheless, it cited Eldredge's (then-recent) seminal paper on trilobites, as well as a draft of the yet-to-be published original Eldredge and Gould paper on punctuated equilibria.

Over the last few decades, Tattersall has repeatedly complained that pa-leoanthropologists laboring under "the dead hand of dogma" have been slow to pick up the latest in evolutionary thought. But this rhetoric is best inter-preted as a complaint about others not accepting his agenda than a demon-strable retardation in hominid paleobiology. Delson applied the concept of punctuated equilibria to primates almost instantaneously. Indeed, the institu-tional proximity of Eldredge, Tattersall, and Delson centered paleoanthropol-ogy firmly within the protracted 1970s and 1980s debates about evolutionary mode and tempo.

The original 1972 Eldredge and Gould essay on punctuated equilibria chose hominid paleobiology to exemplify the most egregious of the "prob-lems" with phyletic gradualism in evolutionary biology. For their purpose, they chose an "admittedly extreme, example of *a priori* beliefs in phyletic gradualism . . . the work of Brace on human evolution." (98). As noted above, by 1972, Brace provided a convenient straw man. Most other hominid paleo-biologists had long ago abandoned the single-hominid lineage concept.

The 1970s choice of fossil hominids to exemplify evolutionary theory con-tinued a long tradition. Most of the key paleontological protagonists who followed Eldredge and Gould (Eldredge, Fortey, Gould, Gingerich, Mayr, Stanley, Vrba) would repeatedly appeal to hominid paleobiology in debates about punctuated equilibria and the fossil record. Why do hominid examples continually find their way into key works on general evolutionary theory, par-ticularly when other paleontologists regularly show a certain disdain for work done on such a small clade with such shallow temporal roots?

Perhaps it is because high-profile examples are more likely to appeal to a wider audience. And perhaps it is because the plethora of workers focused on this small clade has managed to produce such an unusually broad spectrum of interpretations, from aquatic apes to hobbits. With that kind of breadth, hominids provide many good and/or bad examples.

By the mid-1970s, Gould's *Natural History* essays were widely read and hugely influential. His original 1976 article "Ladders and Bushes in Human Evolution" extended the hominid paleobiology theme that he and Eldredge had exploited from the beginning. Gould asserted: "We are merely the surviv-ing branch of a once luxuriant bush" (31). Gould took the liberty of predicting when and what paleoanthropologists would find next: "We know about three coexisting branches of the human bush. I will be surprised if twice as many more are not discovered before the end of the century" (31). Part III evaluates what actually happened.

Hominid paleobiologists outside of Manhattan did not waste time in examining the initial claims of Eldredge and Gould. These ideas were tested repeatedly by new fossils during the 1970s. When *Nature* published a 1981 cover article claiming that phyletic gradualism with variable rates was the most parsimonious interpretation of hominid evolution, it allowed its authors to attempt a point-by-point rebuttal to Gould's popular article on bushes and ladders, thereby reifying the latter's place in the academic *and* popular literature.

Eldredge and Tattersall aimed at both popular and scientific audiences with their 1982 book, *The Myths of Human Evolution*, a volume that claimed: "We have debunked the myth that evolutionary change is gradual and progressive." (p. 175). This all occurred in a climate of creationist resurgence, prompting Delson to urge paleoanthropologists to draw on paleontology and evolutionary theory in their work (which they had already been doing, at least since the Synthesis).

The debates of the 1980s centered on whether any hominid species lineage exhibited stasis. For example, Rightmire squared off against Wolpoff in an argument about whether *Homo erectus* showed stasis or phyletic change. The impact of cladistics and punctuated equilibria were felt strongly in subsequent debates about *Homo erectus*. As early as the 1950s, Louis Leakey employed basically cladistic arguments to argue that this taxon was a "highly specialized offshoot of the human stock." But during the 1960s, most workers conceptualized the taxon as the geographically widespread, direct phyletic ancestor of *Homo sapiens*. In their 1975 debut paleoanthropological cladistics paper, Eldredge and Tattersall resurrected Leakey's views, suggesting that *H. erectus* was not the most parsimonious ancestor of *Homo sapiens*. This tradition was continued throughout the 1980s and 1990s by Andrews, Wood, and others.

Into that debate was thrown a calvaria from Ethiopia's Middle Awash, recovered by our research team in 1997. It is very similar to the classical specimens of *Homo erectus* from Asia. We used it to examine the claim for a deep cladogenetic split between Asian and African Pleistocene *Homo*. We ran an experiment with Hennigian parsimony analysis to investigate clustering of "demic" operational taxonomic units (OTUs) defined by a third party. When our characters (also adopted from other workers) failed to support the hypothesis of divergent geographic species lineages, we previewed the poorly resolved cladogram to Berkeley graduate students in vertebrate paleontology (they were better versed in parsimony cladistics than we were). They suggested that we "fiddle" with character inclusion and coding in order to obtain a higher consistency index (i.e., a "better result").

Meanwhile, Tattersall and Eldredge kept up a steady campaign of popular and professional contributions to focus attention on the controversies about the number of hominid taxa and their evolutionary mode and tempo. They took every opportunity to heap abuse on the Synthesis and on unenlightened "destructively minimalist" paleoanthropological "followers," whom they accused of "slavishly" following its "dictates" (such as phyletic evolution). The arguments about evolutionary mode and tempo extended throughout the realm of primate paleontology. For example, Gingerich, Bown, and Rose squared off against Eldredge, Gould, and others by insisting that evolution as revealed by the Paleogene primate fossil record frequently required arbitrary boundaries between species and even genera.

Hominid fossils, accumulating during these debates, contributed expectedly little to the resolution of the general question of evolutionary mode and tempo. In their *Paleobiology* paper in 1977, Gould and Eldredge had already declared the issue to be one of relative frequency rather than any single example (which the hominids were, even by their reckoning). Nevertheless, paleoanthropology continued to play a central role in Gould's writings. By 1997, he was still engaged in Brace-bashing, coupled with a pronouncement that "a virtual explosion of hominid species occurred on both major branches of the hominid bush" (69). By the time Gould published *The Structure of Evolutionary Theory* in 2002, the study of human evolution had been, for him, "recast" in "speciational terms" (910).

PART III. POST-SYNTHETIC ASSEMBLY
OF THE HOMINID FOSSIL RECORD

Without early hominid fossils, Darwin could only triangulate from highly derived living apes and humans in his attempts to conceptualize early hominids. Fossils were subsequently found, and by 1960 it was generally held that the earliest hominid was the middle Miocene *Ramapithecus. Australopithecus* was, by then, widely recognized as ancestral to *Homo*. By the 1960s, it was widely accepted that a limited amount of cladogenesis had occurred among hominids, but most 1960s workers followed the Synthesis, interpreting *Homo erectus* as a chronospecies of *Homo sapiens*. The simplicity of that view would soon be challenged.

The knowledge claims and practices comprising contemporary hominid paleobiology developed within the dynamic technological, methodological, and theoretical milieu are outlined in parts I and II. As new hominid fossils were discovered, they played roles in this development. The current debate about Pliocene hominid species diversity may be evaluated in the context of these interactions.

Fossil Recoveries and Interpretations

During the 1950s the Piltdown fraud was exposed and *Australopithecus* was entrenched as a hominid. Robinson interpreted the South African early hominids in the ecological, adaptive context of the Modern Synthesis, clearly differentiating robust and nonrobust hominids. Attention turned to eastern Africa in 1959, where the focus remained through the mid-1960s. Nonrobust hominids from Olduvai were christened *Homo habilis* in 1964. The ensuing debate about whether the Olduvai form was separate from South African *Australopithecus* endured for the rest of the decade.

Paleobiological fieldwork involving early hominids became truly international and multidisciplinary beginning in the 1960s, most prominently with work in southern Ethiopia under the late F. Clark Howell and French colleagues. The Omo Research Project extended what started on a local scale during the early 1960s at Olduvai Gorge in Tanzania. A full battery of paleobiologists and earth scientists joined with archaeologists and others in collaboration to unravel early hominid evolution in eastern Africa. The decades that followed saw that model extended to many similar efforts in Africa and beyond. Indeed, a major spinoff was the early work in Ethiopia's Afar that produced the bonanza of "Lucy" and conspecifics. These 1970s' Ethiopian fossils represented the beginning of the end of the dominant Leakey dynasty in paleoanthropology.

Richard Leakey, having grown up witness to the heated debates over Olduvai's fossils, began his own project in northern Kenya in the late 1960s. In a misconceived effort to evade debate, he adopted a peculiar classificatory procedure, identifying his fossils at only the generic level, as either *Homo* or *Australopithecus*, usually in *Nature*. He conceptualized *Australopithecus* to be extinct and followed his father's belief that *Homo*'s roots were Miocene. Detailed descriptions of the fossils were presented without species designations in the *American Journal of Physical Anthropology*.

When *H. habilis* was found at Olduvai, I was ten years old. Like many others, I was drawn to the subject by the *National Geographic*'s coverage. As a biology major at the University of California at Riverside, I took my first anthropology class in 1970—just as young Leakey was making his name. I read textbooks by Simpson and Le Gros Clark in my upper division courses before entering Michigan in 1972. A fringe of paleoanthropologists there persisted in interpreting Africa's burgeoning hominid record in a single lineage framework inherited from Dobzhansky and Mayr. In 1974 the "Lucy" fossil was found in Ethiopia, and my "single-species hypothesis" professors (Brace and Wolpoff) took exception to a lack of party-line loyalty expressed in my written qualifying exams. The experience left no doubt about how indoctrination can proceed in the academy.

I joined Richard Leakey's team in northern Kenya between 1974 and 1976. In 1975, on one of his rare visits to the field site, Leakey had admonished Kamoya Kimeu and me for covering an exposed hominid fossil before a rainstorm. We were banished from Koobi Fora for spoiling a photo-op, but shortly thereafter Bernard Ngeneo was blowing his whistle to beckon me to what he thought was a baboon freshly eroding from 1.6 M sediments. It turned out to be a *Homo erectus* cranium lithostratigraphically contemporary with the most derived robust species of *Australopithecus*. This evidence terminally falsified the "single species hypothesis." Alan Walker kindly transmitted the news to Michigan, and I was freed to finish a dissertation. I met Don Johanson in 1975, on what he called his annual "Richard Leakey eye-pop" tour of Nairobi. He had a bonanza of newly found Ethiopian fossils, enroute to Cleveland. It had been a good year.

Paleoanthropology in eastern Africa in the early 1970s was conducted under the influence of the Leakeys and was very much in the public eye. Mary Leakey's team had recovered a few fossils from the Laetoli site south of Olduvai. She asked me to describe them according to the bigeneric, dichotomous model that her son had established (*Homo* or *Australopithecus*). The Laetoli jaws and teeth, dated to c. 3.5 M, were much older than any other hominids then known (except those from Hadar). The 1976 *Nature* announcement of the Laetoli fossils (in which I regrettably coauthored the mistaken attribution to *Homo*) was the basis of Gould's 1976 *Natural History* article on bushes and ladders, which he called "one of the greatest fossil discoveries of the decade" (26). Gould reasonably concluded that divergence between hominid lineages was very deep, opening up potential for diversity: "Based on the fossils as we know them, *Homo* is as old as *Australopithecus*" (29).

My subsequent late-1970s work with Johanson suggested that the *Homo/ Australopithecus* dichotomy was systematically unsatisfactory, there being insufficient evidence for hominid species lineage diversification before 3.0 M. Our methods were explicitly cladistic, but without excluding phyletic evolution. Our conclusions were published in *Science* in 1979, and included both a cladogram and a phylogenetic tree. I had crossed paths with Gould early in my Berkeley career when he came to lecture on punctuated equilibria, and we debated phyletic evolution among African suid lineages. He wrote almost immediately to congratulate me on the *Science* paper: "So I'll be rooting for your conclusion, even though I rather suspect from my own biases (well known to you) that *Au. robustus* speciated rapidly from *Au. africanus* and that *Au. africanus* persisted—not, as you show in your figure, that one evolved directly into the other." He went on to say how excited he was in the primitive nature of *Au. afarensis* and its implications for the timing of the ape-hominid split.

Debates about *Au. afarensis* continued throughout the 1980s. They focused on whether the species was distinct from *Au. africanus*, whether the Laetoli and Hadar specimens were conspecific, and about the locomotor mode revealed by the Hadar remains and Laetoli footprints. Early in these debates, Skelton and McHenry itemized characters from our 1979 *Science* paper and attempted a formal Hennigian parsimony analysis. Such exercises are, more than twenty years later, a nearly annual event. Nevertheless, a well-resolved early hominid cladogram remains elusive, despite intensive atomization of characters and juggling of OTUs. The stridency and rigidity of early hominid Hennigians has been tempered by a growing appreciation for rampant homoplasy, lack of hard tissue character independence, and the fact that many operational taxonomic units do not represent real biological species.

Constructing Early Hominid Species Diversity

Nearly a decade after the rise of cladistics and punctuated equilibria, Mayr mentioned neither in his 1982 history of human paleontology. Here, even as he took another swipe at human paleontologists for being typological, Mayr was underestimating the degree to which the Modern Synthesis had been perturbed by the rise of cladistics and punctuated equilibria.

Gould showed that it was possible to be a noncladist punctuationist, but there is an obviously intimate relationship (and high correlation) between the two character states. Nowhere is this truer than in hominid paleobiology, a pursuit in which reside some of the most prominent, prolific, and vocal practitioners of cladistics and punctuationistics. With theoretical expectations prescribed by punctuated equilibria, Gould's metaphor of a hominid bush has been forcibly imposed on hominid paleobiology for the last thirty years.

By the 1980s, as Delisle notes (2006, 116), "key elements of the wider paradigm (the theory of punctuated equilibria, species as individual entities, species selection, the notion of hierarchy, and cladistics) were all being applied to human evolution." "Hominid diversity" came into vogue in paleoanthropology during the 1980s, fueled by institutional and editorial rewards for something "new," and abetted by the climate of "diversity politics" that had risen to the forefront of the academy. Ironically, as historians of science uncritically adopted the species-diversity model for hominid evolution (perhaps on the strength of Gould's predictions), famed *New York Times* science journalist John Noble Wilford appreciated the linkage between the two movements: "This may be a reflexion of political as well as scientific currents. Just as Darwin's ideas on evolution by natural selection were congenial to Victorian England's belief in progress through gradual improvement, anthropologists concede that many of them have been steeped in multiculturalism and diversity, and see them in a favorable light" (March 25, 2001).

For many workers, Gould's 1976 prediction of bushiness became a paleoanthropological presumption, if not an obsession. For example, Wood's repeatedly published hominid phylogenetic trees include twenty-seven species lineages. But fully ten of these are unnamed and for good reason, denoted by embedded question marks—they are entirely imaginary, included as "a reminder that in the relatively unexplored period between 6 and 2 myr ago the number of taxa will probably increase" (Wood and Richmond 2000). By 2002 Gould would write: "I don't think that any leading expert would now deny the theme of extensive hominid speciation as a central phenomenon of our phylogeny" (910).

I began working in the Middle Awash of Ethiopia with the late J. Desmond Clark in 1981, but research there was soon halted for most of that decade. When reinitiated in the 1990s, the Afar work began to yield fossil hominids that pushed knowledge of hominid evolution into the upper Miocene. By this time, the interpretive climate of paleoanthropology had been radically transformed from the days of the Synthesis by punctuationism, parsimony cladistics, and an academy enamored of diversity politics. The result has been a return to the "(good or bad)-old-days" of hominid splitting. Many "species" of early hominids have been created in the last fifteen years. Two examples will illustrate the changes.

Meave Leakey and colleagues announced a new genus and species of 3.4 M hominid from northern Kenya, *Kenyanthropus platyops*, via a 2001 *Nature* article. The title was assertive and revealing: "New Hominin Genus from Eastern Africa Shows Diverse Middle Pliocene lineages." Not two lineages, but "diverse" lineages. Given its prominent publication on the cover of a high-profile journal coupled with global electronic media, the interpretation received widespread and uncritical acceptance. After all, by the mid-1990s, obviously inspired by the prodiversity climate of the times, Kappelman and

Fleagle had already pronounced in *Nature:* "It thus appears that the phylogeny of hominids, like that of many other mammalian groups is very bushy at its base" (1995, 559). It seemed that early hominid evolution had moved beyond bushes to hedges, even *before* the distorted cranium came to light.

For Wood and Tattersall, Leakey's christening of *Kenyanthropus* was a vindication. *Nature's* pundit Lieberman pronounced: "We can now say with confidence that hominin evolution, like that of many other mammalian groups, occurred through a series of complex radiations, in which many new species evolve and diversify rapidly." (2001, 420). Gould's 2002 opus would use the fossil as follows: "multiple events of speciation now seem to operate as the primary drivers of human phylogeny (see Leakey et al. 2001, for a striking extension to the base of the known hominid bush in the fossil record" [909]).

The new genus is represented by only *two* fossils. The holotype is an intensely fractured cranium suffering from expanding matrix distortion. The paratype (singular) is a tiny fragment of maxilla. Most workers overlooked the fact that the holotype's poor condition did not allow unambiguous differentiation from the contemporary *Au. afarensis*. This prompted me to observe: "Confusing true biological species diversity with analytical mistakes (15, 16), preservational artifacts, diachronic evolution, or normal biological variation grossly distorts our understanding of human evolution. Past hominid diversity should be established by the canons of modern biology, not by a populist zeal for diversity" (White 2003, 1996).

For another example of how hominid diversity mania permeates contemporary paleoanthropology, consider the case of the earliest hominid fossil, a cranium found by Brunet in Chad and published on *Nature's* cover in 2002. Brunet called the cranium *Sahelanthropus tchadensis*, but made no claims for early hominid diversity. How could he, since he was announcing the very *first* hominid fossil from a *previously unknown* time period (c. 6–7 M)? This time, *Nature* chose Wood for commentary. He did not disappoint diversity aficionados, suggesting that the new fossil belonged to an African ape diversity-equivalent of the Burgess Shale! Not to be outdone, Lieberman likened the impact of the discovery on the field to a "small nuclear bomb." His op-ed in the *New York Times* was entitled "Upending the Expectations of Science." The fossil had been predicted for decades.

In the 1990s and continuing until today, our Ethiopian research has revealed the time-successive series *Ar. kadabba* and *Ar. ramidus* at 5.7 and 4.4 M, interpreted by us as chronospecies along a lineage, but by others as different generic clades. In superimposed strata in this single Ethiopian depository we have found the taxa *Au. anamensis* (4.1 M), *Au. afarensis* (3.5 M), and

Au. garhi (2.5 M). Again, we interpret these as chronospecies, whereas others draw bushier conclusions. Yet Szalay has recently called me a "paleoanthropologist disciple of punctuationism" (2000, 146). Go figure.

Szalay discusses early hominid species diversity as follows: "This term is commonly used, however, in a sense that diversity is the number of existing species at a given moment in time. This is problematical for the fossil record in general as asynchronous morphological diversity is not necessarily lineage diversity, and synchrony becomes contestable with increasing geographical separation in paleontology" (1999, 24). Applying this logic to the hominid fossil record in its "bushiest" contemporary interpretation is an important step in deconstructing species lineage diversity.

Deconstructing Hominid Species Diversity

The last four decades of the twentieth century witnessed an avalanche of molecular, fossil, and contextual data pertaining to hominid origins and evolution. The data came from a variety of sources and continue to be interpreted in different ways. Basic in all interpretations is the species unit. It is worth remembering that the components of the metaphorical hominid bush are its biological species lineages, and that an abundance of names is not necessarily a good barometer of species diversity in zoology or paleobiology.

But hominid paleobiologists have often been vague about whether they conceptualize their "species" as the evolutionary species lineages of the Synthesis, or as static or dynamic Hennigian segments between branching points. Wiley states (as a logical corollary of the evolutionary species concept): "No presumed separate, single, evolutionary lineage may be subdivided into a series of ancestral and descendant 'species.'" (1978, 21). It is ironic that if Wiley's admonition is applied to contemporary hominid trees of twenty-six species, their diversity is considerably pruned.

Tattersall has been the most prolific advocate of hominid "diversity" during the last two decades. From a pulpit at the American Museum of Natural History, he and Eldredge have railed against the Synthesis in both professional and popular media, describing the "dead weight" of a "legacy of linear thought" that "preached a strict linearity in evolutionary pattern." Tattersall maintains "There *is* diversity out there in the fossil record, and there is plenty of evidence for it. If we continue to ignore that diversity in the service of a misbegotten 'antitypological' view of biology, we risk peripheralizing paleoanthropology within science even more than it is already" (Tattersall 2001,

8). Because the difference between science and science fiction is evidence, I have classified this kind of practice and rhetoric as "X-Files paleontology" (White 2000).

Close inspection of the popular contemporary twenty-six-species bushy hominid phylogeny (the one shorn of purely hypothetical lineages) reveals three major causes of nonbiological (i.e., artificial) species diversity. These are: arbitrary names given to chronospecies along lineages, inaccurate species appearance and extinction dates, and invalid species names. When invalid taxa (such as *Au. bahrelgazeli*, a junior synonym of *Au. afarensis*) are eliminated, when chronospecies of species lineages are accounted for, and when overextended first and last appearance data for the remaining lineages are corrected, the resultant hominid phylogeny is left with a pitiful paucity of branches that no gardener would call a bush. This is not species diversity in any biologically meaningful sense. And it is not a signal of adaptive radiation in the sense of contemporary biological science.

As Szalay has observed for hominids: "the corroboration of more than two [or possibly three] bushy lineages through lithosympatry does not exist. Hominids are a perfect and high-profile example of the failure of cladistic, OTU-driven [i.e., operational taxonomic unit-driven] attempts at the reconstruction of evolutionary history. . . . Ongoing and repeated arm waving not only about how speciose the family was, but also by implication [sic] about how many independent lineages came to be during hominid evolution have repeatedly failed to demonstrate the synchronous co-occurrence of acceptable species-level lineages other than the robust and gracile hominids" (2000, 24).

The fossil record has already revealed a modicum of cladogenesis in the hominid clade. There is now universal agreement that *Au. robustus* and *A. boisei* arose during the Pliocene and went extinct in the Pleistocene, coexisting in at least partial sympatry with *Homo*. There is increasing evidence that Neanderthals were a European Pleistocene clade that persisted until about 30 Ka. And there is the hint that hominids were, at least on one occasion, subjected to insular dwarfing in the late Pleistocene. These are valid and valuable examples of evolutionary process.

As large terrestrial generalists, hominids were spread across Africa by three million years ago. At least one lineage arising from these early hominids further specialized in intelligence. Adopting technology, this bipedal primate would become us—2.5 million years later—and was then able to dramatically expand its range, habitats, and niche by 2 M. Surely there were peripheral isolates consequent to these expansions. We might even be able to recognize some of them as valid species if we are lucky enough to find them. But the overall hominid fossil record is remarkable for its *lack* of diversity, whether judged by lemurian, trilobite, or even hominoid standards.

From the perspective of biology, the "diversity" issue is distraction. The questions that we should be asking about hominid evolution do not include misplaced notions of "adaptive radiation." Rather, we should be asking why, compared to other mammals, did species diversity among hominids remain moderate? And for readers who don't think that the diversity *was* moderate, consider the forty-two extant genera comprising 173 species of Old World fruit bats. Or take squirrels, rats, or the five contemporary species of three-toed pygmy jerboas of genus *Salpingotus*.

Might the lack of hominid species diversity roughly appreciated by Dob-zhansky be related to Eldredge's characterization of an ecological "eurytope?" Haven't hominids, whether judged by their teeth or by their technology, been predominantly the "broadly adapted generalists" of Eldredge's definition? To what extent was culture the motor of phyletic evolution in real hominid species lineages? There are good reasons to think that hominid evolution may provide the exceptions that prove the rule of punctuated equilibria. But get-ting there will require more empirical research, exactly the kind of research that gives value to paleontology's seat at the "high table."

CONCLUSIONS

Hominid Paleobiology as Paleontology

Post-Darwinian assembly of the hominid fossil record has been a highly suc-cessful endeavor, aided by factors ranging from taphonomic to academic. Hominids have internal skeletons and robust heterodont dentitions. Many of their available fossil remains are calibrated by trusted radioisotopic methods. The clade has limited time depth (only about six to eight M) and is accessible by virtue of its position atop the geological column. There are close extant apes and baboons for outgroups and analogs. High public interest maintains adequate research funding. And, abundant workers study the clade, including archaeologists who can provide behavioral and distributional data not acces-sible in other areas of paleontology. As a result, in some ways at least, hominid paleobiologists are advantaged relative to workers interested in other more speciose but less-sampled clades. How many times have dinosaur paleontolo-gists debated demes? Human paleontologists do it all the time, illustrating the difference in timescales and completeness of the records.

There are, however, disadvantages to paleontological work with hominids. The clade's fossil record is terrestrial and therefore full of breaks. Hominoid primates are, in general, highly variable as judged by any of their living representatives. This makes the delineation of valid taxa in the fossil record very difficult. All workers agree that there is rampant homoplasy within the clade. Hominids have always lived fairly high on the food chain. Relative to many other mammals, they are k-selected and therefore rare as fossils. And most unfortunately, hominid paleobiology is often characterized as an endeavor short on data and high on emotion.

For example, Gould (2002) tells us: "no true consensus exists in this most contentious of all scientific professions. . . . A field that features more minds at work than bones to study" (910). Measuring contention is difficult, but surely there are now far more bones than practitioners of hominid paleobiology, even if you include the trilobite and lemur specialists, and throw in an editor or two. But perhaps I'm biased and old-fashioned for taking the view that simply being a human is insufficient qualification for pontificating about hominid paleontology.

Gould, however, was on to something important about the structure of contemporary hominid paleobiology. Relative to other sciences, even other paleontological sciences, field fossil recovery efforts are limited compared to numerous laboratory and armchair analysts demanding immediate access to recovered hominid fossils. I have called the structure an "inverted pyramid" of productivity (White 2000). Simpson called general primate studies "covertly or overtly emotional." That may be the case. And even among the rare primates, hominids represent a tiny clade under intense public scrutiny. I suspect that any branch of paleontology with this prominence and these attributes would be characterized as "out of the mainstream" of paleontology, regardless of its practitioners' antics or interpretations. How do these unique aspects of hominid paleobiology affect its place, if any, at the "high table" of evolutionary studies?

The Evolution of Hominid Paleobiology

John Maynard Smith's 1984 premise that there exists a "high table" of evolutionary theory is both pompous and debatable. But how could Gould and Eldredge reject the metaphor when it signaled "mission accomplished" (and provided a good title and organizing principle for Eldredge's 1995 book)? Realistically, however, any "high table" of evolutionary biology that excludes paleontology is hardly worth aspiring to. Furthermore, the "high table" of evolutionary biology has included hominid paleobiology ever since Darwin.

Maynard Smith's "welcome" to the high table was extended consequent to Gould's Tanner Lectures on evolution at Cambridge. By this time, debates about mode and tempo in hominid evolution had been underway for nearly a decade, and parsimony cladistics was firmly entrenched in paleoanthropology. Was there ever a place at the table for specialists narrowly dissecting a bizarre clade of bipedal primates? The paleontologist Simpson sat there, but not the paleoanthropologists whom he employed as examples of bad vertebrate systematics.

Pre-Synthesis paleoanthropologists (many with backgrounds in anatomy) routinely used species names as convenient labels for the specimens they found, a practice that Mayr and Simpson had effectively terminated in hominid paleobiology by the 1960s. It seemed that populationist thinking had carried the century, when Simpson wrote in 1957: "typologists are now in the old guard fighting a lost battle." Indeed, when Mayr wrote about hominid evolution in 1982, he seemed oblivious to the violent rattling of the Synthesis well underway: "As one who has observed this field for about 50 years, I do not hesitate to express my extreme satisfaction with the current state of research in hominid evolution and the multiplicity of approaches that have developed toward the solution of the remaining problems." (237).

In reality, what Szalay (2000) has called a "formidable operational concoction" of "punctuationism, parsimony cladistics, and a strict OTU approach" had emerged by the 1980s. Combined with the deliberate maligning of the Synthesis, a radically different environment for hominid paleobiology had been structured. When diversity politics of the academy, mediaphilic journals, biology envy, new fossils from previously unplumbed periods, and the inevitable public interest inherent in hominid paleobiology were added to the mix during the 1990s, it was obvious that the simple 60s would never return. It had become permissible, indeed fashionable, to split hominid fossils into the species diversity predicted by Gould.

Today it seems that if you broke certain hominid specimens in half, you could easily find a Hennigian paleoanthropologist willing to tally the pieces as representing two hominid species, create two OTUs for a parsimony analysis, and proclaim that predictions of species diversity had been met again. Indeed, contemporary hominid paleobiology features prominent nonspecialists hailing minor hard-tissue differences between fossils as indicators of species lineage differentiation (if not "adaptive radiation").

The hominid fossil record has expanded dramatically, and this enormously successful recovery operation throws light on six million years of human ancestors and their closest relatives. But the species diversity proclaimed by some contemporary paleoanthropologists is an artifact of systematic techniques and an accommodation to preconception. Misrepresentation of reality in the "service" of theory (or as a jobs program for laboratory analysts) should never be the mission of hominid paleobiology.

The hominid fossil record has proven exceptionally hardy, having withstood rough handling by a long line of theorists eager to twist it to conform to their view of how evolution proceeds. For the nonspecialist, cutting through the "spin-doctoring" of contemporary paleoanthropological diversity advocates is difficult, but worth doing, if an appreciation of the underlying biology is the goal. The rhetoric of this movement, in both scientific and popular arenas, has rivaled that of a political campaign, complete with Rovesque smear tactics directed against the Synthesis.

Why the urgency and strength of conviction of what Eldredge has called the "taxic" approach (as opposed to the "transformational") to hominid paleobiology? Hominids have been used as exemplars ever since Darwin. If this tiny but potent clade could be accommodated to a punctuationist diversity view of evolutionary biology, then the advocates' purposes would be well served. But does this not edge hominid paleobiology toward what Brace has termed a "legacy of Medieval Neoplatonism?"

Hominid Paleobiology and Evolution

Hominids are routinely used by evolutionary biologists to exemplify their general themes. Early hominids are fascinating, not just because we evolved from them (or T. H. Huxley found them interesting). In their anatomies, physiologies, and behaviors, their wide geographic distribution, and their trophic heterogeneity, hominid ancestors and close relatives fascinate by being peculiar. But how capable are they of underpinning evolutionary generalities? By itself, hominid paleobiology seems destined to make little real contribution to the understanding of the general principles of evolutionary biology, except as oft-used and popular exemplars. For the paleobiologists who study them, it has been useful having heavyweight evolutionary biologists paying inordinate attention. History suggests that this will continue.

We hominids have a proximate fossil record that is better than those of most frogs, but worse than those of cave bears. Any test of evolutionary mode and tempo that hominids might someday (with a better fossil record) be able to make will be largely irrelevant to the issue of the frequency of punctuated equilibria among primates, mammals, or vertebrates. In many ways, hominid primates are probably one of the *worst* possible clades to choose in generalizing about mammals, vertebrates, or even evolution. Can evolutionary lessons really be gleaned from this bizarre, geographically widespread, ecologically generalist, weirdly technological, multitrophic clade? Can any general pattern be extrapolated from the fossil record of such a tiny clade? Will hominids be the exceptions to evolutionary "rules" of mode and tempo?

The presently inadequate understanding of our immediate evolutionary history has been structured by the Darwinian paradigm, translated through the Synthesis and modified by theoretical and methodological advances of the last thirty years. It has also been structured by the order of various discoveries and their interplay with prevailing method and theory, and a myriad of other contingencies.

To borrow Gould's metaphor, what if we could "rewind" the historical tape of hominid paleobiology? What would have happened had Eldredge and Gould not pursued paleontology? Would cladistics have inevitably forced a similar reconsideration of the Synthetic version of evolutionary mode and tempo (albeit presented by other parties in a far less literate, articulate, coherent, or effectively promoted form)? I don't know the answer to that, or thousands of other "what-if" questions. But for hominid evolution, I suspect that we would probably discern much the same current phylogeny. And we would probably still face the same outstanding questions about it—questions that troubled T. H. Huxley as well as the architects of the Synthesis. Paleoanthropology is often thought of as driven by fossil data. As Mayr remarked in 1982 (232), "Nothing, of course, has shed as much light on the history of the hominids as new fossil discoveries." Such discoveries—while conditioned by theory and interpretation—do have evidentiary value independent of prevailing academic fashions.

It seems to me that theoretical and methodological advances of the last two centuries have established a well-grounded and well-balanced framework for the work yet to be done in hominid paleobiology. The integration of hominid paleontology into paleobiology—and the integration of paleobiology into the biological sciences—are positive, ongoing developments. Whether punctuated equilibria extended the Synthesis (as claimed by Szalay), overthrew it (as claimed by its zealots), or just corrected it, evolutionary biology is better off for it because the ideas and empirical work it generated have challenged and enriched our understanding of origins and evolution—hominid and otherwise. Much the same is true of other developments in evolutionary theory. The generation of variation is just as important for intellectual evolution as for biological evolution. So thank you, Steve and Niles; Elisabeth, Steven, and

Phil. Thank you Theodosius, George Gaylord, Ernst, and Willi. And thank you Charles and T. H.—we needed all of that.

ACKNOWLEDGMENTS

Thanks to all those who have invited me to their laboratories, field study areas, and museums over the years. Thanks to the late F .C. Howell, and to L. Hlusko, D. DeGusta, and H. Gilbert for review and editorial suggestions. Thanks to the editors for inviting me to participate.

REFERENCES

Bowler, P. J. 1986. *Theories of human evolution: A century of debate 1844–1944.* London: Basil Blackwell.

Delisle, R. G. 2001. Adaptationism versus cladism in human evolution studies. In *Studying human origins: Disciplinary history and epistemology,* ed R. Corbey and W. Roebroeks, 107–22. Amsterdam: Amsterdam University Press.

———. 2006. *Debating humankind's place in nature, 1860–2000.* Upper Saddle River, NJ: Pearson/Prentice Hall.

Eldredge, N., and Gould, S. J. 1972. Punctuated equilibria: An alternative to phyletic gradualism. In *Models in paleobiology,* ed. T. J. M. Schopf, 82–115. San Francisco: Freeman, Cooper and Co.

Eldredge, N., and I. Tattersall. 1975. Evolutionary models, phylogenetic reconstruction, and another look at hominid phylogeny. In *Approaches to primate paleobiology: Contributions to primatology,* ed. F. Szalay, 218–42. Basel, Switzerland: Karger.

———. 1982. *The myths of human evolution.* New York: Columbia University Press.

Gould, S. J. 1976. Ladders, bushes, and human evolution. *Natural History Magazine* 85:24–31.

———. 1997. Unusual unity. *Natural History Magazine* 106:20–23.

———. 2002. *The structure of evolutionary theory.* Cambridge, MA: Belknap.

Hull, D. 1988. *Science as a process.* Chicago: University of Chicago Press.

Kappelman, J., and J. G. Fleagle. 1995. Age of early hominids. *Nature* 376:558–59.

Jolly, C. J. 2001. A proper study for mankind: Analogies from the papionin monkeys and their implications for human evolution. *Yearbook of Physical Anthropology* 44:177–204.

Leakey, M. G., F. Spoor, F. H. Brown, P. N. Gathogo, C. Kiarie, L. N. Leakey, and I. McDougall. 2001. New hominin genus from eastern Africa shows diverse middle Pliocene lineages. *Nature* 410:433–40.

Le Gros Clark, W. E. 1955. *The fossil evidence for human evolution*. Chicago: University of Chicago Press.

Lieberman, D. E. 2001. Another face in our family tree. *Nature* 410:419–20.

Marks, J. 2005. Phylogenetic trees and evolutionary forests. *Evolutionary Anthropology* 14:49–53.

Mayr, E. 1950. Taxonomic categories in fossil hominids. In *Cold Spring Harbor Symposia on Quantitative Biology 15*, 109–18. New York: Cold Spring Harbor, L.I.

———. 1963. The taxonomic evaluation of fossil hominids. In *Classification and human evolution*, ed. S. L. Washburn, 332–45. New York:Viking Fund Publications in Anthropology 37.

———. 1982. Reflections on human paleontology. In *A History of American physical anthropology 1930–1982*. Ed. F Spencer, 231–37. New York: Academic Press.

Szalay, F. 1999. Paleontology and macroevolution: On the theoretical conflict between an expanded Synthesis and hierarchic punctuationism. In *African biogeography, climate change, and human evolution*, ed. T. G. Bromage and F. Schrenck, 35–56. Oxford: Oxford University Press.

———. F. 2000. Function and adaptation in paleontology and phylogenetics: Why do we omit Darwin? *Palaeontologia Electronica* 3:21–25.

Tattersall, I. 1997a. *The fossil trail*. Oxford: Oxford University Press.

———. 1997b. Paleoanthropology and evolutionary theory. In *Research frontiers in anthropology*, volume 3, ed. C. R. Ember, M. Ember, and P. N. Peregrine, 325–42. Englewood Cliffs, NJ: Prentice Hall.

———. 2001. Diversity in paleoanthropology. In *Homo-Unsere Herkunft und Zukunft: Proceedings 4. Kongress der Gesellschaft fur Anthropologie* e. V. (GFA). Ed. M. Schultz et al., 6–8. Gottingen: Cuvillier.

White, T. D. 2000. A view on the science: Physical anthropology at the millennium. *American Journal of Physical Anthropology* 113:287–92.

———. 2003. Early hominids—Diversity or distortion? *Science* 299:1994–96.

Wiley, E. O. 1978. The evolutionary species concept reconsidered. *Systematic Zoology* 27:17–26.

Wood, B. A., and B. G. Richmond. 2000. Human evolution: Taxonomy and paleobiology. *Journal of Anatomy* 196:19–60.

CHAPTER EIGHT

Punctuated Equilibria and Speciation: What Does It Mean to Be a Darwinian?

Patricia Princehouse

I well remember how the synthetic theory beguiled me with its unifying power when I was a graduate student in the mid-1960s. Since then I have been watching it slowly unravel as a universal description of evolution. The molecular assault came first, followed quickly by renewed attention to unorthodox theories of speciation and by challenges at the level of macroevolution itself. I have been reluctant to admit it—since beguiling is often forever—but if Mayr's characterization of the synthetic theory is accurate, then that theory, as a general proposition, is effectively dead, despite its persistence as textbook orthodoxy.
— Stephen Jay Gould[1]

Evolution is not changes in gene frequency. That's a really stupid definition.
—Bob Bakker[2]

The hopeful monster is an extreme formulation of something that becomes much more acceptable if you do not call it that. —Adolf Seilacher[3]

Introduced by Niles Eldredge and Stephen Jay Gould in 1971–1972, punctuated equilibria is by far the best-known manifestation of the paleobiology revolution. But its high profile sometimes gives a mistaken impression of its centrality as an organizing principle—leading the public to conflate it with mass extinction, Goldschmidtian hopeful monsters, and X-men mutants. Meanwhile, its truly revolutionary elements are much more subtle—in particular, its macroevolutionary consequences, such as the need to explain evolutionary trends in terms of species sorting. These consequences and their implications call for a more nuanced construction of evolutionary theory in the Darwinian mode—that is, in the tradition of Charles Darwin himself, rather than the convention established by the Modern Synthesis.

TRANSFORMATIONS

An invasion of strange creatures marked the late twentieth-century scientific landscape—cloned sheep, two-headed snakes, six-legged frogs, antennapediacs, fruit flies sprouting eyes all over their bodies, innumerable chimeras. Weird new fossils appeared: fish with fingers, six-million-year-old ape-men walking upright in trees, twenty-thousand-year-old hobbits, zillion-toothed water monsters that were somehow half wolf, half pig, and half porpoise, salamander-fish hybrids, giant carnivorous shrimp, and a petrified nightmare with five eyes and a nozzle on its head.

And that was just the peer-reviewed literature.

The popular imagination also exploded with mutants, monsters, and dinosaurs. The blockbuster *Jurassic Park* movies transformed the public's view of dinosaurs from slow, lumbering obsolescence to vibrant, sharp-witted athletes, ready to rumble with anything the Cenozoic threw their way. The public image of the scientists who studied them underwent similar changes— no more nerdy, white-coated four-eyes from central casting. Instead, we find leather-jacketed chaos theorists, long-legged blond paleopalynologist girls, and galactic hitchhikers. The movie called *Evolution* introduced a 10-nucletide system capable of evolving far faster than our terrestrial norm.

Pop culture took this trajectory not due to stochastic processes, but because amateurs picked up on excitement within the scientific community. They especially grabbed hold of discussions of rates of change. The dinosaurs ran fast and went extinct suddenly, and new species emerged rapidly. Theoretic clarity was not an issue; the public psyche latched onto speed, and ran with the terms *mass extinction* and *punctuated equilibria*.

EVOLUTION BY JERKS

The public is not wrong to be interested in patterns and rates of change over evolutionary history. These issues have "revolutionized the study of Natural History, and carried away captive the best men" of Darwin's day and our own.[4] The mechanisms of such change resonate in deep and personal ways— even among professionals. It is not surprising, then, that when the theory of punctuated equilibria appeared on the scene in 1971–1972, the response was swift and acrimonious. Promulgators of punctuated equilibria as a form of species formation were derided as Darwinian heretics; it was evolution by jerks.

The jerk-minded averred that both speciation and major morphological change could and regularly did happen hand in hand and very quickly—with new species typically needing perhaps as little as 5,000 years to form before settling into a period of prolonged morphological stasis that might endure millions of years. The proposed mechanisms attracted rather less attention than the structural and sociological concomitants. Biologists were not accustomed to seeing major theory emanate from paleontologists, especially from young upstarts not ten years out of graduate school. Eldredge was well aware of the social relations involved, and may well have been drawn to speciation—that "mystery of mysteries"— to buck the establishment as much as for the sheer intellectual headiness of addressing the fountainhead of biological diversity:

> I confess that as a student I disliked the prospect of spending my life rediscovering the wheel . . . merely documenting the fact of evolutionary history, applying a rote interpretation to my fossils' reconstructed evolutionary histories . . . so I frankly admit I started to look for something more, some other way to approach the problem. I stumbled on evolutionary patterns in the grand scale of geologic time.[5]

Eldredge and Gould's enthusiasm was further fueled by the plate tectonics revolution of the mid 1960s. Gould was "fascinated to watch it develop. As a graduate student with no stake, I was inclined in its favor. Who wouldn't be at that stage of a career?" Further, he found it "interesting to watch people react. I ended up admiring quite a few people like Marshall Kay who was such a crusty old bigot about it at first, but then once it was proved, changes his mind and spent the last years of his life redoing a lot of his Newfoundland field work."[6] Others "were so rooted in their own views of continental stability that they would never fall for that, like Bernie Kummel."

The self-proclaimed young Turks steeled themselves to establish and defend paleontology as a source of theory. Crucial to positioning paleo as theory-generating was an influence I call the German Synthesis, which had embraced paleo's theoretical importance in '20s and '30s Germany and given it the name *Palaeobiologie.* I argue this previously unrecognized German Synthesis was thoroughly theorized and well organized, and played a much larger role in the later paleobiological revolution than previously realized by philosophers and historians, particularly in this formative period of the 1960s. The patterns of stasis and sudden change were major themes of the German Synthesis. And although Gould and Eldredge did not accept all mechanisms of the German Synthesis (or of the Modern Synthesis, for that matter), with support from their adviser at Columbia, Norman Newell, both did theses documenting stasis and sudden change in the fossil record. But the sixties ended without seeing them develop a coherent explanation of the mechanisms behind these patterns.

Newell, curator of paleontology at the American Museum of Natural History, was a field colleague of the German Synthesis' Otto Schindewolf, and championed all efforts at synthesis, encouraging his students to communicate with Dobzhansky and Simpson and their students, including Richard Lewontin. Newell's paleoecological work on the Permian of West Texas posited an environmental disturbance for the Permian extinction. His own work supported the idea of widespread stasis in the morphology of animal species over vast periods of time. In 1952, 1963, and 1967, Newell wrote articles emphasizing the punctuated and episodic nature of extinction.[7]

Eldredge's 1971 foray into speciation theory attempted to explain how known mechanisms might reasonably be expected to play out over geologic time.[8] The paper was ambitious but attracted relatively little interest, except from Gould. Nonetheless, Eldredge and Gould were but two among a generation of young paleontologists looking to upend the apple cart.

The previous generation had been marked by enthusiasm for the Modern Synthesis' melding of subfields under the aegis of neontologically oriented mechanisms. David Raup rode the wave of the post-WWII Modern Synthesis, working with Ernst Mayr as much or more than with his geologist adviser Bernard Kummel to earn his PhD in paleontology from Harvard in 1956 with a thesis on modern and fossil echinoids. But as time went by, he found himself continuing to work on neontological theory, only using fossils to check that they conformed to expectations.[9] John Imbrie tried to apply speciation theory to fossil lineages, but instead of the gradual change expected by Mayr's 1942 Modern Synthesis "dumbbell allopatry" speciation model, he found cases after case after case of apparent stasis. Frustrated, he turned to straight sedimentology and left investigation of evolutionary dynamics to the neontologists.[10] Eldredge and Gould were very taken with Imbrie and were struck by his frustration with Devonian brachiopod sequences from the Michigan basin. Gould recalls: "Out of fifty lineages, all but one had stasis. . . . That's when it started. Niles and I kept in pretty close touch after that with John."

As Eldredge and Gould witnessed these developments, they committed even more firmly to raising the profile of paleontology among the evolutionary sciences. The Germans provided a useful counterweight. The 1930s and 40s saw two alternative syntheses—the Modern Synthesis, and the slightly earlier incarnation I call the German Synthesis,[11] which beautifully integrated genetic and paleontological data and hypotheses into a major theory encompassing a plurality of causes acting on levels from the gene to major radiations of plants and animals seen in the fossil record. Primary spokesmen were the geneticist and systematist Richard Goldschmidt and paleontologist Otto Schindewolf. In this view, mutations occurred on several levels. Very minor mutations happened with some frequency, such as Morgan's white-eyed and red-eyed flies, and were responsible for local geographic or "racial" variation and could be adaptive. These microevolutionary phenomena were of interest to microtaxonomists, lab geneticists, and fieldworkers, and could give insight into the mechanisms of physiological genetics, but had nothing to do with speciation or higher taxa. Speciation was the result of chance—systemic mutations that rearranged the entire genome, often in one generation. Small changes to the system produced a new but similar species; macromutations affecting more aspects of the organism's organization might result in the creation of a new genus or family. New orders or classes were the result of even more major rearrangements, usually involving not only physiological genetics, but also regulatory genes, such as those governing heterochrony; that is, changes in the timing of events during embryological development. Higher-level systemic mutations were chosen among by natural selection at a gross level: those suitable enough to the environment would eke out an existence. But once a system was established, its major features changed very little, and primarily due to the playing out of potentialities in its genome, not to the exigencies of environment. Genetic rearrangements that resulted in the formation of new taxa at the family level or above carried certain genetic tendencies that were generally expressed in most or all of the descendents in that clade. Hence the fossil record's testimony of rampant parallelism among sister taxa during given periods of time. The morphologic potential was limited, as shown not only in the frequency of parallelism, but also in the fossil record's revelation that clades tended to run out of steam at a certain point. Again and again the rocks show that many closely related lineages tend to go extinct at about the same time, with only one or two bloodlines squeaking through. The shelf life of particular clades was built into the genomic rearrangements that produced the clade in the first place. The lines that continued did so only because of new systemic mutations that occurred within them. The new macromutations reset the clock and allowed those new taxa to diversify and radiate. The German Synthesis view took the mystery out of puzzling phenomena such as parallelism, mass extinction, sudden radiations of new morphotypes, and the persistence of morphological types across different environments and over long periods of time. It united field and lab genetics and embryology with biogeography, systematics, and paleontology, and regarded all these fields as vitally important.

The German Synthesis work of Goldschmidt and Schindewolf was widely derided in America, but that only made it more compelling to Gould. Julian Huxley considered many of Goldschmidt's mechanisms wrong, but he was never as unfriendly to Goldschmidt as Mayr and Simpson were. Reading Huxley on allometry gave Gould access to some elements of the German Synthesis. Another pointer toward the Germans came when Newell remarked that he considered Schindewolf "the world's greatest living paleontologist."[12] So although the German Synthesis was not referred to as an entity, and its characteristic features were rejected by the Modern Synthesis, its elements were around for the young Turks to draw on. The perceived extremism of the German Synthesis provided a backdrop against which this new generation of theory-oriented young paleontologists could work. They could venture outside the confines of the hardened Modern Synthesis while still seeing themselves as moderates compared with the excesses of the German Synthesis. It also helped that Schindewolf student Adolf Seilacher was circulating in the United States. Beginning in the mid-fifties, Preston Cloud, chief of the paleontology and stratigraphy branch of the U.S. Geological Survey, promoted his work among Americans and urged Seilacher to integrate Modern Synthesis ideas into his work. Cloud "was a very astute politician and he for several years took every opportunity he could to push Dolf."[13] By the late sixties, Seilacher's knack for understanding whole organisms, especially his trace fossil work, fit right in with paleobiology's ecological trend, but retained its German Synthesis flavor very strongly, particularly his work on *Konstruktionsmorphologie*. Newell was greatly impressed also: "Seilacher is a great man. I was interviewed by the Harvard faculty about him, and I urged them to hire him on any basis."[14]

In the mid-sixties, Julian Huxley provided a fateful boost to the young Turks when he recommended that the prestigious Cambridge Philosophical Society journal *Biological Reviews* request a review article on allometry from Gould, who was still a grad student.[15] Fallout from this review made Gould's early reputation and helped secure his professorship at Harvard and tenure at age 30. Eldredge became Newell's heir apparent at the American Museum of Natural History. Gould and Eldredge together had the platform from which to catapult paleontology to new fortune via a newly mutant speciation theory—punctuated equilibria.

PUNK EEK EVOLVING

As students at Columbia, Gould and Eldredge "saw a lot of each other... the main thing was that we had both been trained in statistical methods with John Imbrie." Unlike most of their geology professors, Imbrie was prepared to discuss evolutionary theory with his younger colleagues. "I guess our belief, well Imbrie's belief, which was suggested to us and we accepted at the time was that gradualism was an expectation. It was certainly true that it was almost never found." Attempts were made to fit everything that was found into the pattern of gradual evolution. Any deviations from this pattern were seen to be imperfections of the fossil record. Imbrie "didn't call it gradualism, he called it evolution."[16]

But gradual change was not widely found in the fossil record. This was understandable for terrestrial vertebrate species whose fossils are few and far between due to large body size, small populations, destruction by heat and cold, and transport by streams. But the reasons were less clear why this would be the case for some small, hard-shelled marine organisms with good fossil records. At the time, students thought perhaps "the reason that so few cases had been discovered was that maybe gradual change was very subtle, and needed statistical methods." This prompted increased enthusiasm for quantitative methods. Gould recalls that at the time this made sense, but "it came back on me several years later that that's prima facie rather ridiculous. Because if it's so subtle that you absolutely can't really see it, then it's stable . . . if the only gradualism that exists is so subtle nobody ever sees it, then it's not the stuff of major evolutionary [change.]" But he and Eldredge threw themselves into it: "We learned the statistical methods. We didn't find any more gradual change that way—neither did Imbrie."[17]

Gould had become interested in allometry as an undergrad at Antioch. His geology professor John White studied allometry of hill slopes—"match to exponential equations, convex slope of a hill . . . power function. It actually worked pretty well. And I looked at this dome of this snail shell, and I said hey let's see if that works." This started Gould on land snails, his career specialty. It also led him to D'Arcy Thompson's 1917 classic *On Growth and Form*.

Thompson did not use numerical methods, but rather a transformed coordinate analysis, which, like power functions, were "a way of measuring change, whether it be ontogenetic or phylogenetic or static within a species."[18]

Because of Gould's sustained interest in allometry, while still a graduate student he made bold to send Huxley manuscripts of his first papers. "I was fairly naive about the politics of science. . . . I was just awestruck by Huxley and so I wrote him a long letter saying how much his work had meant to me." Much to Gould's surprise, "about a month later, I got back a letter from the editor of the *Biological Reviews* . . . who said he'd wanted a review article on allometry, and he'd been talking to Huxley. . . . And I guess Huxley had just got my letter and papers, so he had suggested me." This exciting opportunity nevertheless put Gould in a dilemma: "I felt that I could do it. For one thing, the great advantage that I had is I could read French and German, which most of the literature was in, and very few Americans could. So I really felt I could do it. I was very up on the literature, but I was only a graduate student." So, with both hope and trepidation, Gould wrote back: "I said look, I'm sure I can do this [and] I'm very pleased that Huxley has confidence in me. But I have to tell you I don't yet have my PhD. I put it that way." To Gould's amazement the editor "wrote back the sweetest card. . . . He said 'Mister or doctor makes no difference to us. I myself am the former.'" Writing this review article "really got me into allometric studies 'cause to do that I had to read a couple hundred pages." Gould finally met Julian Huxley in person about six months before Huxley died.[19]

The allometry paper was a watershed. It sparked renewed interest in morphology and helped touch off the movement that is now called "Evo-Devo," or evolutionary developmental biology. Today it is seen not only as a classic, but still plays an active role in theory formation. "The '66 paper is still cited now," stresses developmental biologist Rebecca German.[20]

Gould's allometry work attracted Seilacher's attention. In 1968, "Seilacher . . . invited me to that meeting in Tübingen, which was the only time I ever met Schindewolf. . . . That was really the first time, outside my own preliminary interest in ontogeny and phylogeny that I came in touch with that alternative—structuralist, formalist views." Gould recalls it was "a little awesome to meet Schindewolf, whom Newell had called the world's greatest paleontologist." Gould was puzzled by the German students, "they were totally deferential. I don't know if that's what they really felt."[21]

The allometric work on snails and Raup's quantitative approach to bivalves led Gould to better appreciate mollusks in general and land snails in particular. "I came to see there was this Darwin's finch among land snails, a genus that nobody'd really ever studied much. And so I did my thesis on the general paleobiology and evolution of the most prominent subgenera of *Poecilozonites*."[22] Newell guided Eldredge and Gould toward "pretty orthodox synthetic theory." Aside from his youthful browsing of all Simpson's work, Gould did not read *Tempo and Mode*, "which was a book with more radical suggestions," until around 1971. "Newell had told me to read the *Major Features of Evolution*, because, well, naturally you would because that's the update, the rewrite, so to speak. And that's a much more hardline synthetic book." Newell, and thus Gould, saw *Tempo and Mode* as "just the first edition of *Major Features*."

The thesis project was typical of students of the Modern Synthesis, but Gould's unique background, informed by the continental structuralist work of the German Synthesis, made him take note of aspects that did not interest most of his fellow students. But he did not dwell on the inconsistencies while writing the dissertation: "I found a lot of lineages that I could link up. I found transitions in the sense that you could infer based on character changes." But Gould had a hard time making gradual evolutionary sequences out of the data: "I'm not saying that that's the foundation of punctuated equilibria, but it's one of those things that stays on your mind. . . . Meanwhile Niles was finding the same thing looking at trilobites." Gould says that as he was writing his dissertation, "I remember Bobb Schaeffer telling me when I was developing some doubts about certain things, that *Tempo and Mode* was a very different book. And I should really read what Simpson had originally said about quantum evolution, and that I would be very surprised. And he was right."[23]

Tempo and Mode had such a strong effect on Gould, when he finally read it, because it was so different from the gradualist *Major Features*, which had come to represent Simpson's thought. The magnitude of Simpson's pluralism, especially with respect to variable rates of evolution in different lineages at different times, was brought home to Gould with particular force, because he turned to *Tempo and Mode* while preparing to collaborate with Eldredge on what became the 1972 punctuated equilibria paper. Coming to *Tempo and Mode* with explicit attention to questions of ecological process during speciation sharply pointed up the differences between Simpson's early thinking, and his later, hardened Modern Synthesis stance on how speciation was supposed to happen and how quickly it proceeded. Gould then turned to the remnants of the German Synthesis to flesh out his understanding, not only of the factors affecting speciation, but of the nature of biodiversity and its variable genesis with respect to selection in different ecological settings.

Gould found it especially intriguing that genetic theory seemed to reinforce some of the German ideas, particularly Seilacher's *Konstruktionsmorphologie*. Central among recent genetic work was the Neutral Theory—the controversial suggestion developed by Motoo Kimura and endorsed by Sewall Wright that most mutations are neutral—that is, that they are not "visible" to (and thus not acted on by) natural selection when they first arise, and that the course their gene frequencies chart through a population is effectively a random walk.[24] Aside from the role of allometry in generating morphological diversity, structuralist issues had not garnered Eldredge and Gould's concerted attention while they were working on their dissertations. Serious questions about the process of speciation came only later, when Tom Schopf assigned the topic to Gould for the 1972 edited volume *Models in Paleobiology*. By this later date, Eldredge and Gould were able to approach these questions, not as young graduate students writing dissertations largely within the Modern Synthesis, but as an ambitious Harvard professor with tenure, and as Newell's hand-picked successor at the AMNH.

Gould lucked into his job at Harvard before he had even finished his dissertation. Although he would certainly have been a viable candidate in an open competition, the old-boy network made him the hire. Harry Whittington was leaving Harvard to become professor of geology at Cambridge. "Bernie Kummel thought it was good riddance of this old fuddy duddy. Of course then Harry went there and did the greatest work of his career on the Burgess Shale." Kummel decided to look for someone with new ideas. "Bernie was a crusty old guy in various ways, but he had a good heart and a good general vision. He wasn't always able to see through his prejudices, but he really could have just hired a clone of himself." However, Kummel was looking for something different. "He wanted to get a young guy who did the opposite of what he did which to him meant, well, the hot stuff was quantification. He wanted someone who could do quantitative evolutionary work . . . stratigraphic of course" So Kummel approached his own thesis advisor, Norman Newell. "He was Norman's first student. And [Kummel] said do you have a good young man, I'm sure he said 'man.' And Norman said 'as a matter of fact I do.'" Gould's creativity from the beginning and self-confidence for the rest of his career reflect this early lack of struggle to achieve a solid academic job. Rather than rest on these easy laurels, Gould took full advantage of the opportunities it offered to work and publish like a madman. "I came by the job somewhat dishonorably by modern standards, or at least by my own." But, he says, "I kept it totally honorably."[25]

While Gould got along well with the senior members of the profession, he had yet to prove himself to his peers. He attended the Geological Society of

America meeting in New Orleans in 1967, but his situation was so unusual that he did not find much cohesiveness with other young paleontologists. Richard Bambach first met Gould at the eleventh annual North American paleontology convention in 1969. They spent some time together on a field trip to the Essex fauna in Mason Creek to see the Tully monsters. Gould was like no one Bambach had ever met. "I didn't know if Steve was just talk and flash or if there was some substance . . . Steve was sort of an *enfant terrible*" with a "studied effort to be more cultured than everybody else." Sara Stuart Bretsky voiced her opinion that Gould was a "four-flusher." In retrospect, though, Bambach feels Gould was "more successful than most in forcing people to improve their thinking."[26]

Raup notes, "Steve ruffled a lot of feathers in the early days because, you know, the pretentious way he wrote. That was his biggest sin." He also notes that Gould "was using a style in writing research papers that nobody'd ever used before in our experience." More than anything, Gould was "very much younger than one was supposed to be. Certainly it was unspoken but clear as I was coming up that you didn't generalize until you were much farther along. And he was generalizing from the word go."[27] Gould's early tenure put him in a position to take risks at a very early stage in his career. The most significant limb he went out on turned into the theory of punctuated equilibria.

As with so many important events of this era, the catalyst was Tom Schopf, organizer extraordinaire. "Tom was proselytizing; he was trying to get paleontologists out of their lethargy, to take models from the biological sciences." Schopf wanted to do a book based on a symposium at the 1971 Geological Society of America meeting in Washington. "Everyone would get one biological topic and discuss how models from that topic might help paleontology." Schopf asked Gould to address speciation, "and I looked at him and I said why don't you let me do morphology, or evolutionary trends, because I don't know all that much about speciation, but he wanted to give morphology to Raup. So, if I wanted to be in on this thing, I would have to do speciation." Although Gould had some interest in the topic, he did not feel any particular expertise. "I said I don't have any good ideas about speciation except that I think Niles Eldredge's paper in *Evolution* [was] interesting . . . so if you'll let me take him on as a collaborator, we'll do it."[28]

The book, *Models in Paleobiology*, achieved the result Schopf was looking for. It contained many key papers, including the English translation of Seilacher's triangle paper. But by far the greatest effect has been from the Punc Eq paper, which was ironic, since Schopf never really accepted punctuated equilibria.

Like many influential ideas, the formulators of punctuated equilibria feel the theory has often been misunderstood:

A lot of people think it came out of high-falutin theory, but it didn't. . . . General theory came later when we started thinking about its implications. What it mostly came from is the frustration of ambitious, young paleontologists. We'd spent a lot of years studying evolutionary theory with statistical methods. And then to be told this terrible paradox, that gradualism is the way of the world, but you never see it, or only rarely see it because the fossil record's so imperfect. Now that works logically . . . it's a logical statement. But science is about doing, it's not about cogitating. If you spend all this time studying evolution, and you think it means gradual change, and that's what you want to do, and then you can't ever find it, or finding it is exceedingly rare. That's a counsel for despair."[29]

This frustration rankled the two young paleontologists as they went over the latest views on speciation of the towering figure of the Modern Synthesis, Ernst Mayr—particularly his 1963 book, *Animal Species and Evolution*. They began to make progress when Eldredge explained his idea that part of the widespread expectation of gradual evolutionary change might be a scaling mistake. "Maybe what the biologists call slow, they only mean slow by the passage of their lives. Ten thousand years is a bedding plane. [If] you take stasis seriously, ten thousand years is a fraction of one percent of most stabilities in most species. Scaled properly to geological time, the slowness of speciation in ecological time is punc eq. It's not gradualism." When they worked out the consequences, they knew they were on to something big. Taking a step back, they realized that this might make their careers. Gould recalls: "Niles had the idea, though I did all the math." When it came to "the touchy issue of priority in publication, we were going to toss a coin. Well, I won the toss but I [said] 'hey, I can't do this.'" Although completely and enthusiastically committed to Punc Eq, Gould has consistently maintained Eldredge's priority. "It's really his idea."[30]

The seminal 1972 paper revolves around several interrelated theses:[31]

1. The expectations of theory color perception to such a degree that new notions seldom arise from facts collected under the influence of old pictures of the world. New pictures must cast their influence before facts can be seen in different perspective.

2. Paleontology's view of speciation has been dominated by the picture of "phyletic gradualism." It holds that new species arise from the slow and steady transformation of entire populations. Under its influence, we seek unbroken fossil series linking two forms by insensible gradation as the only complete mirror of Darwinian processes; we ascribe all breaks to imperfections in the record.

3. The theory of allopatric (or geographic) speciation suggests a different interpretation of paleontological data. If new species arise very rapidly in small, peripherally isolated local populations, then the great expectation of insensibly graded fossil sequences is a chimera. A new species does not evolve in the area of its ancestors; it does not arise from the slow transformation of all its forebears. Many breaks in the fossil record are real.

4. The history of life is more adequately represented by a picture of "punctuated equilibria" than by the notion of phyletic gradualism. The history of evolution is not one of stately unfolding, but a story of homeostatic equilibria, disturbed only "rarely" (i.e., rather often in the fullness of time) by rapid and episodic events of speciation.

This vision does not sound so revolutionary in retrospect, but the response at the time was deafening. Like Imbrie, most evolutionary biologists in most subfields had internalized the Modern Synthesis notion that Darwinian meant slow, steady, gradual change in large populations. Even folks who fully understood and supported Mayr's 1963 peripatric model of speciation had not realized the implications for the fossil record; they clung to their cognitive disequilibria. Bucking the hegemony of the Modern Synthesis seemed tantamount to attacking Darwin.

But it was hard to deny the power of the data. The paper showcased Eldredge's painstaking stratigraphic work on the trilobite *Phacops rana*, detailing a fossil record that showed new morphologies appearing at the edge of an ancestor's home range, then expanding geographically. Eldredge brilliantly documents variation in a trait appearing in a very small region peripheral to the home range of the ancestor, quickly transforming to stabilize a new standard morphology, then spreading out to inhabit part of the ancestor's previous territory and remaining in morphological stasis thereafter.

A heated debate ensued throughout the 1970s. Punctuated equilibria was the focus, but it also stood in for issues of age, class, and subfield. Although Eldredge and Gould presented it as the natural extension to paleontology of the Modern Synthesis view of speciation, "evolution by jerks" provoked a passionate outcry. The contested terrain ranged across the entire field of evolutionary study—from how the biophysical basis of heredity should be understood in an evolutionary context, to implications of the evolutionary history of dinosaurs for the future of humankind, to the limits and nature of science as a creative human enterprise.

"Punk Eek" asserted the major data of paleontology—widespread stasis in the fossil record—as of utmost theoretical importance, and constructed a space for normal science around it. Stasis became "a theoretically meaningful and interesting phenomenon, and not just an embarrassing failure to detect 'evolution.'"[32] Eldredge and Gould's 1972 assertion of punctuated equilibria as the primary mode of species formation inspired a generation of young paleontologists to not only apply the neontological Modern Synthesis to fossils, but also create theory and interpretive apparatus within paleontology and apply it to the entire field of evolutionary biology. The following decade witnessed the establishment of the movement's flagship journal *Paleobiology*, and saw British geneticist and J. B. S. Haldane student John Maynard Smith welcome paleontology to the high table of evolutionary theory.[33] Subsequent years saw a transition within genetics, as young lab researchers investigated developmental processes and genetic/genomic structures that could produce the macroevolutionary change required or predicted by the paleontologists.[34]

SORTING SPECIES

The resulting flurry of activity among young paleontologists included among its most important work that of Steven Stanley on species selection.[35] Stanley explored the ramifications of ideas only briefly addressed by Eldredge and Gould. They found it "great to see someone developing the implications." Gould says "I remember saying to Niles this is the most important part of the paper but I'm not quite sure why. What I had seen was this implication, or the explanation that turns on species sorting."[36]

Stanley had been trained in the middle of the developing paleobiology movement, receiving his PhD in 1968 from Yale. He quickly became a major player in the macroevolution debate through his analysis of extinctions, his consideration of the role of species in evolution, and his work on functional morphology and adaptation, especially shell shape in mollusks. Like Raup, Eldredge, and Gould, he wrote his dissertation in the thrall of the Modern Synthesis, believing "that the fossil record can best be brought to life by injecting it with large doses of conceptual serum from the field of biology." But he quickly found that approach less satisfying that he had expected: "Like many other young paleontologists, I embarked on a career with this idea in mind and, though trained primarily in geology, put the belief into practice by undertaking a dissertation on living animals." He says his efforts "left no cause for regret," yet during the mid-70s he found himself "aiming more and more in the direction of bringing fossil data to bear on biologic questions—in particular, questions relating to the process of evolution."[37]

While on the faculty at Rochester, he coauthored with Raup the influential textbook *Principles of Paleontology*, first published in 1971. He then moved to Johns Hopkins to pursue macroevolution as a "point of departure from the Modern Synthesis."[38] In a dramatic moment at the Geology Society of America meeting in Miami Beach in 1974, Stanley substituted his species selection paper at the podium, causing quite a sensation. Stanley's work laid out the consequences of punctuated equilibria for long-term trends in the history of life: "my strategy was to identify two potential loci for evolution and then show that one of these (generation-by-generation descent within established species) was generally characterized by such slow evolution that we were compelled to look to the other site (speciation) to account for the large majority of evolutionary innovations and higher taxonomic transitions—that is, most total evolution."[39] Species that last longer and generate more daughter species will have a greater effect on the composition of species on earth. The differential in speciation and extinction thus will determine in part the nature of overall biodiversity. In contrast to the Modern Synthesis emphasis on individual selection as the main patterning force in the history of life on earth, Stanley asserted it was simply the differential between speciation and extinction that determined the direction of trends. It is in many ways more analogous to sexual selection than individual natural selection; it is a differential in reproduction.

Elisabeth Vrba considers the term *selection* a misnomer here. She and Gould preferred the term *species sorting*. To a large extent, the agents of extinction are also the agents of species sorting—competition, predation, disease, habitat change, and random changes in population size. But for Stanley, the causative factor is the rate of origination of species: "It's differential speciation rates; it's not different rates of extinction."[40] Stanley explains: "If phyletic evolution were to prevail, large-scale evolutionary trends would simply track environmental change." But Stanley feels paleontologists have adequately demonstrated that "large-scale transitions in the physical environment cannot be expected to produce simple phylogenetic trends." Stanley and many other paleontologists are convinced that the punctuated equilibria model predominates, and thus puts a "macroevolutionary premium on the survival and splitting of lineages," since most morphological change happens during spe-

ciation. The descendents of species that possess "a behavioral or morphogenetic feature that promotes the divergence or reproductive isolation of small
populations" will be disproportionately represented in future biota. Thus,
resurrecting shades of Schindewolf, Stanley argues that in higher taxa: "The
presence of such a trait can be viewed as something analogous to the life history 'strategy' of a single species."[41]

Stanley used Punc Eq as a springboard. Even more important to him than
the expectation of quick speciation was the demonstration of pervasive stasis: "There's a whole lot of stasis. It's something we hadn't anticipated." He
feels the jury is still out on how much of it is ecological and how much is
morphogenetic (as the German Synthesists claimed), but he is convinced that
stasis is a major feature of evolution that would not have been addressed, or
even noticed, without paleontology: "Quite simply, I have come to believe
that paleontologic data tell us things about evolution that have not gained
general acceptance through the collective biologic effort known as the Modern Synthesis."[42] Stanley is energized by the ways paleobiology can extend
and inform the study of ecology:

> I think that probably the most important implication of all this is for
> ecology, because I grew up in an era when people believed that ecosys-
> tems were finely tuned and that species were coadapted and that there
> was enough genetic variability and general wherewithal for things to
> evolve in the right direction when the environment changed. This is
> saying no, a little bit of adjustment, change your clutch size, change your
> body size somewhat, but really turn into something different, no. And it
> changes our view of ecology—given the temporal dimension of ecology.
> It's saying that things are going to be pretty much making it as they are,
> with some fine-tuning, or they're going to be out of there. I think that's
> very important.

The evolutionary sciences are now more integrated, more cohesive, more
synthesized, than ever before. Yet many practitioners weaned on the Modern
Synthesis question whether the emerging theory is, properly speaking, Darwinian.

DARWIN FLINCHES?

The outcry in the 70s was against the incursion of what many thought were
anti-Darwinian theories. Some elements of them are still controversial. Is the

resulting post-Modern Synthesis no longer entirely Darwinian? Stephen Jay Gould came to that conclusion in his last major work, *The Structure of Evolutionary Theory*. He explained that current evolutionary theory has "an intact Darwinian foundation, but with a general form sufficiently expanded, revised or reconstructed to present an interestingly different structure of general explanation."[43] Gould held that aspects persist of Darwin's original framework for evolutionary theory, but that:

> substantial changes, introduced during the last half of the twentieth century, have built a structure so expanded beyond the original Darwinian core, and so enlarged by new principles of macroevolutionary explanation, that the full exposition, while remaining within the domain of Darwinian logic, must be construed as basically different from the canonical theory of natural selection, rather than simply extended.[44]

Eldredge doesn't go quite so far. In his *Unfinished Synthesis* he locates the new evolution within the continuing tradition of the Modern Synthesis. "When the masters of the synthesis—Dobzhansky, Mayr, and Simpson—were writing, they were well aware of the notion that evolution, meaning change, is somehow inevitable . . . in one passage Simpson was openly enthusiastic about the inevitability of change" in the theory.[45] This view was largely endorsed by Mayr in 1991, who then considered punctuated equilibria just a stage in the natural development of the Modern Synthesis.[46] Ledyard Stebbins, the fourth American architect of the Modern Synthesis, concurs: "the proposed new 'themes' are part and parcel of the modern synthesis."[47]

By 1997 Mayr had embraced punctuated equilibria to such an extent that he was taking credit for the idea well beyond the Mayrian grounding Eldredge and Gould jockeyed for in 1972:

> According to a model I proposed in 1954, evolution progresses rather slowly in large, populous species, while most rapid evolutionary changes occur in small, peripherally isolated founder populations. . . . Eldredge and Gould (1972), using the phrase "punctuated equilibria" accepted this model and proposed that the developmental stasis of populous species may last through millions of years. Subsequent research has confirmed that this is indeed true for many species.[48]

Nevertheless, Mayr asserts that the punctuated equilibria understanding of evolution "is almost the exact opposite of the one proposed by Fisher and

Haldane in the early 1930s."[49] In walking this fine line, Mayr bolsters the importance of American and nongeneticist contributions to the Modern Synthesis against a more Anglo- and genetics-centered view that places the most important work of the Modern Synthesis among geneticists primarily in England in the 20s and 30s.

ON THE ORIGINS OF SPECIATION

In 1859, Darwin made his case that "a naturalist, reflecting on the mutual affinities of organic beings, on their embryological relations, their geographic distribution, geological succession, and other such facts, might come to the conclusion that each species had not been independently created, but had descended, like varieties, from other species."[50] By 1868, most naturalists had been won over to that conclusion. Darwin remarked to Hooker, "This now almost universal belief in the evolution (somehow) of species, I think may be fairly attributed in large part to the *Origin*."[51] But just what processes caused speciation were less clear, as was the relationship between species formation and patterns in the diversity of life.

A century later, questions of macroevolution emerged as a distinct locus of evolutionary theory. Did prevailing evolutionary theory need substantial revision to explain macroevolution, and especially to explain the paleontological data? The controversy moved paleontology from a marginal role of compiling a photo album of the history of life on earth, to a central role as a source of evolutionary theory and of challenges toward further theory. The new ideas seemed, to many theorists, to diverge from what Julian Huxley called the "true blue Darwinian stream." Huxley claimed this stream was "eclipsed" for a time before it was established in the Modern Synthesis.[52] Such rhetoric was useful to Huxley and Mayr in the 40s and to Gould later on, but as history or philosophy, the eclipse model is inadequate. There was no one uniquely Darwinian stream to eclipse, but numerous interacting streams, all legitimately descended from Darwin's thought.

Evolutionary theory as it descended from Darwin had invoked several competing causal mechanisms. Certainly natural selection was one. So was blending inheritance of continuously variable traits, influential among animal breeders and late nineteenth-century British biometricians. Blending inheritance was Darwin's basic understanding of heredity. Other candidate causes were called "Lamarckian" or "neo-Lamarckian," but were closer to strands in Darwin's thought than Lamarck's. They postulated both heritable direct effects of the environment on organism and heritable effects of the organism's actions. Darwin believed the second was the more important of the two, but Lamarck believed the first could not happen in animals (though it did in plants). Darwin gave inheritance of acquired characters increased elaboration in successive editions of the *Origin* and developed his quasi-particulate theory of pangenesis to explain both it and blending inheritance. The late nineteenth century saw laboratory evidence accumulate supporting both these kinds of inheritance.[53]

There were also proposed mechanisms of directed evolution, including Darwin's laws of growth and the various kinds of orthogenesis derived from them. These were often defended by appeal to the fossil record and to embryology. A great deal of normal science was conducted in all these Darwinian causal fields for the better part of a century, and many of them were represented in the German Synthesis, but much of this work was elided or labeled anti-Darwinian in the triumph of the Modern Synthesis.

Modern Synthesists expected phyletic gradualism and supported it as Darwinian by referring to Darwin's argument that there must be many gaps in the fossil record. But they tended to ignore other statements that showed Darwin had a more sophisticated understanding of the fossil record than many twentieth-century neontologists. Discussions of speciation and patterns of diversity in the *Origin of Species* cohere with the expectations of punctuated equilibria. Darwin maintained that "migration has played an important part in the first appearance of new forms in any one area or formation" and that "widely ranging species are those which have varied most, and have oftenest given rise to new species; and that varieties have at first often been local."[54] Darwin explains that "we have no reason to believe that forms successively produced necessarily endure for corresponding lengths of time: a very ancient form might occasionally last much longer than a form elsewhere subsequently produced."[55] Darwin asserts widespread stasis: "Consider the prodigious vicissitudes of climate during the pleistocene period, which includes the whole glacial period, and note how little the specific forms of the inhabitants of the sea have been affected."[56] Periods of stasis are interrupted by narrow zones in which transitions occur rather quickly:

> we have no just right to expect often to find intermediate varieties in the intermediate zone. For we have reason to believe that only a few species are undergoing change at any one period . . . I have also shown that the intermediate varieties which will at first probably exist in the intermedi-

ate zones, will be liable to be supplanted by the allied forms on either hand; and the latter, from existing in greater numbers, will generally be modified and improved at a quicker rate.[57]

This is not to say that Darwin concocted punc eq in 1859. Darwin's proposed causes differ in important ways from those underlying the theory of punctuated equilibria (just as those of the German Synthesis are not isomorphic with those of late twentieth-century paleobiology). But claims that evolution must be gradual and continuous in order to be Darwinian simply do not hold water.

The current evolutionary synthesis can be construed as a revised Modern Synthesis. But it is just as easily construed as a revised German Synthesis—especially as regards pattern, process, continuity, and the hierarchical role of selection. Gould chose to regard the German Synthesis as anti-Darwinian and so to see the current view as only partly Darwinian. But to make this case Gould takes only part of Darwin. If we recognize the German Synthesis as a legitimate Darwin descendent then we gain a more accurate historical understanding. Evolutionary theory today is a synthesis of two alternate but fully Darwinian syntheses.

PALEONTOLOGY AT THE HIGH TABLE

The paleobiology movement rose from its roots in '50s paleoecology, gathered steam via a dynamic group of students—especially at Yale—in the '60s, was focused sharply by James Valentine's 1967 landmark paper in mathematical models, and was organized by Tom Schopf's work at Woods Hole and Chicago in the '70s and his founding of the journal *Paleobiology*. Yet the icons for this movement have consistently been punctuated equilibria (itself also the result of Schopf's organizational genius) and the K-T mass extinction. But for the young Turks themselves, perhaps no event better signaled that they had arrived than the 1980 Chicago Macroevolution Conference, where folks from all subfields confronted what theoretical developments in paleobiology meant for the evolutionary sciences.

As a consequence of the success of the Chicago Macroevolution conference, English geneticist John Maynard Smith welcomed paleontology to the high table of evolutionary science. No more suitable host could have welcomed the paleontologists. Maynard Smith pioneered the application of game theory to evolutionary interpretations of animal behavior, and epitomizes English respectability. He has been perhaps the most severe, but also the most useful, critic of the potential for paleontology and macroevolution to serve as sources of evolutionary theory.

John Maynard Smith (1920–2004) was a "natural history fanatic since childhood" and spent hours on end poking about in the forest. Like most upper-class English boys, young Maynard Smith spent most of his time at boarding school. But he passed his holidays in Exford on Exmoor in the West of England. Like a character from Thelwell, he ranged across the moor on his own Exmoor pony, very much at home with the birds and beetles.[58] Maynard Smith feels his naturalist grounding combined with his years of laboratory genetics to give him an approach to theory quite different from Haldane and Fisher: "In practice, I start with a question about a particular organismic problem; I don't start with a general problem."

In classic Fisherian fashion, Maynard Smith feels nature cannot select for a coadapted set of genes on different chromosomes, but can only add one at a time. He thinks there is a continuum between microevolution and macroevolution, and that speciation as perceived by Mayr is correct overall. Mendelian mutations are not in and of themselves responsible for speciation, but interact with biogeography and other population-level phenomena to produce new races and species: "I can't really say anything constructive about macroevolution," but "there's no reason to think there's anything funny going on." He is nevertheless willing to accept a role for such things as Goldschmidtian hopeful monsters, exaptation as a prelude to adaptation, and the importance of chance: "A random walk can take you into new adaptive possibilities."

Maynard Smith regretted the lack of open exchange between subfields: "Since George Gaylord Simpson, there's been a real lack of communication between paleontology and population genetics." He hoped the rise of more sophisticated understanding of developmental genetics would lead the two fields to understand one another better. As developmental biology learns more about the genetic basis of large-scale morphological change, Maynard Smith looked forward "to a stage when a bridge will be built between the two fields."

Raup and the quantitatively minded paleontology group at the University of Chicago impressed Maynard Smith. He appreciated Raup's grounding in the Modern Synthesis, and so was willing to entertain what he saw as Raup's more far-out ideas about pattern in the history of life, mass extinction, and macroevolution: "You can go to a pub and sit down and talk with Raup." But he could not find much common ground with Gould, "so muddled, so inconsistent . . . dreadful."

Maynard Smith had "the highest regard for Lewontin," but found the spandrels paper "fundamentally beside the point." Randomness is normal. It is what one would expect. For Maynard Smith, the really difficult issue that evolutionary theory must address is how adaptation comes about. Any question asked "is assumed to have an adaptive answer, otherwise we cannot proceed to establish whether a given adaptive process can generate the correct solution."[59] However, he admitted that the disputation brought about by Gould and Lewontin's criticism has been fruitful. His famous paper on optimization was "stimulated largely by the spandrels paper."[60]

Maynard Smith regarded the primary importance of randomness as its ability to move phenotype to a different potential genetic space. He was not wedded to gradualism: "The only way I'm a gradualist is genes have to be added one by one to an adaptive complex." Though many traits do not seem adaptive at first glance, he did not think there are very many "things which don't make sense when you look at them closely." He and Lewontin "don't disagree about how the world works," just about "what it's important to convince people of."

Gould noted in 1998: "Even John Maynard Smith might admit he welcomed us to the high table once. He might be reassessing that now."[61] Gould felt a major way that macroevolution has become an important force informing all evolutionary theory was through hierarchical selection theory: "The main thing that people don't understand about punc eq and macroevolution is that so many of them can't get it out of their heads that if there's anything radical it has to say it must be a claim to new macroevolutionary mechanism." But Eldredge and Gould have not proposed new mechanisms; rather, a different way of understanding evolutionary forces, and the development of more rigorous predictions based on that understanding. Gould complains that when he and his collaborators "tell them that it's just ordinary allopatric speciation properly scaled, then they say well then you're just giving us bombast; you're just selling us a bunch of words. They don't get it that what's radical about it is the need to explain trends in terms of species sorting or species selection." Although punctuated equilibria and the new macroevolutionary theory are extensions of the Modern Synthesis, they were not predicted by it. "The radical part is at a higher level, considering it's not macroevolutionary mechanisms being called upon but . . . that hierarchical selection theory is hierarchical causation in general." Another major contribution, in Gould's eyes, was the realization that at times selection operates by "different rules imposed by rare events like mass extinction." These make any adaptive aspect to trends "effectively invisible."[62]

Gould considered the most dramatic moment from the 1980 Chicago Macroevolution conference to be when "Maynard Smith got up and made his little announcement, as people always do, 'oh you guys have presented nothing new here. We've said that all the time.' And I'll never forget George Oster standing up and saying 'You know, John, we always hear things like this. Well, John, you know. You may have had the bicycle. But you didn't ride it.'"

Of all the innovations contributed by paleobiology, reinterpretation of the expectations for empirical evidence of speciation in the fossil record met the most resistance, especially early on. After the initially stormy reception—both within and outside paleontology, punctuated equilibria gained wide acceptance in the scientific community, even to the point of routine inclusion in biology textbooks by the late 1980s.[63] The National Center for Science Education even featured a bumper sticker that read "Honk if you understand punctuated equilibria."[64] Species selection enjoys similar acceptance by specialists and textbooks, though discussion is usually more abbreviated, and it is more often left out of the most popular literature.

Sapp characterizes a scientific field as "a system of objective relations involving a competitive struggle between positions already won." Bourdieu calls it "the objective space defined by the play of opposing forces in a struggle for scientific stakes."[65] These objects, relations, and spaces relate in this case to positions first won by Darwin himself. As the long argument has continued, the spaces have expanded. Places won in the Modern Synthesis have been challenged anew—particularly by paleobiologists and evolutionary-developmental biologists. And at no spot has competition been fiercer than that of speciation. The origin of species was the Holy Grail in Darwin's day, and remains the prized position today. The theory of punctuated equilibria gained power for its subfield by virtue of its ambition, and gained even more by its staying power. Punc eq demonstrated that paleobiology had the data and theoretical power to be taken seriously as a top player in the Darwinian sciences. Paleo had the nomothetic goods to compete at and even remap the top levels of the biological sciences. This brash positioning, more than any one specific element of the complex, is what got paleobiology recognized at the high table and served up widespread recognition for the paleobiological revolution.

NOTES

1. Stephen Jay Gould (1980). "Is a New and General Theory of Evolution Emerging?" *Paleobiology* 6, 120.

2. Bakker interview, 2/20/99.

3. Seilacher interview, 12/9/96.

4. A. R. Wallace; letter to Darwin from Wallace, May 29, 1864. Charles Darwin, Francis Darwin, and Alfred Seward, *More Letters of Charles Darwin, a Record of His Work in a Series of Hitherto Unpublished Letters* (New York: D. Appleton and Company, 1903), vol 2, Letter number 406.

5. Niles Eldredge, *Reinventing Darwin: The Great Debate at the High Table of Evolutionary Theory* (New York: Wiley, 1995), 56.

6. Gould interview, 11/2/98.

7. Norman D. Newell (1952). "Periodicity in Invertebrate Evolution." *Journal of Paleontology* 26:371–85; Newell (1963). "Crises in the History of Life." *Scientific American* 208:76–92; Newell, "Revolutions in the History of Life," in *Uniformity and Simplicity* (Boulder: Geological Society of America, 1967), 63–91.

8. Niles Eldredge (1971). "The Allopatric Model and Phylogeny in Paleozoic Invertebrates." *Evolution* 25:156–67.

9. For example, see Raup's seminal work on bivalve coiling parameters. David M. Raup (1962), "Computer as Aid in Describing Form in Gastropod Shells," *Science* 138; David M. Raup (1966), "Geometric Analysis of Shell Coiling: General Problems," *Journal of Paleontology* 40.

10. PhD, 1951, adviser Carl Owen Dunbar. Thesis: John Imbrie, "Protremate Brachiopods of the Traverse Group 'Devonian' of Michigan" (Yale University, 1951). *Neontologist* is a word coined by G. G. Simpson to denote biologists who work primarily on extant species.

11. The German Synthesis and other ideas here are developed more extensively in Princehouse, *Mutant Phoenix*. Unpub. diss., Harvard University, 2003; and in "Evolution: The German Synthesis," forthcoming.

12. Stephen Jay Gould. "Foreword," in Otto H. Schindewolf, *Basic Questions in Paleontology: Geologic Time, Organic Evolution, and Biological Systematics.* Translated by Judith Schaefer. (Chicago: University of Chicago Press, 1993).

13. Raup interview, 5/11/98 and 5/12/98.

14. Newell interview, 7/16/98.

15. Stephen Jay Gould (1966), "Allometry and Size in Ontogeny and Phylogeny," *Biological Reviews* 41.

16. Gould interview, 11/2/98.

17. Gould interview, 11/2/98.

18. Gould interview, 11/2/98.

19. Gould interview, 11/2/98. Julian Huxley was born in 1887 and died 1975.

20. Gould, "Allometry and Size in Ontogeny and Phylogeny." Interview, May 14, 2002.

21. Gould interview, 11/2/98. After Schindewolf's retirement, German paleontology came much more into line with the Modern Synthesis, and some of Schindewolf's later students feel his effect on the profession in Germany was stifling.

22. Gould spent most of the rest of his career working on the Bahamian land snail

Cerion. "*Cerion* is probably the most diverse snail in the world in terms of form, and my main interest is the evolution of form."

23. Gould interview, 11/2/98.

24. Motoo Kimura (1968), "Evolutionary Rate at the Molecular Level." *Nature* 217.

25. Gould interview, 11/2/98.

26. All quotes from Bambach interview, 5/16/95.

27. Raup interview, 5/11/98 and 5/12/98.

28. Gould interview, 11/2/98. Eldredge, 1971.

29. Gould interview, 11/2/98.

30. Gould interview, 11/2/98.

31. See also: Albert Somit and Steven A. Peterson (1992), *The Dynamics of Evolution : The Punctuated Equilibrium Debate in the Natural and Social Sciences* (Ithaca: Cornell University Press), and F. S. Szalay (1999), "Paleontology and Macroevolution: On the Theoretical Conflict between an Expanded Synthesis and Hierarchic Punctuationism," in *African Biogeography, Climate Change, and Human Evolution*, ed. T. Bromage and F. Schrenk (Oxford: Oxford University Press. F. S. Szalay and W. J. Bock (1991), "Evolutionary Theory and Systematics: Relationships between Process and Patterns," *Zeitschrift Zoologische Systematic und Evolutions-Forschung* 29.

32. Stephen Jay Gould (2002), *The Structure of Evolutionary Theory* (Cambridge, MA: Belknap Press of Harvard University Press), 26.

33. See J. Maynard Smith (1984), "Paleontology at the High Table," *Nature* 309. See also the reply, Eldredge, *Reinventing Darwin*.

34. The change in paleontology complemented an expansion of genetics beginning in the late 1960s to include mechanisms and phenomena not expected by the Modern Synthesis, especially rampant genetic variation in normal populations, the neutral theory and transposition. See, e.g. , Motoo Kimura (1968), "Evolutionary Rate at the Molecular Level," *Nature* 217. Richard C. Lewontin and John L. Hubby (1966), "A Molecular Approach to the Study of Genic Heterozygosity in Natural Populations. II. Amount of Variation and Degree of Heterozygosity in Natural Populations of Drosophila Pseudoobscura," *Genetics* 54. John L. Hubby and Richard C. Lewontin (1966), "A Molecular Approach to the Study of Genic Heterozygosity in Natural Populations. I. The Number of Alleles at Different Loci in Drosophila Pseudoobscura," *Genetics* 54. Jack King and Thomas Jukes (1969), "Non-Darwinian Evolution," *Science* 164.

35. Steven M. Stanley (1975), "A Theory of Evolution above the Species Level," *Proceedings of the National Academy of Sciences* 72.

36. Gould interview, 11/2/98.

37. Steven M. Stanley (1988), *Macroevolution, Pattern and Process*, (Baltimore: Johns Hopkins University Press), ix.

38. Ibid., xi.

39. Stanley, *Macroevolution, Pattern and Process*, xxiii.

40. Interview with Steven Stanley, 10/5/98.

41. Stanley, *Macroevolution, Pattern and Process*, 205.

42. Ibid., ix.

43. Gould, *The Structure of Evolutionary Theory*, 19.

44. Ibid., 3.

45. Niles Eldredge (1985), *Unfinished Synthesis: Biological Hierarchies and Modern Evolutionary Thought* (New York: Oxford University Press), 213.

46. Ernst Mayr (1991). *One Long Argument: Charles Darwin and the Genesis of Modern Evolutionary Thought, Questions of Science*. (Cambridge, Mass.: Harvard University Press), 162–64.

47. G. Ledyard Stebbins and Francisco J. Ayala (1981), "Is a New Evolutionary Synthesis Necessary?" *Science* 213:1967. Stebbins' Modern Synthesis effort brought botany into the fold: G. Ledyard Stebbins (1950), *Variation and Evolution in Plants* (New York: Columbia University Press).

48. Ernst Mayr (1997), *This Is Biology* (Cambridge: Harvard University Press), 172.

49. Ibid., 173.

50. Charles Darwin (1964), *On the Origin of Species. A Facsimile of the 1st Edition* (Cambridge, MA: Harvard University Press), 3.

51. Darwin to Hooker, July 28, 1868. In: *On the Origin of Species*, 304.

52. Bowler follows Huxley's eclipse model closely: Peter J. Bowler (1983), *The Eclipse of Darwin* (Baltimore: The Johns Hopkins University Press), Peter J. Bowler (1988), *The Non-Darwinian Revolution* (Baltimore: Johns Hopkins University Press). Smocovitis questions the word "eclipse" by consistently using it in scare quotes but does use this interpretation exclusively. Smocovitis, *Unifying Biology*, 44.

53. Much laboratory evidence for the inheritance of acquired characters is indicated in Vernon Kellogg (1907), *Darwinism To-Day*, 290–326. Sources include: Charles E. Brown-Sequard (1875), "On the Hereditary Transmission of Effects of Certain Injuries to the Nervous System," *Lancet* 2, J. T. Cunningham (1893), "The Problem of Variation," *Natural Science* 3, Alpheus Hyatt (1893), "Phylogeny of an Acquired Characteristic," *Proceedings of the American Philosophical Society* 32, Félix Alexandre Le Dantec(1903), *Traité De Biologie* (Paris: Alcan), Heinrich Obersteiner (1875), "Zur Kenntnis einiger Hereditaetsgesetze," *Medizinische Jahrbuch* 2, George Romanes (1892–1897), *Darwin and after Darwin an Exposition of the Darwinian Theory and a Discussion of Post-Darwinian Problems*, 3 vols. (London: Open Court). See also, T. H. Morgan *Animals*, and Thomas Hunt Morgan (1907), *Experimental Zoölogy* (New York: Macmillan). And we should not dismiss Kammerer, e.g., Paul Kammerer and A. Paul Maerker-Branden (1924), *The Inheritance of Acquired Characteristics* (New York,: Boni and Liveright), simply because it was later determined that his experiments were finagled. The data was taken seriously for a time and Gliboff has demonstrated that Kammerer believed himself to be an heir to Darwin, Sander J. Gliboff, "The Case of Paul Kammerer," (forthcoming).

54. Darwin, *Origin*, 342.

55. Ibid., 335.

56. Ibid., 336.

57. Ibid., 463.

58. All quotes from interview with Maynard Smith, July 30, 1996, unless otherwise attributed.

59. Geoffrey Parker and John Maynard Smith (1990), "Optimality Theory in Evolutionary Biology," *Nature* 348:27.

60. J. Maynard Smith (1978), "Optimization Theory in Evolution," *Annual Review of Ecology and Systematics* 9.

61. Gould interview, 11/2/98.

62. Gould interview, 11/2/98.

63. E.g., Douglas J. Futuyma (1986), *Evolutionary Biology*, 2nd ed. (Sunderland, MA: Sinauer Associates). Kenneth R. Miller and Joseph Levine (1999), *Biology: The Living Science.* (Needham, MA: Prentice Hall).

64. NCSE began making Punc Eq bumper stickers in 1987.

65. Jan Sapp (1983), "The Struggle for Authority in the Field of Heredity, 1900–1932: New Perspectives on the Rise of Genetics" *Journal of the History of Biology* 16, no. 3: 312. Pierre Bourdieu (1975), "The Specificity of the Scientific Field and the Social Conditions of the Progress of Reason," *Social Science Information* 14.

CHAPTER NINE

Molecular Evolution vis-à-vis Paleontology

Francisco J. Ayala

Biological evolution is a time-dependent process: by and large, change is unidirectional over time. Some degree of correlation is, therefore, expected between the biological differentiation of two organisms and the time elapsed since their separation, be the comparison between an organism and its ancestor, or between two organisms sharing a common ancestor. The correlation, however, need not be exact, if only because organisms evolve in response to the vagaries of environmental change in time and space. It is well known that some organisms have morphologically evolved quickly, at least with respect to some traits, whereas others have changed but little over millions of years.

Molecular biology, a discipline that emerged in the second half of the twentieth century, nearly one hundred years after the publication of *The Origin of Species*, has provided what many scientists consider the strongest evidence yet of the evolution of organisms. Molecular biology proves evolution in two ways: first, by showing the unity of life in the nature of deoxyribonucleic acid (DNA) and the workings of organisms at the level of enzymes and other protein molecules; second, and most important for evolutionists, by making it possible to reconstruct evolutionary relationships that were previously unknown, and to confirm, refine, and time all evolutionary relationships from the universal common ancestor up to all living organisms.

DNA and proteins have been called *informational macromolecules*, because they are long linear molecules made up of sequences of units—nucleotides in the case of nucleic acids, amino acids in the case of proteins—that embody evolutionary information (Zuckerkandl and Pauling 1965). Comparing the sequence of the components in two macromolecules establishes how many components are different. Because evolution usually occurs by changing one unit at a time, the number of differences is an indication of the recency of common ancestry.

The degree of similarity in the sequence of nucleotides, or of amino acids, can be simply quantified. For example, in humans and chimpanzees, the protein molecule called cytochrome-c, which serves a vital function in respiration within cells, consists of the same 104 amino acids in exactly the same order. It differs, however, from the cytochrome-c of rhesus monkeys by one amino acid, from that of horses by eleven additional amino acids, and from that of tuna by twenty-one additional amino acids.

The degree of similarity reflects the recency of common ancestry. Thus, the inferences from comparative anatomy and other disciplines that study evolutionary history can be tested in molecular studies of DNA and proteins by examining the sequences of nucleotides and amino acids. The authority of this kind of test is overwhelming: each of the thousands of genes and thousands of proteins contained in an organism provides an independent test of that organism's evolutionary history. In what follows, I'll sometimes refer to proteins and other encoded molecules; other times, to the genes that code for them or other DNA sequences. In most relevant respects, what I say about DNA and genes applies to the encoded molecules, and vice versa.

Molecular evolutionary studies have three notable advantages over paleontology, comparative anatomy, and other classical disciplines. One is that the information is readily *quantifiable*. The number of units that are different is easily established when the sequence of units is known for a given macromolecule in different organisms. The second advantage is that comparisons can be made between *very different sorts of organisms*. There is very little that comparative anatomy or paleontology can say when, for example, organisms as diverse as yeasts, pine trees, and human beings are compared, but there are numerous DNA and protein sequences that can be compared in all three. The third advantage is *multiplicity*. Each organism possesses thousands of genes and proteins, which all reflect the same evolutionary history. If the investigation of one particular gene or protein does not satisfactorily resolve the evolutionary relationship of a set of species, additional genes and proteins can be investigated until the matter has been settled.

Moreover, the widely different rates of evolution of different sets of genes opens up the opportunity for investigating different genes in order to achieve different degrees of resolution in the tree of evolution. Molecular evolutionists rely on slowly evolving genes for reconstructing remote evolutionary events, but increasingly faster-evolving genes for reconstructing the evolutionary history of more recently diverged organisms.

Genes that encode ribosomal RNA molecules are among the slowest-evolving genes. (Ribosomes are complex molecules that mediate the synthesis of proteins; each ribosome consists of several proteins and several RNA molecules.) They have been used to reconstruct the evolutionary relationships between groups of organisms that diverged very long ago: for example, among bacteria, archaea, and eukaryotes (the three major divisions of the living world), which diverged more than two billion years ago, or among kingdoms or phyla, such as the microscopic protozoa (e.g., *Plasmodium*, which causes malaria), compared to plants and to animals—groups of organisms that diverged more than one billion years ago. Cytochrome-c evolves slowly, but not as slowly as the ribosomal RNA genes. Thus, cytochrome-c is used to decipher the relationships between phyla or between large groups of organisms within phyla, such as comparisons of humans, fishes, and insects. Fast-evolving molecules, such as the fibrinopeptides involved in blood clotting, are appropriate for investigating the evolution of closely related animals, the primates for example: macaques, chimps and humans. Molecules with intermediate rates of evolution, such as hemoglobin, are useful for comparisons within phyla, such as among the vertebrates.

LINEAGE EVOLUTION AND SPECIES DIVERSIFICATION

DNA and proteins provide information not only about the branching of lineages from common ancestors (*cladogenesis*) but also about the degree of genetic change that has occurred in any given lineage (*anagenesis*). Molecular evolutionary trees are models or hypotheses that seek to reconstruct the evolutionary history of taxa—that is, species, genera, families, orders, and other groups of organisms. Moreover, the trees embrace information about both dimensions of evolutionary change, cladogenesis and anagenesis.

It might seem at first that quantifying anagenesis for proteins and nucleic acids would be impossible, because it seems to require comparison of molecules from organisms that are now extinct with molecules from living organisms or from other extinct organisms. Organisms of the past are sometimes preserved as fossils, but their DNA and proteins have largely disintegrated. Nevertheless, comparisons between living species provide information about anagenesis.

Consider again, for example, the protein cytochrome-c. The sequence of amino acids in this protein is known for many organisms, from bacteria and yeasts to insects and humans; in animals, cytochrome-c consists of 104 amino acids. When the amino acid sequences of humans and rhesus monkeys are compared, they are found to be different at position 58, but identical at the other 103 positions. When humans are compared with horses, twelve amino acid differences are found, and when horses are compared with rhesus monkeys, there are eleven amino acid differences. Even without knowing anything else about the evolutionary history of mammals, we would conclude that the lineages of humans and rhesus monkeys diverged from each other much more recently than they diverged from the horse lineage.

Moreover, it can be concluded that the amino acid difference between humans and rhesus monkeys must have occurred in the human lineage after its separation from the rhesus monkey lineage. This conclusion is drawn from the observation that, at position 58, monkeys and horses (as well as other animals) have the same amino acid (threonine), while humans have a different one (isoleucine), which therefore must have changed in the human lineage after it separated from the monkey lineage. The amino acid sequences in the cytochrome-c of twenty very diverse organisms were ascertained in 1967. Counting the amino acid differences between the twenty species resulted in the evolutionary tree shown in figure 9.1 (Fitch and Margoliash 1967).

THE MOLECULAR CLOCK

One conspicuous attribute of molecular evolution is that, as pointed out, differences between homologous molecules can readily be quantified and expressed as, for example, number of nucleotides or amino acids that have changed. Rates of evolutionary change can therefore be fairly precisely established with respect to DNA and proteins. Studies of molecular evolution rates have led to the proposition that macromolecules may serve as evolutionary clocks.

It was first observed in the 1960s that the number of amino acid differences between homologous proteins of any two given species seemed to be nearly proportional to the time of their divergence from a common ancestor. If the rate of evolution of a protein or gene were approximately the same in the evolutionary lineages leading to different species, proteins and DNA sequences would provide a molecular clock of evolution. The sequences could then be used to reconstruct not only the sequence of branching events of a phylogeny but also the time when the various events occurred.

Consider, for example, figure 9.1. If the substitution of nucleotides in the

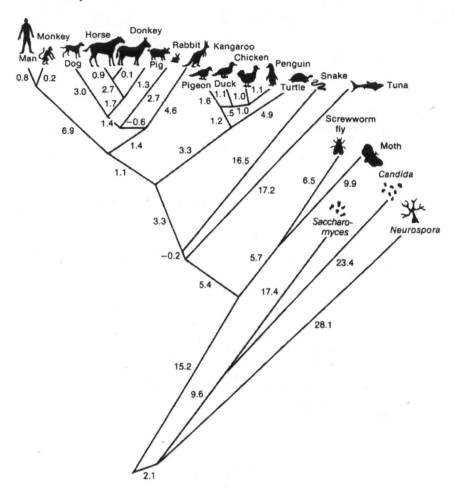

Figure 9.1 Phylogeny of 20 organisms, based on differences in the amino acid sequence of cytochrome c. The phylogeny agrees fairly well with evolutionary relationships inferred from the fossil record and other sources. The minimum number of nucleotide substitutions required for each branch is shown. Although fractional numbers of nucleotide substitutions cannot occur, the numbers shown are those that best fit the data. (After Fitch and Margoliash 1967.)

gene coding for cytochrome-c occurred at a constant rate through time, we could determine the time elapsed along any branch of the phylogeny simply by examining the number of nucleotide substitutions along that branch. We would need only to calibrate the clock by reference to an outside source, such as the fossil record, that would provide the actual geologic time elapsed in at least one specific lineage or since one branching point. For example, if the time of divergence between insects and vertebrates is determined to have occurred seven hundred million years ago, other times of divergence can be determined by proportion of the number of amino acid changes.

The molecular evolutionary clock is not expected to be a metronomic clock, like a watch or other timepieces that measure time exactly, but a stochastic (probabilistic) clock, like radioactive decay, where the *probability* of a certain amount of change is constant, although some variation occurs in the actual amount of change. Over fairly long periods of time, a stochastic clock is quite accurate. The enormous potential of the molecular evolutionary clock lies in the fact that each gene or protein is a separate clock. Each clock ticks at a different rate—the rate of evolution characteristic of a particular gene or protein—but each of the thousands and thousands of genes or proteins provides an independent measure of the same evolutionary events.

Evolutionists have found that the amount of variation observed in the evolution of DNA and proteins is greater than is expected from a stochastic clock—in other words, the clock is overdispersed, or somewhat erratic. The discrepancies in evolutionary rates along different lineages are not excessively large, however. So it is possible, in principle, to time phylogenetic events with considerable accuracy, but more genes or proteins must be examined than would be required if the clock were stochastically constant in order to achieve a desired degree of accuracy. The average rates obtained for several proteins, taken together, become a fairly precise clock, particularly when many species are studied.

This conclusion is illustrated in figure 9.2, which plots the cumulative number of nucleotide changes in seven proteins against the dates of divergence of seventeen species of mammals (sixteen pairings) as determined from the fossil record. The overall rate of nucleotide substitution is fairly uniform. Some primate species (represented by the points below the line at the lower left of the figure) appear to have evolved at a slower rate than the average for the rest of the species. This anomaly occurs because the more recent the divergence of any two species, the more likely it is that the changes observed will depart from the average evolutionary rate. As the length of time increases, periods of rapid and slow evolution in any lineage will tend to cancel one another out.

Figure 9.2 conveys the prevailing point of view among molecular evolutionists concerning the reliability of the molecular clock. The reliability of the clock is the matter to which I will now turn. I will proceed in three steps. First, I will point out an intrinsic bias in the methodology of the molecular clock, which therefore yields time estimates of past events that are systematically

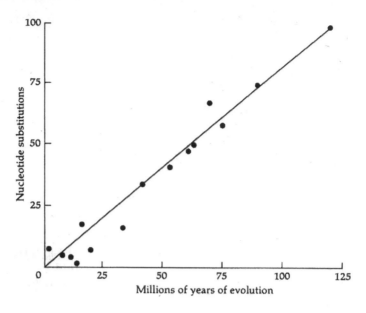

Figure 9.2 Nucleotide substitutions versus paleontological time. The minimum numbers of nucleotide substitutions for seven proteins (cytochrome c, fibrinopeptides A and B, hemoglobins α and β, myoglobin, and insulin C-peptide), sequenced in 17 species of mammals, have been calculated for comparisons between pairs of species whose ancestors diverged at the time indicated on the abscissa. The line has been drawn from the origin to the outermost point and corresponds to a rate of 0.41 nucleotide substitution per million years for all seven proteins together. Most points fall near the line, except for some representing comparisons between primates (points below the line at lower left), in which protein evolution seems to have occurred at a lower-than-average rate.

older than the correct time. (The mathematical details in this section can be ignored without missing the gist of the argument.) Second, I will review a set of experiments designed to test the accuracy of the molecular clock. We'll see that simple hypotheses to account for the larger-than-expected variances of the clock are not generally valid. These hypotheses include the proposals that for a given gene the rate of evolution is faster for organisms with shorter generation time, or for organisms with larger population sizes. Third, I will examine, as a case study, efforts to use molecular clocks to determine the time of origination of the major animal phyla. We'll see that large errors can easily be introduced, but they often can be by and large corrected with appropriate statistical methodologies.

A SYSTEMATIC BIAS OF THE MOLECULAR CLOCK

Molecular time estimates display asymmetric distributions, with a constrained younger end but an unconstrained older end. A typical plot of age estimates is right-skewed, with a large number of values at the left-hand (younger) end, and a long tail of ever-older values to the right. This is because rates of evolution are constrained to be nonnegative (that is, the lower boundary is nonelastic, it cannot be less than zero), while the rates are unbounded above zero (the upper boundary is elastic). Simply taking an arithmetic average of the estimated divergence times based on all possible calculated rates of evolution will consistently overestimate the true date. This overestimation problem becomes more marked as the rate of molecular evolution decreases and/or the sequences become shorter. The overestimates also grow as target times become increasingly remote, so this could be an especial problem for estimates of dates in the Precambrian; for example, for the diversification of life, or for the plant-fungi-animals splits, or for the radiation of animal phyla. I will now analyze this bias in greater detail (see Rodriguez-Trelles, Tarrio, and Ayala 2002).

Suppose we have three orthologous protein sequences related as in figure 9.3, which have passed some molecular clock criterion (usually a relative rate test—see the following). We seek to determine the date when lineages C and AB split (denoted as t_T, or target time in figure 9.3). Let us assume that the average number of amino acid replacements per site between A and B is $K_{AB} = 1$, and that C differs from either A or B by $K_{AC} = K_{BC} = 10$. Also, it is known from the fossil record that A and B split from a common ancestor one hundred My (million years) ago (denoted as t_C, or calibration time). If we assume that the rate of evolution is constant, so that r_{AB} (hereafter denoted as r_R, or reference rate) is equal to the rate between C and AB (hereafter denoted as r_U, or unknown rate), then the unknown (target) date would be placed at $t_T = 1,000$ My ago. (Saying it without symbols: given that the differences between C and A or B is 10, or ten times larger than between A and B, we expect the time of divergence between C and A or B to be also ten times greater, or 1,000 My rather than one hundred My.) In actual practice, after conducting analogous calculations separately for each of n independent, putatively rate-constant protein regions, conventional molecular dating approaches would set the time of the split between lineages C and AB as the arithmetic mean across the ensuing n t_T values.

Note, however, that (a) even if rate constancy holds, r_R and r_U represent different realizations of a stochastic process, subject to sampling variation such

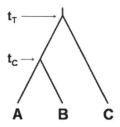

Figure 9.3 Tree topology for lineages A, B, and C. Calibration and target times are represented, respectively, by t_C and t_T.

that they are not expected to be identical; indeed, the dispersion of the rate of molecular evolution has proved to be much larger than expected if the probability of change were constant; and (b) because of its definition as a quotient of (often nonindependent, gamma-distributed) rates, time since divergence is an asymmetrically bounded random variate: constrained to be nonnegative as pointed out (the lower boundary is nonelastic) but unbounded at the upper boundary (elastic boundary). Equivalent random deviations around target times scale divisively forward (i.e., to the present), but multiplicatively backward (i.e., to the past) on their target times. As a result of this reciprocal scaling of under- and overestimates, the frequency distribution of time-since-divergence estimates is squashed up near the origin with a long tail to the right, yielding arithmetic averages that are upwardly biased with respect to the true times.

Suppose that in figure 9.3 one hunded and 1,000 My are, respectively, the true divergence times between A and B, and between either of them and C. Now consider two protein sequences with an observed r_R that is two times r_U for one protein, but r_R is half r_U for the other protein. The first protein would date the split between C and AB 500 My later than it happened (i.e., 500 My ago), whereas the second one would set the split 1,000 My earlier (i.e., 2000 My ago). The arithmetic average across the two proteins is 1,250 My, which still overestimates the true time by 250 My. These numbers become increasingly disparate as the ratio r_R/r_U deviates from 1.

To evaluate the extent of the overestimation that results from equating target times to arithmetic means across multiple-gene data, Rodriguez-Trelles, Tarrio, and Ayala (2002) simulated the evolution of an ancestral amino acid sequence along the topology of figure 9.3 under different sets of conditions.

For each condition set, the rate of replacement was fixed throughout the tree (i.e., $r_R = r_U$). Amino acid changes were generated using the discrete gamma distribution with shape parameter α (the JTT + dG model; Yang et al. 1998) to accommodate among-site rate variation. Three different, biologically meaningful replacement rates were considered to represent slow (one replacement per site per 10^{10} years), intermediate (five replacements per site per 10^{10} years), and fast (ten replacements per site per 10^{10} years) evolving genes. Each replacement rate was combined with a specific value of α (0.5, 1.0, and 2.0, respectively), to take into account that slowly evolving proteins tend to have a high level of rate variation among sites, and vice versa. In all cases t_C was set to 300 My, and for each rate class three total tree lengths were considered by setting, alternatively, t_T at 600 My, 1,100 My, and 3,000 My, four sequence lengths (75, 150, 300, and 500 amino acids) that span the typical lengths of proteins or DNA sequences used in experiments. For each set of conditions 1,000 simulations were performed. Each simulation produced three amino acid sequences related as in figure 9.3.

The simulation results show that, as expected, owing to the distributional asymmetry of divergence times, even under a uniform rate model of evolution, arithmetic averages across molecular clock projections consistently overestimate the true date of divergence (Rodriguez-Trelles, Tarrio, and Ayala 2002). The overestimation problem becomes aggravated as the rate of replacement decreases and/or the sequences become shorter. Both circumstances are expected to result in enhanced sampling variation of estimates, thus yielding increasingly right-skewed distributions.

Figure 9.4 illustrates the frequency distribution of 1,000 time-estimates for the case of a short (seventy-five residues long), slowly evolving (five replacements/site/10^{10} years) protein used to date an episode 3,000 My old. The distribution is highly skewed to the right, giving an arithmetic average that places the event 4,084 My ago, i.e., more than 1,000 My earlier than actually happened. Overestimates grow as target times become increasingly remote. This pattern results because, when the rate of replacement is low enough such that the sequences being handled become too short for accurately reflecting the expected number of variable sites, evolutionary rates become consistently underestimated. Underestimation is most acute for the reference rate, because it involves the shortest time span (i.e., reference times are more recent than the times we want to estimate), and diminishes as the rate to be ascertained involves an increasingly remote divergence. Because of these systematic differences in sampling error between the calibration and extrapolation rates, the

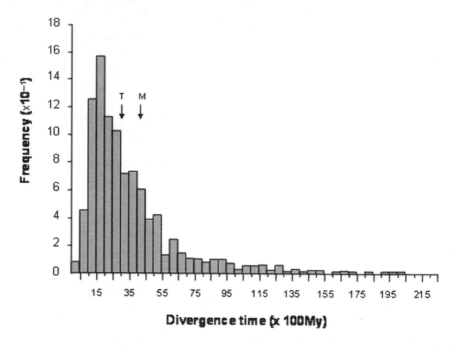

Figure 9.4 Frequency distribution of 1,000 estimates of the divergence time between lineages C and AB in figure 9.3, set to have occurred 3,000 My ago, obtained using a short (75 residues long), slow-evolving (one replacement per site per 10^{10} years) protein, and using the split between A and B, set to 300 My ago, as the calibration point. T and M represent target (i.e., 3,000 My) and estimated mean (i.e., 4,084 My) times, respectively.

least-related sequences will often appear to have diverged more, leading to inflated divergence times. This methodological bias becomes enhanced as a consequence of the multiplicative scale of overestimates.

With real-world sequences, overestimates of divergence times are expected to be larger than suggested by the simulations, particularly, because relative rate tests (see the following) used to identify and exclude sequences that violate the rate-constancy assumption have limited statistical power. Relative rate tests typically neglect levels of rate variation between lineages where the rate of one lineage is as much as four times the rate of the other. In addition, the power of relative rate tests decreases with the length of the sequences and the number of variable sites, which are precisely the conditions where sampling error differences between calibration and extrapolation rates become more pronounced.

ADDITIONAL METHODOLOGICAL BIASES

Three additional pervasive biases that can make molecular dates too old are the following:

1. If calibration (reference) dates are too old, then all other dates estimated from them will also be too old. More generally, biased calibration rates will impact all estimates dependent on the calibration, yielding estimates that are proportionally older or younger than would be correct, depending, respectively, on whether the calibration is older or younger than the correct date. Molecular clock estimates often involve circularity, in that the calibration date has been obtained from molecular data as well. The effect is a compounding of ever-greater errors. Some commonly used calibration dates (see references in Benton and Ayala 2003) are incompatible: the nematode-chordate date (1,177 My) cannot be older than the fungal-metazoan date (1,100 My), since the first branching point is higher (more recent) in the tree than the second. The choice of maximal dates such as these promulgates maximal estimates, all of which become too old. Circularity makes statistical estimates of the robustness of a tree misleading, since the statistical error estimates depend on the calibration time being correct.

2. Undetected fast-evolving genes can bias estimates of timing. Some statistical analyses of molecular data for vertebrates and other animals claim that such nonclocklike genes may be detected (Kumar and Hedges 1998). But, in addition to the problem of circularity pointed out earlier, statistical tests commonly used to exclude molecular sequences that are susceptible to variation in rates of evolution have unacceptably low power and can produce consistent overestimation of dates of divergence (Ayala et al. 1998; Soltis et al. 2002; see the following). These tests generally cannot reliably reject short molecular sequences that show higher-than-normal rates of evolution, and hence the calculated time since divergence is higher than it should be. One way to avoid this particular problem is using longer concatenated sequences and appropriate correction factors (Benton and Ayala 2003).

3. Polymorphism is a third source of bias that inflates molecular dates. Two species often become fixed for alternative alleles (DNA sequences) that existed as a polymorphism in their ancestral species. If so, the divergence time estimated from the DNA sequences would time the origin of the polymorphism, which predates the divergence of the species. In cases of bal-

anced polymorphisms, millions of years could be at stake. The HLA/MHC genes are an extreme case of this, where polymorphisms may be millions of years older than related speciation events (Ayala 1995).

FIXING THE CLOCK

The neutrality theory of molecular evolution asserts that most amino acid substitutions in a protein (or nucleotides in a gene) are neutral—that is, functionally equivalent and, thus, not subject to the vagaries of natural selection. This is the conceptual core of the neutral theory of molecular evolution, which in turn is the theoretical foundation for the construct known as the *molecular clock* of evolution. If molecular evolution is neutral with respect to adaptation, the rate of evolution is expected to occur with a constant probability, because the rate of amino acid or nucleotide replacement along evolving lineages would be determined by mutation rate and time elapsed, rather than by natural selection. Natural selection is rather fickle, subject to the vagaries of environmental change and organism interactions, whereas mutation rate for a given gene is likely to remain constant through time and across lineages. The number of amino acid replacements (or nucleotide substitutions) between species would, then, reflect the time elapsed since their last common ancestor. The time of remote events, as well as the degree of relationship among contemporary lineages, could be thus determined on the basis of amino acid (or nucleotide) differences (Zuckerkandl and Pauling 1965).

Early investigations showed that the evolution of the globins in vertebrates conformed fairly well to the clock hypothesis, which allowed reconstructing, for example, the history of globin gene duplications (Zuckerkandl and Pauling 1965). Fitch and Margoliash (1967) would soon provide a genetic distance method that was effectively used for reconstruction of the history of twenty organisms, from yeast to moth to human, based on the amino acid sequence of a small protein, cytochrome-c. A mathematico-theoretical foundation for the clock was provided by Kimura (1968), who developed a "neutral theory of molecular evolution," which was formulated with great mathematical simplicity. Notably, the theory states that the rate of substitution of adaptively equivalent (neutral) alleles, k, is precisely the rate of mutation, u, of neutral alleles, $k = u$. The neutrality theory predicts that molecular evolution behaves like a stochastic clock, such as radioactive decay, as stated earlier, with the properties of a Poisson distribution, in which the mean, M, and variance, V, are expected to be identical, so that $V/M = 1$. The index of dispersion, measuring the deviation of this ratio from the expected value of 1, is a way to test whether observations fit the theory.

As pointed out earlier, experimental data have shown that often the rate of molecular evolution is "overdispersed," that is, that the index of dispersion is often significantly greater than 1, as expected. Deviations from rate constancy occur between lineages, say between rodents and mammals, as well as at different times along a given lineage, both factors having significant effects. Consequently, several modifications of the neutral theory have been proposed, seeking to account for the excess variance of the molecular clock.

Four subsidiary hypotheses that have been proposed to fix the clock are: (1) that most protein evolution involves *slightly deleterious* replacements rather than strictly neutral ones; (2) that certain *biological properties*, such as the effectiveness of the error-correcting polymerases, vary among organisms; (3) the *population size* hypothesis, which proposes that organisms with larger effective population size have a slower rate of evolution than organisms with smaller population size, because the time required to fix new mutations increases with population size; (4) the *generation-time* hypothesis. Protein evolution has been extensively investigated in primates and rodents, with the common observation that the number of amino acid replacements is greater in rodents. In plants, the overall rate at the *rbc*L locus is more than five times greater in annual grasses than in palms, which have much longer generations (Gaut et al. 1992). These rate differences could be accounted for, according to the generation-time hypothesis, by assuming that the time rate of evolution depends on the number of germ-line replications per year, which is several times greater for the short-generation rodents and grasses than for the long-generation primates and palms. The rationale of the assumption is that the larger the number of replication cycles, the greater the number of mutational errors that will occur.

From a theoretical, as well as operational, perspective, these and other supplementary hypotheses have the discomforting consequence that they invoke additional empirical parameters, often not easy to estimate. It is of great epistemological significance that the original proposal of the neutral theory was highly predictive ($k = u$, and $V/M = 1$) and, therefore, eminently testable. The supplementary hypotheses lead, nevertheless, to certain predictions that can be tested. The generation-time, population size, and biological properties hypotheses uniformly predict that rate variations observed between lineages or at different times will equally affect (in direction and magnitude) all genes of any particular organism, since these attributes are common to all genes of the same species. The "slightly deleterious" hypothesis predicts that the rate of evolution will be inversely related to population size, and thus reduces to the population size hypothesis.

In the following section I will summarize an extensive investigation undertaken as a test of these four supplementary hypotheses, as well as of the more general, or null hypothesis underlying the molecular clock hypothesis (Rodriguez-Trelles, Tarrio, and Ayala 2004, and references therein). The conclusion reached is that inferences about the timing of past events (and about phylogenetic relationships among species) based on molecular evolution are subject to sources of error not altogether disparate from inferences based on anatomy, embryology, or other phenotypic characteristics.

EXPERIMENTAL TESTS OF THE MOLECULAR CLOCK

I will now summarize a set of experiments designed to test the molecular clock and to assess whether the subsidiary hypotheses proposed to fix the clock, can indeed fix it, even if at the expense of additional parameters. The experiments that I want to review have been summarized by Rodriguez-Trelles, Tarrio, and Ayala (2004). The experiments encompass nine protein-coding genes (*Adh, Amd, Ddc, Gpdh, G6pd, Pgd, Sod, Tpi,* and *Xdh*) and ninety-three species (Rodriguez-Trelles, Tarrio, and Ayala 2004). Comparisons are made at multiple levels of phylogenetic differentiation. This becomes possible in these studies because the organisms represented include all three multicellular kingdoms (plants, fungi, and animals) and within the animals, four phyla, different classes, orders, families, and genera, down to twenty-nine different species of the broad genus *Drosophila*, some very closely related, others belonging to different subgenera. The particulars can be found in the reference cited (Rodriguez-Trelles, Tarrio, and Ayala 2004), as well as in the references given therein.

Table 9.1 lists (left column) the nine levels of increasing phylogenetic divergence examined in the tests. The times of divergence given (in million years) with standard errors are, approximately, commonly accepted ones. As we shall soon see, changing any, several, or all of these times will not alter the relevant conclusions derived from the analysis. The table gives the estimated rates of evolution for each level of taxonomic divergence in each of six proteins (out of the nine proteins studied) for which the available data are most extensive, as well as the average rate when all six proteins are jointly considered. Figure 9.5 displays the average rate of evolution for species from three different groups: the genus *Drosophila*, the order Mammalia, and the kingdom Fungi. (The graph

TABLE 9.1 Normalized rates of evolution of GPDH, G6PD, PGD, SOD, TPI, and XDH for increasingly remote lineages.

Comparison	My	Amino acid replacements per 100 My						
		GPDH	G6PD	PGD	SOD	TPI	XDH	Average
1. Within *Drosophila* groups	25–30	0.0–1.9	44.1	19.7	4.8–46.0	0–8.8	20.3–36.7	25.0
2. Between *Drosophila* groups	55 ± 10	1.5	—	—	25.7	26.1	30.4	25.9
3. Between *Drosophila* subgenera	60 ± 10	2.0	44.0	20.1	30.7	29.6	29.2	28.9
4. Between drosophilid genera	65 ± 10	4.4	—	—	34.9	—	31.7	27.1
5. Between mammalian orders	70 ± 10	11.6	8.5	12.4	46.0	6.8	17.1	15.2
6. Between dipteran families	100 ± 20	9.3	21.2	16.1	33.7	43.5	28.9	23.6
7. Between Fungi	300 ± 50	40.0	35.3	13.8	24.9	40.5	13.7	24.8
8. Between animal phyla	600 ± 100	13.2	13.4	9.7	19.2	17.8	19.2	15.7
9. Between kingdoms	1100 ± 200	13.0	11.7	11.7	12.6	19.9	11.5	12.3

Note: The ninety-three species compared are listed in figure 1.1 of Rodriguez-Trelles, Tarrio, and Ayala (2004). The plus/minus values are crude estimates of error for My. Rate values are expressed in units of 10^{-10} substitutions per site per year. Averages across loci are obtained by weighing the rate of each gene by the length of its sequence; i.e., 0.15, 0.23, 0.13, 0.07, 0.05, and 0.37, corresponding to 241, 367, 208, 107, 78, and 599 residues of GPDH, G6PDH, 6PGDH, SOD, TPI, and XDH, respectively.

for the *Sod* gene includes one additional rate, namely for comparisons within each of two species groups, *Drosophila melanogaster* and *D. obscura*, which have a rate of evolution much lower than the rate observed when other *Drosophila* species are compared: 4.8 versus 46.0).

The results of the studies summarized in table 9.1 and figure 9.5 make it eminently clear that the rate of evolution is not constant for any of the six genes or even for the average (see table 9.1) of all six genes. The large heterogeneity of rates has been discovered making comparisons at different taxonomic levels, which is a way to disentangle the variation of rates in any given lineage as time proceeds along the lineage. Notice in this respect that there is no consistent pattern as the time span changes. Thus, for example, the rate of *Gpdh* is 9.3 for comparisons between dipteran families, which diverged one hundred My ago, jumps to 40.0 for comparisons between fungi species diverged 300

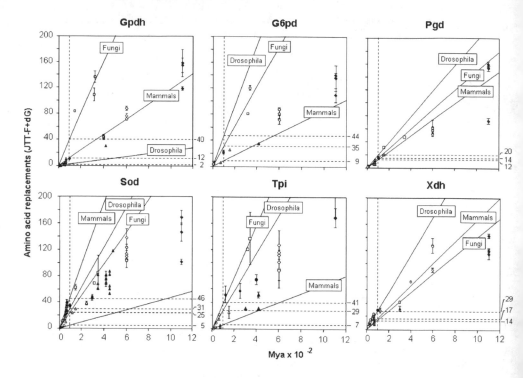

Figure 9.5 Rates of amino acid replacement in six genes. The time unit (abscissa) is 100 million years. The rates on the right are for replacements × 10^{-10} per site per year. These rates correspond to comparisons between *Drosophila* subgenera, mammal orders, or fungi (rows 3, 5, and 7 in table 9.1). The comparisons between *Drosophila* subgenera and mammal orders correspond to roughly contemporary time lapses (60 My and 70 My, respectively). Other points in the figure are for other comparisons, such as between kingdoms (1,100 My) or animal phyla (600 My; see also table 9.1). For SOD the rate of 5 on the right is for comparisons between species within the *Drosophila melanogaster* and *D. obscura* groups.

My ago, but decreases to 13.2 or 13.0 for comparisons between animal phyla diverged 600 My ago or between kingdoms (animals versus plants versus fungi) diverged 1,100 My ago.

Notice also that rates of evolution may vary considerably for a given gene within the same span of astronomic time. Thus, the *Gpdh* rate is 2.0 or 4.4 for comparisons between *Drosophila* subgenera or between *Drosophila* and closely related genera, such as *Chymomyza*, which diverged sixty to sixty-five My ago; but it is 11.6, three or more times faster, for comparisons between mammalian genera diverged about seventy My ago. Figure 9.5 makes this lack of pattern in the change of rates clear. The rates are displayed for two comparisons encompassing about the same spans of astronomic time: between *Drosophila* subgenera and between mammalian orders (rows 3 and 5 in table 9.1), as well as between fungi (row 7). In the case of *Gpdh*, the rate of evolution is much greater for mammals than for *Drosophila*, but for *G6pd* the opposite is the case.

The heterogeneity of evolutionary rates as displayed in table 9.1 and figure 9.5 hides a much greater heterogeneity that becomes apparent when different branches of the tree encompassing the species are analyzed separately. Thus the *Gpdh* rate of 4.4×10^{-10} for comparisons between *Chymomyza* and *Drosophila* (row 4 in table 9.1) needs to be decomposed into a rate of $\leq 2.0 \times 10^{-10}$ for most of the overlapping period of *Drosophila* and *Chymomyza* evolution and a rate $> 10 \times 10^{-10}$ for the twenty-five million years after their divergence (Ayala 1997).

The same lack of pattern obtains for other genes. Consider, for example, *Tpi*. The average rate within *Drosophila* groups is 4.4×10^{-10}/site/year. This rate becomes six times greater between *Drosophila* groups (26.1×10^{-10}/site/year), seven times greater between *Drosophila* subgenera (29.6×10^{-10}), and ten times greater between dipteran families (43.5×10^{-10}). Notice also the mammalian rate of 6.8×10^{-10}, which is several times slower than the rate of the contemporaneously evolving *Drosophila* subgenera (29.6×10^{-10}).

The enormous range of rate variation observed in these studies does not depend on the times of divergence assumed (see My column in table 9.1). Consider the two bottom rows, where the comparisons made are between some animal phyla (row 8) or between different multicellular kingdoms (row 9). The rates are fairly similar for four genes (*Gpdh, G6pd, Pgd,* and *Tpi*). For two other genes (*Sod* and Xdh) and for the average, however, the rate is considerably faster for comparisons between phyla. If the assumed dates were changed so that these three estimates would be more nearly similar, the other rates would become proportionally more disparate. This erratic behavior of rates from one to another gene is even larger for comparisons at other taxonomic levels.

Do we know what causes the disparate rates among different taxonomic levels and their erratic behavior across genes? The answer is "no" in many cases, but there are reasonable explanations in some cases. In other cases, the disparities may inspire hypotheses and additional research. Consider GPDH, the nicotinamide-adenine dinucleotide (NAD)-dependent cytoplasmic glycerol-3-phosphate dehydrogenase. GPDH plays a crucial role in insect flight metabolism because of its keystone position in the glycerophosphate cycle, which provides energy for flight in the thoracic muscles of *Drosophila*. In *Drosophila melanogaster* the *Gpdh* gene is located on chromosome 2 and consists of eight coding exons. It produces three isozymes by differential splicing of the last three exons (Cook, Bewley, and Schaffer 1988). The GPDH polypeptide can be divided into two main domains: the NAD-binding domain and the catalytic domain. The NAD-binding domain is typically more highly conserved than the catalytic domain. The essential role of GPDH in generating energy for the *Drosophila* flight muscle makes it subject to strong purifying selection, so that few or no amino acid replacements are tolerated.

The gene loci just reviewed serve well for the purpose of illustrating the biases that may be associated with the choice of reference for calibrating the clock. Table 9.2 is based on the data in table 9.1, but using two different calibrations. The rates have been normalized by reference (1) to the *Drosophila* subgenera rate and (2) to mammal orders. The molecular clock dates that are obtained for the various phylogenetic events are nearly twice as old when the mammal rate is used as reference rate.

ORIGIN OF METAZOAN PHYLA: MOLECULAR CLOCK VERSUS PALEONTOLOGY?

The time of origin of the metazoan phyla has been a persistent subject of controversy. A common view is that the first coelomates appeared in the late Neoproterozoic, some 700 My ago, and the divergence between protostomes (arthropods, annelids, mollusks, and other invertebrates) and deuterostomes (echinoderms and chordates) occurred somewhat more than 600 My ago. The divergence between the deuterostome phyla, echinoderms and chordates, is generally thought to have occurred during the Vendian, somewhat before the beginning of the Cambrian 544 My ago. Fossil remains of nearly all readily fossilizable animal phyla have been recovered from Cambrian rocks.

The previous interpretations have been challenged on the grounds that they rely on negative evidence, namely the scarcity of fossil remains preceding the Cambrian, followed by the relatively sudden appearance during the Cambrian of many diverse phyla, classes, and orders. This Cambrian explosion might simply reflect the difficulty of preservation and discovery of soft-bodied and perhaps tiny animals.

Resolution of the controversy concerning metazoan origins has been sought in DNA and protein sequence data and the theory of the molecular clock. One investigation that sought to settle the issue was conducted by Wray, Levinton,

TABLE 9.2 Normalized rates of evolution across the loci, normalized to the rates of *Drosophila* subgenera and mammal orders, and corresponding estimates of divergence times.

Comparison	Normalized rates		Clock estimates (My)	
	(1)	(2)	(1)	(2)
1. *Drosophila* subgenera	1.0	1.9	60	111
2. Drosophilid genera	0.9	1.6	60	110
3. Mammal orders	0.5	1.0	38	70
4. Dipteran families	0.8	1.6	179	330
5. Fungi	0.9	1.6	276	509
6. Animal phyla	0.5	1.0	337	621
7. Kingdoms	0.4	0.8	485	893

Note: The normalized rates are derived from the averages in table 9.1: (1) normalized to the *Drosophila*-subgenera average; (2) normalized to the mammal-orders average. The clock estimates assume that the divergence times are sixty My for the *Drosophila* subgenera and seventy My for the mammal orders.

and Shapiro (1996), who concluded from a seven-gene analysis that the divergence of protostomes and deuterostomes occurred more than twice as early as the Cambrian, that is, about 1,200 My ago, and that chordates diverged from the echinoderms about 1,000 My ago, several hundred million years earlier than common paleontological estimates. The number of taxa analyzed ranged from twenty-five to eighty-five, depending on the gene, but in all cases included several representatives of at least five different animal phyla.

A critical point in the argument of Wray, Levinton, and Shapiro (1996) was the demonstration that the rates of evolution were constant across lineages for the genes in their investigation. This they sought to achieve by means of the relative-rate test, in which the taxa under investigation are compared to a taxon outside their phylogeny. Thus, the independent comparison of all the animal taxa to a plant (*Arabidopsis thaliana*) gave an average genetic distance of 1.40 ± 0.62, which the authors considered reflected rates that were fairly constant across animal taxa, given the relatively small standard error. Indeed, when *Arabidopsis* was compared to only the vertebrate taxa, the average genetic distance was 1.44 ± 0.66; when compared to the invertebrate taxa, the average genetic distance was 1.22 ± 0.44. These two estimates are not statistically significantly different.

Wray, Levinton, and Shapiro (1996) sought to confirm rate constancy by another set of relative-rate tests; in this case, comparison of the animal taxa to the bacterium *Rhodobacter capsulatus*. The average genetic distance between the bacterium and all animal taxa was 1.51 ± 0.50. The average genetic distance between the bacterium and the vertebrate taxa was 1.49 ± 0.50 and to the invertebrate taxa 1.53 ± 0.50, statistically not different. Thus, they affirmed that the molecular rates were reasonably constant and proceeded with their analysis, reaching the conclusion that the time of origin of the animal phyla was several hundred million years older than commonly accepted by paleontologists.

But Wray, Levinton, and Shapiro (1996) failed to notice a remarkable feature of the results I have just summarized, which demonstrate that the rates of evolution in their study were extraordinarily heterogeneous, precisely the opposite of what they thought to have confirmed. This is simply shown by pointing out that the bacterium is a prokaryote, while the plants and the animals are multicellular eukaryotes. The divergence between prokaryotes and eukaryotes occurred many hundred million years, perhaps as much as two billion years, before the divergence between plants and animals. Yet the genetic distance between the prokaryote and the animals was about the same as between the plant and the animals (1.51 ± 0.50 and 1.40 ± 0.62, respectively). This, at face value, would indicate that the genes they were examining have *not evolved at all* during the two billion years or so elapsed between the divergence of eukaryotes from prokaryotes and the divergence between plants and animals, since the two distances, from animals to plants and from animals to bacteria, are nearly identical. Evidently the rates of evolution have not been nearly constant across lineages and through time.

Ayala, Rzhetsky, and Ayala (1998) reexamined the data of Wray, Levinton, and Shapiro (1996) by using rate-constancy tests, such as the branch-length test, which are more sensitive to rate variation. The rate of evolution was significantly heterogeneous across lineages for every one of the genes in Wray, Levinton, and Shapiro's (1996) investigation. Indeed, the rates were disparate, enormously heterogeneous. Further, Ayala, Rzhetsky, and Ayala (1998) analyzed eighteen protein-encoding genes that included those in the study of Wray, Levinton, and Shapiro (1996). Their methods included a statistical process that eliminates from consideration, for each gene, all branches in the tree that evolved significantly faster or slower than the average. Their molecular clock estimates are 670 My for the divergence between protostomes (arthropods, annelids, and mollusks) and deuterostomes (echinoderms and chordates); and 600 My for the divergence between echinoderms and chordates (Ayala, Rzhetsky, and Ayala 1998). These estimates are consistent with prevailing paleontological estimates.

CODA

There can be little doubt, in my view, that there is no molecular clock in the general sense that we can assume that any given gene evolves at a nearly constant rate over time and across lineages. But I will argue that my statement does not imply that the timing of evolutionary events cannot be determined using molecular data. It rather means that caution should be used in assuming that molecular evolution is proceeding in any particular case in a clocklike manner. Severe tests of constancy should be used, as well as statistical procedures to discard outliers (branches evolving significantly faster or slower than the average).

As I pointed out earlier, molecular investigations have three obvious advantages, in degree if not completely in kind, over phenotypic traits and paleontological data: namely, easy quantification, breadth of comparisons, and multiplicity. Every one of the thousands of genes in the makeup of each organism provides information about the evolutionary history of any taxon, and differences can be precisely quantified, measured as they are in terms of distinct units, such as amino acids or nucleotides.

There are many evolutionary issues concerning both timing and phylogenetic relationships between species for which molecular sequence data provide the best, if not the only, dependable evidence. The large-scale reconstruction of the universal tree of life is a case in point: the phylogenetic relationships among archaean and bacterial prokaryotes and between them and the eukaryotes have best been determined with DNA sequences encoding ribosomal RNA genes. The multiplicity of genes opens up the possibility of combining data for numerous genes in assessing the time of particular evolutionary events, or the phylogeny of species. Because of the time dependence of the evolutionary process, the multiplicity of independent results is expected to tend to converge (by the so-called "law of large numbers") on average values reflecting with reasonable accuracy the time elapsed since the divergence of taxa.

REFERENCES

Ayala, F. J. 1995. The myth of Eve: Molecular biology and human origins. *Science* 270:1930–36.
———. 1997. Vagaries of the molecular clock. *Proceedings of the National Academy of Sciences, U.S.* 94:7776–83.
Ayala, F. J., A. Rzhetsky, and F. J. Ayala. 1998. Origin of the metazoan phyla: Molecu-

lar clocks confirm paleontological estimates, *Proceedings of the National Academy of Sciences, U.S.* 95:606–11.

Benton, M. J. 1990. Phylogeny of the major tetrapod groups: Morphological data and divergence dates. *Journal of Molecular Evolution* 30:409–24.

Benton, M. J., and F. J. Ayala. 2003. Dating the tree of life. *Science* 300:1698–1700.

Cook, J. L., G. C. Bewley, and J. B. Schaffer. 1988. *Drosophila sn*-glycerol-3-phosphate dehydrogenase isozymes are generated by alternate pathways of RNA processing resulting in different carboxyl-terminal amino acid sequences. *Journal of Biological Chemistry* 263:19858–10864.

Fitch, W. M., and E. Margoliash. 1967. Construction of phylogenetic trees. *Science* 155:279–84.

Gaut, B. S., S. V. Muse, W. F. Clark, and M. T. Clegg. 1992. Relative rates of nucleotide substitution at the *rbc*L locus of monocotyledonous plants. *Journal of Molecular Evolution* 35:292–303.

Kimura, M. 1968. Evolutionary rate at the molecular level. *Nature* (London) 217:624–26.

Kumar, S., and S. B. Hedges. 1998. A molecular timescale for vertebrate evolution. *Nature* 392:917–20.

M. Nei, P. Xu and G. Glazko. 2001. Estimation of divergence times from multiprotein sequences for a few mammalian species and several distantly related organisms, *Proceedings of the National Academy of Sciences, U.S.* 98: 2497–2502.

Rodriguez-Trelles, F., R. Tarrio, and F. J. Ayala. 2002. A methodological bias toward overestimation of molecular evolutionary time scales. *Proceedings of the National Academy of Sciences, U.S.* 99:8112–15.

———. 2004. Molecular clocks: Whence and whither? In *Telling the evolutionary time: Molecular clocks and the fossil record*, ed. P. C. J. Donoghue and M. P. Smith, 5–26. London: CRC Press.

Soltis, P. S., D. E. Soltis, V. Savolainen, P. R. Crane, and T. G. Barraclough. 2002. Rate heterogeneity among lineages of tracheophytes: Integration of molecular and fossil data and evidence for molecular living fossils. *Proceedings of the National Academy of Sciences, U.S.* 99:4430–35.

Wray, G. A., J. S. Levinton, and L. H. Shapiro. 1996. Molecular evidence for deep precambrian divergences among Metazoan phyla. *Science* 274:568–73.

Zuckerkandl, E., and L. Pauling. 1965. Evolutionary divergence and convergence in proteins. In *Evolving genes and proteins*, ed. V. Bryson and H. J. Vogel, 97–166. New York: Academic Press.

* II *

The Historical and Conceptual Significance of Recent Paleontology

CHAPTER TEN

Beyond Detective Work: Empirical Testing in Paleontology

Derek Turner

EMPIRICAL TESTING WITHOUT EXPERIMENT?

Like all scientists, paleontologists need to subject their hypotheses and theories to rigorous empirical tests. In many areas of science, empirical testing involves experimentation: researchers carry out a series of trials while varying the initial conditions a little bit each time in order to see how those changes affect the outcomes. For example, Charles Darwin became interested in the question of whether seeds could germinate after prolonged exposure to saltwater. He wanted to understand more about seed dispersal: how could a plant species that started out on the mainland end up flourishing on an island hundreds of miles away? To test the hypothesis that seeds could have survived transportation by ocean currents, he placed large numbers of seeds from many different plant species in small bottles of saltwater for varying periods of time (two weeks, three weeks, etc.) Then he planted the seeds to see which ones sprouted. Darwin knew that if none of the seeds sprouted, he would need to find some other explanation of seed dispersal. This type of biological experimentation—and indeed much of the research that biologists do today—involves manipulation and control. Although it may sound so obvious as to be hardly worth pointing out, *no one can manipulate or experiment with the past*. This simple fact creates a major methodological problem for paleontologists: how can they test hypotheses about prehistoric life without being able to experiment on the things they wish to study?

Contrary to the impression that one might get from textbook presentations of the scientific method, it really is possible to carry out rigorous empirical tests without actually performing any experimental manipulations on the objects of interest. In what follows, I describe two techniques that paleobiologists have come to rely on over the last few decades. The first technique, which Michael Ruse (1999) has called *crunching the fossils,* involves the compilation of huge amounts of data from the fossil record in order to test hypotheses about large-scale evolutionary processes. A second involves *virtual experiments:* in lieu of actual manipulation of the past, the next best thing is to manipulate a computer model of some past entity or process. The development of these techniques for carrying out nonexperimental empirical tests is a major—and largely unsung—scientific achievement.

Within paleontology, it is possible to distinguish at least two very different kinds of research. One kind of research aims to reconstruct organisms from fossil remains, and to answer such questions as: what did they eat, and what ate them? Were they solitary or sociable? In what sort of environment did they live? How did they die? How did they reproduce? A second kind of research, which is more closely associated with the subfield of paleobiology that emerged during the 1970s, aims to reconstruct large-scale evolutionary processes from the fossil record. What can the fossil record tell us about evolution, about speciation and extinction, and about evolutionary trends? One plausible suggestion is that paleontologists earned their place at the high table of evolutionary theory in part by shifting the focus of their work from organismal reconstruction to the reconstruction of evolutionary processes. I will argue, however, that the really crucial development has been the introduction of rigorous methods of nonexperimental testing into both of these branches of paleontological research. I will also show, by examining two case studies having to do with the evolution of body size, that the two distinct kinds of research—reconstruction of prehistoric organisms versus reconstruction of past evolutionary processes—can link up in surprising and fruitful ways.

COPE'S RULE

Much work in paleobiology involves testing ideas about large-scale evolutionary trends, and one example of a possible evolutionary trend is *size increase*. According to "Cope's rule," named in honor of the nineteenth-century American paleontologist, E. D. Cope, "the size of individuals tends to increase in most evolutionary groups" (McShea 1998, 306). Cope himself, though he did make some tantalizing claims about body size and evolution, never quite asserted the rule that has come to bear his name (Polly 1998). He came close, though. In a discussion of "the phylogeny of the horse," Cope did remark that "the species have all the while been growing gradually larger" (1896, 148). Indeed, for a long time, horses served as the classic illustration of Cope's rule. Existing horses and zebras evolved from *Hyracotherium,* a small animal that was only about fifteen to twenty inches high at the shoulder, and that lived during the Eocene, about fifty-five to fifty-seven million years ago. Horses really do seem to have gotten bigger as they have evolved.

Today, Cope's rule has some ardent and able defenders within the pale- ontological community, such as John Alroy (1998). Yet there are also some well-documented exceptions to it. The story of the evolution of horses turns out to be rather complicated, with animals in some lineages getting bigger and bigger over time, but animals in a few (now extinct) lineages getting smaller (McFadden 1986; Gould and McFadden 2004). The horse genus *Nannipus,* which went extinct in North America several million years ago, was much smaller than today's horses, and evolved from earlier horses that were much larger than it was. Exceptions like this help explain why no scientists these days are willing to talk anymore about Cope's "law" (which is what it used to be called). A law of nature is supposed to be a generalization that holds true at all times and places, without any exceptions.

Still, even if we demote the idea from a law to a rule, why would anyone persist in defending Cope's rule in the face of such clear counterexamples? The answer to this question is that Alroy and others have put Cope's rule to the test by *crunching the fossils*—that is, by searching for interesting patterns in huge collections of fossils. For some time, paleobiologists have relied on large databases to facilitate this work.

In his 1998 study, for instance, Alroy looked at 1,534 different species of mammals that have lived in North America during the Cenozoic. He then asked a very straightforward question: if we examine *pairs* of species that belong to the same genus, one of which shows up in the fossil record earlier than the other, will the organisms in the newer species be bigger than those in the older one? He found that on average, the newer species are about 9.1% larger than the older species in the same genus. This is perfectly compatible with there being some cases, like that of *Nannipus,* in which the older species are actually bigger. But Alroy takes his results to provide a strong confirmation of "the most narrow and deterministic interpretation of Cope's rule; namely, that there are directional trends within lineages" (1998, 732).

This last reference to trends within lineages is important. In this study, Alroy was not focusing on whether the average size of mammals has increased during the Cenozoic. An increase in the average size of mammals over time would be a trend within the clade. Whether the average size of mammals has increased could be affected by the differential extinction and persistence of lineages. For example, if (for whatever reason) lineages with smaller body sizes are more likely to go extinct, that could contribute to an increase in the average size of organisms within the entire clade—that is, to an increase in the average size of mammals. But Alroy designed his test to determine whether size increase has occurred within mammalian lineages rather than among lineages.

In recent years, a number of other scientists have carried out fossil-crunching tests of Cope's rule in different groups of animals. In one of the most important such studies, David Jablonski (1996; 1997) found that the rule does not hold up very well for marine mollusks that lived during the Cretaceous period. Other recent fossil-crunching studies have focused on dinosaurs (Hone et al. 2005), the earliest reptiles (Laurin 2004), deep-sea ostracodes (Hunt and Roy 2006), as well as pterosaurs (Hone and Benton 2007). It is not too much of a stretch to see this as a larger research program using fossil-crunching techniques to determine where Cope's rule applies and where it does not. Although this type of quantitative work is typical of the new paleobiology, fossil-crunching tests are not the only way to approach questions about evolution and body size.

VIRTUAL EXPERIMENTS

In his entertaining and influential book, *The Dinosaur Heresies* (1986), Robert Bakker sought to overthrow the traditional view of dinosaurs as cold-blooded, dim-witted, slow-moving giants. He deployed a series of biomechanical arguments in defense of the view that *Tyrannosaurus rex* was an agile predator, capable of pouncing, lunging, and running down prey at speeds over forty miles per hour. Bakker's view caught on, and in the movie *Jurassic Park,* a tyrannosaur runs fast enough to keep up with a speeding jeep. How could scientists possibly test the hypothesis that *T. rex* could run that fast? After all, no one can observe a living tyrannosaur in action or experiment with a creature that has been extinct for sixty-five million years. Was the portrayal of *T. rex* in *Jurassic Park* based on solid, well-tested science, or merely on educated (but untested) guesswork?

Bakker's biomechanical arguments involved adaptationist reasoning: he argued that tyrannosaurs have at least three distinct adaptations for fast running: first, unlike earlier predatory dinosaurs, tyrannosaurs had relatively small claws on their hind feet, which "were thus adapted for running and dodging, avoiding counterattacks from the spikes, tail clubs, and horns of their prey" (1986, 272). Second, *Tyrannosaurus* had an "absolutely huge" cnemial crest—a bony projection at the top of the shin that served as an attachment point for muscles used in running. Living animals with big cnemial crests also tend to be fast runners. Third, tyrannosaur leg bones had thick shafts relative to their length, which Bakker took to mean that they could have withstood the tremendous stresses associated with fast running, without fear of fracture. Bakker concluded that "at full speed, a bull *Tyrannosaurus* could easily have overhauled a galloping white rhino—at speeds above forty miles per hour, for sure" (1986, 218). Although he did not put it in quite this way, it is fair to interpret Bakker as arguing that the hypothesis that *T. rex* could run forty mph affords the best adaptation explanation of some otherwise puzzling features of *T. rex* hindlimb morphology (for a more detailed account of this type of reasoning, see Turner 2000). The challenge was figuring out how to subject this idea to some sort of predictive test.

Bakker knew that one way to test his hypothesis would be to look for tyrannosaur tracks. There is a well-established mathematical technique for estimating the speed of an animal based on its stride length and the height of its hip (Alexander 1976). The basic idea behind this approach is that an animal's *relative stride length* (which is defined as the ratio of stride length to hip height) increases as its speed increases. This is because the stride length increases while the hip height remains constant. Thus, given the stride length and the hip height, it is possible to come up with an estimate of speed.

In developing this technique, Alexander (1976) borrowed a concept from shipbuilding known as the *Froude number*. The Froude number (say, of a boat) is u^2/gl, where u is the velocity, g represents the gravitational force, and l is the length of the hull. Arguing that the Froude number "applies to any situation where inertia and gravity interact," he substituted hip height for hull length, and began calculating the Froude numbers of animals. Empirical studies of living organisms (humans, horses, and jirds—rodents that are closely related to gerbils) then showed that an animal's relative stride length is a function of its Froude number. With a little bit of algebra, Alexander used this finding to demonstrate that speed is a function of hip height and stride length. Bakker was well aware of Alexander's work, and he knew that this technique might provide one way of testing the hypothesis that tyrannosaurs could run forty mph.

There are, however, two problems with this approach. First, no one has ever found any well-preserved trackways that can be assigned to *T. rex* (Farlow et al. 2000). Recently, scientists found an intriguing theropod trackway in Oxfordshire, England. One section of the trackway is wide-gauge, with the feet spread farther apart and the toes pointed inward. That section also has a relatively short stride length. Another section of the same trackway is narrow-gauge, with the feet landing right on the midline and the toes angled slightly outward; in this narrow-gauge section, the stride length is much longer (Day et al. 2002). The trackway seems to show a theropod dinosaur breaking from a walk into a fast run. Using Alexander's technique, Day and colleagues estimated that the animal was running at about twenty-nine kilometers per hour, or about eight meters per second. But this rare trackway does not help Bakker's case much. It was made during the mid-Jurassic period, long before the evolution of tyrannosaurs, and the animal that made it was considerably smaller than *Tyrannosaurus*. Finally, the estimated running speed of eight meters per second is still much slower than Bakker's high-end estimate of forty mph, or about twenty meters per second.

Second, this technique only tells scientists how fast an animal was going when it happened to make a given set of footprints; it cannot reveal the animal's maximum speed. Since animals do not run at maximum speed very often, it is likely that if anyone ever did find a *T. rex* trackway, the footprints would indicate that the animal was moving much more slowly than the maximum speed proposed by Bakker. When *Jurassic Park* was released in theaters, the hypothesis that a tyrannosaur could have kept up with a speeding jeep was simply not well tested.

One general strategy for conducting rigorous empirical tests without experimental manipulations is to take advantage of the fact that the fossil record contains different kinds of evidence. A hypothesis formed with the aim of explaining one kind of evidence (such as skeletal remains) can be tested by making a prediction about some other kind of evidence (such as footprints). The problem is that scientists pursuing this strategy often get stymied by the incompleteness of the fossil record. The failure to find a trackway left by a speeding tyrannosaur does not really disconfirm Bakker's forty-mph estimate, because even if tyrannosaurs could have run that fast, it is not very likely that anyone will find a trackway made by a tyrannosaur that was running forty mph at the time. The prediction, in short, is not a terribly risky one.

Ironically, one early sign of trouble for Bakker's sprinting tyrannosaurs may have come during the making of *Jurassic Park*. It is true that the jeep in the famous chase scene appears to be going very fast, but when John Hutchinson and Mariano Garcia—two scientists who have challenged Bakker's view—looked closely at the tyrannosaur, they noticed that "it always has one foot on the ground, and the cadence of its footfalls is rather slow, less than two steps (one stride) per second" (2002a). From a biomechanical perspective, it would be impossible for the animal to run at forty mph with normal stride lengths while taking only one stride per second. When the scientists contacted the special effects engineers at Industrial Light and Magic, the movie makers admitted that when they had the digitally animated *T. rex* move at forty mph, the scene did not "look right." So they reduced the animal's speed to ten to fifteen mph while using other visual tricks to make it look like the animal was running faster (Hutchinson and Garcia 2002a).

Another early sign of trouble was an argument made by Farlow, Smith, and Robinson (1995), who claimed that if an animal as big and as heavy as *Tyrannosaurus*—weighing in at more than five tons—were to trip and fall while sprinting at forty miles per hour, the impact would probably be lethal.

Twenty years ago, when he wrote *The Dinosaur Heresies,* Bakker probably did not suspect that his speed estimates would ever be tested using computer models. In 2002, Hutchinson and Garcia created a simple two-dimensional biomechanical model of *T. rex*. (Since then, Hutchinson et al. 2005 have developed a fancier three-dimensional model, but I will focus on the simpler one here.) Essentially, they created a virtual tyrannosaur that they could then experiment with on a computer.

Hutchinson and Garcia's (2002b) computer model of *Tyrannosaurus* is a good example of a *numerical model*—a mathematical representation of some real system or process. Their virtual tyrannosaur is little more than a stick figure that represents a snapshot of a running tyrannosaur. (The virtual tyrannosaur does not run at all.) They tested Bakker's speed estimate by asking how large the extensor muscles in each leg would have to be in order to support the animal during a single instant of fast running. It is probably easiest to think of their model as a system that takes certain inputs and produces a single output. The output in this case is the minimum extensor muscle mass required to support the animal. The inputs included information about the angles of the hip, knee, and toe joints, the lengths and weights of the leg segments, the animal's total mass, and the ground reaction force. They found that if *Tyrannosaurus* were a fast runner, as Bakker speculated, no less than 86% of its total body mass would have to be taken up by its leg extensor muscles. *T. rex* certainly could not have run as fast as Bakker suggested, and quite possibly could not have run at all.

One might reasonably question my description of Hutchinson and Garcia's work as virtual experimentation. It is fair to say that all they did was use a computer to carry out some biomechanical calculations—calculations that could, in principle at least, have been done with pencil and paper. Nevertheless, their work was experimental in the sense that they manipulated the model in systematic ways in order to see what would happen. They performed two distinct kinds of empirical tests: tests *of* the model and tests *with* the model.

First, they applied their model to other organisms—in particular, a chicken, an alligator, and a human—in order to see how realistic it is. This was a test *of* the model. They used it to calculate the minimum extensor muscle mass needed to support each animal during fast running, and compared that to the actual extensor muscle mass. It turns out that humans and chickens have more extensor muscle mass than is required. For example, the model yielded the result that a chicken needed at least 4.7% of its total body mass in the extensor muscles of each leg in order to support itself while running, but in fact 8.8% of a chicken's total body mass is in the muscles of each leg. Both chickens and humans have more muscle mass than is needed. On the other hand, the model yielded the result that alligators would need to have 7.7% of their total body mass taken up by the muscles of each leg in order to run, but in fact only 3.6% of an alligator's body mass is in the muscles of each leg, and this conforms to the observation that alligators do not run.

Hutchinson and Garcia also performed a series of numerical experiments *with* the model to test the sensitivity of their results to other parameters. One interesting finding was that assumptions about posture make a big difference to the model output. When they reconstructed *T. rex* with its legs held straight, in columnar fashion, the minimum extensor muscle mass needed to support the animal during running was much *less* than would have been needed if the animal ran in a crouching pose. They obtained this result simply by varying the inputs to the model and letting the computer solve the biomechanical equations. This result also runs counter to Bakker's view, and to the portrayal of *T. rex* in *Jurassic Park*. Bakker had *T. rex* running fast in a crouching pose.

Although Hutchinson and Garcia (2002b) and Hutchinson et al. (2005) are the first studies involving a computer model of a dinosaur's musculoskeletal system, numerical experiments involving computer models of evolutionary, geological, and climatic processes have for some time been a mainstay of research in paleobiology and earth science. (See Huss 2004 for an illuminating discussion of virtual experiments in paleobiology, as well as Oreskes, Shrader-Frechette, and Belitz 1994 for an assessment of policy-relevant numerical modeling in earth science.) Numerical experiments have enabled scientists to test their ideas even in cases where they cannot manipulate the things they really want to study. Geologists, for example, have used computer models to study the ebb and flow of glaciers, and such models have also been indispensable to scientists who are trying to understand the complexities of global climate change. Hutchinson and Garcia's work also builds on and extends a paleontological tradition of using living organisms as experimental models (Padian and Olsen 1989).

NO SMOKING GUNS

As these two sketches show, scientists really can test predictions without actually experimenting on the things they want to study. But that is not the only reason why these two cases involving virtual experiments and fossil-crunching tests are interesting.

It is natural to suppose that since paleobiologists cannot directly experiment with the past, their historical research will have a lot in common with detective work. An investigator may visit a crime scene in search of a clue—the smoking gun, the testimony of an eyewitness, a DNA sample, anything—that will indicate whether suspect A or suspect B committed the crime. Similarly, one might think that paleontologists begin by formulating different hypotheses about whodunit—say, about what caused the extinction of the dinosaurs. Then they head out into the field in search of some fossil or geological clue that will solve the mystery. At least one philosopher of science has recently argued that this is exactly how historical scientists typically proceed (Cleland 2002).

Carol Cleland argues that just as a detective might try to solve a crime by searching for the smoking gun, historical scientists try to test their hypotheses by searching for a trace or a collection of traces "that unambiguously discriminates one hypothesis from among a set of currently available hypotheses as providing 'the best explanation' of the traces thus far observed" (Cleland 2002, 480–81). Cleland is so persuaded by the analogy with detective work that she makes "smoking gun" into a term of art, defining it as "a trace (or subcollection of traces) that (so-to-speak) cinches the case for a particular causal story" (2002, 482).

Cleland is right that paleontologists often do proceed in this way. Yet her description of "prototypical historical science" as a kind of detective work can also help to show what is so significant about virtual experimentation and fossil crunching as methods of empirical testing. One remarkable feature of the cases I have described here is that there simply are no proverbial smoking guns—no particular clues in the fossil record that answer scientists' questions. In the first case, Cope's rule is not even the kind of scientific hypothesis that could be supported or disconfirmed by any particular clue—or even by any small collection of clues—from the fossil record. In the second case, the problem is that even if *T. rex* could have run forty mph, it is highly unlikely that anyone would ever discover the smoking gun—a set of footprints made by a tyrannosaur that was running that fast. *T. rex* tracks are hopelessly rare, and animals seldom run at their top speeds. Not only do these cases show that paleobiologists have devised ways of testing hypotheses in spite of the obvious fact that they cannot experiment on the past, but they also show that research in paleobiology often goes beyond detective work.

Earlier, I considered the hypothesis that paleobiologists have earned their place at the high table mainly by downplaying organismal reconstruction in favor of the reconstruction of past evolutionary processes. While there may be some truth to this suggestion, it cannot be the whole story. In recent years, both of these types of research have benefited from the new methods of empirical testing. In the next section, I argue that the two distinct styles of paleontological research, as illustrated by these two case studies, can actually complement one another.

HOW ARE THE TWO CASES RELATED?

If we ask *why* the average size in mammals tends to increase within lineages over time, as Alroy found, one plausible answer is that natural selection generally favors bigger organisms. There are a number of different ways in which size might give one an edge in the Darwinian struggle for existence: bigger animals could be better able to intimidate rivals, more desirable as mates, more effective predators, more difficult for predators to kill, better able to regulate their body temperatures, and so on (Hone and Benton 2005). We might call this the " bigger is fitter" explanation of Cope's rule. It turns out, however that the bigger is fitter explanation is rather difficult to test. (For one interesting attempt to test it by looking at living organisms, see Kingsolver and Pfennig 2004.)

In an important and widely cited paper published in 1973, Steven Stanley argued that there is another equally good explanation of Cope's Rule. He contrasts the following two ways of thinking about natural selection and body size:

1. Traditional "bigger is fitter" view:
   Smaller size $\rightarrow$ Larger size
2. Alternative "bigger is not necessarily fitter" view:
   Starting size $\rightarrow$ Optimum size

Rather than supposing that natural selection always favors larger size, Stanley invites us to suppose that selection will drive a lineage from its starting size (whatever that happens to be) in the direction of the optimum size for whatever ecological niche it happens to occupy.

Stanley then borrows an idea from Cope himself, which is that smaller species are likelier to survive mass extinction events. Cope wrote:

> Changes of climate and food consequent on disturbances of the earth's crust have rendered existence impossible to many plants and animals [including the dinosaurs]. . . . Such changes have been often especially severe in their effects on large species, which required food in large quantities. The results have been degeneracy or extinction. On the other hand, plants and animals of unspecialized habits [mammals, perhaps?] have survived . . . species of small size would survive a scarcity of food, while large ones would perish (1896, 173–74).

If this reasoning is correct, it would explain why most lineages happen to start out small. The average size of the organisms within each lineage would then increase over time as a result of natural selection—until, that is, the organisms within the lineage reach the optimum size (whatever that happens to be). According to this view, we need not suppose that larger body size necessarily confers any advantage in the struggle for existence. If lineages happened to start out bigger than their optimum size, then natural selection would in that case lead to size *decrease*. Thus, Stanley argues that if we want to explain why Cope's rule holds true, the important thing is to explain why lineages typically start out small. Cope's rule "is more fruitfully viewed as describing evolution from small size rather than toward large size" (1973, 22).

McShea (1994) draws an important distinction between *passive* and *driven* evolutionary trends. A passive trend results from a random walk away from some lower boundary. Computer models have shown that if there is a fixed lower bound with respect to some variable—such as body size—and subsequent increases and decreases with respect to that variable are equally likely, the resulting random walk can give rise to a real trend. A trend is driven when there is a directional bias—for example, when increases in body size are likelier than decreases (see also McShea 2001). Thus, when Stanley says that Cope's rule "is more fruitfully viewed as describing evolution from small size rather than toward large size," it is natural to take him to be saying that Cope's rule is more fruitfully viewed as a passive rather than a driven evolutionary trend. McShea (1994) proposes a number of different empirical tests to determine whether a given trend is passive or driven. One of these is the ancestor-descendant test—roughly, the style of test that Alroy used in the study previously described. (Alroy's was a bit different, because he looked at earlier and later congeneric species, without necessarily knowing that the later species was a descendant of the earlier one.)

Even if Alroy's work shows that size increase in mammals is a driven trend, one could still ask: *what is the best way to describe the underlying cause of the directional bias?* This is where Stanley's argument causes trouble. Is natural selection driving lineages from smaller to larger sizes (the bigger is fitter view)? Or is it driving lineages from their starting sizes—which happen to be small— toward the optimal sizes for the niches they occupy? Even if fossil-crunching tests can tell us whether a trend is passive or driven, it is not clear that they can tell us how best to describe what is driving a given trend, such as size increase. Is there an empirical test that could discriminate between the bigger is fitter view and Stanley's alternative view, which is that bigger is fitter, but only so long as the lineage's starting size is smaller than its optimum size? One problem is that devising an empirical fossil-crunching test that could discriminate between these two rival views would seem to require that scientists somehow identify the optimal sizes for the ecological niches occupied by extinct lineages, and it is by no means clear how to do that, especially since optimal size is a moving target, in the sense that the optimal size is liable to change over the course of evolution. (See, however, Alroy 1998, who tries to address this issue.)

Hutchinson and Garcia's work is highly relevant to the assessment of the bigger is fitter view. Although they focus on *Tyrannosaurus,* their work also explores the biomechanical costs and benefits of large body size in a more general way. Indeed, *Tyrannosaurus,* one of the largest terrestrial predators that has ever lived, is a good test case for the bigger is fitter view. Hutchinson and Garcia have shown that larger body size has clear biomechanical costs in terms of reduced speed and mobility. Of course, these costs may be out-weighed by other fitness benefits, but at a certain point, the biomechanical costs will become so great that further body size increases are no longer worth it—that is, they no longer confer any net fitness advantage.

To illustrate the biomechanical costs of size increase in a colorful way, Hutch-inson and Garcia (2002b) produced a computer model of a *Tyrannosaurus*-sized chicken weighing 6,000 kg. Their model yielded the absurd result that if the giant chicken were to keep pace with a speeding jeep, 99% of its total body mass would have to be taken up by the extensor muscles in each leg. A chicken the size of *Tyrannosaurus* could not even walk. Although this ex-ample is obviously extreme, it would be interesting to perform a more system-atic set of trials along these lines. How much bigger could *T. rex* have gotten before walking became biomechanically impossible?

In this case, two lines of paleontological research—reconstruction of organisms and reconstruction of evolutionary processes—represent two different and complementary approaches to the problem of understanding the evolution of body size. By quantifying the biomechanical costs of larger body size, Hutchinson and Garcia's work poses a serious challenge to the bigger is fitter explanation of Cope's rule. In this case, both lines of research have produced significant results, and both have benefited immensely from the introduction of new, nonexperimental methods of empirical testing. Both go beyond historical detective work.

REFERENCES

Alexander, R. McN. 1976. Estimates of speeds of dinosaurs. *Nature* 261:129–30.
Alroy, J. 1998. Cope's Rule and the dynamics of body mass evolution in North American fossil mammals. *Science* 280(5364): 731–34.
Bakker, R. T. 1986. *The dinosaur heresies.* New York: Kensington.
Cleland, C. E. 2002. Methodological and epistemic differences between historical and experimental science. *Philosophy of Science* 69(3): 474–96.
Cope, E. D. 1896. *The primary factors of organic evolution.* Chicago: Open Court.
Day, J. J., D. B. Norman, P. Upchurch, and H. P. Powell. 2002. Dinosaur locomotion from a new trackway. *Nature* 415:494–95.
Farlow, J. E., S. M. Gatesy, T. R. Holz, Jr., J. R. Hutchinson, and J.M. Robinson. 2000. Theropod locomotion. *American Zoologist* 40:640–63.
Farlow, J. O., M. B. Smith, and J. M. Robinson. 1995. Body mass, bone 'strength indicator,' and cursorial potential of Tyrannosaurus rex. *Journal of Vertebrate Paleontology* 15:713–25.
Gould, G. C., and B. J. McFadden. 2004. Gigantism, dwarfism, and Cope's Rule: Nothing in evolution makes sense without a phylogeny. *Bulletin of the American Museum of Natural History* 285:219–37.
Hone, D. W. E., and M. J. Benton. 2005. The evolution of large size: How does Cope's Rule work? *Trends in Ecology and Evolution* 20(1): 4–6.
———. 2007. Cope's rule in the Pterosauria, and differing perceptions of Cope's rule at different taxonomic levels. *Journal of Evolutionary Biology* 20:1164–70.
Hone, D. W. E., T. M. Keesey, D. Pisanis, and A. Purvis. 2005. "Macroevolutionary trends in the dinosauria: Cope's Rule." *Journal of Evolutionary Biology* 18:587–95.
Hunt, G., and K. Roy. 2006. Climate change, body size evolution, and Cope's Rule in deep-sea ostracodes. *Proceedings of the National Academy of Sciences* 103(5): 1347–52.
Huss, J. E. 2004. Experimental reasoning in non-experimental science: Case studies from paleobiology. Unpublished PhD diss., University of Chicago. UMI number 3125619.

Hutchinson, J., and M. Garcia. 2002a. Popular depictions of dinosaur running: Did *Jurassic Park* get it wrong? Available online at http://www.rvc.ac.uk/AboutUs/ Staff/jhutchinson/ResearchInterests/TRex/JurassicPark.cfm.

———. 2002b. Tyrannosaurus was not a fast runner. *Nature* 415:1018–21.

Hutchinson, J. R., F. C. Anderson, S. S. Blemker, and S. J. Delp. 2005. Analysis of hindlimb muscle moment arms in Tyrannosaurus rex using a three-dimensional musculoskeletal computer model: Implications for stance, gait, and speed. *Paleobiology* 31(4): 676–701.

Jablonski, D. 1996. Body size and macroevolution. In *Evolutionary Paleobiology,* ed. D. Jablonski, D. H. Erwin, and J. H. Lipps, 256–89. Chicago: University of Chicago Press.

———. 1997. Body-size evolution in Cretaceous molluscs and the status of Cope's Rule. *Nature* 385:250–52.

Kingsolver, J. G., and D. W. Pfennig. 2004. Individual-level selection as a cause of Cope's Rule of phyletic size increase. *Evolution* 58(7): 1608–12.

Laurin, M. 2004. The evolution of body size, Cope's rule, and the origin of amniotes. *Systematic Biology* 53(4) : 594–622.

McFadden, B. J. 1986. Fossil horses from 'Eohippus' (Hyracotherium) to Equus: Scaling, Cope's Law, and the evolution of body size. *Paleobiology* 12(4): 355–369.

McShea, D. W. 1994. Mechanisms of large-scale evolutionary trends. *Evolution* 48(6): 1747–63.

———. 1998. Possible largest-scale trends in organismal evolution: Eight 'live hypotheses.' *Annual Review of Ecology and Systematics* 29:293–318.

———. 2001. Evolutionary trends. In *Palaeobiology II*, ed. D. E. G. Briggs and P. R. Crowther, 206–10. Oxford: Blackwell.

Oreskes, N., K. Shrader-Frechette, and K. Belitz. 1994. Verification, validation, and confirmation of numerical models in the earth sciences. *Science* 263(5147): 641–46.

Padian, K. and P. Olsen. 1989. Ratite footprints and the stance and gait of Mesozoic theropods. In *Dinosaur Tracks and Traces,* ed. D. D. Gillette and M. Lockley, 231–42. Cambridge: Cambridge University Press.

Polly, P. D. 1998. Cope's Rule. *Science* 282(5386): 51.

Ruse, M. 1999. *Mystery of mysteries: Is evolution a social construction?* Cambridge, MA: Harvard University Press.

Stanley, S. M. 1973. An explanation for Cope's Rule. *Evolution* 27(1): 1–26.

Turner, D. 2000. The functions of fossils: Inference and explanation in functional morphology. *Studies in History and Philosophy of Biology and Biomedical Sciences* 31(1): 193–212.

CHAPTER ELEVEN

Taxic Paleobiology and the Pursuit of a Unified Evolutionary Theory

Todd A. Grantham

INTRODUCTION

In 1973, Jack Sepkoski was hired as a graduate assistant to compile data on the first appearance, growth, and decline of all classes of organisms from the Cambrian to the present. So began Sepkoski's lifelong devotion to producing, refining, and analyzing large taxonomic databases as a means of studying the major patterns of life, from the Cambrian to the present. In 1982, he published his first database: the *Compendium of Marine Fossil Families*. Later, he produced a similar (posthumously published) compendium of marine animal genera. Sepkoski's compendia were not the first taxonomic databases to be developed. As Benton (1999) and Miller (2000) note, earlier efforts by Newell and Valentine were particularly important. However, the taxonomic and temporal breadth of Sepkoski's data, the rise of computer technologies for both compiling and analyzing the data, and Sepkoski's own challenging interpretations of the data all contributed to their impact. The family- and genus-level databases stimulated the production and testing of hypotheses about macroevolutionary patterns and encouraged others to develop largescale databases (e.g., Benton 1993). Collectively, these methods for compiling and analyzing taxonomic databases can be called "taxic paleobiology" (Eldredge 1979, Smith 1994).

Taxic paleontology grew up amid the controversies that roiled evolutionary biology in the late 1970s and 1980s. Disputes over sociobiology, adaptationism, units of selection, and mass extinctions raged. Despite these controversies, some evolutionists optimistically aspired to develop an integrated hierarchical understanding of evolutionary processes (e.g., Eldredge 1985). In particular, Gould (1980) argued that paleontology was poised to become a more "nomothetic" discipline that could make vital contributions to a hierarchically expanded theory of evolution. It is somewhat ironic, then, that twenty-five years later, paleobiology remains only poorly integrated with neontological evolutionary biology (Grantham 2004). The goal of this chapter is to examine the ways in which the growth of taxic methods affected interaction between neontological and paleontological approaches to evolutionary biology.

The central and defining difference between neontology and paleontology concerns the objects of study: paleontologists study fossils, whereas neontologists study living organisms. Obviously, both fossils and living organisms are studied for a variety of reasons, many of which do not concern evolution but, for the purposes of this chapter, I will focus on how neontologists and paleontologists study evolution. The defining difference between these approaches leads to a variety of other differences (see table 11.1; Grantham 2004 describes these fields in more detail). For example, neontology and paleobiology typically study phenomena on different timescales, rely on somewhat different bodies of theory, and utilize very different methods.

I believe that the evolutionary sciences are a complex array of partly overlapping and only loosely integrated fields. These various fields—that is, population genetics, ecology, physical anthropology, paleontology, and systematics, as well as fields devoted to the study of particular taxonomic groups—are all loosely guided by the synthetic theory of evolution, but often employ significantly different methods to address rather different questions. Given this diversity, some measure of methodological and conceptual pluralism is prudent; a single overarching synthesis is neither likely nor desirable (Burian 2005). However, the theoretical and methodological pluralism should be constrained: the data and methods of one field often conflict with the data and methods of closely related fields, requiring that the fields be made compatible. In some cases, conceptual connections between fields provide the basis for active integration. Thus, I am interested in how and why evolutionary biologists pursue partial integration across disciplinary boundaries, though I suspect that paleobiology is only loosely integrated with neontological approaches.

This paper examines how the development of taxic paleobiology affected the paleobiology/neontology interface during the 1980s. I begin by reminding the reader that taxic paleobiology was often linked to the hierarchical expansion of evolutionary theory. One might, therefore, expect that the rise of the taxic approach would promote cross-disciplinary interaction between paleontologists and neontologists. But this does not appear to be the case. The

TABLE 11.1 Principal differences between neontology and paleobiology

	Neontological evolutionary biology	Evolutionary paleobiology
Focus of study	Living organisms	Fossil remains of organisms
Temporal perspective	Often shorter term: 10^{-2}–10^3 years; though comparative and phylogenetic can be used to probe patterns on longer time scales.	Typically longer term (10^3–10^7 years), though the fossil record occasionally provides higher temporal resolution.
Theory	Models typically emphasize natural selection and speciation (though other processes such as random drift are acknowledged); generally articulated in terms of population or quantitative genetics (though ESS modeling does not make any assumptions about the underlying genetics).	Relies on broader neo-Darwinian theory; rarely uses population genetic theory. Some distinctively paleobiological theory (e.g., taphonomy and models of cladogenesis).
Methods	Greater emphasis on laboratory and field experiments; lab methods for studying genes, development, behavior, population structure.	Less emphasis on experiments; special methods for addressing sampling problems in using fossil data.
Data	Emphasizes genetic data and population structure.	Extremely limited access to genetic data and population structure.

third section examines patterns of cross-disciplinary interaction (as expressed in the journals *Paleobiology* and *Evolution*). This survey shows that although taxic methods quickly became influential among paleontologists, they had little impact in *Evolution*. Furthermore, the rise of taxic methods did not seem to promote closer collaboration across disciplines. The fourth and fifth sections then explore the issue of why taxic paleobiology did not (in the 1980s) realize its promise to promote an integrated hierarchical theory of evolution. More specifically, I review the protracted debate over the legitimacy of taxic methods. Tracing out this debate, we will see that taxic paleobiology pushed paleontology to focus on patterns among higher taxa, to use specialized paleontological methods, and to emphasize problems that are substantially different from those that are most central to microevolutionary (and neontological) evolutionary biology. All of these factors acted to maintain the conceptual distance between these research traditions. Although these communities are unified by their common interest in some evolutionary phenomena (e.g., diversification, speciation, functional morphology, how developmental systems can channel morphological change), the process of developing taxic methods led paleobiologists (in the short run) to focus on technical problems associated with paleontological data, thereby emphasizing the distance between fields.

The basic idea of taxic paleontology is quite simple. One simply records the first and last appearance of taxonomic groups in the fossil record. The time from first to last appearance is the *stratigraphic range* of the taxon. In principle, this can be done at any taxonomic level: species, genera, families, classes, and so on. While the basic idea is simple, creating a good database is not. Different authors have used different nomenclature, both for the taxonomic units and for the stratigraphic intervals. Organizing this mass of data—as Sepkoski did for all the known families of marine taxa, from the Cambrian to the present—requires considerable effort. Sepkoski began from other published compendia but updated and corrected the data by consulting the primary literature and taxonomic specialists. Once the data are compiled, it is easy to calculate the standing diversity at any time: the standing diversity is simply the number of taxa (in this case, families) that occur during a taxonomic interval. (Fig.11.1, column A illustrates Sepkoski's original method; column B illustrates an alternative and more phylogenetic approach. For the purposes of this paper, I will regard both approaches as alternative versions of taxic paleobiology because both use large taxonomic databases and stratigraphic ranges.) Based on the data gathered in his *Compendium*, Sepkoski graphed the diversity of marine families—the now iconic Phanerozoic diversity curve.

The advent of large databases and the development of (often computerized) means of analyzing the data provided one means to rigorously address large scale patterns of macroevolution (Benton 1999, Miller 2000). As researchers began to examine the compendium more closely, intriguing patterns emerged. Paleontologists quickly realized database research could also shed light on origination and extinction rates. One of the first attention-grabbing results was Raup and Sepkoski's (1984) claim of periodicity in mass extinctions. Jablonski used taxic data to argue for species selection (1986), the heritability of species-level geographic ranges (1987), and a tendency (in marine invertebrates) for higher taxa to preferentially originate in on-shore environments, even though there was no corresponding bias in the origination of genera (Jablonski and Bottjer 1991). In short, paleobiologists quickly came to see that these methods could be used to probe complex macroevoltuionary patterns.

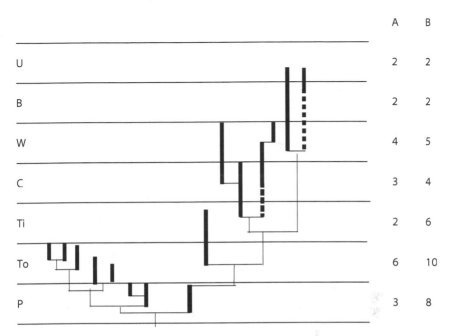

Figure 11.1 Two taxic methods. Originally, Sepkoski simply counted the number of lineages, whether species or higher taxa, documented to occur in a given stratigraphic interval (shown by heavy solid lines). This procedure produces standing diversity counts (shown in column A). An alternative (phylogenetic) method places the observed taxa in a phylogenetic tree and counts all observed and inferred taxa. (The thin vertical lines denote *ghost taxa*—taxa that are never observed, but which are inferred to exist. The thick dotted lines denote *range extensions* of taxa that are found in the fossil record.) For example, direct fossil evidence only shows two species during the Tiffanian (Ti), but phylogenetic analysis suggests that at least six lineages must have been present. This figure is based on a phylogeny presented in Archibald's (1993) study of Paleocene mammals.

The excitement of this period is well captured in Gould's "The Promise of Paleobiology as a Nomothetic, Evolutionary Discipline" (1980). One of Gould's many taking-stock-of-the-field essays, this piece was written in celebration of the fifth anniversary of *Paleobiology*. Gould's main theme was that although paleontology had deservedly been criticized for merely describing, sometimes in excruciating detail, the history of life without any attempt to uncover deeper evolutionary laws (processes), paleontology was now poised to become a more theoretical discipline. In Gould's view, paleontologists had

been too willing to passively receive evolutionary theory from below without actively generating and testing new hypotheses about macroevolutionary
processes. Paleontology has traditionally been idiographic (simply historical). Gould argued that paleontology should retain its traditional emphasis
on history, but should also expand to include nomothetic approaches (i.e.,
develop and test new theoretical hypotheses about the mechanisms and laws
of evolution). While Gould praised *Paleobiology* for attempting to integrate
paleontology and evolutionary theory, he pushed paleontologists to be more
daring in proposing new higher-level processes based on fossil data.

Recall the historical setting within which Gould was writing. The intense
work on the units of selection problem in the 1970s led some to propose a
hierarchical expansion of evolutionary theory that would incorporate both
micro- and macroevolution (e.g., Eldredge 1985). Although taxic paleontology dates from the 1960s—well before the rise of interest in hierarchy in the
1970s—the two ideas grew up together, and it is reasonable to believe that
scientists recognized taxic methods as one means to advance the cause of hierarchy. It was Gould, after all, who asked Sepkoski to develop his first taxonomic database. And Gould (1980) treats Sepkoski's work as an exemplar of
the newer nomothetic approach. It is reasonable to believe that paleontologists
such as Valentine, Gilinsky, McKinney, and Jablonski didn't just happen to
advocate both taxic methods and hierarchy; rather, the two ideas were closely
intertwined. Taxic methods were attractive, in part, because of their potential
to foster a more nomothetic and hierarchical paleontology. This reorientation
would help paleontology become a full partner in the project of developing
a hierarchically expanded evolutionary theory. But taxic paleontology did
not, at least in the short run, promote much productive interaction between
fields.

DID TAXIC PALEOBIOLOGY FOSTER CROSS-DISCIPLINARY COOPERATION?

As we have seen, taxic paleobiology emerged into a charged environment in
which some authors aspired to develop a more mechanistic (nomothetic) paleontology as part of an integrated hierarchical account of evolution. In an effort to quantify the rise of taxic paleobiology and to explore its influence on
the paleontology-neontology (henceforth: paleo-neo) interface, I examined the
contents of *Evolution* and *Paleobiology* during the 1980s.[1] These two journals
seem to be appropriate places to look for signs of cooperation. *Paleobiology* was
launched with the intent of encouraging closer interaction between paleontol-

ogy and biology, and the Society for the Study of Evolution (which sponsors *Evolution*) was established to unify the various evolutionary sciences.

I examined all the primary articles from three years: 1981, 1985, and 1989. Nineteen-eighty-one is a sensible year to begin the analysis because it includes two important landmarks in the history of the taxic approach: Sepkoski's (1981) three faunas paper, and the influential consensus paper (Sepkoski et al. 1981), which argued that the Phanerozoic diversity curve reflects a real signal. Thus, the 1981 sample occurs just as the taxic method is taking off; it provides a kind of baseline just prior to the publication of Sepkoski's *Compendium* (1982). By the close of the decade, these methods were yielding challenging results and were being challenged (e.g., Patterson and Smith 1987). Thus, my sample from the 1980s reflects the first wave of results from taxic methods, as well as criticisms and refinements of those methods. (In future work I hope to broaden the scope of my analysis to cover the 1990s as well.) How were these developments expressed in these two journals? Did these controversies lead to a more robust pattern of cross-disciplinary cooperation?

My analysis focuses on several measures of interfield interaction. First, I determined the proportion of papers in *Evolution* that utilize fossil data and/or taxic methods. Second, I examined patterns of coauthorship. Scientists generally cooperate because they have differing areas of expertise that, when combined, allow them to address an important question that would otherwise be very hard to address. If paleontology is providing theories, data, or methods that are useful to neontologists, one would expect members of these two communities to collaborate. Finally, I examined the extent to which authors published their results to an audience other than their home discipline.

My analysis includes only full-length articles (not book reviews or short commentaries). Departmental affiliation was used as the primary guide to disciplinary affiliation. For the purposes of this analysis, I focused on traditionally defined disciplines (e.g., mathematics, biology, geology, psychology, physics, anthropology). Thus, a paper cowritten by a geneticist and an ecologist was regarded as occurring within a single discipline (biology), whereas a collaboration between a physical anthropologist and a behavioral ecologist was regarded as cross-disciplinary. People who worked in museums were often hard to classify: two people in the same museum department might work in very different traditions. In such cases, I conducted quick Internet searches and was generally able to locate the author's home page, which provided indications about disciplinary education and publications. A few people who straddle the boundary between neontological and paleontological research were classified as half neontologist and half paleontologist (e.g., Karl Niklas, Michael Bell).

In addition to noting cross-disciplinary coauthorship, I also noted cases of presentation to an audience across disciplinary boundaries (e.g., a paper by a neontologist published in *Paleobiology*, or a paper by an anthropologist published in *Evolution*). Finally, I read the abstract (and descriptions of methods, when necessary) in order to determine whether fossil data or taxic methods were used. The results are summarized in table 11.2.

Consider, first, how frequently fossil data and taxic methods appear in these two journals. Within *Paleobiology*, taxic methods rise quite dramatically, from 5% (1981) to 22% (1989) of the primary articles. In contrast, taxic methods are almost invisible within this sample of *Evolution:* given the low frequency of papers using fossil data, it is not surprising that very few articles featuring taxic methods appeared in my sample. The poor representation of fossil data (and taxic methods specifically) seems to indicate considerable intellectual isolation between these communities. Two different and opposing factors need to be weighed when interpreting these results. On the one hand, my study focused on only three years' worth of publications from two journals. It is possible that more robust mutual influence might be observable with a larger database (e.g., including book reviews and commentaries, additional journals, or additional years in the study). For example, articles discussing taxic methods did appear in influential generalist journals, suggesting an alternative vehicle for cross-disciplinary communication (e.g., Raup and Sepkoski 1984, Jablonski 1987, Patterson and Smith 1987).[2] On the other hand, researchers using taxic methods were documenting challenging patterns in the fossil record (e.g., species selection, mass extinctions); one would have expected that these findings would merit serious attention in the pages of *Evolution*. While the data presented here cover only a small portion of the available literature, the paucity of fossil data presented in *Evolution* strongly suggests that paleontology was poorly integrated into evolutionary biology during the 1980s.

A second set of observations concern interdisciplinary coauthorship and collaboration. Multiple authorship (two or more authors) is more common in *Evolution* (58%) than in *Paleobiology* (30%). Despite this, the frequency of cross-disciplinary coauthorship is much higher in *Paleobiology*. Of the 183 multiauthored papers in *Evolution*, only eight involved interdisciplinary teams (4%); by contrast, *Paleobiology* had only twenty-eight multiauthored papers, but seven of these were interdisciplinary (25%). Furthermore, when we examine the specific disciplines involved, we find only two genuinely paleo-neo collaborations in the entire sample (both occur in 1981 in *Paleobiology*). Biologists publishing their work in *Evolution* are more likely to collaborate

TABLE 11.2 Coauthorship and Taxic Methods in *Evolution* and *Paleobiology*

Journal	year	co-author	X-D team	X-D audience	Fossil/Taxic
Paleobiology	1981	14/38	4/38	8/38	2/38 T (5.3%)
	1985	6/31	1/31	4.5/31	3/31 T (9.7%)
	1989	8/25	2/25	5/25	5.5/25 T (22%)
	total	28/94 (30%)	7/94 (7%)	17.5/94 (18.6%)	10.5/94 T (11%)
Evolution	1981	43/94	2/94	3.5/94	1/94 F; 1/94 T
	1985	62/102	4/102	7/102	2/102 F; 0/102 T
	1989	78/117	2/117	4.5/117	1.5/117 F; 0/117 T
	total	183/313(58%)	8/313 (2%)	15/313 (4.8%)	4.5/313 F (1.5%)
					1/313 T (0.5%)

Note: Based on a study of all the primary articles published in these journals during three years. The table shows: the proportion (percentage) of: papers with two or more authors (column 3), papers that involve interdisciplinary (X-D) teams of authors (column 4), and papers in which authors present their research to an audience outside of their home discipline (column 5). The final column reports the proportion (percentage) of papers that use fossil data (F) or taxic methods (T).

with physicists, mathematicians, or psychologists than with paleontologists (table 11.3)! Having now completed this initial phase of research, it is clear that collaboration is a demanding standard. The levels of cross-disciplinary collaboration are too low to providing much meaningful data beyond the observation that paleontologists are not chosen as collaborators any more than mathematicians or psychologists. I hope that subsequent studies of citation patterns will provide a more fine-grained analysis of interfield interaction. While the rates of interdisciplinary coauthorship are low, scientists did present results to audiences beyond their primary discipline with somewhat higher frequency—particularly in *Paleobiology*. There is no evidence that the rise of taxic paleontology led to higher rates of cross-disciplinary presentation in either journal.

In sum, while taxic methods were clearly rising in significance within *Paleobiology*, they did not receive much attention within the pages of *Evolution*. Nor did the rise of taxic paleobiology promote cross-disciplinary coauthorship or cross-disciplinary presentation of results. This (admittedly limited) sample suggests that neontologists and paleontologists operate in fairly distinct spheres. Some neontologists published their work in *Paleobiology*, very few paleontologists published in *Evolution*, and even fewer attempted to collaborate across the neo-paleo divide. If this sample is representative, then the

TABLE 11.3   Fields of authors engaging in interdisciplinary publication

	X-D coauthors	Author presenting to x-d audience
Paleobiology	3 math/computer 2 neontology (both 1981!) 1 geophysics (not paleo) 1 medicine	14.5 neontology 2 psychology 1 math/statistics
Evolution	2 Psychology 4 Physics/Engineering 2 Math/Statistics 0 paleontology	5.5 geology/paleo 4 neo-paleo interface 4.5 anthropology 1 medicine

Note: This table shows that home discipline of authors who either worked in interdisciplinary (X-D) teams (column 1) or who presented their research (either as single author or within single-discipline teams) in a journal outside of their home discipline. Some people who regularly worked at the interface of two fields (e.g., physical anthropology/behavioral ecology or neo- and paleo- work with a single taxonomic group) were scored as 0.5 in each discipline.

rise of taxic methods did not contribute to closer working relations—at least not within this first decade. In fact, the situation is arguably worse than this; when we examine the various controversies surrounding taxic paleontology, there are reasons to think that the development of taxic methods may have, at least in the short run, discouraged cross-disciplinary collaboration.

METHODOLOGICAL OBJECTIONS
TO TAXIC PALEOBIOLOGY

One reason taxic paleobiology was not quickly adopted by neontologists was a sense of methodological caution: the taxic approach was perceived (rightly or wrongly) to suffer from methodological problems. I suspect that awareness of methodological objections made relying on these methods appear risky. Furthermore, as taxic methods evolved to address these objections, the knowledge needed to apply them may have made the cost of learning these techniques too high for most neontologists. Finally, as I will argue in the next section, taxic methods did not help neontologists to directly address their central research questions. The following discussion of objections to taxic methods is fairly detailed. These details will, I hope, clarify why taxic methods did not immediately produce cross-disciplinary cooperation, and will demonstrate that these problems do not pose any permanent obstacle to interdisciplinary co-operation.

After Sepkoski and colleagues (1981) showed that several independent data sets identify a similar Phanerozoic diversity pattern, paleontologists began using large databases with greater confidence to address a wider variety of questions. The publication of controversial and surprising conclusions (e.g., Raup and Sepkoski's [1984] claim of periodicity) led to increased critical examination of taxic methods. For example, Patterson and Smith (1987; Smith and Patterson 1988) argued that reliance on nonmonophyletic taxonomic groupings can bias our perception of extinction patterns. Furthermore, because large databases are generated by nonspecialists compiling published literature, they inevitably contain many errors. Specifically, Patterson and Smith (1987) found that 75% of the *Compendium* data they examined contained errors, and that it is this "noise" that generates the appearance of periodicity. Before turning our attention directly to the problem of whether paraphyletic taxa bias our perception of evolutionary patterns (their first criticism), it is useful to address the concern about error.

The Problem of Error

Taxonomic specialists argued that Sepkoski's method of culling data from the literature without specialist knowledge generated a data set full of error. Patterson and Smith (1987) lumped a variety of sins under the heading of "noise." Some examples (e.g., incorrectly dated first appearances, failure to recognize two names as synonyms, inclusion of polyphyletic taxa) are universally regarded as errors. But Patterson and Smith also classified monotypic and paraphyletic taxa as noise, even though some systematists would defend their use. To begin, I will focus on the unproblematic kinds of errors: are these errors so pervasive as to undermine the raison d'etre of taxonomic compilations?

Although the evidence put forward by the critics appeared quite damning, Sepkoski (1993) found that extensive correction of the *Compendium* hardly changed the overall pattern of standing diversity in the oceans. Similarly, Benton (1993) organized a team of specialists to produce a database of both marine and terrestrial Phanerozoic diversity. Although this newer database is largely independent of Sepkoski's database and relied on strictly cladistic classifications whenever possible, *Fossil Record 2* confirms, in broad outline, Sepkoski's family-level diversity curve (Benton 1995). Recent efforts at taxonomic standardization support this conclusion. For example, using a standardized[3] and largely cladistic classification of trilobites, Adrain and Westrop (2000) checked the validity of the taxonomic groupings and the accuracy of the range endpoints in Sepkoski's data. Although roughly 70% of Sepkoski's genus-level data contained some kind of error, Adrain and Westrop's diversity curve is almost identical to Sepkoski's! (See fig. 11.2.)

In sum, a substantial body of evidence now suggests that the errors in taxonomic databases do not invalidate the conclusions of taxic paleobiology. As long as one relies on large data sets (e.g., of the temporal and taxonomic scope of the Adrain and Westrop study), we can safely set the problem of error aside. Andrew B. Smith, one of the most vocal critics of taxic methods, concedes that "stratigraphical and taxonomic biases in large data sets are effectively random in distribution and thus cannot be responsible for creating trends in diversity patterns" (2001, 353).

A somewhat different problem concerns the nature of stratigraphic ranges themselves. Observed stratigraphic ranges typically underestimate the true ranges. When, as often happens, species originate as small populations, it is unlikely that the earliest members of the species will be fossilized; similarly, as a species undergoing extinction dwindles in geographic range or abundance, it is less likely to enter the fossil record. As a result, the observed range will typically underestimate the true range—on both ends. To address this problem, Marshall (2001) developed methods for putting confidence intervals on the observed stratigraphic ranges. In the following, I will argue that the development of such specialized methods is one factor that makes interdisciplinary cooperation more difficult.

Do Higher Taxa Provide Good Estimates of Species-level Diversity?

Sepkoski's database of marine *families* was often interpreted as revealing the underlying patterns of *species* diversification. That is, if one finds more *families* of snails, then there are probably more *species* of snails. But is this assumption reasonable? Two main lines of evidence have been offered to support the proposition that higher taxon counts are reliable proxies for species-level standing diversity: data from conservation biology and computer simulations.

Conservation biologists have suggested that higher-taxon diversity may be a good proxy for hard-to-obtain species-level data (Williams and Gaston 1994). If higher taxon diversity is strongly correlated with species-level diversity, then, since it is easier to determine the presence or absence of higher taxa, assaying the diversity of larger groups might be an efficient means of setting conservation priorities. The reliability of higher taxon estimates has now

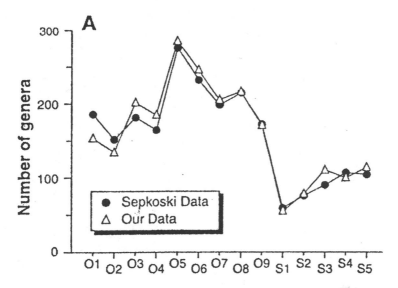

Figure 11.2 Diversity of Ordovician and Silurian Trilobites. Adrain and Westrop re-analyzed trilobite diversity based on a standardized and corrected database. Although 70% of Sepkoski's data contained some kind of error, the resulting curve closely mirrors Sepkoski's, suggesting that the problem of error does not undermine taxic methods. From Adrain and Westrop (2000), *Science* 289:110–12. Reprinted with permission from AAAS.

been studied in a wide variety of organisms and at several different geographic and taxonomic levels. For example, Balmford, Jayasuriya, and Green (1996) found that species-level diversity is strongly correlated with genus- and family-level diversity. As one would expect, genus-level data provide a more sensitive measure of species-level diversity than family-level data. Although higher taxon diversity is strongly *correlated* with species diversity, higher taxa remain relatively poor *predictors* of species-level diversity. Because higher taxa vary in the number of species they contain, a given number of higher taxa can correspond to a wide range of species-level diversity values. The prediction that higher taxon data accurately represent geographic patterns of species-level diversity is supported by Roy, Jablonski, and Valentine (1996), who found that family data accurately represent the global latitudinal gradient in species diversity but do not reveal the structure of biogeographic provinces.

Computer simulations have also been used to explore how well higher-taxon counts capture species-level diversification and extinction patterns. Sepkoski and Kendrick (1993) and Robeck, Maley, and Donoghue (2000) argue that higher taxa are useful proxies for species-level diversity. Under conditions of poor sampling, higher taxon diversities actually provide better estimates of diversity patterns than species-level data. (This point is discussed more fully in the following.)

While these two lines of evidence show that higher taxa are reasonable proxies for species-level standing diversity in global databases, let me add one cautionary note. The number of taxa found depends on sampling intensity, and the fossil record does not provide even sampling over time. During times of poor preservation, the record will systematically underestimate true diversity. This problem is currently the focus of much work (e.g., Smith 2001; Bush, Markey, and Marshall 2004; Kowalewski et al. 2006); further work is necessary to determine how extensively the observed diversity patterns are shaped by fluctuations in sampling intensity. (Note: major fluctuations in the sampling intensity pose significant problems for phylogenetic methods, too.)

The Problem of Paraphyletic Taxa

Sepkoski's *Compendium* simply counted the recognized taxa in existing classifications—a method that allowed paraphyletic and even polyphyletic taxa to enter the data set. (Paraphyletic taxa include some, but not all the descendants of a common ancestor. For example, *dinosaur* is paraphyletic, because it excludes the birds that evolved from dinosaur ancestors. In contrast, polyphyletic taxa lump species together based on phenetic resemblance, while excluding more closely related, but phenotypically different, taxa.) Everyone agrees that the inclusion of polyphyletic taxa is an error, but opinions on paraphyletic taxa are divided. One particularly important worry is that including paraphyletic taxa can skew our perception of extinction rates. While counting traditional taxa may provide a reasonable estimate of standing diversity, counting the extinctions of paraphyletic taxa is, critics contend, fundamentally confused. The disappearance of a paraphyletic taxon does not require the extinction of a monophyletic taxon. Rather, systematists terminate paraphyletic taxa by definitional fiat (i.e., the stem group is said to be extinct even though descendants of the clade persist). Imagine, for example, a family of trilobite species that undergoes a period of poor preservation lasting 10 million years. During this interval considerable morphological evolution occurs, so that we can now recognize two morphologically distinct groups of species: the stem group (before the gap) and the derived group (after the gap). Now suppose that many different taxonomic groups undergo periods of poor preservation at roughly the same time (say, because few fossiliferous rocks from that interval survive). Under these circumstances, the apparent extinctions of paraphyletic taxa can accumulate just before an interval of nonpreservation. However, since the paraphyletic taxa are not (according to cladist orthodoxy) ontologically real, we should not count their extinction as biologically significant events. In short: when an interval of poor preservation and taxonomic practices of constructing paraphyletic taxa combine in just the right (perhaps one should say "wrong") way, taxic paleontology can mislead us about extinction and origination rates (Patterson and Smith 1987, Smith and Patterson 1988, Edgecombe 1992). This argument led many authors to develop alternative phylogenetic methods.

At present, it is not feasible to undertake the investigation of large-scale diversity patterns using species-level phylogentic data: we do not have species-level phylogenies for many groups and, even if we did, the species-level fossil record is generally less reliable than the record of higher taxa. Thus, advocates of phylogenetic methods claim that strictly cladistic *higher taxa* provide better estimates of species-level diversity, extinction rates, and origination rates than traditional higher taxa. Both Sepkoski and his critics used higher taxa as proxies for species-level diversification patterns; they simply differ in the kinds of higher taxa they use. Following Robeck and colleagues (2000), I refer to traditional classifications as "mixed classifications," because they include a mixture of paraphyletic, polyphyletic, and monophyletic taxa. Thus, one key methodological question is: do strictly cladistic classifications provide better estimates of species-level patterns than mixed classifications? Many authors assume that strictly cladistic data will provide more reliable estimates. This assumption may turn out to be correct, but the few studies that bear on the question suggest that, under some conditions, mixed classifications can outperform cladistic classifications.

An ideal empirical investigation would compare three datasets: (1) a standardized mixed classification at the genus or family level, (2) a standardized cladistic classification at the same level, and (3) a full species-level phylogeny. One could then assess how well the mixed and cladistic classifications capture the species-level patterns of diversification. (I propose using standardized databases to eliminate errors in the databases. The possibility that a nonstandardized database contains more errors than a newer cladistic database introduces a confounding variable.) Some influential criticisms of taxic methods are rather far from this ideal. Consider Edgecombe's (1992) study of turnover among Cambrian and Ordovician trilobites. Edgecombe shows that the genus-level turnover patterns seen in a revised (cladistic) database differ from the turnover patterns of traditional genera. But simply showing that analyses based on mixed and cladistic databases *differ* does not demonstrate that cladistic higher taxa are *more reliable* proxies for species-level patterns. Edgecombe correctly points out that many of the "extinctions" in taxic analyses are not true extinctions; rather, they are pseudo-extinctions in which a paraphyletic taxon gives rise to a descendant taxon. Further, if we assume that taxa always originate through branching and extend the range of a taxon back to the earliest appearance of its sister taxon, then taxon originations are pushed back in time. (Fig. 11.1, column B shows a simple example of this.) But to assume that these points undermine taxic methods is to confuse metaphysics and methodology. Even if taxic methods fail to make the metaphysical distinction between real and psuedo-extinctions (and therefore overestimate the number of higher-taxon extinction events), this fact does not address the methodological question of how well cladistic classifications function as proxies for *species-level* diversity.

Consider the role of ghost lineages (taxa that are not found in the fossil record but are inferred to exist based on phylogenetic analysis—see fig. 11.1, column B) and range extensions. Suppose one tallies all the genera (including ghost taxa) in an interval. This pushes the origination of the *higher taxa* further back into the past. But when did *species-level* diversity increase? If ghost taxa generally contain few species and enter the fossil record only as their species richness increases, then neglecting ghost taxa might actually provide a more accurate depiction of the underlying species-level diversification. Similarly, since the extinction of a large paraphyletic taxon would involve many species-level extinctions, classifications containing paraphyletic taxa may provide more sensitive measures of species-level extinction rates. The considerations offered here do not show the superiority of mixed classifications; they only make the negative point that many criticisms of taxic methods are not decisive.

Wagner (1995) offers one of the most direct attempts to empirically assess the relative reliability of mixed and phylogenetic classifications. He compared a species-level phylogeny of Paleozoic gastropods with three higher-taxon analyses: a mixed classification and two somewhat different phylogenetic classifications. One phylogenetic analysis used monophyla (i.e., the smallest clade whose sister taxon is a multispecies clade, considering only present and past species). The second analysis uses Hennigian taxa—taxa that are monophyletic relative to the entire cladogram. (Cladists typically use Hennigian taxa.) The mixed classification was significantly different from the phylogenetic classifications: fewer than 30% of the traditional taxa are monophyletic. Wagner then compares how well traditional taxa, monophyla, and Hennigian taxa capture the species-level diversification patterns. In this study, monophyla were the best proxy for lineage-level diversity patterns, followed by mixed classifications, with Hennigian taxa coming in third. Thus, mixed classifications can, in real examples, outperform Hennigian classifications. However, the alternative method using monophyla outperformed the mixed classification.

Computer simulations provide another tool for assessing the reliability of mixed classifications. Simulations conducted by Sepkoski and Kendrick (1993), Robeck and colleagues (2000), and Lane, Janis, and Sepkoski (2005) show that higher taxa are useful proxies for species-level diversity. Under conditions of poor sampling, both cladistic and mixed higher taxa provide better estimates of diversity patterns than the raw lineage-level data. Further, both studies found that mixed classifications can outperform cladistic classifications under some conditions.

Do mixed or strictly monophyletic classifications provide better estimates of species-level diversity patterns? Critics of taxic methods point out that relying on paraphyletic taxa can (in principle) bias our perception of extinction and origination rates. A number of studies have been offered to support this conclusion (e.g., Patterson and Smith 1987, Smith and Patterson 1988, Edgecombe 1992). However, the common argument that cladistic and mixed classifications display *different* patterns simply does not demonstrate that monophyletic taxa provide a *better* estimate of species-level diversity. The data that bear most directly on the issue (e.g., Wagner 1995 and computer simulations) suggest that mixed classifications can outperform cladistic classifications under some conditions. This limited range of evidence suggests that in many circumstances mixed classifications perform satisfactorily.

DISCUSSION: WHY TAXIC PALEOBIOLOGY DID NOT FOSTER INTEGRATION

Although taxic paleobiology was shaped by the ideal of an expanded evolutionary theory that integrated paleontology and neontology, authorship patterns in *Evolution* and *Paleobiology* suggest that taxic paleobiology did not stimulate paleo-neo interaction during the 1980s. Based on a review of criticisms of taxic methods (in the fourth section), this concluding discussion offers three reasons why taxic paleontology did not promote much neo-paleo cooperation.

1. The idea that paleontology could make important contributions to evolutionary theory always contained the seeds of interfield conflict.

2. Neontologists may have been cautious about utilizing a method that elicited such vigorous controversy within paleontology.
3. The research agendas of the neontological and paleobiological approaches remain largely (though not completely) distinct.

Gould's program

Gould (1980) hoped that paleontology would make a *distinctive* contribution to evolutionary theory. He criticized paleontologists' tendency to passively receive and apply (micro-)evolutionary theory to fossil data. In contrast, he hoped that paleobiology would offer its own unique contributions to evolutionary biology—but this, of course, requires that the macroevolutionary hypotheses be different from (or independent of) microevolutionary theory. Thus, I suggest that some measure of independence (if not outright conflict) between these fields was always implicit in the program for the hierarchical expansion of evolutionary biology. For example, if group or species selection is to be genuinely distinct from organismic selection, it must, under at least some circumstances, make predictions that differ from models of organismic selection. Although the ultimate aim was to produce a genuine synthesis of micro- and macroevolution, paleobiology first needed to establish a robust realm of macroevolutionary phenomena that are not adequately captured by microevolutionary approaches—an idea that was sure to ruffle some feathers. (In my view, paleobiology has provided some challenging data, including evidence for mass extinction selection regimes that differ from background selection (Jablonski 1986), species selection (Jablonski 1986, 1987), passive diffusion as an explanation for evolutionary trends (McShea 1994), a tendency for higher taxa to preferentially originate in on-shore environments (Jablonski and Bottjer 1991), and developmental constraints (e.g., Eble 2000). All of these findings challenge the idea that we can smoothly extrapolate microevolutionary processes to explain macroevolutionary patterns.)

Methodological caution

Some of the initial worries about taxic paleobiology would have been highly visible to nonspecialists (e.g., Patterson and Smith's [1987] critique was published in *Nature*). If the paleontologists are in a state of internal dispute, a sensible neontologist might wait until the dust settles before relying on methods that are perceived to be controversial.

Although this cautious approach was reasonable during the 1980s, the previous section argued that these methodological concerns have largely been answered. Empirical and simulation data show that errors in Sepkoski's database do not undermine the large-scale patterns he documented. Further, higher taxa (whether strictly monophyletic or not) are often adequate proxies for species-level diversity. Finally, the development of strictly cladistic databases increasingly provides an alternative to relying on mixed classifications. Thus, methodological caution does not seem to provide a compelling argument for continuing to ignore taxic methods (especially if this is understood to include work with large cladistic databases). It is worth noting, however, that as taxic methods developed throughout the 1980s and 1990s, paleobiologists began using increasingly sophisticated and specialized methods (e.g., Marshall's technique for placing confidence intervals on stratigraphic ranges). Paleontology's emphasis on improving its methods—while crucial to the advancement of the field—did not provide a basis for much interdisciplinary cooperation. It is natural that people working with living organisms would largely ignore the development of increasingly sophisticated methods for handling problems associated with fossil data. This leads to the most central obstacle to closer integration: I suspect that few neontologists are convinced that the methods or findings of taxic paleontology provide important resources for solving the problems that are most central in their research.

Division of labor

I believe there is a rough division of labor between paleobiology and neontological evolutionary biology. It is important to state this division with some care, however. Some authors have suggested that paleobiology is merely idiographic, and unconcerned with mechanisms; this is certainly an overstatement (Grantham 2004). Instead, it is more accurate to say that paleobiology focuses on somewhat different (typically large-scale) mechanisms, including Vermeij's "arms races," passive diffusion models, or species selection. Thus, the general sense of methodological caution was reinforced by the fact that it is hard to directly apply taxic methods to the central problems that were gripping neontological evolutionary biology during the 1980s. To make this idea clearer, let us review some of the primary topics discussed in the 1980s in *Paleobiology* and *Evolution*.

In addition to the more formal study presented in the third section, I also reviewed the content of articles in my sample to identify research hot spots. In *Evolution*, several topics were the focus of sustained discussion (i.e., discussed in at least five articles): the evolution of behavior (e.g., cannibalism, mate

choice, feeding preferences), hybrid zones, reproductive isolation and speciation, hierarchy and group selection, life history evolution, phylogenetic methods, the evolution of sex, diversification/adaptive radiation, developmental
constraints, as well as studies of natural selection and functional morphology.
Several of these topics are very hard to address with fossil data (e.g., evolution
of behavior and life-history traits, genetics of hybrid zones and speciation).
Further, some of the most hotly discussed topics in *Paleobiology* received little
attention in *Evolution* (e.g., evolutionary rates, gradualism/punctuated equilibrium, taphonomy, and extinction). This is not to say that there are no areas
of overlap. Common interests include speciation, influence of development
on evolution, diversification, functional morphology, and the development of
phylogenetic methods. However, even in these cases we often find more subtle
differences between fields. For example, neontologists and paleontologists approach phylogeny reconstruction in somewhat different ways, using different
kinds of data and (as a result) different methods (Grantham 2004). Neontological analyses of speciation often focus on molecular and behavioral mechanisms that are largely invisible to paleontologists. When viewed in this light,
the fact that taxic methods forced paleontologists to analyze diversification
patterns at the level of families or genera may have reinforced neontologists'
judgment that paleontological data are rather far removed from their research
interests. Although these observations are still rather coarse grained, they support the idea that taxic methods pushed (some) paleobiology further away
from areas of common interest, actually making interdisciplinary cooperation
less likely.

Here is another way to see how taxic paleontology may, in the short run,
have made collaboration less likely. I imagine that many readers were startled
by the contrast between the promise of a unified evolutionary theory and the
discussion of objections to taxic methods. What, you might ask, do all of these
technical questions have to do with the fate of an integrated theory of evolution? That's exactly the point. During the 1980s, paleobiologists focused
intently on the development and refinement of taxic (and, more generally
quantitative and nomothetic) methods. At least until the 1990s, paleobiology was, quite naturally, turned inward to develop its own methods. Although
paleontologists pursued these methods as a means to use fossil data to address
evolutionary questions, the big evolutionary questions often took a back seat
until the methodological issues could be resolved. (In fact, the discussion of
the fourth section understates the depth of the problem. It illustrates only
one of a number of thorny methodological issues that gripped paleontology
between 1980–2000.) In retrospect, then, it was naturally difficult to develop

neo-paleo collaborations at a time when taxic paleobiology generally did not bear directly on the most central issues motivating neontological research.

Even in this environment, some fields did manage to work fairly closely with paleobiology. For example, macroecology and conservation biology both interacted significantly with paleobiology (e.g., Brown 1995, Balmford et al. 1996) and systematists grappled with the problem of integrating fossil, molecular, and character data to assess phylogenetic hypotheses (Grantham 2004). Why didn't more active integration occur at the micro/macroevolution interface? In the case of conservation biology, the compelling practical need to study biodiversity before the taxa go extinct (combined with significant paleontological expertise in the use of higher taxa as proxies) provided a strong basis for cooperation. Similarly, phylogenies based on fossil data were often incompatible with molecular phylogenies, suggesting an immediate need to coordinate research efforts. In the case of micro- versus macro-evolution, there was not, it seems, such a strong, immediate need to coordinate research. My aim has not been to criticize neontologists or paleontologists for failing to deliver on the promise of a more unified theory of evolution. But I do hope that as paleobiology matures and stabilizes, we can return our attention to the longstanding problem of developing a hierarchical framework that is adequate to explain the challenging macroevolutionary patterns documented by paleobiology.

ACKNOWLEDGMENTS

A remote ancestor of this paper was presented at a symposium honoring Jack Sepkoski, sponsored by the Field Museum of Natural History (Chicago) in 2001. Thanks to Peter Wagner for encouraging my initial exploration of this topic. The research for that initial presentation was supported by NSF (Grant #SES-9818397). This very different presentation was written with the support of a College of Charleston Faculty Research and Development Grant. David Sepkoski, David Jablonski, and Steven Orzack all reviewed an earlier draft and saved me from various stylistic and substantive gaffes.

NOTES

1. This paper presents the results of a preliminary survey. The data cover only two journals for a short time period (the 1980s) and focus on a demanding measure of interdisciplinary cooperation—collaborative authorship. In the future, I hope to extend this analysis to cover a longer time period and to also examine citation patterns in these (and other) journals.

2. Thanks to David Jablonski for this observation.

3. *Taxonomic standardization* means systematizing the various different classifications to insure consistency (e.g., by eliminating synonyms and competing/overlapping taxonomic groupings). While Adrain and Westrop (2000) offer standardized and largely cladistic classifications, they did not offer a full cladistic revision (Adrain, personal communication 2000).

REFERENCES

Adrain, J. M., and S. R. Westrop. 2000. An empirical assessment of taxic paleobiology. *Science* 289:110–12.

Archibald, J. D. 1993. The importance of phylogenetic analysis for the assessment of species turnover: A case history of Paleocene mammals in North America. *Paleobiology* 19:1–27.

Balmford, A., A. M. H. Jayasuriya, and M. J. B. Green. 1996. Using higher taxon richness as a surrogate for species richness: I Regional Tests. *Proceedings of the Royal Society of London B* 263:1571–75.

Benton, M. J., ed. 1993. *Fossil Record 2*. New York: Chapman & Hall.

———. 1995. Diversification and the history of life. *Science* 268:52–58.

———. 1999. History of life: Large databases in palaeontology. In *Numerical Palaeobiology*, ed. D.A.T Harper, 249–83. Chichester, UK: Wiley & Sons.

Brown, J. H. 1995. *Macroecology*. Chicago: University of Chicago Press.

Burian, R. M. 2005. *The epistemology of development, evolution, and genetics*. Cambridge: Cambridge University Press.

Bush, A. M., M. J. Markey, and C. R. Marshall. 2004. Removing bias from diversity curves. *Paleobiology* 30:666–86.

Eble, G. J. 2000. Contrasting evolutionary flexibility in sister groups: Disparity and diversity in Mesozoic atelostomate echinoids. *Paleobiology* 26:56–79.

Edgecombe, G. D. 1992. Trilobite phylogeny and the Cambrian-Ordovocian "event": Cladistic reappraisal. In *Extinction and phylogeny*, ed. M. J. Novacek and Q. D. Wheeler, 144–77. Oxford, UK: Oxford University Press.

Eldredge, N. 1979. Alternative approaches to evolutionary theory. *Bulletin of the Carnegie Museum of Natural History* 13:7–19.

———. 1985. *Unfinished synthesis*. Oxford, UK: Oxford University Press.

Gould, S. J. 1980. The promise of paleobiology as a nomothetic, evolutionary discipline. *Paleobiology* 6(1): 96–118.

Grantham, T. A. 2004. The role of fossils in phylogeny reconstruction, or why is it difficult to integrate paleontological and neontological evolutionary biology? *Biology and Philosophy* 19:687–720.

Jablonski, D. 1986. Background and mass extinctions: The alternation of macroevolutionary regimes. *Science* 231:129–3.

———. 1987. Heritability at the species level: Analysis of geographic ranges of Cretaceous mollusks. *Science* 238:360–63.

Jablonski, D. and D. J. Bottjer. 1991. Environmental patterns in the origins of higher taxa: The post-Paleozoic fossil record. *Science* 252:1831–33.

Kowalewski, M., W. Kiessling, M. Aberhan, F. T. Fúrsich, D. Scarponi, S. L. Barbour Wood, and A. P. Hoffmeister. 2006. Ecological, taxonomic, and taphonomic components of the post-Paleozoic increase in sample-level species diversity of marine benthos. *Paleobiology* 32:533–61.

Lane, A., C. M. Janis, and J. J. Sepkoski. 2005. Estimating paleodiversities: A test of the taxic and phylogenetic methods. *Paleobiology* 31:21–34.

Marshall, C. R. 2001. Confidence limits in stratigraphy. In *Palaeobiology II*, ed. D. E. G. Briggs and P. R. Crowther, 542–45. London: Blackwell.

McShea, D. 1994. Mechanisms of large-scale evolutionary trends. *Evolution* 48:1747–63.

Miller, A. I. 2000. Conversations about Phanerozoic diversity. Pp. 53–73 In *Deep time: Paleontology's perspective,* ed. D. H. Erwin and S. L. Wing, 53–73. Lawrence, KS: Paleontological Society.

Patterson, C., and A. B. Smith. 1987. Is the periodicity of extinctions a taxonomic artifact? *Nature* 330:248–51.

Raup, D. M., and J. J. Sepkoski. 1984. Periodicity of mass extinctions in the geological past. *Proceedings of the National Academy of Sciences, USA* 81:801–5.

Robeck, H. E., C. C. Maley, and M. J. Donoghue. 2000. Taxonomy and temporal diversity patterns. *Paleobiology* 26:171–87.

Roy, K., D. Jablonski, and J. W. Valentine. 1996. Higher taxa in biodiversity studies: Patterns from eastern Pacific molluscs. *Philosophical Transactions of the Royal Society of London B* 351:1605–13.

Sepkoski, J. J. 1981. A factor analytic description of the Phanerozoic marine fossil record. *Paleobiology* 7:36–53.

———. 1982. Compendium of Marine Fossil Marine Families. *Milwaukee Public Museum Contributions in Biology and Geology* 51: 1-125.

———. 1993. Ten years in the library: New paleontological data confirm evolutionary patterns. *Paleobiology* 19:43–51.

Sepkoski, J. J., R. K. Bambach, D. M. Raup, and J. W. Valentine. 1981. Phanerozoic marine diversity and the fossil record. *Nature* 293:435–37.

Sepkoski, J. J., and D. C. Kendrick. 1993. Numerical experiments with model monophyletic and paraphyletic taxa. *Paleobiology* 19:168–84.

Smith, A. B. 1994. *Systematics and the fossil record: Documenting evolutionary patterns.* Oxford: Blackwell Scientific.

———. 2001. Large-scale heterogeneity of the fossil record: Implications for Phanerozoic biodiversity studies. *Philosophical Transactions of the Royal Society of London, B* 356:351–67.

Smith, A. B., and C. Patterson. 1988. The influence of taxonomic method on the perception of patterns of evolution. *Evolutionary Biology* 23:127–216.

Wagner, P. J. 1995. Diversity patterns among early gastropods: Contrasting taxonomic and phylogenetic descriptions. *Paleobiology* 21(4): 410–39.

Williams, P. H., and K. J. Gaston. 1994. Measuring more of biodiversity: Can higher taxon richness predict wholesale species richness. *Biological Conservation* 67:211–17.

CHAPTER TWELVE

Ideas in Dinosaur Paleontology: Resonating to Social and Political Context

David E. Fastovsky

INTRODUCTION

Nonavian dinosaurs (dinosaurs) are compelling beasts to young children as buttresses for the myths and stories that characterize and enrich their developing psychologies. For adults, by contrast, dinosaurs epitomize failure and outdatedness. Among professional paleontologists, this symbolism is presumably left behind, and science is supposed to be the driving force for discovery and interpretation. Yet, discoveries and interpretations don't occur in a vacuum, and here I track the ways in which interpretations of dinosaur paleobiology appear to be conditioned by social context.

The influence of dinosaurs on social climates has been well documented (e.g., Mitchell 1998). Its converse—the effect of culture on dinosaur paleontology—has not been as thoroughly investigated. Interestingly, much is known about the relationship between culture and science in Victorian times. Studies of this subject include Adrian Desmond's *Hot-Blooded Dinosaurs* (1976) and *Archetypes and Ancestors* (1982), Martin Rudwick's *Bursting the Limits of Time* (2005), Deborah Cadbury's *Terrible Lizard* (2000), *Georges Cuvier, Fossil Bones, and Geological Catastrophes* (1998), *Scenes from Deep Time* (1995), *The Great Devonian Controversy* (1988), and *The Meaning of Fossils* (1985), "Politics and Paleontology" by Hugh Torrens in *The Complete Dinosaur* (1997), and Christopher McGowan's *The Dragon Seekers* (2001). And this is very short of a complete listing, leaving one wondering whether perhaps the distance of 125 years makes the subject somewhat more acceptable.

Treatments of the more recent culture of paleontology run more to whodun-
its (*The End of the Dinosaurs* [1999]; *Night Comes to the Cretaceous* [1998]);
personalities (*Kings of Creation* [1992]); voyages of discovery (*T. rex and
the Crater of Doom* [1997]; *The Nemesis Affair* [1986]; *Digging Dinosaurs*
[1988]; *Quest for the African Dinosaurs* [1993]) and personal memoirs (*Time
Traveler* [2002]). How the various discoveries and interpretations laid out in
these works may or may not have been influenced by their cultural context is
left generally unstated.

Here I consider three case studies: the paleobiology of the large theropod
T. rex, the discovery of dinosaur maternity, nests, eggs, and embryos, and the
dinosaur extinction. In each case, my thesis is that the work gained a foothold
not only because the interpretations were supported by discoveries, but because the social climate was ripe for these kinds of inferences.

PALEOBIOLOGY OF *TYRANNOSAURUS REX*

Tyrannosaurus rex is a dinosaur at once completely familiar and paradoxically
utterly unfamiliar. Ever since its discovery in 1902, it has exerted a ubiquitous hold on imaginations as the ultimate icon of terrestrial carnivory. Yet, its
uniqueness precludes satisfactory comparison with the modern analogs that
might provide robust insights into its behavior. Among the features of *Tyrannosaurus* that leave many paleontologists shaking their heads in bafflement are
(a) large size (about 15 meters from stem to stern), (b) powerful, stocky legs
that either allowed it to run fast or not; (c) short, stout, powerfully muscled
arms whose robust, clawed, two-fingered grasping hands could nonetheless
not reach even its massive jaws; and (d) banana-like teeth whose bulbous
cross-sections are too inflated to be solely a consequence of allometry.

The context in which the discovery of *T. rex* occurred provides clues
about the early interpretations. The first specimen was described in 1905
at a time of burgeoning European and North American global imperialism.
Nationalism, imperialism, domination, and cultural and military hegemony
were themes that were commonly sounded as European nations and the
United States scrambled to exert control over underdeveloped regions in Africa and Asia (via spheres of influence [see Hay 1899] and/or colonization).
Early twentieth-century imperialism has been well studied; here I offer but
an example to lend the era a flavor. The following are opening remarks to a
conference convened by King Leopold II of Belgium, who hoped underwrite
his colonial activites in Africa:

To open to civilization the only part of our globe which it has not yet penetrated, to pierce the darkness which hangs over entire peoples, is, I dare say, a crusade worthy of this century of progress . . . (quoted in Hochschild 1998, 44)

The overriding image here is the imposition of, and domination by, one culture over another. The righteous establishment of cultural hegemony (the "crusade") was generally accompanied by a need to build and maintain physical hegemony, generally by naval strength and by forcibly obtaining parcels of land for colonization and/or refueling and resources.

The nationalism of the age is echoed in early interpretations of *T. rex*. Osborne's own names for the first specimens (he thought he had two genera) reveal his mental imagery: *Tyrannosaurus rex* ("tyrant king") and *Dynamosaurus imperiosus* ("imperial power"). Themes of power and domination again reverberate in American Museum of Natural History (AMNH) expedition leader Roy Chapman Andrews' (1953) lurid description of *T. rex* in action. Andrews was fully imbued with the spirit of his time, and revealingly titled his report on his paleontological expeditions in the Gobi Desert in the 1920s *The New Conquest of Central Asia* (1933). He clearly harbored no doubts as to how *T. rex* went about its business:

> *Tyrannosaurus rex*, the King of Tyrants, rises on its two powerful hind legs and looks about. . . . Nothing about *Tyrannosaurus* is weak. It is the most terrible creature of destruction that ever walked upon the earth!
>
> Then it settles to the feast. Huge chunks of warm flesh . . . slide down the cave-like throat . . . He stretches out beneath a palm tree. For days, or perhaps a week, he lies motionless in a death-like sleep. When his stomach is empty, he gets to his feet and goes to kill again. That is his life—killing, eating, and sleeping (Andrews 1953, 64–67).

There is much in this scenario with which modern paleontologists would disagree. The model is the stereotypical crocodilian one, in which an ectothermic predator kills robotically. Moreover, palm trees weren't the vegetation of choice in the temperate habitats frequented by *T. rex*. Nonetheless, the description is redolent with the sense of individual domination and power that was so important a part of the aggressive nationalism of the early twentieth century. The chauvinism of the time suggests that it is perhaps not strictly serendipitous that as the description progresses, Andrews (presumably unconsciously) slides into use of the male pronoun.

Figure 12.1   Charles R. Knight's Field Museum of Natural History (Chicago) mural of *Tyrannosaurus* and *Triceratops* (1926–1930), claimed by Czerkas and Glut (1982) to be "unequivocally Knight's most influential work" (p. 81). Reproduced from Czerkas and Glut.

Images are important in paleontology. As noted by Harris (1987), paleontological art directly helps "refine our understanding of [ancient] animals" (1). Perhaps no artist made a more significant contribution in this respect than C. R. Knight, whose most famous creation is arguably his Field Museum mural of the classic confrontation between *T. rex* and *Triceratops* (fig. 12.1). While dinosaur art often verges toward overpopulation, in this evocative mural there are just three dinosaurs. *Tyrannosaurus* dominates the foreground, while in the background a second image of *T. rex* echoes and reinforces the first. *Triceratops* is shown defensively, low and off to the side, in darker hues. *T. rex* is the focal point of the painting and the brightest hues in the work highlight the aggressor's head. There is no doubt in the viewer's mind as to the ultimate victor of this contest. The mural brilliantly renders the psychology of early twentieth-century thinking about *Tyrannosaurus rex*.

THE TIMES THEY WERE A-CHANGIN'

In the time since Andrews wrote his description of *T. rex* in action, social roles significantly changed; notably, those of women, as well as the status of children. Feminism again became a significant force in the 1960s and '70s, explicitly with the 1963 publication of *The Feminine Mystique*. By 1966, the National Organization of Women was founded, and the importance of women's rights and gender equality had once again become part of the national debate. In perhaps an even more fundamental, but related, cultural reorganization, the 1960s and '70s also brought with them heightened social consciousness (and public discussion) of issues historically restricted to the privacy of the home and within the purview of women. Such discussions included the importance of family and the shared parental responsibilities of child-rearing. A culmination of the national discourse about the growing sense of family, community involvement in, and responsibility for, shared childrearing was the 1996 book *It Takes a Village,* by Hillary Clinton.

Other well-documented changes took place during these decades as well. There were popular, generalized calls for "revolution" against "the establishment" in all its guises, as reflected in the political sphere by *The Greening of America* [Reich 1970], including, most notoriously, conventional approaches to sexuality. Antiestablishment heroes (e.g., John Lennon, Bob Dylan, Angela Davis, Abbie Hoffman, Jerry Rubin, Che Gevara, Janis Joplin) tended to be young and energetic, and affected informality in speech and dress when compared with establishment icons (e.g., Richard Nixon, Julius Hoffman, Hubert Humphrey, J. Edgar Hoover, Lyndon Johnson, and William Westmoreland).

Within the bio- and geosciences, episodicity supplanted hoary notions of gradualism: in sedimentology, Lyellian gradualism was replaced by episodic pulses of sedimentation; in evolutionary biology, the gradualist New Synthesis was replaced by the theories of punctuated equilibria and macroevolution; likewise, mass extinctions were reevaluated as episodic rather than gradual events. The timing of these "revolutions" paralleled the social and political upheavals.

In lockstep with related disciplines, paleontology mirrored these social changes, as revolutionary ideas pervaded the field. Emblematic of the times was the 1975 establishment of the journal *Paleobiology,* a self-conscious rebellion against paleontology as it had been traditionally practiced. The journal was started by a group of talented, ambitious young paleontologists, including Thomas J. M. Schopf, Jack Sepkoski, Ralph G. Johnson, Steven J. Gould, Steven Stanley, and Niles Eldredge. They dubbed themselves "young turks," a term harkening back to Mustafa Kemal's 1919–1923 populist overthrow of the Ottoman caliphate. Miller (2000), writing on the twenty-fifth anniversary of the founding of *Paleobiology,* offered this perspective:

Paleontology has been transformed [by paleobiologists] from a science that was once largely descriptive to a more synthetic enterprise in which information about fossils is assembled into databases, and the data are then analyzed to address large-scale questions that could not possibly have been evaluated exhaustively by our pre-1950 forebears, because of the lack of computers. (Miller 2000, 55).

Like many academics, paleontologists donned the uniform of youth and antiestablishment rebellion, some verging on caricature by wearing, at national indoor conferences, jeans, boots, and other informal wear clearly best suited to fieldwork. It was without a trace of irony that R. T. Bakker titled his 1986 book *The Dinosaur Heresies*, even if by 1986, the views in it were no longer particularly heretical.

T. REX GETS AN EXTREME MAKEOVER

Ideas about dinosaurs changed, too, notably ideas about their metabolism, their descendents, and their extinction, and *T. rex* was not immune to these changes. In the century or so since *T. rex* was first identified and named, some thirty-three plus specimens had been found, most of them in the last third of the twentieth century and after; the anatomy of the beast was thus far better understood than when it was first described. Consider, then, this late twentieth-century description of a *T. rex* kill by "dinosaurologist" G. S. Paul:

> The *T. rex* is a monster of 10 tonnes . . . suddenly she and her consorts launch themselves into a horse-speed run, panicking the . . . elephant-sized *Triceratops* into a galloping stampede. . . . The tyrannosaurs' 5-foot-long jaws and 7-inch teeth rip a gaping hole in the herbivore's belly . . .dazed and wobbly, [*Triceratops*] slows . . . The pack of titans [*T. rex*] moves in, yanking, pulling . . . squabbling over the bits. One of the grown-ups leads in the youngsters. Having been hiding in the bush, they now chirp in excitement as they join the feast (Paul 1988, 27).

The breathless, lurid imagery remains, but now here are all the icons of post-1950s social change. In contrast with the lone monsters of an earlier generation, these dinosaurs are highly social and communicate vocally. Childrearing has become an important role of the now-maternal *T.rex*; indeed, the entire pack behaves maternally. These new beasts are characterized by levels of activity modeled after mammals, and the gender assignment, unlike that of R. C. Andrews, is assuredly conscious.

How much of this is pure invention? Discoveries suggest that some of it is at least probable. The antagonistic pairing of *Tyrannosaurus* and coeval large herbivores (*Edmontosaurus* and *Triceratops*) got a boost in the form of a coprolite, referred by its size to *T. rex,* bearing ornithischian bone fragments. Discoveries reinforcing Paul's scenario include bite marks, inferred to originate with *T. rex*, on *Triceratops* and *Edmontosaurus* bones. The idea of pack behavior, likewise, is not strictly outré: enough discoveries containing multiple *T. rex* fossils have been made to hint that the animal did not behave solitarily. More significantly, several of these discoveries have contained a size range of animals—suggesting that juveniles as well as adults constituted part of the pack. Vocal communication is important in the closest living relatives of *T. rex,* birds, and is inferred in other dinosaurs (hadrosaurs); is it too great a stretch to infer that it was used by tyrannosaurids, too? Maternal behavior is immanent to modern birds and is therefore reasonably inferred in nonavian dinosaurs.

Gender determination, however, is more problematic. Despite a popular tendency to refer to all dinosaurs as "he," female dinosaurs obviously existed and, if the bird model is valid, were likely caregivers. In that context, Larson (1997; Larson and Donnan, 2002) claimed that the Field Museum's famous *T. rex,* dubbed "Sue," was female, based upon the position and morphology of the chevron closest to the pelvis. Ultimately, however, Larson rejected his own diagnostic character (Erickson, Lappin, and Larson 2005).

It is true that with new discoveries come new insights, and certainly new discoveries contributed to modern interpretations of tyrannosaur behavior. These interpretations are, indeed, *modern,* but not necessarily *timeless.* Had the cultural ground not been fertile, the significance of these discoveries and insights might not have been appreciated. Why were the first assessments of *T. rex* that of a solitary, dominant predator? Why are the later assessments so different? The change resulted partly from a paradigm shift from the crocodile model of dinosaur metabolism to something more birdlike. But it is not co-incidence that the biggest advances in our understanding of *Tyrannosaurus* parallel the cultural milieu in which those advances were made.

With a broader ethological palette than that which was available to an earlier generation, Horner and Lessem (1993), and Horner (1994), resurrected a longstanding supposition that *T. rex* might be a scavenger. They cited several lines of evidence for their proposal, including (a) large size (and presumably weight) precluding extremely fast running; (b) stout long-bone morphology, a type not associated with extreme cursoriality; (c) bulbous teeth that do not look like the narrow blades that equip unambiguous hunters such as deinonychosaurs; and (d) small hands that are atypical of active theropodan predators.

The *T. rex*-as-scavenger idea elicited the following sputtering response from Larson (1997):

Much ado has been made . . .proposing that *T. rex* was a scavenger. Nothing more than an overgrown garbage disposal. What does the evidence show? *T. rex* had a large brain with a large olfactory bulb. It was capable of recognizing and tracking prey. Even I can smell a dead cow from a mile away downwind. *T. rex* was built for speed and could easily run down *Edmontosaurus* . . . acting in concert, a family group could probably even bring down the formidable *Triceratops* by attacking from the rear while another family member held the attention of the horned and armored front end. *Tyrannosaurus rex* was an endotherm which possessed . . . a large brain, a highly developed sense of sight and keen senses of smell, touch, and hearing. The skull . . . was ideal for capturing, killing, dismembering, and swallowing large chunks of its prey. This largest of all land carnivores possessed surprising speed, balance, and agility. (69).

These views were not likely derived from a dispassionate view of the data. The "large brain" of *T. rex* was larger than some dinosaurs, but well within the encephalization quotient (EQ) range of modern living reptiles. Some deinonychosaurs had significantly larger EQs. The claim that "*T. rex* was built for speed" survives neither modeling, nor trackway evidence, nor anatomical observation. We still know next to nothing of the "balance and agility" of *T. rex*, although it is reasonable to assume that the animal was functionally well integrated. That it may have been exceptional in these characteristics is unsubstantiated. Finally, we have seen that the possibility that *T. rex* exhibited some social behaviors has some support.

A less-impassioned take on *T. rex*-as-scavenger was provided by G. M. Erickson (1999). Asking the rhetorically "Hawk or Vulture?" Erickson answers:

Within *T. rex*'s former range exist bone beds consisting of hundreds and sometimes thousands of edmontosaurs that died from floods, droughts and causes other than predation. Bite marks and shed tooth crowns in these edmontosaur assemblages attest to scavenging behavior by *T. rex*. Jacobsen has found comparable evidence for albertosaur scavenging. Carpenter, on the other hand, has provided solid proof of predaceous behavior, in the form of an unsuccessful attack by a *T. rex* on an adult *Edmontosaurus*. The intended prey escaped with several broken tailbones that later healed. The only animal with the stature, proper dentition and biting force to account for this injury is *T. rex*. (47).

"Dinosaurs," notes Mitchell (1998, 149), "symbolize the dominant master race that commands a global empire. . . ." And culturally, at least, *T. rex* is surely the dominant dinosaur. Larson's (1997) outraged reaction to the proposal that *T. rex* scavenged suggests that the cultural appeal of *Tyrannosaurus*-as-alpha-beast transcends mere evidence. Although we may hope to be far removed from the colonization-hungry time of early twentieth-century nationalism, identification—either national or personal—with the "ruling" beast can be hard to avoid.

DINOSAUR MATERNITY, NESTS, EGGS, AND EMBRYOS

As part of the R. C. Andrews-led, AMNH expeditions to the Gobi Desert in the 1920s, Walter Grainger discovered nests of eggs in south central Mongolia. Andrews famously and incorrectly attributed these to the small ceratopsian *Protoceratops*. Attribution notwithstanding, this was the first definitive demonstration that dinosaurs laid eggs, and justifiably generated considerable excitement at the time. Ironically, Andrews was at something of a loss to explain the attraction:

> I have often wondered why the dinosaur eggs hold such interest for the layman. I suppose it is because of their great age. Ninety-five million years is the estimated age of the strata in which the eggs were embedded. Probably that is not far wrong (Andrews 1933, 664).

Dinosaur eggs remained extremely rare until 1978, when Jack Horner and Bob Maleka discovered the remains of what they interpreted to be a fossilized nesting ground in eastern Montana replete with hadrosaur eggs, juveniles, and adults. Based upon these remains, Horner developed a series of stunning ethological hypotheses that provoked extraordinary media attention. The title of the initial publication on the subject signaled the brave new world: "Nest of juveniles provides evidence of family structure among dinosaurs" (Horner and Maleka 1979). Later publications revisited and elaborated the idea that dinosaurs cared for their young. The work and related studies conclusively demonstrated something that paleontologists had long suspected—namely, that dinosaurs functioned in social herds. Even more tantalizing, it breathed promise of dinosaur endothermy, an idea that was generating much excitement at the time.

There were many reasons for the idea of altricial behavior in dinosaurs to be attractive. For one, it fit well into the newly resurrected idea of dinosaur endothermy. Moreover, the growing abundance of juvenile dinosaur fossils— and the study of their growth in the context of their bone histology—opened up major, unforeseen avenues of research in dinosaur ontogeny and physiology. The research that underpins current interpretations of dinosaur physiology and behavior arguably represents the single most important contribution to our understanding of Dinosauria since the late 1800s.

Still, the timing is somewhat suspect. It had been believed since before the turn of the twentieth century that dinosaurs laid eggs. Babies (or embryos) were a very rare commodity prior to Horner and Maleka's discoveries of the early 1980s, but were nonetheless known. The microscope-based histology that was used for the new behavioral inferences was not particularly new. Parental care was suggested for dinosaurs as early as 1978, and while it was implied by the discoveries since the 1980s, demonstrating it unambiguously continues to be elusive.

So what changed? Why was there a sudden increase in the discovery of fossil eggs and babies? And why was their meaning so long overlooked? Was it fortuitous that brooding was only finally identified in *Oviraptor* in the late 1900s, after the discoveries of Horner and colleagues?

Obviously, the ideas were themselves mutually reinforcing. The discovery of herding, eggs, babies, and nests provoked thought about behaviors that reflected on dinosaur physiology, itself a subject of considerable interest. But above and beyond this, the ideas resonated strongly because of the contemporary heightened consciousness of maternal roles, families, offspring, and socialization. A clue is found in Horner's name for the new juvenile-bearing hadrosaur: "*Maiasaura.*" *Maiasaura* uses the conventional dinosaur suffix (*saurus*) in the female gender. "*Maia-*" refers to motherhood. Interestingly, Horner translated *Maiasaura* in *Digging Dinosaurs* (1988) as "good mother dinosaur." *Maiasaura* wasn't just a mother dinosaur; she was a "good" mother dinosaur! And Roy Chapman Andrews and two subsequent generations of highly competent paleontologists knew that dinosaurs laid eggs, knew of (rare) dinosaur juveniles, knew of presumed dinosaur nests, and believed that they knew the identities of the mothers—but were unable to recognize the real significance of the discovery. It would appear that they were not culturally ready.

DINOSAUR EXTINCTION

Within less than fifty years of Owen's coining the word *Dinosauria* (1842), it was known that the group did not persist past the Cretaceous-Tertiary (K/T) boundary. Since then, the compelling question has been the cause of the dinosaur extinction. Theories exclusively explaining the dinosaur extinction are abundant; theories that meet the twin criteria of testability and explaining the full range of what is known about the breadth and pace of the K/T extinctions, however, are extremely rare.

A recent theory that met those criteria was proposed by Alvarez et al. (1980). In the context of the many absurd antecedent theories, the Alvarez et al. theory appeared to have little to recommend it. The idea was based upon only three localities (two in Europe, one in New Zealand) where the Alvarezes (*père et fils*) and coworkers recorded elevated levels of the Platinum-Group metal iridium (Ir) at the K/T boundary, as identified by marine microorganisms. As Ir was supposed to have an extraterrestrial source, the inference was that the elevated levels must represent extraterrestrial input. An asteroid was thus proposed as the vector for the iridium; its explosive collision with Earth would then distribute the iridium globally. Based upon that supposed global Ir distribution (the three localities), the theoretical size of this asteroid was calculated. As for the dinosaur extinction that the asteroid was presumed to have caused, no data were presented. Instead, a scenario was offered:

> A second food chain is based upon land plants. Among these plants, existing individuals would die, or at least stop producing new growth, during an interval of darkness, but after light returned they would regenerate from seeds, spores, and existing root systems. However, the large herbivorous and carnivorous animals [e.g., dinosaurs] that were directly and indirectly dependent upon this vegetation would become extinct (Alvarez et al. 1980, 1106).

In short, despite the absence of any real data about dinosaurs and the pace of their extinction, the theory invoked a deus ex machina ending for dinosaurs (and other organisms) at the end of the Cretaceous. On the face of it, it was absurd—so why did it catch fire?

The most important answer to this question is that it withstood tests, subsequently adduced, that might have falsified it. But it also worked because the geosciences as a discipline were culturally ready for an idea like this. As we have seen, the geosciences were undergoing a revolution paralleling the societal one. The growing recognition of the importance of episodic events as dominant contributors to the sedimentary record was becoming manifest. Episodicity implied the unusual rather than the commonplace. And acceptance of the unusual as Earth-shaping forces helped make conceivable extraterrestrial influences on earthbound events. As we have seen, the hypothesis that evolution proceeds by punctuated equilibria had recently gained adherents, and asteroids were obvious, if unanticipated, potential punctuating agents.

The theory played well within the culture of science as well. It was multi-disciplinary, at a time when the importance of multidisciplinary approaches to scientific questions was coming to be appreciated. The team was headed by a youthful scientist and supported by outsiders to the paleontological establishment although, ironically, coauthor Luis Alvarez was a highly entrenched member of the U.S. scientific establishment.

Finally, the idea resonated within the general social and political context of the times. In 1977, the movie *Star Wars* hit the theaters and rapidly attained cult status. "Star Wars" became the popular name of the antimissile defense program instituted by Ronald Reagan for the protection of the United States from intercontinental missile attacks. The idea, therefore, that destruction could come from above—even space—had reached popular radar as of the 1980s.

It went further, however. Western politicians analogized national security with the precariousness of the position of dinosaurs at the end of the Cretaceous. Was the conventional military arsenal a *dinosaur*—a word signifying large, dated, clueless, and ultimately extinct? More tellingly, was the U.S. population itself as vulnerable as the dinosaurs?

It was not coincidental that NASA's Lunar and Planetary Institute (LPI) took a keen interest in the apparently arcane question of the extinction of the dinosaurs. Indeed, it sponsored a series of conferences on the effects of large-body impacts with the Earth, starting with the convening of the first Snowbird Conference of 1981. Likewise, it was not a coincidence that estimates of the explosive force of the impact, presented at LPI-sponsored events, were measured and presented in megatons, a metric not particularly familiar to most scientists. Nor was it coincidence that in 1983, shortly after the Alvarez hypothesis was unveiled, Carl Sagan and colleagues invented the term *nuclear winter*, a term used to describe the climate of a postatomic Earth. The "winter" was supposed to be engendered by atmospheric aerosols blocking sunlight, just as the Alvarez et al. hypothesis proposed. Finally, it was no coincidence that in 1984, when astronomer Richard Muller first proposed that there exists an invisible companion star to the sun, which was responsible for the K/T asteroid, he called that companion star, again without irony, a "Death Star"—a term lifted directly from *Star Wars*. Reagan was said to have blurred the line between reality and the movies, but it would seem that he was not the only person in the 1980s to do so. And early in that decade, the connection between the Cold War politics of the United States and end-Cretaceous dinosaurs became explicit.

CONCLUSION

Science is conventionally portrayed as distinct from its cultural context. The implication is that an objective truth exists apart, and that science and scientists are all about uncovering it. More recent portrayals of science, however, suggest that, as trenchantly stated by K. M. Parsons (2001), "'objective knowledge' is an oppressive illusion, that all knowledge is inevitably political" (80). He dichotomizes these two viewpoints as "rationalism" and "constructivism." Rationalists, he says,

> Affirm the existence of an external, independent (of human wishes or concepts), non-socially constructed physical world which is at least partially knowable; that is, we can observe (either directly or with instruments), measure, and experiment with that world and thereby ascertain certain facts about it (81).

Constructivists, on the other hand, claim that

> the 'nature' that scientists pretend to study is a fiction cooked up by the scientists themselves—that, as Bruno Latour puts it, natural objects are the *consequence* of scientific work rather than its *cause* (Cartmill 1999, quoted in Parsons, 81).

"A corollary," he notes, of the constructivist claim is that "any putative non-social physical reality has negligible bearing on the formation of our beliefs" (Parsons, 2001, 82). Parsons' question, then, was, "Are dinosaurs social constructs?" This is an interesting existentialist problem; however, because it is fundamentally untestable, it is not within the purview of science.

My conclusion that paleontology is highly influenced by its times may be disheartening for those who would see it as divorced from apparently unrelated social and political influences. But science is a human endeavor and is bound to reflect the humanity of those who carry it out. Given that fact, Parsons asks whether science is simply a suite of social or political agendas, new ones supplanting older ones. In a sense, should we view T. S. Kuhn's scientific paradigms as, when stripped to their bare essentials, sequential social and/or political agendas?

My preference is that we should not, although the claim, as I have noted earlier, is effectively untestable. Barring existential considerations, fossils at least *exist* regardless of social and political context, and it is striking that in many cases, nineteenth-century descriptions, illustrations, and even interpretations of fossils are still valid. Parsons (2001) suggests that methodological innovations permit hypotheses to be tested more rigorously; however, this may beg the question, since the methodological innovations may themselves be the result of social and/or political agendas.

Regardless, insofar as humans can test human hypotheses, the social revolutions that appear to have so strongly influenced the scientific ones have produced hypotheses that more robustly withstand falsification than their antecedents. The freeing of all the Earth sciences from the tightly clenched grip of Lyell's extreme uniformitarianism can only be viewed as a blessing regardless of whether some of the driving forces were not actually scientific. It should be acknowledged that the social and political context of fossil discoveries are a necessary part of the process of understanding them, but it is equally clear that as observations and interpretations resist falsification in widely divergent social and political climates, context isn't the whole story.

ACKNOWLEDGMENT

This manuscript results from a thoughtful invitation to address the College of Wooster on the occasion of the endowment of the Lewis M. and Marian Senter Nixon Professorship of Natural Sciences in the Department of Geology. An abbreviated version of it was presented orally at the 2004 annual meeting of the Geological Society of America in Denver. Finally, it has greatly benefited from careful readings by Drs. R. H. Dott, Jr., and M. A. Wilson; however, any errors of fact or interpretation are, obviously, solely my responsibility.

REFERENCES

Alvarez, L. W., W. Alvarez, F. Asaro, and H. V. Michel. 1980. Extraterrestrial cause for the Cretaceous-Tertiary extinction. *Science* 208, p. 1095-1108.
Andrews, R. C. 1933. Explorations in the Gobi Desert. *National Geographic* 63:653–716.
———. 1953. *All about dinosaurs.* New York: Random House.
Bakker, R. T. 1986. *The dinosaur heresies.* New York: William Morrow.
Cartmill, M. 1999. Review of *Mystery of mysteries: Is evolution a social construct? Report of the National Center for Science Education* 19:49–50.
Czerkas, S. M., and D. F. Glut. 1982. *Dinosaurs, mammoths, and cavemen—The art of Charles R. Knight.* New York: E. P. Dutton.

Erickson, G. M. 1999. Breathing life into *Tyrannosaurus rex*. *Scientific American* 281:42–50.

Erickson, G. M., Lappin, K., and Larson, P. L. 2005. Androgenous rex: The utility of chevrons for determining the sex of crocodilians and non-avian dinosaurs. *Zoology* 108:277–86.

Harris, J. M. 1987. Introduction. In *Dinosaurs past and present* (vol. 1)., ed. S. J. Czerkas and E. C. Olsen, 1–6. Seattle, WA: Natural History Museum of Los Angeles County and University of Washington Press.

Hay, J. 1899. First open door note: Letter from John Hay to Andrew D. White: Papers relating to the foreign relations of the United States, 1899, 129–30, accessed at http://www.let.rug.nl/~usa/D/1876-1900/foreignpolicy/opendr.htm.

Hochschild, A. 1998. *King Leopold's ghost*. Boston: Houghton Mifflin.

Horner, J. 1994. Steak knives, beady eyes, and tiny little arms (A portrait of *T. rex* as a scavenger). DinoFest Proceedings, *Paleontological Society Special Publication* 7:157–64.

Horner, J. R., and J. Gorman, J. 1988. *Digging dinosaurs*. New York: Workman.

Horner, J. R., and D. Lessem. 1993. *The complete* T. rex: *How stunning discoveries are changing our understanding of the world's most famous dinosaur*: New York: Simon and Schuster.

Horner, J. R., and R. Maleka. 1979. Nest of juveniles provides evidence of family structure among dinosaurs. *Nature* 282:296–98.

Larson, P. L. 1997. The king's new clothes: A fresh look at *Tyrannosaurus rex*. In *Dinofest international: Proceedings of a symposium held at Arizona State University*, ed. D. L. Wolberg, E. Stump, and G. D. Rosenberg, 65–71. Philadelphia: The Academy of Natural Sciences.

Larson, P. L., and K. Donnan. 2002. *Rex appeal*. Montpelier, VT: Invisible Cities Press.

Miller, A. I. 2000. Conversations about Phanerozoic global diversity. In *Deep time— Paleobiology's perspective*, ed. D. H. Erwin and S. L. Wing, 53–73. *Paleobiology* (suppl.) 26:53–73.

Mitchell, W. T. J. 1998. *The last dinosaur book*. Chicago: University of Chicago Press.

Parsons, K. M. 2001. *Drawing out Leviathan: Dinosaurs and the science wars*. Indianapolis: Indiana University Press.

Paul, G. S. 1988. *Predatory dinosaurs of the world*. New York: Simon and Schuster.

CHAPTER THIRTEEN

Reg Sprigg and the Discovery of the Ediacara Fauna in South Australia: Its Approach to the High Table

Susan Turner and David Oldroyd

On 16 April 2005, a metal marker was hammered into the rocks at a site in the Flinders Ranges by Mike Rann, premier of the state of South Australia. Other big shots were there, too, such as the well-known writer Tim Flannery, then director of the Museum of South Australia, and John Hill, minister of the environment for South Australia.[1] A set of Australian commemorative stamps was also issued (on 21 April) to mark the occasion. The marker is intended to serve as a global reference point for the base of a new geological period: the "Ediacaran," the first to have been established for 120 years. This major subdivision of the stratigraphic column is the oldest to be defined on the basis of fossils, and is the only one, thus far, to be established in the Precambrian. It is also the only one that has received its definition in the southern hemisphere, though some stages have been defined in that part of the world. This chapter tells something of the events that led to the placement of the golden spike at the base of the Ediacaran and how those ancient fossils were slowly and sometimes painfully admitted to the High Table of paleontology and stratigraphy. We do not dwell here on the debates about their dating, taxonomy, anatomies, and manner of preservation, or their status in the evolutionary tree of life.

The story of the investigations of the Burgess Shale fauna by Charles Doolittle Walcott and others is known worldwide and has attracted much attention (Gould 1989; Conway Morris 1998; Yochelson 1998). Generally less well known but almost as important to paleontologists and to understanding of the history of life on Earth, was the discovery in 1946, in semidesert outback South Australia, of a remarkable macrofossil site with well-preserved remains of soft-bodied animals, some resembling jellyfish; others primitive worms;

and others resembling sea-pens. They occurred in clayey laminae in ancient sandstones/quartzites, which were subsequently regarded as Precambrian (where most people did not expect to find macrofossils). Five representative examples are shown in figures 13.1A–E.

In his work on the Burgess Shale, Walcott was already recognized as a major paleontologist whose ideas certainly counted. By contrast, the Australian discovery was made by a young geologist working with the Geological Survey of South Australia. His ideas were initially ignored or rejected, and were not accepted until they were taken up (or taken over) by a distinguished European paleontologist who accepted an appointment at The University of Adelaide after World War II, partly as a way of escaping from the torments of Europe.[2]

The young geologist was Reginald Claude Sprigg (1919–1994), a remarkable man who had an extraordinarily varied career, making contributions to paleontology, mineralogy, economic geology (especially uranium, nickel, and petroleum prospecting, on land and under water), specimen collecting (primarily for academic but also for commercial purposes), and conservation (Cooper 1995). During his later career, he acquired a large pastoral lease at Arkaroola about 600 km north of Adelaide, not far from where he had first found his "jelly-fish," and successfully turned it into what would today be called an eco-resort. He was also a prolific writer and popularizer of geoscience. But perhaps because of his polymathic interests Sprigg was not generally regarded as a serious paleontologist, and it was the aforementioned European-trained paleontologist who was chiefly responsible for getting Sprigg's work known and accepted, though Sprigg had already written it up in the *Transactions of the Royal Society of South Australia* and in his little privately published journal, *The Australian Amateur Mineralogist*. The fossils found by Sprigg were named and described by him, and drawings and photographs published, well before the European paleontologist became involved in the investigations (Sprigg 1947a, 1949). It was only in 1983, after the period of Sprigg's work described in the present chapter, that his contributions were formally recognized by the award of an Order of Australia (the equivalent of a British knighthood). He was then cited for "service to industry, particularly in the fields of geology and petroleum exploration." He also received honorary doctorates from the University of Adelaide and the Australian National University.

The background to Sprigg's paleontological work is important. Sir Tannatt William Edgeworth David (1858–1934), professor of geology at The University of Sydney, believed (like some others) that animal fossils would one day be found below the Cambrian, for the simple reason that such complicated creatures as (say) Cambrian trilobites must have had simpler evolutionary

Figure 13.1A *Ediacaria flindersi* Sprigg (Sprigg 1947a: plate V, figure 1). Reproduced by courtesy of the Royal Society of South Australia.

Figure 13.1B *Dickinsonia costata* Sprigg (Sprigg 1949: plate XVIII, figure 2). Reproduced by courtesy of the Royal Society of South Australia.

Figure 13.1C *Spriggina floundersi* Glaessner (Glaessner 1958: figure 1). Reproduced by courtesy of the Royal Society of South Australia.

Figure 13.1D *(left) Charniodiscus arboreus* Glaessner (Cloud and Glaessner 1982: figure 2A, p. 783).[3] Reprinted with permission from AAAS.

Figure 13.1E *(right) Mawsonites spriggi* Glaessner and Wade (Glaessner and Wade 1966: plate 99, figure 1) holotype. Photo provided by the fossil's finder, Dennis Walter, and published with his permission.

ancestors, (but microbial fossils—stromatolites—had long been known in Precambrian rocks.) Edgeworth David had found structures, well down in the "Lower Division" of the Neoproterozoic (Precambrian) rocks, which he thought were "giant Annelids and large Arthropods," near Adelaide (in a quartzite at Tea Tree Gully). David's ideas were published posthumously, with the entomologist Robin Tillyard, as a 122-page memoir with a commercial publisher (David and Tillyard 1936), having failed acceptance by the Royal Society of London in 1932. The general opinion was that these (unconvincing) fossils were sedimentary structures seen with an overly imaginative eye. This was the view of David's former student, the famous explorer and professor of geology and mineralogy at The University of Adelaide, Sir Douglas Mawson (1882–1958; with whom David had had a long and somewhat strained relationship since their expedition to the South Magnetic Pole back in 1909[4]).

In 1936, aged seventeen and studying at the Adelaide Technical High School but also employed as a "part-time Acting Curator of the Mineral and Fossil Museum at the South Australian School of Mines," Sprigg felt there were prospects for finding Precambrian macrofossils. With advice from the doyen of South Australia geology, Walter Howchin, he searched below the Cambrian archaeocyathid beds at Sellick Hill south of Adelaide, and at Ardrossan on Yorke Peninsula northwest of Adelaide, finding what he took to be a eurypterid. Sprigg later described the specimen quite minutely. It had a:

> mould of part of a 4 cm. wide, and squarish, cephalon of a plainly eurypterid-type animal. In the upper right corner was a half-moon-shaped "eye," opposite which a 4-segmented tufted antenna protruded, and below it was a four segment "pleopod" [i.e., an abdominal appendage of a crustacean] or swimming arm. Three well preserved thoracic segments continued below the head shield each approximately one centimetre deep. The whole impression was about 7 cm. long and would have been part of an animal of 15–20 cm. in length (Sprigg 1988, 48).

He was sure in his own mind that it was a eurypterid, but showing it to Adelaide's geological big shots (Sprigg was already attending meetings of the Royal Society of South Australia) his claim was discounted, particularly by Mawson, who, as said, had rejected David's claims about the Tea Tree Gully fossils, but was nonetheless looking for Precambrian fossils himself. Thus young Sprigg may have been a victim of disagreements or rivalries at the Australian High Table between Mawson and David.

But undeterred, Sprigg made some plaster casts and presented the original specimen to the Tate Museum of Adelaide University's geology department, where it remained for several years before being transferred to the Third Year Geology teaching cabinet. In 1939, a student geological club was formed at the University, where Sprigg was by then studying geology. At one of its meetings, he gave a talk on David's fossils and exhibited his claimed eurypterid, which he dubbed *Sellicksia*. Mawson didn't normally attend the club's meetings but he did so on this occasion, and attacked Sprigg's claim about the fossil, saying that he was "headed for the madhouse and would possibly never be allowed to graduate if he . . . [did] not change his attitudes" (Sprigg 1988, 48). Feelings were running high.

The matter ended deplorably. In 1948, Sprigg was visiting the United States and showing people his recently found Ediacara specimens (see the following). He also showed his claimed eurypterid cast to Otto Haas (b. 1887) (an ammonite specialist) at the American Museum of Natural History in New York (said by Sprigg to have been at the Smithsonian, though this does not mesh satisfactorily with other records). Anyway, Haas was apparently impressed and asked to see the original specimen. But on returning to Adelaide, Sprigg found that it had been thrown out, on Mawson's instructions, without photos taken or drawings having been made. Hmm!

It was later determined that the strata where the specimen had been found were Cambrian, not Precambrian. But be that as it may, early in his career Sprigg learned about the way things can happen in paleontology, and how egos may influence what happens at high tables! Advanced forms of life were not expected below the Cambrian archeocyathid strata. And after the fiasco of David's Tea Tree Gully claims, the Adelaide establishment seemingly wanted no more of such things.

After a period of war service (in Australia), Sprigg gained employment (1944–1954) as an assistant geologist with the Geological Survey of South Australia, where, with the post-war boom, there was great interest in exploring for mineral deposits, including uranium, copper, and other metals. In 1946, he was mapping near an old lead–silver mine at a place called Ediacara (meaning, in the local Aboriginal language, a veinlike spring of water) near Lake Torrens, about 600 km north of Adelaide. There he found Cambrian archaeocyathids in limestone, in an area previously thought unfossiliferous.

The limestone was underlain by argillaceous beds, containing metallic ores. And below them were flaggy rocks that overlay a widespread sandstone/ quartzite unit, important in the area and regarded as Precambrian. The flags had some clay partings that Sprigg thought looked promising for fossils; and sure enough he recorded in his notebook (for 31 March, 1946[5]) his discovery of "queer markings suggestive of jellyfish, and also tracks suggesting other creatures" (Sprigg 1988, 49).[6]

Sprigg collected a specimen of the "jellyfish" but was prevented from doing much further work because of a foot injury that required him to return to Adelaide. He showed his specimen (which he named *Ediacaria flindersi*) to colleagues, who were apparently unimpressed. Then Sprigg exhibited his specimen at the meeting of the Australian and New Zealand Association for the Advancement of Science at Adelaide in 1946 and gave a paper about his finds at the Perth meeting in August 1947, with Mawson and the paleontologists Curt Teichert (1905–1996) and Martin Glaessner (1906–1989; the previously mentioned "distinguished European geologist") present; but only the title of his paper was printed in the published *Report* of the meeting (Sprigg 1947b). Again the big shots were skeptical, but Teichert encouraged Sprigg to publish a description of the specimen.

At the end of 1946, Sprigg had shown a different group of big shots, this time politicians and bureaucrats, around the area of the mine workings at Ediacara. The group included the premier of South Australia, Thomas Playford, the highways commissioner, David Fleming (who had a background in engineering geology), and Sprigg's boss, the director of mines, Ben Dickinson. Sprigg judiciously steered them to a spot near his fossil locality for a "billy break," where he was able to collect some thirty good specimens. Dickinson was impressed, whereas politician Playford couldn't see how South Australia could benefit from "a lot of old fossils."

Back in Adelaide, Mawson was eventually converted, and encouraged Sprigg to publish an account of the fossils, which he did in December 1947 in the *Transactions of the Royal Society of South Australia* (paper read 8 May that year), introducing new names (*Ediacaria flindersi, Beltanella gilesi,*[7] *Cyclomedusa davidi, Dickinsonia costata,* and *Papilionata eyri*). Descriptions, drawings, and photographs were provided. We note how both David and Dickinson became immortalized! Sprigg specifically acknowledged the assistance of Teichert, who had accompanied him on a visit to Ediacara (but he did not follow up on his work there). Further work was published in the *Transactions* in 1949, and Sprigg's specimens were deposited in the Museum.

In his 1947 paper, Sprigg was uncertain as to the age of the strata where the fossils were found. He placed them at the top of the Adelaide Series (then regarded as Proterozoic–Lower Cambrian), which underlay the agreed-upon Cambrian, but hesitantly called them Eo-Cambrian ("dawn of," or earliest Cambrian). When used today (as it is occasionally), this term is taken to refer to the Riphean (of Russian geologists) or late Precambrian rocks; but as used by Sprigg it tended to suggest earliest Cambrian. Indeed, he stated that he thought his specimens were "probably Lower Cambrian." In his subsequent recollections, he said that he originally called them "Upper Proterozoic cum Eo-Cambrian" (Sprigg 1988, 50), which was more suggestive of Precambrian. However, he could find no angular unconformity between the Cambrian limestones and the underlying sediments containing the Ediacara fossils, so there was some reason for not placing them in the Precambrian. But this is the paradox, for if one wants to find the boundary between Cambrian and Precambrian one would like a continuous succession in which the macrofossils first appeared, and place a stratigraphic "golden spike" there, at a time-determinable point. An angular unconformity could represent a clear distinction between one system and the next, but also an indefinite period of time. So what to do? (We are speaking of the 1940s, when absolute-age determinations by radiometric methods were not well developed and spikes, golden or otherwise, had not been dreamed of.)

There was a further special problem when looking for a marker or definition of the Precambrian–Cambrian golden spike. Was it to be the first appearances of macrofossils? If so, then Sprigg's fossils could properly be called Cambrian, according to traditional usage. But if the criterion was the first appearance of shelly macrofossils then his specimens could reasonably be allocated to the Precambrian. It was, in a sense, a semantic question. And it was not one that was settled in the late 1940s. (One of us, Oldroyd, recalls it being raised as an issue for undergraduate tutorial discussion at Cambridge in 1956–1957.)

Now comes the really sad part of this story. Between the reading and publication of his 1947 Royal Society paper Sprigg, "feeling that the discovery was epochal" (as in fact it was to become so for paleontologists, if not politicians), prepared a letter to *Nature* about his discoveries and posted it on 15 October, 1947 (Sprigg 1988, 50). But it was turned down. This document has not been discovered and may no longer be in existence; and according to Henry Gee, of the present editorial staff of *Nature,* the journal does not have records from that period. So we don't know who may have been responsible for blocking what should indeed have been an epochal paper. (Also, we don't know why Sprigg didn't offer his paper to a somewhat less-prestigious international journal.) Things were no better in 1948, for Sprigg was unable to present his findings at the International Geological Congress held in London that year, despite the support of a certain "Dr Shackelton," said by Sprigg (1988, 50) to have been at the British Museum.[8] Thus it would appear that the world was not yet ready to accept, or even seriously consider, Sprigg's ideas.

It is interesting to speculate who may have been Sprigg's *Nature* referee. It seems likely to us that the journal would have consulted some Australian authority who knew the geology of South Australia. That being the case, it may well have been Mawson. He would have been regarded as "Mr. Geology" in South Australia at that time, and his opinion could have been sought. He was definitely close to the High Table as a result of his celebrated Antarctic work. We also know that Mawson had been antipathetic to Sprigg's earlier eurypterid claim and had talked (rhetorically, not seriously) about failing him for his degree, and he had caused Sprigg's prized specimen to be discarded. Also, we know that Mawson was himself looking for signs of Precambrian macrofossils in South Australia, for Sprigg actually wrote that he felt sorry for Mawson, since he pipped him to the post in the discovery of such organisms (Sprigg 1989, 201). There is, however, no evidence in Mawson's papers in Adelaide that he refereed Sprigg's paper, and the editor of *Nature* (L. J. F. Brimble, a botanist) may have put it straight in the bin without consulting anyone. Moreover, Sprigg recorded that following the return to Adelaide after the 1946 expedition with the bureaucrats Mawson "agreed this time as to . . . [the fossils'] momentous importance" (Sprigg 1989, 201).

As for Glaessner, he did not (at least publicly) take a serious interest in Sprigg's work until the mid-1950s, after he had moved from Melbourne to Adelaide University in 1950, despite the fact that Sprigg's specimens had been well described and illustrated back in the 1940s. And when Glaessner did take up the study of Ediacara fossils, he initially worked largely on materials collected by Sprigg and his coworkers. It may be noted that Mawson died in 1958. Possibly he had to depart from the High Table before Sprigg's specimens could be placed thereon (in print) by an expert. (In addition, fossils analogous to those at Ediacara were discovered by schoolboys in Precambrian rocks at Charnwood Forest in Britain and were written up by Trevor Ford [1958] of Leicester University, to whom the boys reported their discovery, which event may have made the Ediacara specimens more palatable![9] However, Ediacara-type fossils were in fact previously known from Namibia (see note 20).

So in the event, it was Glaessner, not Sprigg, who belatedly put the Ediacara fossils on the Table. They failed to gain a place there earlier, despite Sprigg's publications in the *Transactions of the Royal Society of Australia*, his attempt to publish his findings in *Nature*, his presentations at meetings of the Australian and New Zealand Association for the Advancement of Science, and his discussions with scientists in the United States and Britain.

Nevertheless, there were developments in Adelaide in the decade following the first publication of Sprigg's work. Late in 1956 and in 1957, two amateur workers, Hans Mincham (a primary-school teacher, but a member of the Royal Society of South Australia and subsequently the information and education officer at the South Australia Museum) and another schoolteacher/mineral and fossil collector, Ben Flounders, visited the Beltana and the Ediacara sites, looking for further specimens. (It seems that they were prompted by Sprigg's [1956] article in his amateur magazine.) They were highly successful. Good photographs were taken and shown to Glaessner's student, Mary Wade (1928–2005), and specimens were presented to the South Australian Museum. Returning from overseas, Glaessner became seriously interested in the Ediacara fauna, but the timing (1957–1958) may (or may not) be coincidental. Sprigg was by then out of the way, having left government service to set up his own geo-consulting company in 1954 (Geosurveys of Australia Pty) and (in the same year) establishing the now-powerful oil exploration company SANTOS (South Australia Northern Territory Oil Search: later "Santos"), of which Mawson was one of the founding directors. However, Sprigg did not forget the fossils at Ediacara, and through his subsidiary company, Specimen Minerals, and the support of its manager and his partner Dennis Walter, he determined to collect Ediacara fossils and donate them to academic institutions (Sprigg 1958), while selling some specimens to collectors in order to finance the fieldwork. Walter (see note 12) recalls that Mawson and Glaessner visited the company's specimen store and selected every specimen they wished for Adelaide University, so that in fact little was left for sale.

Soon Glaessner began directing his full research effort toward the Ediacara fossils, being assisted particularly by Mary Wade (who completed her PhD under him in 1959 on a topic in micropalaeontology)[10] and another student, Brian Daily (1931–1986).[11] A party of six, organized by the South Australian Museum and led by Daily, including Wade, Mincham, and Flounders, went north in March 1958 to examine the Ediacara site and collect further specimens (some of which, according to Walter, had previously been deliberately exposed to weathering by himself and a fellow collector [a Specimen Minerals employee] to facilitate collecting). In four days, according to Glaessner, their finds filled two small trucks and a trailer!

But in May that year the Ediacara site was designated a fossil reserve under the control of the State Minister of Education and the South Australian Museum, and was closed to unauthorized collectors, including Sprigg. In fact, the proclamation of the reserve is believed to have been aimed primarily at Sprigg and his commercial activities with Specimen Minerals, for he had plans to quarry the site.[12] Needless to say, the closure only served to attract collectors: it could not be protected in its remote location. So, as Sprigg later put it, the area was "literally rape[d]" and an illegal international market in the ancient fossils quickly sprang up. A further University collecting expedition was organized in October 1958, and authorized collecting occurred for several years thereafter.

So, despite the activities of both amateur and academic collectors (as a PhD student and research curator in the 1960s, one of us (Susan Turner) recalls the haul brought back to the University of Reading Geology Departmental Museum by Roland Goldring), some 1,500 specimens were assembled in Adelaide and the Ediacara fossils were subjected to professional scrutiny by Glaessner and Daily (1959), and Glaessner and Wade (1966). Glaessner also published single-authored articles on the fossils in 1958, 1959, 1961, and 1971, and in a book: *The Dawn of Animal Life: A Biohistorical Study* (1984). Wade eventually did even more, publishing at least nine single-authored works, culminating in her contribution on the fossil Scyphozoa (marine coelenterates) in *Traité de Zoologie* (Turner 2007).

It should be noted that in his first publication, which he wrote without knowledge of Trevor Ford's work in Britain, Glaessner (1958), like Sprigg, supposed that the fossils were Cambrian. The term *Ediacarien*, as the oldest paleontologically characterized stage (*sic*), with outcrops worldwide, was introduced by Henri and Geneviève Termier (1960, 82) as part of the *sous-système Eocambrien* and Glaessner's 1961 article in *Scientific American* did much to bring the attention of the remarkable South Australian fossils to the world at large. The paleontologists at the High Table had become interested, and the Ediacara fossils were giving a shake to the whole stratigraphic column.

Glaessner accepted a fair amount of Sprigg's original taxonomic work but in his study with Daily he emphasized that careful comparative work would require many years of work, which was perhaps an unjustified suggestion that Sprigg's original descriptions and taxonomic suggestions were somewhat hasty. Also, in his *Dawn of life* (43), Glaessner suggested that it was "[t]o his [Sprigg's] surprise" that the strange fossils were first found near the old mine at Ediacara. But Sprigg later rejected this suggestion, saying that his "search and ultimate success in discovering the world's first Precambrian animal fossils was deliberate and represented the culmination of a decade of serious search" (1988, 50).

Be that as it may, after the 1950s Sprigg largely dropped out of frontline research so far as the Ediacara fossils were concerned. And why not? By that time, he had committed himself to earning a living from the petroleum industry and other commercial activities and was not funded by the University, the

Museum, nor the Geological Survey. He recorded his pleasure, however, that Glaessner named an annelid family after him: Sprigginidae. Sprigg also had a genus and species named after him: *Spriggina floundersi* (interpreted by Glaessner as a kind of segmented worm) and *Mawsonites spriggi* (a multilobed disc-shaped fossil impression, thought then to be the "bell" of a kind of jellyfish). That was nice.[13] Sprigg's former chief, Ben Dickinson, was honoured by having a genus, again purported segmented flatworms, named after him (by Sprigg): *Dickinsonia*. Fossils resembling those found at Charnwood Forest (United Kingdom) were also discovered: *Charniodiscus opposites* (a sort of fronded colonial animal like a sea-pen, with a disc that might have served as its anchor, which could have been the same sort of thing as *Mawsonites* or Sprigg's original *Ediacaria flindersi*).

At the time of Sprigg initial work it appeared to him (as previously mentioned) that the sandstones containing the Ediacara fauna were conformable to the overlying Cambrian limestones containing archeocyathids, even though there had been a suggestion of a disconformity in mapping in the region, published by Ralph W. Segnit of the Geological Survey of South Australia back in 1939. This was later confirmed by wider regional mapping by Robert Dalgarno (1964), showing again that there was a disconformity, not a smooth transition or conformable relationship. Subsequent detailed mapping of the strata containing Ediacara fossils in the Flinders Ranges by Mary Wade (1970) revealed a significant unconformity between them and the overlying Lower Cambrian rocks. Therefore, the case for placing the Ediacara fossils in the Precambrian was greatly enhanced. Incidentally, the clay partings adjacent to which Sprigg found his fossils at Ediacara had been noticed by Segnit, but no fossils were observed at that time. One can assume that he was not on the lookout for them. It may be remarked that Sprigg had a low opinion of Segnit's mapwork and his ability to comprehend geological structures,[14] so he may not have taken his indication of a break below the Cambrian beds very seriously. Therefore the Ediacara fossils could reasonably have been basal (or Eo) Cambrian for Sprigg.

But, to revert to a question previously raised, how should the bottom of the Cambrian be defined? Should it be pushed down so as to accommodate the soft-bodied Ediacara fauna, or were fossils with shelly hard parts essential to the characterization of the Cambrian? Since the nineteenth century, the Cambrian had been the home of the first macrofossils, and all the rest was Pre-Cambrian (or Precambrian). Back in the 1930s, the paleontologist George Halcott Chadwick (1876–1953) had suggested the terms "Phanerozoic" and "Cryptozoic" (meaning, respectively, "evident/visible/manifest life" and "hidden life"), with the bottom of the Phanerozoic placed at the bottom of the Cambrian. The first of these terms gained wide acceptance, but there was a problem when the Ediacara fossils were discovered. They were evident, visible, and manifest, but lay below the Cambrian. Should the bottom of the Cambrian be pushed downward to accommodate them, absorbing some of the Precambrian in the process? Or should the traditional base of the Cambrian be retained (though it was not fixed by a golden spike in the 1950s), so that the well-established term—Precambrian—could also be retained? Having regard to the meaning of Phanerozoic, logic dictated the first of these alternatives, as was urged by the influential American paleontologist Preston Cloud (e.g., Cloud and Glaessner 1982, with Glaessner dissenting). But tradition won out. The Cambrian was reserved for strata containing shelly organisms and the idea of evident but soft-bodied organisms in the Precambrian was accepted, even though this spelled the demise of the Cryptozoic as a stratigraphic category. The Cambrian–Precambrian boundary was eventually defined by a site chosen in Fortune Head, Burin Peninsula, Newfoundland (Landing 1994).[15]

That issue being established, there followed the possibility of subdividing the Precambrian on the basis of paleontological evidence rather than by lithological criteria or by rather arbitrary divisions according to time measurements based on radiometric evidence.[16] The uppermost part of the Precambrian or Neoproterozoic could be defined in terms of the Ediacara-type fauna, while lower units could be divided by microfossils such as acritarchs[17] or according to the occurrence of algae and various types of single-celled organisms.

But where was the base of the unit with soft-bodied fossils of Ediacaran character to be placed? Such placements entail the definition of Global Stratotype Sections and Points (GSSPs), which task is carried out by subcommissions or working groups of the International Union of Geological Sciences (IUGS), a body that emerged (in 1961) from the work of the peripatetic International Geological Congress (IGC), which was founded in 1878 with the intention of bringing order to the general confusion of names and boundaries that existed in the stratigraphic column. Such work has been ongoing ever since and is still unfinished. Worldwide research projects in the International Geoscience Programme (formerly the International Geological Correlation Program [IGCP]), initiated in 1964 by an Australian and begun by the IUGS and UNESCO in 1972 (see Turner 2006) often complement the work of the IUGS's subcommissions as, for example, does the current IGCP 493 ("The rise and fall of the Vendian[18] biota"). The major theme of Precambrian dating was emphasized by Glaessner when he was a member of the first IGCP scientific board in 1973.[19]

The task of establishing a faunally based subdivision for the top of the Proterozoic (Precambrian) was entrusted to a "Working Group on the Terminal Proterozoic Period" (WGTPP) of the IUGS's International Commission on Stratigraphy, and set up at the Washington IGC in 1989. Proposals to establish a lower boundary for the unit had been made previously in South Australia by another of Glaessner's students, Richard Jenkins (1981) of Adelaide University (and now at the South Australian Museum) and by Cloud and Glaessner (1982). Also, many other sites had been found worldwide with fossils similar to those at Ediacara: in Namibia,[20] the Ukraine, China, Newfoundland, Morocco, India, the United Kingdom, various sites in the United States, Finnmark (Norway), the Urals, the Mackenzie Mountains (Canada), and in places in Australia other than the Flinders Ranges.

The WGTPP[21] drew up a list of candidate countries and sites for the establishment of the GSSP and then embarked on a series of field excursions to compare and contrast the sites and determine the most suitable candidate. Criteria for acceptance included accessibility, continuity of deposition, completeness and coherence of sections, worldwide reference possibilities (including identifiable fossils or lithologies and geomagnetic and isotopic fingerprints) and rocks that might be dated radiometrically with accuracy and precision. Bore-hole data could be considered, but an acceptable GSSP could not be located in a bore-hole! Also, a GSSP cannot be located at an unconformity, which could represent a significant time-gap. After the WGTPP had completed its fieldwork, decisions were reached by a series of ballots among the voting members.

Here we may remark that GSSPs are *constructs* that are *created* socially *by scientists*. They are not just "out there" waiting to be found. We know of no procedure in science for arriving at knowledge or scientific definitions that is more social or conventional in character. First there is the establishment of a committee of knowledgeable experts. Then comes empirical work, often in the field, and theoretical discussion. Then a series of ballots is held (with voting rules codified by the IUGS) to select a suitable site, preferably in an accessible location. Then ratification by the IUGS is needed. And finally a spike may be hammered into place and a paper published setting out the Working Group's rationale for its decision. Given the social character of all this, it is unsurprising that politics may enter into the story, and on occasions votes have split along Cold War lines, and in the case under consideration the Russian members of the group (Boris Sokolov and Mikhail Fedonkin), as well as one of the Australians (Jenkins), dissented from the final decision.[22]

The WGTPP labored for fourteen long years. The Russians had a strong case for supporting a site in the Podolia region of Ukraine where relevant stratigraphic work went back at least to 1952 (Sokolov 1952[23]); and the Vendian Period had been proposed and extensively used and accepted for the upper part of the Proterozoic, though it included both the Ediacaran and a lower unit: the Varanger Epoch (named after the Varanger Fjord in northern Norway) and the Povarovian, named after Povarova in the Ukraine (Sokolov and Fedonkin 1984). The Chinese, by contrast, could claim priority for the Sinian System, characterized by good exposures in the Yangtze Gorge of a largely undeformed unit at the top of the Precambrian, with the name having been introduced by Bailey Willis, Eliot Blackwelder, and Harvey Sargent (1907) and later used by Amadeus Grabau (1922) in relation to localities in northeastern China.

Among the several contenders, the US sites were never seriously considered suitable. Namibia could provide no continuous section whose level could be determined. A proposed Canadian section was good, but was in difficult and inaccessible country. The sites in India and Morocco had received insufficient research compared with Australia. A possible site in Siberia was excluded as the WGTPP's expedition there got lost, after running into difficulties with its helicopter, and the proposal was abandoned. A Norwegian site was remote and had a poor fossil record as well as metamorphosed rocks. The Ukrainian site was, as mentioned, well established in the literature, but had no continuous section, though the acritarchs there (important as well as the jelly-fish!) had been closely studied. The Chinese location was the last left standing before a vote was taken to select South Australia as the locality for the GSSP. A special issue of *Precambrian Research* appeared in 1995, setting out the pros and cons of the different sites.

An important issue was whether to place the spike as close as possible to the bottom of the strata actually containing macrofossils, or whether one should look for some clearly defined, worldwide and unambiguous boundary that could be accurately dated. The first option seems the appropriate one if one is looking for macrofossils to define a new geological system, and this was the view taken by the dissident Australian voice, Richard Jenkins. But he was outvoted, on the grounds that such a site could not be defined with precision, and there was the risk that further fieldwork might reveal macrofossils below the chosen horizon. So, in the event, it was decided to look for indications of worldwide glaciation well *below* the horizon of the fossil outcrops.

Global late-Precambrian glaciations had previously been recognized by others (including Edgeworth David), and the last of these, preceding the appearance of Ediacara fauna, was selected. Helpfully, the rocks of each Precambrian glaciation are marked by a fairly thin distinctive band of cap carbonate.[24] Such rocks are widely distributed and have characteristic geochemical signatures, in that the measurable carbon isotope ratios form a distinctive graphical pattern upward through the caps (due to changes in the ambient temperatures at the times of deposition). So cap carbonates of different dates can be distinguished by such patterns. Geomagnetic signatures (patterns of geomagnetic reversals) can also be useful for correlation purposes.

While reluctantly accepting the utility of the glacial argument, Jenkins maintained that there were other glaciations between the one that was recommended and the horizon where Ediacara fossils first appeared. But his fellow geologists did not accept that what he regarded as higher glacial sediments were in fact so at all, and cap carbonates were missing.[25] So events moved forward. In December 2000, the decision was taken to have the GSSP at the bottom of the cap carbonate that topped the sediments of the Marinoan Glaciation (the last global Proterozoic or Precambrian glaciation). Then the WGTPP's members were invited to nominate sites that would accord with that decision. The ones most favored were two sites in the Flinders Ranges, one at the Yangtze Gorge sections, and one for a site in the Lesser Himalayas in India. The Russians did not at that stage put forward a proposal (probably because it was realized that the glacials in the Ukraine were diachronous). At the next vote (March 2003) all four proposals received some support, but the one at the base of the Nuccaleena Formation in the Flinders Ranges was preferred (63% of votes). The final ballot (September 2003) had two parts: one to determine the exact location of the GSSP at that horizon, and the second to decide the name of the new geological period being defined.

The preferred site (89% of votes) was one at Enorama Creek in the Brachina Gorge, which cuts east-west across the line of strike of the Precambrian and Cambrian strata in the Flinders Ranges, the road through the Gorge being a signed heritage trail. The GSSP was to be located at the base of the cap carbonate at the top of a red-brown glacial diamictite[26] (the Elatina Formation) underlying the creamy dolomite of the Nuccaleena Formation (see fig. 13.2A and 13.2B).

As to the name, *Ediacaran* was chosen for the newly defined system and period. Ediacarian was also considered, since the Termiers' original terminology (*Ediacarien*) had precedence; and the "ian" suffix accorded with that used for all the Paleozoic systems. However, the name Ediacaran was well

Figure 13.2A General view of the locality of the GSSP for the base of the Ediacaran Period, located in the Brachina Gorge, Flinders Ranges, South Australia.

established in the literature, and so that was the name selected, with Jenkins' approval and advocacy. The decisions were ratified by the IUGS's Executive Committee in March 2004 and were promulgated at the Florence IGC later that year. As previously mentioned, the spike was formally emplaced and blessed by the premier of South Australia in April 2005.[27]

Thus after almost fifty years, the work pioneered by Sprigg finally reached the High Table: at the highest level of the geological community and at the highest level of the South Australian Government. His work achieved its apotheosis, but sadly several years too late for him to enjoy it. Should anyone question the significance of Sprigg's work, we need merely point out that he and his fauna are given honourable (albeit slightly garbled) mention in Bill Bryson's (2003) bestseller *A Short History of Nearly Everything*! Also, the originator of an alternative interpretation of the Ediacaran fauna of "Vendobionta" (Seilacher 1989), Professor Dolf Seilacher, even elevated the fossils to high art by their inclusion in his traveling international display (Seilacher

Figure 13.2B Close-up view of the "Golden Spike" (brass!) marking the base of the Ediacaran Period: being the horizon corresponding with the bottom of the brass disc. Below the marker is the brown diamictite of the glacial Marinoan Period. The disc itself is attached to a stratum of cap carbonate, above which lies an unfossiliferous dolomite of the Ediacaran. The borings seen in the picture are the result of sampling undertaken by an American group for geomagnetic study before the spike was emplaced.

1997). And as a kind of icing on the cake, a fossil site containing Ediacaran fossils was placed on Australia's National Heritage List on 11 January, 2007.[28]

There is a further interesting issue. If one visits the current (2007) Ediacaran display at the Museum of South Australia one can see a nice model of what the resident paleontologists think the sea-floor might have looked like in Ediacaran times. It is represented as being covered by a lawn of green algae, on which various Ediacaran organisms are feeding, among which is the almost elliptical segmented organism (worm?): *Dickinsonia costata*. It is modeled as a bilaterally symmetrical ribbed mass of jelly with almost no difference between its front and rear. There is, as the display shows, great variation in the size of these organisms, from about a centimeter to almost a meter. So these creatures could certainly grow. They had no protective cover and presumably they just browsed on the algae beneath and around them by the secretion of enzymes. They could also apparently move: the display shows a flattened area of algae where a *Dickinsonia* had been located for a while; and then it had moved on to newer and presumably greener pasture. (Such resting or feeding spots can be discerned in the Ediacaran sediments.)

It would appear that the organism had no predators and only the most simple of nervous systems. But *Spriggina floundersi* (see figure 13.1) was more elongated than *Dickinsonia*. Indeed, it had "an indisputable head-end" (Sprigg 1989, 205) and presumably a more advanced nervous system. Also, it would appear to have been more mobile than *Dickinsonia*, more like modern annelids, and at a higher stage of evolutionary development. But neither of them, nor any other organisms of that period, had any sign of either offensive or defensive structures. There was no evolutionary maelstrom in the "Garden of Ediacara" (cf. McMenamin 1986). It was the only time that the (macro) animal kingdom was wholly peaceful. But it was not to last: soon the world was to be filled with spiny trilobites, shelled brachiopods, and all the strange creatures of the Burgess Shale fauna, many of which look rather fierce! The evolutionary arms race had begun. The Garden of Eden/Ediacara was no more. Yet even in that peaceful world an organism has been found that was, it has been suggested (Glaessner and Wade, 1971), possibly an ancestor of trilobites, with their protective covers. Perhaps soldiers were already on their way?

ACKNOWLEDGMENTS

For sharing their memories and their time, we thank Barry Cooper, Richard Jenkins, Brian McGowan, the late Mary Wade, Dennis Walter, and his brother Malcolm Walter. The assistance of librarians and archivists at the Australian

Academy of Sciences Basser Library, University of Adelaide (the Barr Smith Library), the Library of South Australia, and the Queensland Museum are gratefully acknowledged. Gabriele Schneider kindly provided information about Namibian Precambrian fossils and Trevor Ford gave us an account of the discoveries at Charnwood Forest. Henry Gee informed us of the absence of records for *Nature* for the 1940s. Zoya Bessudnova sent from Russia a copy of, and Vladimir Simkin construed, the paper of Sorokin (1952). Thanks also to Patricia Vickers-Rich for funding for Susan Turner in 2004, who received a one-year Australian Research Council Discovery Grant, DP0453155; to research assistant Judy Bracefield; and to Monash University Geosciences and the Board of the Queensland Museum for basic facilities.

NOTES

1. See report by Elissa Doherty in the *Sunday Mail* (South Australia), 17 April, 2005, p. 8.

2. During the war, he had been working for an oil company in New Guinea, before moving to Australia, initially in Melbourne.

3. This species was earlier named *Rangea arboreus*, being the same as the supposed plant referred to by Paul Range for a Namibian specimen. (See note 20.) The specimen illustrated here shows the fronds of a colonial animal, plus its holdfast, which was originally thought to be a distinct fossil species. The two are only rarely found in association.

4. See Corbett (2000); Branagan (2005).

5. In some other publications, Sprigg gave the year as 1947, but this would seem to have been in error. (But he did go to Ediacara again in 1947.)

6. Richard Jenkins has reported that the old miners' dwellings near the Ediacara mine were paved with flagstones containing fine discoidal fossils, suggesting their deliberate collection for curiosity value. So presumably Sprigg was not the first Westerner to *see* such organisms. But their occurrence on the miners' flagstones may have been quite fortuitous.

7. Beltana, "running water," is the name of an old mining settlement near Lake Torrens and not far from Ediacara.

8. The museum had no such person on its staff. Sprigg probably meant Dr. R. M. Shackleton of Imperial College, whom he might have met at the Museum, next door. Sprigg did attend the Congress (meeting his future wife, Griselda, there during a field trip) and presented two papers in Section M (of which Shackleton was Secretary): "Stranded Pleistocene sea beaches of South Australia and aspects of the theories of Milankovitch and Zeuner" and "Unusual thrust structures in the Willouran Ranges, South Australia." The titles reveal Sprigg's eclectic interests and the fact that he was not considered unsuitable as a Congress speaker.

9. Ford (like Range: see note 20) initially thought the object was a plant. Ford has informed us that the "object" discovered by the boys had in fact been observed not long before by a schoolgirl, Tina Negus, whose teachers told her that it *couldn't* be a fossil, as the rock in which she found it was *known* to be Precambrian.

10. Wade went on to become an eminent researcher at the Queensland Museum (Turner, 2007).

11. Daily completed his PhD in 1957 and obtained a lectureship at the university in 1961.

12. The "barring" of Sprigg's company has been recalled by Dennis Walter (conversations with Susan Turner, 7 December, 2004, and David R. Oldroyd, 14 June, 2006, and subsequent e-mails). To obtain protection for the site, they had obtained mining rights (to quarry for quartzite), had prepared corner pegs to mark out the site, and were getting ready to depart from Adelaide when the area was unexpectedly gazetted as a reserve. So their work was stymied.

13. Irony was perhaps involved, however, in that *Spriggina floundersi* was said by Glaessner to be "the lowliest worm that ever lived" (Sprigg 1989, 208), a story that Sprigg repeated verbally and in print on several occasions, often as a joke at his own expense. We may also note that the editorial that Sprigg wrote in the last issue of his expiring *Australian Amateur Mineralogist* reveals that he *was* hurt about his exclusion from excavating the Ediacara site: "A 'reserve' was thrown around the area and now science is hamstrung by red tape. The fact that some of the best fossils in the world are found through quarrying, and the fact that the surface material is now largely collected out, has apparently not occurred to those who rushed in to restrict our efforts. . . . [O]ur learned Royal Society [of South Australia] raced in to support the virtual embargo on fossils, and another enterprise of extreme value to science . . . [was] quashed" (Sprigg 1960, 98).

14. Sprigg: record of interview with Barry Cooper in September 1983 (Mortlock Library of South Australia, J. D. Somerville Oral History Collection OH 89/22). This interview reveals also that Sprigg believed that Mawson did not initially appreciate Sprigg's Ediacara finds *because of* the Mawson/David controversy about the Tea Tree Gully claims.

15. Interestingly, Precambrian fossils were described from New Brunswick by George Frederick Matthew in the 1880s but were forgotten until recently.

16. Modern subdivisions are:

Proterozoic Neoproterozoic (1000–542 Ma)
Mesoproterozoic (1600–1000 Ma)
Paleoproterozoic (2500–1600 Ma)
Archean Neoarchean (2500–2800 Ma)
Mesoarchean (2800–3200 Ma)
Paleoarchean (3200–3600 Ma)
Eoarchean (or Hadean) (> 3600 Ma)
Origin of Earth 4567 Ma

For details, see Plumb (1991) and Gradstein *et al.* (2004). The divisions do in fact have rationales that are not merely arbitrary, but they need not be discussed here.

17. Resistant-walled, unicellular fossils, useful for stratigraphic correlation. There is also a distinctive Ediacaran acritarch fauna.

18. 'Vendian' is a Russian term, more or less equivalent to the newly-defined Ediacaran (see below).

19. See *Geological Correlation*, the official IGCP serial.

20. The case of Namibia (formerly German Southwest Africa) is interesting. Fossils similar to those at Ediacara and Charnwood Forest were found there by the Government Geologist Paul Range as early as 1908, and again in 1929 (Gürich 1929, 1933). Gürich was a professor at Hamburg who did a considerable amount of work in Namibia. According to "tradition," the fossils, found in 1908 were put away and forgotten for several years. The type *Rangea* was thought by Range to be a plant, and was subsequently described by him in a botanical journal (Range 1932). Some of the early material survives in Berlin, but not the original specimens. Glaessner, reading German, undoubtedly had "access" to Gürich's papers, whereas Sprigg did not.

21. Later promoted to the status of a subcommission of the IGCP, it was headed by Andrew Knoll (USA), Malcolm Walter (Australia), Guy Narbonne (Canada), and Nicholas Christie-Blick (USA) and there were initially sixteen other voting Members. It had a somewhat fluctuating membership, but there were representatives from all "interested" countries; and there were also non-voting "corresponding members," who could make submissions and have their views duly considered.

22. With their preference for the Vendian and a Russian site for the GSSP, the Russians abstained from the final ballot rather than voting against the GSSP being in Australia.

23. Sokolov at that time recognized Baltic (upper) and Vendian (lower) complexes, and stated that the former was Cambrian and the latter Precambrian, with analogies to the Sinian System in China, the Eocambrian in northern Europe, and various other localities round the world, including the Adelaide System in South Australia. In fact, he suggested that Sinian-type rocks occurred at Podolia. But he did not then suggest a Vendian System and made no mention of macro-fossils in the Vendian rocks at Podolia, though phyllocarids were recorded from the Baltic strata.

24. During the glacial period, dissolved carbon dioxide is thought to have collected at the bottom of the ocean in the form of bicarbonates. With the melting of the ice cover, the waters get stirred up and the bicarbonate rich waters rise and become warmed, leading to the precipitation of carbonates as a "cap" to the glacial deposits.

25. Jenkins (1981) had earlier placed the bottom of the Ediacaran much higher in the succession than the horizon eventually selected for the GSSP, but he did not then discuss higher glacial horizons. He had also suggested a different locality (Bunyeroo Gorge) as the site for the standard section. This was a little to the south of Brachina Gorge, which was eventually selected for the standard section and "point."

26. Diamictite = poorly sorted, non-calcareous, terrigenous rock with a wide range of particle sizes.

27. Information about the history of the proposal and ratification of the Ediacaran are based chiefly on interviews by DRO with Richard Jenkins and Malcolm Walter, and on Knoll *et al.* (2006). See also Preiss (2005).

28. See: "Ediacara Fossil Site – Nilpena, Parachilna, SA", www.deh.gov.au/cgi-bin/ahdb/search.pl?/mode-place_detail;search=place_name%sDNNilpena%3Bkeyword. The exact locality has not been disclosed by the Government, to prevent its being pilfered by collectors. The heritage honour was announced in a media release by former federal minister Ian Campbell, who, however, gave Reg's surname as Spriggs. (Actually it was his Press Officer that made the mistake!)

REFERENCES

Branagan, D. 2005. *T. W. Edgeworth David: A life. Geologist, adventurer, soldier and 'knight in the old brown hat.'* Canberra: National Library of Australia.
Bryson, B. 2003. *A short history of nearly everything.* New York: Broadway.
Chadwick, G. H. 1930. Subdivision of geologic time (abstract). *Bulletin of the Geological Society of America* 41: 47.
Cloud, P. E., and Glaessner, M. F. 1982. The Ediacaran period and system: Metazoa inherit the Earth. *Science* 217:783–92.
Conway Morris, S. 1998. *The crucible of creation: The Burgess Shale and the rise of animals.* Oxford: Oxford University Press.
Cooper, B. J. 1995. Reg Sprigg (1919–1994): A legend in his own lifetime. *The Australian Geologist* 94:73–74.
Corbett, D. W. P. 2000. A staunch but testing friendship: Douglas Mawson and T. W. Edgeworth David. *Records of the South Australian Museum* 33:49–70.
David, T. W. E., and R. J. Tillyard. 1936. *Memoir on fossils of the Late Pre-cambrian (Newer Proterozoic) from the Adelaide Series, South Australia.* Sydney: Angus & Robertson and the Royal Society of New South Wales.
Ford, T. D. 1958. Pre-Cambrian fossils from Charnwood Forest. *Proceedings of the Yorkshire Geological Society* 31:211–17 and plate.
Glaessner, M. F. 1958. New fossils from the base of the Cambrian in South Australia (Preliminary account). *Transactions of the Royal Society of South Australia* 81:185–88 and plate.
———. 1959. The oldest fossil fauna in South Australia. *Geologische Rundschau* 47:522–31.
———. 1961. Precambrian animals. *Scientific American* 204:72–78.
———. 1971. Geographic distribution and time range of the Ediacara Precambrian fauna. *Bulletin of the Geological Society of America* 82:509–14.
———. 1984. *The dawn of animal life: a biohistorical study.* Cambridge: Cambridge University Press.

Glaessner, M. F., and B. Daily. 1959. The geology and Late Precambrian fauna of the Ediacara fossil reserve. *Records of the South Australian Museum* 13:369–401.

Glaessner, M. F., and M. Wade. 1966. The Late Precambrian fossils from Ediacara, South Australia. *Palaeontology* 9:599–628.

———. 1971. *Praecambridium*—a primitive arthropod. *Lethaia* 4:71–77.

Gould, S. J. 1989. *Wonderful life: The Burgess Shale and the nature of history*. New York: W. W. Norton.

Grabau, A. W. 1922. The Sinian System. *Bulletin of the Geological Society of China* 1:1–11.

Gradstein, F. M., J. G. Ogg, and A. G. Smith. 2004. *A geologic time scale*. Cambridge: Cambridge University Press.

Gürich, G. 1929. Die ältesten Fossilien Südafrikas. *Zeitschrift praktische Geologie mit besonderer Berücksichtigung der Lagerstättenkunde* 37:85.

Gürich, G. 1933. Die Kuibis-Fossilien der Nama-Formation von Südwest-Afrika. *Paläontologische Zeitschrift* 15: 137–55.

Jenkins, R. J. F. 1981. The concept of an 'Ediacaran Period' and its stratigraphic significance in Australia. *Transactions of the Royal Society of South Australia* 105:179–94.

Knoll, A. H., M. R. Walter, G. M. Narbonne, and N. Christie-Blick. 2006. The Ediacaran Period: A new addition to the geologic time-scale. *Lethaia* 39:13–30.

Landing, E. 1994. Precambrian-Cambrian boundary global stratotype ratified and a new perspective of Cambrian time. *Geology* 22:179–82.

McMenamin, M. A. S. 1986. The garden of Ediacara. *Palaios* 1:178–82.

Plumb, K. A. 1991. New Precambrian time scale. *Episodes* 14:139–40.

Preiss, W. 2005. Global stratotype for the Ediacaran System and Period: The golden spike has been placed in Australia. *MESA [Mines and Energy of South Australia] journal* 37(May): 20–25.

Range, P. 1932. Die Flora des Namalandes. I. *Repertorium specierum novarum regni vegetabilis [Feddes Repertorium]* 30: 129–158.

Segnit, R. W. 1939. *The Precambrian–Cambrian succession: The general and economic geology of these systems, in portions of South Australia*. Geological Survey of South Australia, Bulletin no. 18.

Seilacher, A. 1989. Vendoza: Organismic construction in the Proterozoic biosphere. *Lethaia* 22:229–39.

———. 1997. *Fossil art*. Tübingen: Geological Institute, Tübingen University.

Sokolov, B. S. 1952. O vozrastye drevneysgego osadochnogo pokrova Russkoy Platformy (On the age of the oldest sedimentary cover of the Russian Platform). Moscow: *Isvestiya Akademii Nauk SSSR, seriya geologisheskaya (Newsletter of the USSR Academy of Science, geology series)* No. 5: 21–31.

Sokolov, B. S. and Fedonkin, M. A. 1984. The Vendian System as the terminal system of the Precambrian. *Episodes* 7(1): 12–19.

Sprigg, R. C. 1947a. Early Cambrian (?) jellyfishes from the Flinders Ranges, South

Australia. *Transactions of the Royal Society of South Australia* 71:212–24 and plates.

———. [1947b]. Some fossil jellyfishes (?) of early Cambrian age from the Flinders Ranges, South Australia. Report of the twenty-sixth meeting of the Australian and New Zealand Association for the Advancement of Science, Perth Meeting, August, 1947. Perth: Government Printer. (Only the title of this paper was published.)

———. 1949. Early Cambrian 'jellyfishes' of Ediacara, South Australia, and Mount John, Kimberley District, Western Australia. *Transactions of the Royal Society of South Australia* 73:72–99 and plates.

———. 1958. Fossil jellyfish from the Cambrian of South Australia. *The Australian Amateur Mineralogist* 2:22–24.

———. 1960. Idealism, is it worth it? *The Australian Amateur Mineralogist* 4:98.

———. 1988. On the 1946 discovery of the Precambrian Ediacaran fossil fauna in South Australia. *Earth Sciences History* 7:46–51.

———. 1989. *Geology is fun (recollections): Or the anatomy and confessions of a geological addict.* Arkaroola, South Australia: Author.

Termier, H., and G. Termier. 1960. L'Ediacarien, premier étage paléontologique. *Revue générale des sciences pures et appliquées et bulletin de l'Association Française pour l'Avancement des Sciences* 67:79–87.

Turner, S. 2006. The rocky road to success: A new history of the International Geoscience Programme (IGCP). In *Sixty years of science at UNESCO: 1945–2005,* ed. P. Petitjean, V. Zharov, G. Glaser, and J. Richardson, 297–314. Paris UNESCO.

———. 2007. Invincible but mostly invisible: Australian women's contribution to palaeontology. In *The role of women geologists' contributions,* ed. C. Burek and B. Higgs. London: The Geological Society, Special Publication 281, 165–201.

Wade, M. J. 1970. The stratigraphic distribution of the Ediacara fauna in Australia. *Transactions of the Royal Society of South Australia* 94:87–104.

Willis, B., E. Blackwelder, and R. H. Sargent. 1907. *Researches in China 1.* Washington, DC: Carnegie Institution.

Yochelson, E. L. 1998. *Charles Doolittle Walcott: Paleontologist.* Kent, OH: The Kent State University Press.

CHAPTER FOURTEEN

The Morphological Tradition in German Paleontology: Otto Jaekel, Walter Zimmermann, and Otto Schindewolf

Manfred D. Laubichler and Karl J. Niklas

INTRODUCTION

The original working title of this volume, *Paleontology at the High Table*, clearly places this endeavor within the Anglo-American cultural milieu, and more specifically, its academic traditions. Here, the honor of being invited to the High Table generally implies acceptance of one's accomplishments or, less flatteringly, the recognition of one's pedigree. The title therefore serves as an apt metaphor for paleontology's struggles to secure its place within the hierarchy of scientific disciplines. Furthermore, it also suggests that there is a difference between acceptance by the scientific establishment and popularity with general audiences. Because it is undoubtedly true that paleontology has always fueled the popular imagination; there are countless tales of children standing in awe before a fossil dinosaur—with some of them subsequently becoming brilliant scientists and communicators; of hobbyists and avid fossil hunters, who not only made some of the most outstanding discoveries of the last centuries but some of whom went as far as to forge the earliest Englishman; and of the high drama and big egos involved in the hunt for human ancestors. The recent ruckus about a journal devoted to the publication of analyses of privately owned fossils highlights that this mutual relationship between professionals and amateurs continues to be problematic.

However, as the chapters in this volume illustrate, the relationships of paleontologists with their scientific kin have often been equally strained, and thus paleontologists still struggle to be invited to the High Table of the scientific establishment and, once there, to secure their position. Their problem at the High Table is one of methods and theories. In both areas, paleontology does not quite fit the current paradigm of laboratory-based translational research and the underlying model of intervening and its associated notion of causality. It is, therefore, not at all surprising that one of the constant themes in paleontological discussions has been the question of specific macroevolutionary processes and their causes. Others in this volume have discussed these issues in great detail.

This chapter focuses on the history of paleontology within a different cultural context. Formal High Tables do not exist at German universities. But this does not mean that there is not an equivalent cultural value system in place that bestows acceptance to a field in the courts of both scientific and popular opinion. Even the most cursory overview of the history of German biology reveals that the towering figure of Johann Wolfgang von Goethe, as both a cultural and scientific icon, has, to no small degree, influenced the values that have shaped the debates about paleontology, morphology, and evolution in the German-speaking world of the late nineteenth and early twentieth century. The German reception of Darwin during this period can not, for instance, be understood without reference to the remnants of Romanticism and idealistic morphology that have their origin in Goethe's idiosyncratic conception of nature and culture (see also Richards 2002; Laubichler 2003). This specifically German notion of romantic evolutionism is most visible in the life and work of Ernst Haeckel, who did more than any other scientist to promote Darwin's ideas, albeit in his own interpretation, in Germany (Laubichler 2005; Richards 2008). Haeckel's distinct morphological emphasis, his focus on phylogeny, and his materialism set the stage for many of the debates among German biologists and paleontologists, as well as for the intense public and ideological debate about evolution.

In recent years historians have discussed whether mid-twentieth-century German biology had its own equivalent to the Anglo-American Modern Synthesis (Junker and Engels 1999; Reif, Junker, and Hoßfeld 2000; Junker 2004). As a result of these discussions we now have a much more differentiated picture of the many strains within German evolutionary biology. But with respect to paleontology the verdict has not changed much. Wolf-Ernst Reif puts it bluntly: "The modern synthesis did not play a role within German paleontology until way in to the 1970s (Reif 1999, 151)." Which implies that there were other factors that shaped the disciplinary identity and orientation of German paleontology, and, we might add, of German evolutionary biology more generally. If we assume that no scientist would willfully place himself or herself outside of a normative system of recognition—a claim that is well substantiated by countless case studies in the history and sociology of science—then, in order to understand the *Sonderweg* of German paleontology, we have to explore the cultural, educational, and scientific value system that rewarded a particular kind of work and interpretation. In other words, we have to identify what High Table German paleontologists attended. But such a cultural history of German paleontology is well beyond the scope of this chapter. We will therefore explore the main themes of our argument in the form of a few select vignettes that are both representative as well as reflective of the variation in the theoretical foundations that existed within the German paleontology community.

SETTING THE STAGE: THEMES IN GERMAN PALEONTOLOGY

If one reviews the major German-language contributions to paleontology and evolutionary biology in the early twentieth century it soon becomes clear that there was no general agreement about the theoretical foundations of the field, no unifying research program that would embed paleontology within a larger framework of evolutionary biology (which did not really exist as a coherent discipline, either), and that the relationship of paleontology and other areas of biology was, for the most part, one of mutual tolerance. As a consequence of this pluralism, rooted as it was in local and disciplinary traditions of morphology, geology (stratigraphy), and history (*Urgeschichte, Erdgeschichte*), it is rather difficult to establish major themes in German paleontology and evolution during this period. Wolf-Ernst Reif, who more than any other has investigated this history, has, for instance, identified six groups within the German paleontology community of the late nineteenth and early twentieth century (Reif 1986). But even though individual members of these groups—traditionalists, early Darwinians, pluralists, neo-Lamarckians, orthogeneticists, and typostrophists—share certain core assumptions, there are still widespread disagreements even within each group, making it all the more difficult to identify core questions and values in German paleontology.

The three historical actors featured in our vignettes—Otto Jaekel, Walter Zimmermann, and Otto Schindewolf—each represent different orientations within German paleontology in terms of their areas of expertise (plants, vertebrates, invertebrates, respectively) and conceptual orientation (a Darwinian evolutionary morphologist, a neo-Lamarckian, and a typostrophist). They also had different institutional goals. Jaekel and Zimmermann advocated a close connection between paleontology and biology, while Schindewolf treasured his influence and control of the German paleontology community from his base as chair in geology and paleontology at the University of Tübingen. These three scientists also reflect different patterns of recognition—they dined, as it were, at different High Tables. Jaekel was a respected paleontologist based at a provincial university (Greifswald) whose attempts to forge closer ties with biology and to reform the institutional landscape of German biology and paleontology were ignored by all powerful biologists and science administrators. He had more luck with his immediate peers in establishing a new professional society, the German Paleontological Society (Deutsche Paläontologische Gesellschaft). Zimmermann, as a botanist and evolutionary biologist, was respected by his peers and students, but was never invited to join the establishment, even though his main theoretical contribution to plant morphology and evolution, the *telome* theory, was a substantial achievement. He was, however, part of a group of scientists and academics who reached a wider audience. His work in the history of evolutionary biology was included in a popular highbrow series of scholarly works. And Otto Schindewolf, after he finally advanced to a chair in geology and paleontology in 1948 (the Nazis had denied him such a promotion on political grounds), readily became the most powerful German paleontologist of his generation, whose influence and anti-Darwinian views shaped German paleontology well into the 1970s. He, in a way, decided who would dine at the High Table of German paleontologists.

So far we have emphasized the conceptual, scientific, and institutional differences within the German paleontology community. But this still leaves open the question whether there are any common themes that would help us understand the disciplinary development and, despite all its diversity, the coherence of German paleontology. One important common factor certainly is the institutional connection to geology—most German paleontologists worked within geological institutes, where they represented historical geology. This orientation brought with it a certain methodological orientation and a skill set that was conducive to detailed work in stratigraphy as well as local geology. It also contributed to a particular emphasis on history as opposed to a more narrow interpretation of phylogeny as simply phylogenetic systematics. This trend was further reinforced by the role history played within the humanistic education that most German paleontologists received before their university studies. A search for general laws or patterns of history has been a prominent feature within nineteenth century historiography and philosophy (Hagner and Laubichler 2006; Ruehl 2006). Independent of whether these historical patterns were thought to be progressive (Hegel) or to show a cyclical pattern of birth, growth, and decline (Spengler), the prevailing cultural attitudes emphasized the value of recognizable patterns and laws of history that, in turn, severely restricted the importance of chance and contingency in historical developments. (Insofar as these attitudes contribute to the belief in the destiny of a nation, they also had most damaging political consequences.) But for the case of German paleontology, we can easily see how such a value system contributed to ideas about macroevolutionary processes and laws, orthogenesis, and typostrophism.

Another important factor that shaped the disciplinary identity of German paleontology was its reception of Darwinian evolutionary biology. German paleontologists were among the first who enthusiastically embraced evolution, but most struggled with the idea of natural selection as the sole or even the main mechanism of evolutionary change and transformation. Drawing on strong traditions as well as active research programs in morphology and embryology (*Entwicklungsgeschichte*), a majority of German paleontologists emphasized the need for additional, internal factors of evolution. Mostly these were thought to lie within the developmental system of the organism, a tradition that dates back at least to von Baer and Haeckel. In many instances, these proposed internal mechanisms were highly speculative, depending, as Haeckel's law of terminal addition, more on intuition than comparative and experimental data. However, the developmental perspective of many German biologists and paleontologists was embedded within a larger foundation of comparative anatomy and embryology, represented, for instance, by the Gegenbaur-Haeckel program of evolutionary morphology, which provided a framework of empirical observations and a conceptual basis for problems of phylogeny and evolution. By the early decades of the twentieth century these morphological and embryological perspectives had become quite diverse, ranging from developmental mechanics and physiology, on the one hand, to idealistic morphology on the other. This diversity of approaches is also reflected in a corresponding plurality of proposed internal mechanisms, such as orthogenesis, saltationism, typostrophism, and so on. Despite these differences, the common theme behind all these proposals is the view that the internal organization of the organism, represented either by an idealized notion of type or a developmental system, is the real driving force behind evolutionary transformations.

We will see in more detail how Schindewolf's anti-Darwinian explanation of evolutionary transformations is a skillful blend—or rather an idiosyncratic combination—of insights from genetics, developmental physiology, morphology, and historical geology. In this context it is, however, worth noting that we still find a similarly catholic perspective in the more recent work of Schindewolf's successor as chair in Tübingen, Adolf Seilacher. Seilacher's proposal for constructional morphology also emphasizes that any explanation of form (morphology) requires a combination of internal and external factors—namely function, phylogeny, and architectural constraints (Seilacher 1970). While Seilacher has become one of the leading advocates of architectural principles and constraints—an emphasis that has meanwhile led to the rather active field of theoretical morphology (see, e.g., Niklas 1992, 1994, 1997; McGhee 1997, 2007)—the German tradition of developmental physiological genetics, originally most prominently represented by Richard Goldschmidt (on whose work Schindewolf relied rather heavily in the *Basic Questions*) and Alfred Kühn, Schindewolf's colleague in Tübingen after World War II, has in the meantime been transformed in the context of evolutionary developmental biology. Kühn's formulation, in the second edition of his Lectures in Developmental Physiology (Kühn 1965), that different morphological types represent different equilibria of the developmental system (which he called the *Wirkgetriebe der Gene*) not only foreshadows the most recent insights of evolutionary developmental biology, it also is in accordance with Schindewolf's idea that morphological transformations are a consequence of mutations acting during different periods of development. Those that affect fundamental developmental processes will then lead to saltational morphological changes and the emergence of different types.

It is thus interesting to see how some of the peculiar ideas of German paleontology are rediscovered in the context of modern Evo-Devo. While German paleontologists did not dine at the High Table of the Modern Synthesis, it is still an open question which tradition of paleontology will be invited to the High Table of the New Evolutionary Synthesis. But before we come back to this question, let us first investigate some of the elements of the German *Sonderweg* of paleontology.

OTTO JAEKEL: BIONTOLOGY, A SYNTHESIS OF BIOLOGY AND PALEONTOLOGY

Among the idiosyncratic mix of German paleontologists Otto Jaekel (1863–1929) stands out in that he, more than most, tried to connect biology and paleontology and did so, albeit without much success, in a rather interesting way, namely by combining theoretical reflections about the relationship between the sciences with what we would today call forays into "science policy." Jaekel's career path is quite typical for a German paleontologist; he first studied geology in Breslau and paleontology in Munich, where he received his doctorate in 1886. Years as an assistant in Straßburg (then again part of Germany), research at the Museum of Natural History in London and a curatorship at the geological-paleontological museum of the University of Berlin preceded his appointment as professor of geology at the University of Greifswald (Abel 1929a; Sucker 2002). He was an evolutionist with neo-Lamarckian views—he even developed his own mechanical model for the inheritance of somatic traits—who also had a nonteleological and interactive conception of the evolutionary process; arguing, for instance, that organisms actively shape their own niche, an idea that is quite similar to the current concept of niche construction (Jaekel 1901; Odling-Smee, Laland, and Feldman 2003). His empirical work in functional morphology and systematics was highly respected— Othenio Abel called him "the most brilliant paleontologist"—and focused on vertebrate paleontology (Abel 1929a). He was a tireless campaigner for the emancipation of paleontology from geology and was instrumental in establishing the "Paläontologische Gesellschaft" in 1912 (Jaekel 1914). To emphasize the close connections between paleontology and biology Jaekel coined a new term—biontology (*Bioontologie*). Whereas biology focused on the study of currently living organisms, paleontology traditionally dealt with those from earlier periods in the earth's history. Biontology, then, would be the general science of all organisms, living and extinct. Jaekel's model for the biological sciences was interdisciplinary; he realized that understanding organisms requires a variety of methodological approaches. But, based on his understanding of *Entwicklung* (describing both evolution and development) as the common and unifying thread of all biological sciences, he developed a blueprint for conceptual unification and disciplinary reform.

The intense discussion about reforming the sciences in Imperial Germany after 1900 provided the framework for Jaekel's efforts (Sucker 2002; Laubichler 2006). At that time, even though German science had become the model for the world, many in Germany perceived an impending crisis, a loss of their competitive edge, and an inability to respond adequately to new developments within several cutting-edge sciences, such as experimental biology. The establishment of the Kaiser Wilhelm Society was but the most visible response to this recognition of a crisis. In the ensuing discussions biology was on the agenda from the start. Many of the leading German biologists were asked to submit their ideas about promising future directions within the biological sciences and to develop proposals how the Kaiser Wilhelm Society should respond to these challenges. These debates provide a fascinating window into the conceptual and political struggles of German biologists shortly after the turn of the century (Sucker 2002). And they also are the background for Otto Jaekel's efforts to unite biology and paleontology.

Jaekel's ideas about biontology and its possible realization went through several stages of development. However, ultimately, he did not succeed in convincing either his colleagues in the biological sciences or state administrators. But Jaekel's proposals and their reception help us understand the role of paleontology (or more precisely, one specific interpretation of paleontology) within the scientific landscape of Imperial Germany. Jaekel complained that paleontology, as one of the most important phylogenetic disciplines, was still largely ignored by most biologists, including those who considered themselves evolutionists. To rectify this situation he suggested a reorganization of the "biontological sciences." These included "botany, zoology, paleontology, comparative anatomy, physiology, biology, physical anthropology, as well as the philosophical essence of all these natural sciences, die *Entwicklungslehre* (theory of evolution)." In a series of proposals Jaekel suggested various combinations of institutes linked by strong collaborations and interdisciplinary research programs and united by a focus on *Entwicklungslehre*. Only in his last and most pragmatic proposal did he suggest that the Kaiser Wilhelm Society establish two institutes, one devoted to experimental biology and a second to morphology and phylogeny. The latter would also be charged with developing "*Theoretische Entwicklungslehre* (theoretical evolutionary biology)," not the least because of its importance as the philosophical foundation of the natural sciences and its role for unifying all of the biontological sciences.

The scientific establishment, however, remained skeptical and did not endorse Jaekel's proposals. There were many reasons for this rejection—the ones that are of interest to us are connected with: (1) disciplinary separation— Jaekel was perceived as an outsider to biological discussions, and a rather aggressive one at that. He was not considered qualified to suggest sweeping structural changes to the biological sciences and he did not succeed in changing biologists' attitude toward geology and paleontology, which continued to be one of benign neglect; (2) theoretical confusion with regard to evolution—Jaekel's proposal was predicated on evolution (*Entwicklungslehre*) as the unifying thread for all the biological sciences. But which of the many different and often conflicting interpretations of evolution should this be? He himself had clearly neo-Lamarckian views, but others in Germany held neo-Darwinian, Mendelian, or even neovitalist positions, which made a conceptual unification of biology based on evolution that much more difficult;

and (3) methodological rather than conceptual unification—it was easier for biologists to put their theoretical differences aside and agree on new methods, such as those of experimental biology, and the newly developing framework of general biology, than any specific theoretical perspective. Furthermore, general biology was focused on an organismal perspective and was based on the integration of morphological, physiological, developmental, behavioral, and genetical approaches, thus incorporating many of the dominant research traditions within German biology (Laubichler 2006). Evolution, on the other hand, while it was considered to be a fact of nature, was not really discussed within these scientific debates. On the other hand, evolution continued to be featured prominently in often contentious public discussions. This fact did not appeal at all to those Prussian administrators, such as Friedrich Althoff and Adolf von Harnack, who decided the strategic orientation of the Kaiser Wilhelm Society.

Thus Jaekel's attempt to align paleontology with biology initially failed. During a period in the history of German biology that was dominated by the experimental ideal, the evolutionary and morphological perspectives of paleontology did not secure an invitation to the high table of German biology. But Jaekel succeeded in drawing attention to the problem of the relationship between paleontology and biology. One of his supporters, the Viennese paleontologist Othenio Abel, made this connection explicit when he introduced the term *paleobiology* for studies of extinct organisms in their environment (Abel 1911, 1921, 1929b). This concept eventually had a more favorable reception than Jaekel's term, *Bioontologie,* and is still in use today. It also signaled growing support among some German paleontologists for forging closer ties with biology.

WALTER ZIMMERMANN: TELOME THEORY AND THE PHYLOGENY OF PLANTS

If Otto Jaekel provides us with a window into the debates about the theoretical orientation of German paleontology during the first decades of the twentieth century, then Walter Zimmermann's work on plant evolution, though strictly speaking not that of a paleontologist per se, represents an interesting synthesis of evolutionary, paleontological, and morphological perspectives that also connects twentieth-century evolutionary biology with the most revered icon of German culture, Johann Wolfgang von Goethe.

Zimmermann is today best known as one of the founders of the telome theory, a theory explaining the morphological diversification of land plants (more on that in the following). But he was also actively engaged in many of the German evolution debates, and both his morphological work and his contributions to systematics and phylogenetics also connect him with German paleontologists. Zimmermann was born into a bourgeois German family and, as would be expected, received a thorough humanistic education. His subsequent studies at several universities were equally broad and included mathematics, botany, embryology, evolutionary biology, paleontology, geology, history, and literature. His original plan was to teach high school; nevertheless, he began working on his dissertation on *Volvox* in 1914. The outbreak of World War I interrupted his research. Zimmermann served as an officer for more than four years. He was wounded toward the end of the war and finally finished his dissertation at the University of Freiburg in 1920. Five years later he moved to the University of Tübingen (the power base of Otto Schindewolf, see the following) where he stayed until the end of his career (see Junker 2004).

During the late 1920s Zimmermann's research focused increasingly on questions of phylogeny, culminating in his first major publication, *Die Phylogenie der Pflanzen*, in 1930. This work was a comprehensive account of plant phylogeny, drawing on many different resources, such as paleontology, morphology, genetics, and developmental physiology. The theoretical sections of this work include discussions of what Zimmermann called "phylogenetic-historical laws," such as Dollo's law, correlated transformations, or the biogenetic law. One of Zimmermann's main methodological arguments concerned the different contributions of paleontology and experimental approaches to evolutionary theory. For him, paleontology is important for establishing general trends in phylogeny and identifying evolutionary innovations and new adaptations. But it is not suited to settle the question which of the different evolutionary models—neo-Lamarckism or neo-Darwinism—better describes the evolutionary process; in his opinion, this question would have to be decided experimentally. In subsequent publications Zimmermann further developed his evolutionary views as well as his best-known contribution, the telome theory (e.g., Zimmermann 1965).

The telome theory and the problem of plant morphology have an illustrious and intertwining history. The work of Kaspar Fredrich Wolff (1733–1794) and Johann Wolfgang Goethe (1749–1832) marked a new era in the history of comparative plant morphology by establishing a dynamic, ontogenetic perspective on the development of plant organs, thereby denying the utility of the largely static worldview advocated by Carl von Linné (1707–1778). Wolff introduced the concept of the *punctum vegetationis* (the growing point, or apical meristem), and, in 1764, he was the first to argue that the parts of the flower are reproductively modified vegetative leaves—a concept whose discovery history typically credits to Goethe.

Goethe's own doctrine of plant metamorphosis, which is fully developed in his *Versuch die Metamorphose der Pflanzen zu erklären* (1790), is as deceptively simple as it is profound—all appendicular organs, including the reproductive parts of the flower, are developmentally homologous organs. This ontogenetic concept dove deeper into the heart of plant metamorphosis than any of Goethe's predecessors, including Wolff, because it ascribed no "ancient" versus "derived" condition, but rather saw all appendages to be derived from a single fundamental type that he called the *Blatt* (= leaf). Although this word immediately evokes the concept of the vegetative leaf, Goethe's writings make it clear that he considered the vegetative leaf as only one among many appendage-types that are all derivatives of a basic organ-type, which, in today's parlance, would be called a "phyllome." Importantly, the idea of transformism, which is so evident in Goethe's writings on the comparative anatomy of animals, is entirely lacking in his treatment of plants. This difference is understandable because, at the time Goethe was developing his botanical concepts, descriptive morphology and anatomy was totally inadequate for the purpose of resolving plant phylogenetic relationships or evolutionary morphological problems, and paleobotany was only in the most incipient state.

Importantly, Goethe's concept of the *Urpflanze* was not an evolutionary concept or even a genetical one. It was, when fully mature in the poet's mind, a concept of "the type"—in keeping with that later advocated by Troll or Owen. The type concept played an important role in Goethe's thinking largely because it eased the tasks of comparative anatomy and morphology, by permitting homologies to be drawn among the body parts of seemingly disparate organisms.

Although history remembers every minute detail about the life and work of Germany's great poet, Goethe's writings on plants were largely ignored by his contemporaries. Most failed to see that he was the founder of a new perspective, which can be called *ontogenetic plant morphology*. Others saw—correctly—that it was a direct repudiation of the prevailing metaphysics of Linnaeus, and attacked it for that reason. Perhaps worse, many of those that adopted a Goethean perspective did so in the context of a purely idealistic romantic *Naturphilosophie*.

The foundation for the development of evolutionary comparative plant morphology was laid down by Karl Sanio (1832–1891), Philippe van Tieghem (1839–1914), Edward Jeffrey (1866–1952), and others who collectively benefited from the phylogenetic insights gained from the study of fossil plants by paleobotanists like Cotta, Whitham, Brogniart, and Williamson in the late nineteenth and early twentieth centuries. During this active period of research and exploration, which blended insights from the study of the living and the dead, the field of plant organography and anatomy changed from a purely descriptive science into a Darwinian exercise in evolutionary thinking.

This transformation into what Thomas (1932) called the "New Morphology" was almost entirely due to the emergence of the telome theory. Although the development of this theory is now almost universally credited to the work of Walter Zimmermann, its basic ideas are clearly outlined in the writings of Henry Potonie (1857–1913) and Octave Lignier (1855–1916). The former argued, on purely theoretical grounds, that leaves evolved from the lateral fusion of much simpler parts of ancient thallophytes. Accordingly, Potonie was one of the first to reject the ironclad distinctions typically drawn across the leaf/stem/root vascular plant organ triad. Drawing on the work of Potonie, Lignier further developed the telome theory by contributing two important ideas: (1) the most ancient land plants had dichotomously branched sporophytes and (2) sporangia were distal and thus evolutionarily preceded the leaf/stem distinction. Lignier's hypothetical morphology for the most ancient vascular plants received stunning validation with the discovery and meticulous description of the Rhynie Chert flora by Kidston and Lang in 1915–1916, which contained fossils identical in almost every respect to Lignier's ur-plant.

Like Goethe, the ideas of Potonie and Lignier were not appreciated by their contemporaries. It was only after the elapse of nearly two decades that the telome theory was resuscitated by Walter Zimmermann's book *The Phylogeny of Plants* (1930). Here and elsewhere Zimmermann argued that the stereotypical architecture of all modern-day vascular plants evolved from a plexus of ancient morphologically simple plants whose sporophytes consisted of simple bifurcating axes called telomes, derived from the division of apical meristems. In turn, telomes are subtended and joined together by axes called mesomes (equivalent to the internodes of modern-day stems). Telomes may be either sterile or fertile (phylloids or spore-bearing, respectively). According to the telome theory, the large leaves produced by ferns and other megaphyllous lineages evolved from syntelomes—specialized photosynthetic organs resulting from the planation and webbing of adjoining telomes. Like Lignier, Zimmermann firmly rejected the organographic distinction between stem and leaf.

The telome theory of Lignier and Zimmermann was elaborated substantially by F. O. Bower in his book *The Primitive Land Plants* (1935), particularly regarding the evolutionary origins of small scalelike leaves, like those of lycopods. Bower argued convincingly that microphylls evolved from small, originally sterile protuberances or "enations," a concept that was substantiated by subsequent paleobotanical discoveries. Among the many important contributions made by Bower, his work demonstrated that the telome theory could (and does) not provide an absolute and monolithic explanation for plant morphological evolution. Unfortunately, many who followed in the footsteps of Lignier and Zimmermann failed to learn from Bower's example and persisted in explaining all of plant morphology in terms of telomes, even to the point of drawing fantastic scenarios for the evolution of the flower directly from simple bifurcate axes!

In addition to his work on the telome theory Zimmermann also contributed to German discussions on evolution. His book *Die Vererbung erworbener Eigenschaften und Auslese* (Zimmermann 1938) contains a detailed rejection of neo-Lamarckian theories, which were still popular among some German biologists and many paleontologists at that time, and his chapter in Gerhard Heberer's monumental work, *Die Evolution der Organismen* (Heberer 1943) was devoted to "Die Methoden der Phylogenetik." In this chapter Zimmermann presents a concise overview of the concepts, goals, challenges, and methods of phylogeny. With regard to paleontology Zimmermann notes that it is difficult, due to the limitations of the fossil record, to establish clear-cut ancestor-descendant relationships. It is much easier to identify sequences of character transformations. However, as phylogeny is based on the principle of gradual nested hierarchy, it is possible to establish a coherent phylogenetic system. In this context Zimmermann repeats his attacks on idealistic morphology, mostly on methodological grounds, as the emphasis of these scientists on intuition does not comply with scientific standards of objectivity and testability.

For our question of paleontology at the High Table Zimmermann is an interesting case study. Not, strictly speaking, a paleontologist, he did contribute greatly to questions of phylogeny, both through his work on the evolutionary history of plants and in plant morphology, especially the telome theory. Zimmermann also represents the humanistic ideal of a scientist, highly valued within German society. His monumental work on the history of evolutionary thought, beginning with the pre-Socratics and paying, of course, special attention to Goethe, is a classic in humanistic scholarship (Zimmermann 1953). And as it was published in a series of works aimed at the educated public—

the series title *orbis academicus* clearly reveals the humanistic orientation of the whole enterprise—Zimmermann thus became a leading representative of evolutionary biology within the German republic of letters.

OTTO SCHINDEWOLF: BASIC QUESTIONS AND TOTAL CONTROL OF PALEONTOLOGY

In the foreword to the English translation of Otto Schindewolf's (1896–1971) major theoretical work, *Basic Questions in Paleontology*, Stephen Jay Gould recalls the following anecdote (Schindewolf and Reif 1993). He once asked his teacher, Norman Newell, who the greatest living paleontologist was. The answer came immediately—Otto Schindewolf. Gould was surprised, as Schindewolf was not part of the Modern Synthesis, which had transformed evolutionary biology from mid-century onward. Rather, Schindewolf stood for an anti-Darwinian form of evolutionism rooted in the morphological and developmental traditions of continental science and philosophy. Is the fact that Schindewolf was held in such regard, even by those with different views, such as Newell, a testament to his powerful position, the quality of his empirical work (as separate from his theoretical ideas), or the fact that even among converts to a neo-Darwinian view of evolution the old questions of paleontology—those of macroevolution and major transitions, continued to hold a strong appeal? The answer, probably, is yes to all of the above, and it also goes a long way in explaining the different trajectory of German paleontology during the twentieth century.

Otto Heinrich Schindewolf first studied geology under the well-known geologist Hans Stille (1876–1966) in Göttingen, who is today best known for his cyclical theory of mountain folding and his opposition to Wegener's theory of plate tectonics, and then took his PhD in Marburg with Rudolf Wedekind (1883–1961). He stayed in Marburg until he was appointed head of the geological survey in Berlin in the late 1920s and remained in Berlin until the end of World War II. Despite his eminence as a scientist he was denied a full professorship under the Nazis due to his lack of support for the regime. He finally was appointed to the most powerful chair in German paleontology in Tübingen in 1948, a position he held until the end of his career.

What, then, is the importance of Otto Schindewolf and why is he considered to be so important? Like other prominent German paleontologists and biologists, Schindewolf believed that evolutionary theories of his time were woefully incomplete, and that for the most part, the evolutionary discourse was utterly confused, especially the many neo-Lamarckian positions. Schindewolf thought that the simple application of known genetic facts should be able to clarify this matter and demonstrate the inadequacy of neo-Lamarckian positions. In that he followed the lead of his teacher Wedekind, who had argued already in 1916, citing the genetic concepts of the day—especially those of Johannson and Goldschmidt—that neo-Lamarckian positions have no basis in modern science (Reif 1993). Schindewolf expressed his own thoughts on this matter first in a book, *Entwicklungslehre und Genetik*, in 1936. Also relying on Goldschmidt's analysis of the role of genes in development and evolution—Goldschmidt had at this time already suggested several models of how quantitative gene action can change phenotypes through changes of developmental processes—Schindewolf tried to combine the current knowledge of genetics with the paleontological evidence for evolutionary innovations and novelties. Seen from our present-day perspective Schindwolf was on the right track, but the terminology of the day, especially Goldschmidt's concept of macromutation proved, in the end, to be unhelpful. But combining those developmental/morphogenetic processes studied by paleontologists—heterochrony and allometry—with underlying genetic causes—mutations affecting the developmental system—clearly was an innovative strategy. However, in 1936 the available evidence was still inadequate, and Schindewolf concluded that evolutionary theory was incomplete.

In the meantime, detailed empirical studies in paleontology, especially those in stratigraphy by Schindewolf himself and Karl Beurlen, revealed common cyclical patterns and trends in the fossil record of different systematic groups, such as ammonites and crustaceans. It was Beurlen who first interpreted these data along the line of a "life history" of a systematic group, arguing that the "pathway of evolution within a taxon . . . is cyclic (Beurlen 1932)." Each group was considered to go through three stages: saltation and explosive creation of forms, orthogenetic continuity, and, finally, disintegration and extinction. With the widespread emphasis on individual development among German biologists, such a version of an ontogenetic sequence of taxa had a certain appeal. In any case, it provided a framework not only for the interpretation of large stratigraphic datasets but also for further conceptual development of paleontological and evolutionary theory. And, in particular, these ideas allowed Schindewolf to incorporate the remnants of idealistic morphology, which had again become popular after World War I and, largely because of its references to Romanticism and Goethe, also had widespread popular appeal (Naef 1919; Troll 1926, 1944; Troll and Wolf 1944) into his system. Schindewolf, who had no use for metaphysical concepts, simply equated idealistic types with the concrete reality or life of taxa that were visible in the fossil record.

During the same period (late 1930s and early 1940s) as Schindewolf was developing his theoretical framework, paleontology played hardly any role in discussions among German evolutionary biologists. These discussions culminated in the 1943 publication of the volume *Die Evolution der Organismen*, edited by Gerhard Heberer, often referred to as the "German Modern Synthesis" (e.g., Juncker 2004). It is striking that there was no paleontologist among the contributors to Heberer's volume, and only Walter Zimmermann touched in any meaningful way on phylogenetic issues. Heberer successfully organized the German Darwinists, such as Bernhard Rensch, who was working on his book *Neuere Probleme der Abstammungslehre* (only published in 1947, due to war-related problems), and advocated a gradual, or microevolutionary model of evolutionary change. At the same time Schindewolf, facing similar publication delays of his own book, developed a radically different view of the evolutionary process, which came to be know as *typostrophism* and dominated German paleontology during the following decades.

Die Grundfragen der Paläontologie was finally published in 1950; an English translation appeared in 1993. As Schindewolf's theory has recently received some attention, we can be brief here and emphasize more contextual factors rather than providing a detailed account of its themes (see also Reif 1993). Typostrophism has several core elements that are all, depending on one's point of view, either important additions to microevolutionary theory or an anti-Darwinian countermodel. The first of these elements is saltationism, the idea that major morphological changes do not arise gradually but in jumps. Saltationism has been a favorite topic of paleontologists and is one of the core themes of most macroevolutionary theories. For Schindewolf, saltationism was not only supported by the fossil record, it also was in accordance with his reading of the literature in genetics, a topic he pursued earlier on in his 1936 book (see the previous discussion). The second core element is orthogenesis, another venerable topic among paleontologists. Schindewolf's conception, however, differs in important ways from most previous accounts of orthogenesis that emphasized some vital or metaphysical principle governing directed evolutionary change. For Schindewolf, the starting point of a new evolutionary trajectory lies in a mutational (macromutational) change that is not directed in any way. But once a new type has been established, orthogenesis is reflected in a constrained unfolding that has the appearance of directionality but actually represents a limitation on future variation. While this view contradicts important elements of the gradual conception of evolutionary change of the Modern Synthesis it is actually more in line with the recent findings of developmental genetics and discussions about developmental and architectural constraints. The last element of Schindewolf's theory, the cyclical view of evolutionary patterns, does not have any obvious predecessors (other than Beurlen), although it resonates with current thinking about species selection. It assumes that taxa have a life cycle similar to that of individual organisms (birth/spontaneous diversification, growth and differentiation, and degeneration and extinction). Schindewolf's proposal is predicated on some sort of continuity and objective existence of taxa—a view similar to the idea of species as individuals, which has recently been proposed in the context of discussions about systematics and evolution. As we have already seen, among the conceptual resources Schindewolf could draw upon were his own materialistic interpretation of idealistic morphology and the widespread emphasis within German biology that development is the core process that characterizes biological systems. Integrating all these themes, *Basic Questions in Paleontology* provided a tightly integrated conceptual and interpretative framework for detailed empirical studies in stratigraphy, Schindewolf's own empirical forte. And seen this way we can more easily understand how typostrophism could become such a force within German paleontology.

Schindewolf continued to defend his approach until his death in 1971, and insisted that the Modern Synthesis is an incomplete theory of evolution. According to Ernst-Wolf Reif this had a damaging effect on younger generations of German paleontologists, who were not exposed to or consciously avoided discussions of evolutionary problems, as typostrophism was not to be challenged (Reif 1993). From today's international perspective the power that Schindewolf had accumulated within German paleontology is difficult to imagine. As a consequence, in the context of evolutionary discussions many younger German paleontologists felt like second-class citizens. But, as the example of Seilacher shows, there is also virtue to be found in different perspectives. Furthermore, today the stronghold of the Modern Synthesis is clearly under attack. Schindewolf's insistence on the incompleteness of mid-twentieth century evolutionary theories has now become almost commonplace, and the current emphasis on the origins of variation as an equally important evolutionary problem as selection has led to a resurgence of discussions of internal factors of evolution. Clearly, the schematism of the typostrophic theory is no longer a viable alternative, but neither is an evolutionary theory solely based on random mutation and subsequent selection. Schindewolf's earlier attempts to reconcile paleontology and genetics might, in this context, be a more productive model. The idea that mutations can affect different parts of developmental processes and therefore have a wide range of phenotypic consequences finds its modern equivalent in the different elements of regulatory gene networks, such as kernels, plug-ins, switches, and differentiating gene batteries (Davidson 2006). Would Schindewolf have accepted modern developmental genetics as completing an otherwise incomplete evolutionary theory? Of course, we don't know, but it is tempting, albeit somewhat ahistorical, to speculate. But we cannot help but wonder what Schindewolf would have said about today's leading developmental geneticists teaming up with paleontologists to address questions of macroevolution. At the very least, we submit, he would have been intrigued.

CONCLUSION

Returning to the question about paleontology's place in twentieth-century science, our brief vignettes in the history of German paleontology explain why, until recently, German paleontologists would not have been welcome at the high table of Darwinian evolutionary biology. Their conceptual orientation was simply too different, and not at all focused on the extension of gradual, population-based mechanisms of evolutionary change to questions of cladogenesis and the emergence of higher taxa that has been characteristic of the Darwinian orthodoxy. Much of the work was anchored in a variety of local traditions, and even though one can distinguish certain clusters of topics and people, German paleontology has not really had a coherent research program. Rather, many of its themes have been rooted equally within strong cultural and academic values—the morphological, historical, geological, and organismal paradigms that have linked a good deal of German biology over the last two centuries with German culture at large—as within any recognizable disciplinary matrix.

German paleontology is thus a perfect topic for a cultural history of science that places the development of scientific theories and concepts clearly within the framework of cultural references, values, and transformations. As Wolf-Ernst Reif put it: "Until WWII almost all leading German paleontologists presented their personal views [sic!] on evolutionary theory in textbooks, journal articles and lectures" (Reif 1999, 151–52). These personal statements reveal a good deal about the kind of cultural, historical, and philosophical factors that shaped the opinions of these scientists. And even though a complete account of all these influences is still missing, based on our preliminary review of the material and the interpretations of others working in this area, we could clearly identify several cultural/scientific paradigms that established the conceptual topology of German paleontology. Among those is an emphasis on form (ranging from morphology to art), history (seen here less as phylogeny and more along the lines of an evolutionary/developmental narrative), geology (especially stratigraphy), and the organism (as the locus of biological processes) clearly stand out. Each of these themes has its own interpretative framework and logic of concepts (Cassirer 1969, 2000). But, as none of these topics fits particularly well within the Modern Synthesis, German paleontologists did not join the high table. They fared somewhat better within their own cultural reference system, but even in the context of German science they were caught between the rock of geology and the hard place of biology.

The question, then, is what does the history of German paleontology mean to us. Is it merely a curious episode, a case study in cultural history (which would be no small accomplishment by itself), or is it still significant today? Even if we take for granted that the history of science can be relevant for science, we still have to demonstrate the significance of these particular traditions. But it seems that the criteria for being invited to the high table of evolutionary biology have changed. New questions are asked, several old problems have made a comeback, and new, sometimes quite unexpected alliances have been formed. Macroevolution is still discussed rather vigorously, the reconstruction of whole ecological histories adds a new dimension to phylogeny, and in the context of evo-devo we not only see the reappearance of traditional morphological concepts (*Bauplan*) we also see a return of an organismal perspective, even at the molecular level. Here we are not saying that Schindewolf or any other German paleontologist was right and that it is our duty to defend misunderstood geniuses. Rather, we are trying to understand what it has meant to be invited to the high table in different countries at different times and what the criteria of admission might be today. The architecture of regulatory gene networks—their kernels, plug-ins, switches, and gene batteries—are thought to correspond to both the observed modular architecture of organisms and the patterns of phylogeny and macroevolution (Davidson and Erwin 2006). Clearly, genomic changes are still the raw material of evolution, but a mutation is not a mutation is not a mutation. The context matters, whether it is molecular, developmental, organismal, or ecological. The theoretical questions that so concerned German paleontologists—the nature of higher taxa and their role in the evolutionary process, the question of possible and impossible transformations of types, and the problem of a seeming directionality in morphological evolution are now again being asked, mostly within developmental genetics and evo-devo. It has long been recognized that the inclusion of paleontology will be crucial for the future success of the evo-devo synthesis (as it was for the Modern Synthesis). And, given the similarity of the current problems to some of those that occupied German paleontologists of the late nineteenth and early twentieth centuries, the history of this period might be a potential resource. It might, it does not have to be. To find out, we first need to know more about this fascinating period.

REFERENCES

Abel, O. 1911. *Paläobiologie.* Stuttgart: Schweizerbart'sche Verlagsbuchhandlung.

———. 1921. *Allgemeine Paläontologie.* Berlin und Leipzig: Walter de Gruyter.

———. 1929a. Otto Jaekel. *Palaeobiologica* 2:143–86.

———. 1929b. *Paläobiologie und Stammesgeschichte.* Jena, Switzerland: Gustav Fischer.

Beurlen, K. 1932. Funktion und Form in der organischen Entwicklung. *Die Naturwissenschaften* 20:73–80.

Bower, F. O. 1935. *Primitive land plants, also known as the Archegoniatae.* London: Macmillan.

Cassirer, E. 1969. *The problem of knowledge.* New Haven, CT: Yale University Press.

———. 2000. *The logic of the cultural sciences.* New Haven, CT: Yale University Press.

Davidson, E. H. 2006. *The regulatory genome : Gene regulatory networks in development and evolution.* Burlington, MA: Academic Press.

Davidson, E. H., and Erwin, D. H. 2006. Regulatory networks and the evolution of animal body plans. *Science* 311:796–800.

Goethe, J. W. 1790. *Versuch die Metamorphose der Pflanzen zu erklären.* Gotha, Germany: C. W. Ettinger.

Hagner, M., and M. D. Laubichler, eds. 2006. *Der Hochsitz des Wissens. Das Allgemeine als wissenschaftlicher Wert.* Berlin and Zürich: Diaphanes.

Jaekel, O. 1901. Über verschieden Wege phylogenetischer Entwicklung. *Verhandlungen des 5. Internationaler Zoologischer Kongress,* 1–60. Berlin.

———. 1914. Die Stellung der Paläontologie zu den Naturwissenschaften. *Paläontologische Zeitschrift* 1:18–50.

Junker, T. 2004. *Die zweite Darwinsche Revolution : Geschichte des synthetischen Darwinismus in Deutschland 1924 bis 1950.* Marburg: Basilisken-Presse.

Junker, T. and E.-M. Engels (1999). *Die Entstehung der synthetischen Theorie : Beiträge zur Geschichte der Evolutionsbiologie in Deutschland 1930–1950.* Berlin: Verlag für Wissenschaft und Bildung.

Kühn, A. 1965. *Vorlesungen über Entwicklungsphysiologie.* Berlin: Springer-Verlag.

Laubichler, M. D. 2003. A pre-modern synthesis: Robert J. Richards. The romantic conception of life. *Science* 299:516–17.

———. 2005. Art forms in nature: David Lebrun. Proteus—A nineteenth century vision. *Science* 308:1746.

———. 2006. Allgemeine Biologie als selbstständige Grundwissenschaft und die allgemeine Grundlagen des Lebens. *Der Hochsitz des Wissens. Das Allgemeine als wissenschaftlicher Wert*, ed. M. Hagner and M. D. Laubichler, 185–206. Berlin and Zürich: Diaphanes.

McGhee, G. R., Jr. 1997. *Theoretical morphology: The concept and its application.* New York: Columbia University Press.

———. 2007. *The geometry of evolution: Adaptive landscapes and theoretical morphology.* Cambridge: Cambridge University Press.

Naef, A. 1919. *Idealistische Morphologie und Phylogenetik (Zur Methodik der systematischen Morphologie).* Jena, Switzerland: Gustav Fischer.

Niklas, K. J. 1992. *Plant biomechanics : An engineering approach to plant form and function.* Chicago: University of Chicago Press.

———. 1994. *Plant allometry : The scaling of form and process.* Chicago: University of Chicago Press.

———. 1997. *The evolutionary biology of plants.* Chicago: University of Chicago Press.

Odling-Smee, F. J., K. N. Laland, and M. W. Feldman. 2003. *Niche construction : The neglected process in evolution.* Princeton, NJ: Princeton University Press.

Reif, W.-E. 1986. The search for a macroevolutionary theory in German paleontology. *Journal of the History of Biology* 19:79–130.

———. 1993. Afterword. *Otto H. Schindewolf: Basic questions in paleontology.* Chicago: University of Chicago Press.

———. 1999. Deutschsprachige Paläontologie im Spannungsfeld zwischen Makroevolutionstheorie und Neo-Darwinismus (1920-1950). In *Die Entstehung der synthetischen Theorie. Beitruage zur Geschichte der Evolutionsbiologie in Deutschland 1930-1950*, ed. T. Junker and E.-M. Engels, 151–88. Berlin: Verlag für Wissenschaft und Bildung.

Reif, W.-E., Junker, T., and Hoßfeld, U. 2000. The synthetic theory of evolution: General problems and the German contribution to the synthesis. *Theory in Biosciences* 119:41–91.

Rensch, B. 1947. *Neuere Probleme der Abstammungslehre.* Stuttgart: Enke.

Richards, R. J. 2002. *The romantic conception of life : Science and philosophy in the age of Goethe.* Chicago: University of Chicago Press.

Richards, R. J. 2008. *The tragic sense of life : Ernst Haeckel and the struggle over evolutionary thought.* Chicago: University of Chicago Press.

Ruehl, M. 2006. Kentaurenkämpfe. Jacob Burckhardt und das Allgemeine in der Geschichte. In *Der Hochsitz des Wissens. das Allgemeine als wissenschaftlicher Wert*, ed. M. Hagner and M. D. Laubichler, 23–72. Berlin and Zürich: Diaphanes.

Schindewolf, O. H. 1936. *Paläontologie, Entwicklungslehre und Genetik; Kritik und Synthese.* Berlin: Gebrüder Borntraeger.

———. 1950. *Grundfragen der Paläontologie; Geologische Zeitmessung.* Stuttgart: E. Schweizerbart.

———. 1969. *Über den "Typus" in morphologischer und phylogenetischer Biologie.*

Mainz: Verlag der Akademie der Wissenschaften und der Literatur; F. Steiner in Kommission.

Schindewolf, O. H., and W.-E. Reif. 1993. *Basic questions in paleontology : Geologic time, organic evolution, and biological systematics.* Chicago: University of Chicago Press.

Seilacher, A. 1970. Arbeitskonzept zur Konstruktions-Morphologie. *Lethaia* 5:325–43.

Sucker, U. 2002. *Das Kaiser-Wilhelm-Institut für Biologie : seine Gründungsgeschichte, seine problemgeschichtlichen und wissenschaftstheoretischen Voraussetzungen (1911–1916).* Stuttgart: Steiner.

Thomas, H. H. 1932. The old morphology and the new. *Proceedings of the Linnaean Society London* 145:17.

Troll, W., ed. 1926. *Goethes Morphologische Schriften ausgewählt und eingeleitet von Wilhelm Troll.* Jena, Switzerland: Eugen Diederichs.

———. 1944. Urbild und Ursache in der Biologie. *Botanisches Archiv* 45:396–416.

Troll, W., and K. L. Wolf. 1940. Goethes morphologischer Auftrag. *Botanisches Archiv* 41:1–71.

Zimmermann, W. 1930. *Die Phylogenie der Pflanzen.* Jena, Switzerladnd: Gustav Fischer.

———. 1938. *Vererbung "erworbener Eigenschaften" und Auslese.* Jena, Switzerland: Gustav Fischer.

———. 1943. Die methoden der phylogenetik. In *Die Evolution der Organismen.* ed. G. Heberer, 20–65. Jena, Switzerland: Gustav Fischer.

———. 1953. *Evolution : Die Geschichte ihrer Probleme und Erkenntnisse.* Freiburg: K. Alber.

———. 1965. *Die Telomtheorie.* Stuttgart: G. Fischer.

CHAPTER FIFTEEN

"Radical" or "Conservative"? The Origin and Early Reception of Punctuated Equilibrium

David Sepkoski

In 2002, only months before his death, Stephen Jay Gould published *The Structure of Evolutionary Theory*, the summation of more than thirty years' grappling with the Darwinian theory of evolution. In the community of evolutionary biologists, Gould is probably best known for authoring, with Niles Eldredge, the theory of punctuated equilibrium. In *Structure*, a lengthy, revisionist history of the punctuated equilibrium debate spans the final third of the volume and attempts to set the record straight about the proper interpretation of his theory in relation to neo-Darwinian evolutionary orthodoxy. That this section of the book runs nearly 300 pages is hardly surprising given the tremendous—and often heated—controversy punctuated equilibrium created during Gould's lifetime. As a number of reviews of *Structure* have noted, Gould attempts to strike a fairly delicate compromise: on the one hand, he argues that punctuated equilibrium does *not* challenge the essential principles of Darwinian natural selection—it is "a speciational theory of macroevolution, with species treated as irreducible Darwinian individuals playing causal roles analogous to those occupied by organisms in Darwinian microevolution."[1] Viewed more broadly, however, it is apparent that the unifying theme of *Structure* as a whole is a reevaluation of what it means to be Darwinian, particularly in the context of the Modern Evolutionary Synthesis of the mid-twentieth century. The chapter on punctuated equilibrium is merely the centerpiece of a broader argument that, in total, spans more than 800 of the book's 1,343 pages, and draws ammunition from macroevolution and species selection, development (ontogeny and phylogeny), adaptationism, and the history of the Modern Synthesis itself. As the capstone to a life's work in evolutionary theory, *Structure* is clearly a frontal assault on neo-Darwinsm; punctuated equilibrium is just as clearly the lynchpin of that attack.

Reviewers of *Structure* have also noted that Gould presents nothing particularly new in the book; nothing, that is, if one has been following his career since the mid-1970s. Over the decades since the publication of *Ever Since Darwin* (in 1977), Gould's public and scientific personae have been somewhat at odds: while to the general public he was the sober defender and promoter of scientific rationality and literacy, to his scientific colleagues he was the ardent proponent of a radical (and perhaps misguided) view of evolutionary change. Yet despite the brashness of many of his claims on behalf of punctuated equilibrium over the years, one is brought time and again back to the reconciliatory, even conservative justifications Gould made for his theory: in *Structure*, for example, he recalls the common "misunderstanding" of his critics that "I proclaimed the total overthrow of Darwinism, and . . . that I intended punctuated equilibrium as both the agent of destruction and the replacement."[2] For the historian attempting to assess the significance of Gould's scientific work, as well as for the scientist or philosopher trying to evaluate it, there is an immediate and important question: just how radical did Gould intend to be?

This essay will attempt to resolve some of the uncertainty surrounding this question. First, ignoring for the most part the literature retrospectively assessing the significance of punctuated equilibrium, I will examine what Gould and his colleagues thought about the theory at the time it was conceived and shortly afterward. Second, I will argue that punctuated equilibrium cannot be separated from the broader development of theoretical paleobiology promoted during the mid-1970s by a group of paleontologists dedicated to revising the goals, agenda, and status of paleontology. Third, I will argue that when viewed in this context, punctuated equilibrium is in fact part of a larger movement that involved not just Gould and Eldredge, but his colleagues David Raup, Thomas Schopf, Steven Stanley, and others as well, which sought to redefine paleontological evolutionary theory largely without the directional causes central to traditional Darwinian evolution via natural selection. This approach made innovative use of kinds of quantitative statistical analysis and computer simulation that were new to paleontology, pioneered modeling of macroevolutionary patterns as stochastic (or random) processes, and also rehabilitated the evolutionary significance of extinction dynamics (particularly mass extinctions), all of which posed questions about the ubiquity of natural selection as the sole causal mechanism in evolution. Punctuated equilibrium was not even necessarily the most iconoclastic or radical product of this movement.

Finally, however, I will argue that a dichotomy between the labels *radical* and *conservative* is not appropriate for describing these theories or the motivations of the authors of punctuated equilibrium. Even among the central figures in the new paleobiology (some of whom cheekily called themselves the "radical fringe" in paleontology) there was significant disagreement concerning the proper theoretical approach with which to challenge neo-Darwinism. In particular, I will focus on the relationship between Gould and Tom Schopf. Schopf was Gould's close friend and collaborator, and he provided important institutional support for Gould's work (in addition to editing the volume *Models in Paleobiology*, where punctuated equilibrium first appeared in print, he was founding editor of the journal *Paleobiology*, where many of Gould's scientific papers were published). He was also the most strident advocate of stochastic modeling and nondirectionality in macroevolution, but at the same time one of the most vocal opponents of punctuated equilibrium. The dynamics of Gould and Schopf's personal friendship and professional rivalry, particularly represented in their correspondence, serves as a microcosm for the broader debates in evolutionary paleobiology during the 1970s. By examining this dynamic I will argue that the paleobiological response to neo-Darwinism involved elements of both radicalism and conservatism, and that in surprising ways, what might be seen as radical by one observer may well have been conservative to another.

THE ORIGIN OF THE THEORY

In light of the great significance punctuated equilibrium was to have in his later career, it is ironic that Gould did not even want to write about speciation theory in the first place. In early 1970, Schopf was organizing a symposium for the Geological Society of America's 1971 annual meeting on the subject of "Models in Paleontology." In soliciting participants, Schopf described the session as an opportunity "to identify and evaluate the theoretical models which are guiding (by accident or design) the development of various parts of our science," since "the theoretical framework . . . dictates where one looks [in empirical data] and how one goes about the descriptive process." "In this way," Schopf continued, "we can encourage the analytical 'problem oriented' approach in Paleontology."[3] Schopf asked Gould to contribute a paper on models in speciation, and Gould's response was decidedly ambivalent: "A damned good idea, your symposium. I'm flattered by your invitation and will gladly accept. My only hesitation is that you have given me a topic that ranks only third on your list in terms of my competence [behind models in morphology and phylogeny]." Nonetheless he agreed, since the other subjects were already taken, but he suggested a joint effort with Niles Eldredge, whom he described as "our best new thinker."[4] The eventual compromise was to coauthor the paper with Eldredge, which Gould presented at the conference but for which Eldredge was first author in the subsequent published proceedings. The result was the now famous essay "Punctuated Equilibrium: An Alternative to Phyletic Gradualism."[5]

The classic Darwinian evolutionary model assumes that species change very gradually over vast amounts of time (tens of millions of years or more), developing in response to equally slow and gradual changes in environment that produce adaptations that ultimately lead to the appearance of new species. A central assumption is that this is a constant, inexorable process, and that the tempo of evolution is unchanging. The major driving force in Darwinian evolution is the mechanism of natural selection, which presses individual organisms to compete with one another and their environments, rewards beneficial adaptations, and punishes less-successful species with extinction. That the fossil record was incomplete—with its intermittent gaps and jumps, and notorious absence of transitional missing link species—was of no concern to evolutionists, who contended that it was simply the result of an imperfectly preserved record.

However, beginning with the unorthodox intuition that the fossil record was in fact a much more accurate record of the history of life than had been previously assumed, Gould and Eldredge proposed a radical revision of this standard narrative. They argued that the pattern of evolutionary history really was composed of fits and starts, consisting of long periods of evolutionary stasis (or "equilibrium") "punctuated" by shorter periods of rapid speciation. This theory presented some very significant revisions of Darwinian evolutionary theory: by suggesting that species can act as independent units in natural selection, Gould and Eldredge upset the orthodox Darwinian assumption that natural selection can only bring about adaptive advantages in single organisms. Punctuated equilibrium instead proposed that entire species have life spans, with a birth, a long, stable period of existence, and a death, followed, in many cases, by offspring species.[6]

This theory was not composed in a vacuum—it was based solidly on previous work done by scientific luminaries like the biologists Ernst Mayr and Theodosius Dobzhansky and the paleontologist George Gaylord Simpson—but it attracted immediate attention for Gould (who became its spokesman) and caused eventual controversy. The theory effectively undercut the traditional understanding of the tempo of evolution by seeing phyletic gradualism as "very rare and too slow, in any case, to produce the major events of evolution," as Gould and Eldredge wrote in a reassessment of their theory in 1977.[7] According to many evolutionary biologists, this was tantamount to heresy: Darwinian evolution required a slow, steady progression over which natural selection could operate, in order for the accumulation of adaptations in individuals to gradually produce sufficient variations to cause speciation. If, as Gould proposed, species remained stable and unchanged for millions of years, only to suddenly branch off as new species or disappear completely, then what mechanism could account for this pattern?

THE IMMEDIATE RECEPTION

From the very beginning, the authors and Schopf were conscious of the potential for misreading punctuated equilibrium: for example, Gould objected to Schopf's draft introduction to the volume, which characterized the Gould-Eldredge paper as a statement of "the way that metaphysics impinges on considerations of speciation."[8] Gould clarified that his paper did not assume "an untestable statement about the world," but rather presented "an a priori testable theory," and was therefore not metaphysical. The theory, Eldredge and Gould argued privately to Schopf, was simply a defense of the significance of allopatric speciation (speciation based on geographic isolation), which was a concept firmly established in current paleontological thought, but whose implications for macroevolution—namely that it challenged traditional phyletic gradualism—were not widely appreciated.[9] Nonetheless, throughout the editing process Schopf worried about possible objections to the paper, and cautioned the authors to moderate their tone: "You self-consciously appear to give the type of case you can easily attack," he warned Eldredge. "That comes through, Niles, and that is precisely what will turn people off to your argument. Rather than educate, you may offend. That was the tone of the review[s]. . . . Given that response, Niles, you cannot be oblivious to the potential general response." "However," Schopf concluded, "the manuscript stays just as you have given it."[10]

As Eldredge recalls, Schopf "really hated the punctuated equilibria paper. He really hated it." As a consequence, Schopf sent the paper out to a dozen reviewers, and forwarded only the reviews that confirmed his own assessment.[11] One of these editorial reviews of the chapter summarized what would characterize many of the later reactions to the published piece. The referee began by commenting on the elegance of the writing, its "quality of introspection," and its "refreshing" emphasis on "the influence of theory on discovery." How-

ever, the review also questioned whether Eldredge and Gould's presentation of the "species problem" was a straw man, noting (quite alliteratively)
that "it has a built-in Platonic paradox for pushing paleontological pundits
toward polemic precipices for purposes of publication. Well written rhetoric,
or sophistry, or something. Not one whit closer to the problem."[12] This reaction was mirrored in the other referees' comments (all of whom urged publication despite criticisms) and in the reception of the paper once published: the
consensus was that it was a flawed but provocative theory, and most agreed
that it was an elegant piece of writing. A few paleontologists became immediate converts, while a few others saw no merit in the idea whatever, but most
simply withheld final judgment.

Over its first few years of life, punctuated equilibrium attracted relatively
little attention. During this time Gould collaborated with Schopf and Raup
on other work involving stochastic simulation of macroevolution, including a
project Schopf organized with Gould and Raup to better understand macroevolutionary patterns through a variety of statistical and computer-generated
models. Schopf suggested that a central goal should be to understand "the
processes underlying [the] patterns" of diversity, morphology, and phylogeny
through time, and to ask "what are their long-term equilibrium consequences."
In outlining the processes to investigate, Schopf made the interdependence
of paleontology and biology clear: "1) speciation theory, including population genetics and the species equilibrium"; "2) the constraints imposed by
size, shape and habitat on organized protoplasm"; "3) the unity (or disunity)
of biochemical pathways, including models of reproduction"; and "4) is
there an equilibrium model of phylogenetic development?"[13] The result of
this initial work was a groundbreaking paper (in 1973) jointly authored by
Gould, Schopf, Dave Raup, and Daniel Simberloff, titled "Stochastic Models
of Phylogeny and the Evolution of Diversity."[14] Essentially, this paper proposed that by generating random phylogenetic trees (using a computer program), one could test whether these would replicate certain aspects of actual
phylogenies—thus demonstrating whether actual patterns of origination and
extinction likely had stochastic variables.

In the fall of 1976, Gould mailed a manuscript to Schopf at *Paleobiology*
titled "Punctuated Equilibrium: The Tempo and Mode of Evolution Reconsidered." This paper, again coauthored with Eldredge, was the authors' first
major reconsideration of their theory. In his covering letter Gould described
the "rather exuberant manuscript" as "a labor of love," and as a defense of the
theory it served several functions.[15] First, more than even the original paper,
the new manuscript made punctuationism the center of a philosophical recon-

ceptualization of evolutionary change. Here the authors were more explicit about the exact nature of the conceptual reconfiguration their theory brought to macroevolution—in particular by adapting Steven Stanley's formulation of the asynchrony of micro- and macroevolution published a year earlier.[16] The paper also gave Gould and Eldredge a chance to address some of the criticisms that had been leveled at their theory, and to clarify what they actually intended to say in their first paper. Here the authors dismissed charges that they were motivated by an a priori disdain for gradualism, and that they took a "defeatist attitude" toward the testability of macroevolutionary claims using the fossil record.[17] They also presented a significant amount of empirical evidence to prove the theory by considering a wider sample of taxa than they had in the 1972 essay, and by answering empirical challenges to their earlier conclusions. Finally, Gould and Eldredge extended their model to propose a new and "general philosophy of change" in the natural world. Here they explored the sociocultural basis for gradualism, and proposed punctuationism as a new "metaphysic," which they suggested "may prove to map tempos of change in our world better and more often than any of its competitors."[18] This was done without any apparent awareness of irony, given Gould's claim in 1972 that the theory was "not metaphysical."

Gould and Schopf had by this time developed a friendship through their collaboration on stochastic modeling (which came to be known as the "MBL model"), but this submission was potentially complicated by the fact that Schopf was an avowed gradualist who had privately expressed his strong reservations about punctuated equilibrium to Gould.[19] While Gould acknowledged their difference of opinion, his letters do not suggest he anticipated difficulty having the paper published, or that he worried Schopf's private beliefs would influence editorial decisions. On the contrary, he invited Schopf to "collect your thoughts on gradualism into a full-scale paper," since "we would certainly welcome a rebuttal at a higher level than has been directed to us so far."[20] In any case, the paper was accepted with only minor revisions, and Schopf closed his letter acknowledging acceptance of the final manuscript by commenting "we are pleased (and proud) to have your article in *Paleobiology*. . . . I feel this is a most remarkable article, one that will merit and require a lot of careful attention."[21]

Leaving aside discussions of its conceptual significance (for the time being), from a sociological standpoint this paper is interesting because of what it did *not* do. Gould has written that the article was inspired, in part, because "enough data, argument, and misconception as well had accumulated by the summer of 1976."[22] This may be true, but it is worth noting how little controversy the 1977 paper provoked, especially compared to the fairly violent reaction the theory saw in the early-to-mid-1980s. As already mentioned, Schopf—who would later publish several fairly harsh criticisms of the theory—was unambiguously pleased with the manuscript. Additionally, one of the paper's two reviewers was Philip Gingerich, who became a staunch opponent of Gould's in the 1980s (whom Gould suggests became intellectually jealous of the theory).[23] Gingerich's review was not especially critical—he gave it a lower rating than the other referee ("good" as opposed to "excellent")—but he also characterized it as "an interesting and important paper."[24]

Why, then, did the 1977 paper fail to cause the uproar that followed later articles (most notably Gould's 1980 "Is a New and General Theory of Evolution Emerging?", which will be discussed in the following)? Certainly not because the paper was timid; if anything, the authors had strengthened their argument for the macroevolutionary significance of punctuation, had clarified the anti-neo-Darwinian implications of the theory, and moreover had made a case for broad "metaphysical" punctuationism. It is difficult to know precisely why the 1977 paper did not excite passions the way later articulations of the theory did; perhaps the explanation Gould gives in *Structure* is most accurate: "The early history of punctuated equilibrium unfolded in a fairly conventional manner for ideas that 'catch on' within a field. The debate remained pretty much restricted to paleontology . . . [and] most discussion, to our delight, arose from empirical and quantitative studies."[25]

THE TURNING POINT

In the three years following publication of the 1977 paper—prior to Gould's next major articulation of the theory, in other words—the article was cited several times in *Paleobiology*, although nearly always favorably or at least neutrally.[26] The single example of criticism to appear in *Paleobiology* was Fred Bookstein, Gingerich, and Arnold Kluge's coauthored paper, "Hierarchical Linear Modeling of the Tempo and Mode of Evolution," which appeared in the spring of 1978. In reviewing the manuscript, Raup called it "the best paper to date on the gradualism argument" and commented "it breaks a lot of new ground and thus should have a seminal effect on future research."[27] While the authors do conclude that "we see little use for further speculation based on the generality of punctuated equilibrium," they do not entirely discount punctuation as an occasional pattern, and overall the criticism is a fairly mild. It is not an attack on Gould and Eldredge's theory.[28]

The turning point seems to have come—as Gould himself acknowledges— when Schopf asked him to contribute an essay assessing the "Status of Paleontology—1980" for the fifth anniversary of *Paleobiology*. Gould happily obliged, and in a single issue of the journal published two reflective papers on the state of the discipline. In the first, which he titled "The Promise of Paleobiology as a Nomothetic, Evolutionary Discipline," Gould both celebrated the recent advances in paleobiology and reiterated his call for further progress toward revision of evolutionary theory based on macroevolutionary modeling. One of Gould's longstanding concerns was to raise the status of paleontology with respect to other evolutionary fields, and he was pleased to report that "our profession now wears the glass slipper and, if not queen of the evolutionary ball, at least cuts a figure worth more than a passing glance."[29] He found the most significant new direction for the field the widespread acceptance of "[G.G.] Simpson's procedure of modeling and testing," and the reduction of emphasis on ideographic studies in the "'empirical law' tradition."[30] This emphasis reflected the goal he shared with Schopf of making paleontology more nomothetic, or law-producing; the title of this essay harkened back to a 1974 paper he coauthored with Raup in *Systematic Zoology*, "Stochastic Simulation and Evolution of Morphology: Towards a Nomothetic Paleontology."[31]

This reference was quite intentional: the aim of Gould's first 1980 essay was to carve out paleontology's unique niche among the evolutionary sciences, and the area where he saw most promise for both originality and nomotheticism was in stochastic modeling of macroevolutionary patterns. Macroevolution clearly offered the best opportunity to emerge from serfdom to biology, since "if evolution works on a hierarchy of levels (as it does), and if emerging theories of macroevolution have an independent status within evolutionary theory (as they do), then paleontology may become an equal partner among the evolutionary disciplines."[32] However, Gould also suggested that this contribution might entail a reevaluation of the received view of Darwinism, as codified in the Modern Synthesis. The power of stochastic simulation and modeling in recent years seemed to suggest, Gould proposed, that macroevolutionary patterns did not follow the same deterministic lines as microevolutionary trends. Here was the decoupling of micro- and macroevolution Stanley had forcefully argued for, but Gould saw the potential for an even more profound shift in beliefs about the metaphysics of natural change: he concluded that "the world's frequent fit to stochastic models might mean that ontological randomness really is an admissible way to encompass part of our universe—and that our preferences for determinism are a cultural prejudice born of the ideographics that prevail at the scale of our short personal existence."[33]

If a hint that Gould was using the occasion of the fifth anniversary of *Paleobiology* to attack the sacred cows of neo-Darwinism was detectable in the first essay, it was overwhelming in the second, "Is a New and General Theory of Evolution Emerging?" Much later, reflecting on the difficulty punctuated equilibrium had over the years, Gould located the origin of much of the controversy here: "the received legend about this paper . . . holds that I wrote a propagandistic screed [claiming] . . . first, the impending death of the Modern Synthesis; and second, the identification of punctuated equilibrium as the exterminating angel (or devil)."[34] Whether this interpretation is fair, it is notable that the two papers, taken together, more explicitly challenge the synthetic account of evolution than either of the Eldredge-Gould papers of the 1970s. Perhaps the most controversial statement in "New and General Theory" is the oft-quoted claim that "if Mayr's characterization of the synthetic theory is accurate"—that is, if evolution is guided solely by the accumulation of small genetic changes and natural selection, and if macroevolution is nothing more than microevolution writ large—"then that theory, as a general proposition, is effectively dead, despite its persistence as textbook orthodoxy."[35]

In *Structure*, Gould protests on a number of grounds that his critics have read this and similar statements unfairly. He argues in particular that punctuated equilibrium "does not occupy a major, or even a prominent, place in my 1980 paper ['New and General Theory']," and he denies that his discussion of "the Goldschmidt break" promoted saltationism.[36] I will not attempt to assess whether Gould is justified in feeling misunderstood, but I will note that the language he used in "New and General Theory" to describe the synthetic account is fairly unequivocal: "I have no doubt that many species originate in this way [i.e., via neo-Darwinian mechanisms]; but it now appears that many, perhaps most, do not."[37] Is it surprising that these statements caused biologists to sit up and take notice of what had formerly been seen as a fairly benign macroevolutionary theory of interest mostly to paleontologists? However, despite the fallout that resulted, this paper (and its companion) must be seen in a very real sense as a triumph for *Paleobiology* (and the discipline): here, finally, biologists (neontologists) were reading the journal and taking its contents seriously enough to become upset, and it is only really after that point that punctuated equilibrium entered general cultural currency.

PUNCTUATION UNDER FIRE

Over the next two years, Schopf finally went public with his own assessment of punctuated equilibrium in a series of highly critical publications. The first was an item in *Paleobiology*'s Spring, 1981 "Current Happenings" section, which was a recently instituted forum in the journal for news and editorial comment about the field. The second was a scholarly, analytic paper in *Evolution*, titled "A Critical Assessment of Punctuated Equilibrium," published in late 1982. The third was a letter to the journal *Science*, coauthored with Antoni Hoffman, published in February of 1983.[38] In each of these publications Schopf attacked the theory from a different angle—as impartial journal editor, then careful empirical paleontologist, and finally concerned scientific citizen. These attacks raise a somewhat perplexing question: given Schopf and Gould's close agreement about the general agenda and goal of paleobiological work, why did Schopf choose to undermine his discipline's most prominent theorist (and close personal friend) in a publicly and potentially embarrassing way?

Schopf's "Current Happenings" piece, "Punctuated Equilibrium and Evolutionary Stasis," is presented as an objective summary of the state of evidence for and against Gould and Eldredge's theory. Schopf opens the piece with the fairly neutral goal that "it seems worthwhile to try to place the paleontological and biological evidence in a 1981 perspective," and the implication is that the essay will consider the pros and cons of the argument.[39] However, Schopf's personal beliefs come quickly to the fore. He notes, for instance, that punctuated equilibrium seems to demand "some strongly deterministic factors in order to account for patterns of speciation and extinction in the fossil record," and shortly thereafter clarifies his intentions with the piece: "the major purpose of this 'Current Happenings' is to encourage the quantitative and qualitative evaluation of these limitations and prediction of the punctuated equilibrium model so that a truer picture of evolutionary history may be obtained."[40] (Schopf was intellectually committed to a nondeterministic, stochastic view of life, so his use of the label "deterministic" in the first quotation should be considered a kind of epithet).

Schopf then systematically enumerates what he interprets as the weaknesses of punctuated equilibrium: (1) The incompleteness of the fossil record: the appearance of suddenness (or punctuation) of speciation is an artifact "almost guaranteed" by the state of fossil knowledge, and "indicates nothing of any meaning about the process of evolution which led to these classes."[41] (2) Artifacts related to the recording of taxa durations: since taxa are commonly assumed to be present for the entire duration of the geologic stages in which they originate, their "book value" is often exaggerated. (3) Insufficient morphological information: "because only the most resistant and most numerous of hard parts can ever be studied, paleontologists must recognize species by recourse to only a small part of an organism's actual evolutionary change."[42] (4) Poor sample population: it is easier to discern species-level evolution in organisms with well-defined hard parts, so we tend to omit organisms with "simple, relatively undifferentiated forms." (5) General limitations of taphonomy: short-lived species are much less likely to be preserved, and hence recorded. The majority of these complaints have to do with *signal* errors— i.e., observations about perceived limitations on generalizing from fossil data. Notably, only the potential weaknesses of the theory are considered, and there is no mention of any of the recent studies that attempted to produce empirical verification of punctuated equilibrium.

It appears that Gould obtained a copy of the piece shortly before it was to be published, and his reaction was surprised and indignant. He did not object to Schopf's critique of the theory per se, but rather questioned the use of his position as editor to present it:

> I must confess—and I expressed this to Jim [Hopson, Schopf's coeditor] when he called me for another reason two weeks ago—that I am not altogether happy with the forum that you have chosen for the piece. If it had been submitted as a regular article, I would have welcomed it entirely (while disagreeing strongly, of course, with its conclusions). In a sense, I am flattered that you consider punctuated equilibrium as a "happening"—and therefore worthy of inclusion in your section. Yet I confess that I do not think it fair for you, as editor, to use this section as a forum for expressing personal viewpoints on issues of the moment.[43]

Gould went on to stress his belief "that accounts of happenings may and indeed even should express a point of view," but added "ideas aren't events— and I would argue that the editor of such a section should not use his prerogative as a platform for expressing personal opinions about theoretical issues." It seems that Gould objected particularly to the ostensibly objective way Schopf had presented the piece: "I don't think I feel angry about this, but I am not unconcerned either. I just don't see how one man's viewpoint can become a kind of official line in one section of a journal."

This disagreement did not cause serious damage to Schopf's friendship with Gould, but it did begin a new, more directly combative, phase of their relationship. For the next several years (until his untimely death in 1984), Schopf devoted considerable energy to attacking punctuated equilibrium in print and elsewhere. His paper in *Evolution*, published in late 1982, added more empirical depth to the arguments he raised in his "Current Happenings" piece, and takes a scholarly, measured approach to arguing against general morphological stasis in favor of gradualism (for example, the paper opens by claiming to "take the position of a devil's advocate" with regard to punctuation). However, this appearance exists only in the final, published version; the manuscript Schopf originally submitted was far more partisan, and often seemed to associate criticism of the theory with its author. One review of the manuscript stated bluntly, "I don't feel that this paper needs to be riddled with *ad hominum* [sic] references to Steve Gould. From a supporter, the term 'Gouldism' would appear laudatory; in this ms it simply seems to be sarcastic."[44] Another reviewer found it "heavily slanted" and "irritatingly polemical and biased," and recommended extensive revisions before publication.[45]

Schopf's motivation is not entirely clear, but a third review—from a paleontologist acquainted with Schopf—sheds some light. Her comments draw attention to Schopf's "overly conciliatory tone" and "evident awe of Steve," and suggest "Tom could probably do him far more honor by sharing his views with him than by monumentalizing his name with an as-yet insufficiently tested theory." The review also expresses the sort of frustration with the faddishness of the theory that was only just then beginning to publicly surface:

> the overly conciliatory tone he [Schopf] takes towards Gould, can only delay the acceptance of his ideas both among those of us who never particularly cared for punctuated equilibrium but found it difficult to voice dissenting views at meetings (I was there, so were you) or even in print and those who knew a bandwagon coming when they saw it and jumped without too critical a look. I do not agree with him that "punctuated equilibrium was put forth in a spirit of conjecture and refutation, utility and testing, thesis and antithesis." My recollection is that punctuation was pretty much forced down our throats—some of us swallowed, others gagged, and, for a time refutation and testing and antithesis received very short shrift indeed. There are fashions in science—punctuation was one of them.[46]

Here the intriguing possibility is raised that Schopf was—perhaps unconsciously—using his criticism of Gould the scientist to promote the reputation of his friend. This is even more apparent in the letter of 1983 he and Hoffman sent to *Science*. The ostensible purpose was to question whether "a static hierarchy [is] a true and correct view of life."[47] Again, however, the published version differs significantly from the original manuscript. The opening paragraph of the first draft begins "S. J. Gould arguably is becoming the most important single force in the shaping of current popular evolutionary thought," and describes him as "a very fine human being who values scholarship," who "for [his] skill at argumentation and conceptual organization is widely and deservedly admired."[48] The second draft of the letter is toned down considerably, but is still (in spite of its criticism) quite effusive: the new opening paragraph opines "we admire very much [Gould's] efforts and readily acknowledge that no paleontologist has contributed as much to the popularity of evolutionary theory."[49]

The point here is not to psychoanalyze Schopf, but rather to seek a nuanced appreciation of his actions—both editorially and otherwise—toward Gould's work. In the published version of the letter to *Science* the phrases quoted above are omitted, and most readers would have been unaware of Schopf's appreciation for Gould's ideas and character. Indeed, Schopf's more casual acquaintances were often surprised by his personal affection for Gould. In one particularly striking instance, Schopf chided the prominent biologist Norman Horowitz after Horowitz had written a harsh review of Gould's *Mismeasure of Man*. In his letter, Schopf upbraided Horowitz for labeling Gould a "Marxist," and in a follow-up described Gould as "one of the most decent and humane humans I have ever met."[50] Schopf's feelings about both Gould and punctuated equilibrium were clearly quite complicated, and even conflicted. In his editorial capacity at *Paleobiology* and elsewhere (as editor of *Models*, for example), Schopf again and again provided a platform for helping Gould launch and establish his evolutionary beliefs—but he was also quite capable of using such positions to attack Gould's theory.

In late 1982, Schopf sent Gould a letter explaining his position toward punctuated equilibrium in very candid terms. His major objection, he wrote, was that the theory had "got taken too far." He worried that "the many biases of the fossil record that needed carefully, and systematically, to be looked into, never got looked into," and that the effect it would have on future work could be harmful: "unless those who are brought into the field learn the rigor of testing, it will be for naught."[51] Still, he explained, "I don't have a campaign against P. E."; rather, "I do have a campaign *for* rigorous testing of these ideas." Part of Schopf's justification for his critical stance—and for his continued warm regard for Gould—was that he felt Gould was not entirely responsible for the excesses connected with the theory:

> I don't think it's entirely your fault that P. E. got out of hand. As I see it, the "press" (Roger Lewin et al.) discovered Steve Gould, and what Steve Gould *happened* to be on was P. E. If it had been some other issue, then

that issue would be well known. The press recognizes personalities. It publicizes what those personalities are saying and doing. Sometimes, those particular sayings, and doings, are beyond their worth. I *think* this is what happened to P. E. The press didn't discover P.E. It discovered S. J. G.—and S. J. G. happened to discover P. E. Pure accident. Five years later, or earlier, it would have been different.

In Schopf's stochastic account of the reception of the theory, events might have been entirely different if the clock were wound back and restarted—just as trilobites might have survived in an alternate evolutionary scenario.[52] Schopf was very clear, however, that his feelings on the matter did not jeopardize his affection for Gould: "So, Steve, I have felt and do feel, very close to you. But, I have to (had to) go my own way on P. E. But I'll defend you as a person as long as I can write."

INTERPRETATIONS OF THE THEORY

Before concluding, I want to return to the more general question about the relation of 1970s (and later) paleobiology to Darwinian orthodoxy. The expected protagonist in this tale is Gould, and indeed there are reasons why his work should be viewed as central to the subversive set of ideas that came out of the "radical fringe" of paleontology to challenge Darwinian evolutionary theory. But despite being the most recognizable character, is Gould the real radical in this story?

Historians, sociologists, and philosophers have turned Gould's openness concerning his political views and their relation to his scientific work into a minor cottage industry. However, I submit that the real radical was not Gould, but Schopf. Consider the circumstances that led to punctuated equilibrium: at the time Schopf was organizing the original models symposium, Gould was a promising, if unspectacular young professor at Harvard known for his work on Bermudan snails and his review paper on allometry. He might truly have never even done his work on punctuated equilibrium had it not been for Schopf—a fact that was not lost on either man. In a very personal letter to Schopf in 1977, Gould reflected on this fact: "I have been dragged literally (once by you and once by Niles) into the two creative things I have done," and he noted "my frame of mind, [was] so correctly identified by you at one of those dinners as conservative scientifically, despite general social and political views."[53] Where, then, does the assumption that Gould was a committed evolutionary radical, and punctuated equilibrium his manifesto, come from?

In 1982, Ernst Mayr published an essay titled "Speciation and Macro-evolution" in the journal *Evolution*. In it, he gave an account of two possible readings of punctuated equilibrium that, I argue, have significantly influenced the historiography of the theory. Mayr acknowledged Gould and Eldredge's debt to him, writing that after being "totally ignored by the paleontologists for almost twenty years," their theory explained "some of the most important findings of the paleontologists . . . in terms of my theory of peripatric speciation."[54] He went on to specify that a "moderate" or "Mayr version" of punctuated equilibrium involves only a "slight translation" of his 1954 theory "into vertical terms." However, he also notes (ominously) that a "drastic or 'Goldschmidtian version'" of punctuated equilibrium was presented in Gould and Eldredge's 1977 paper, which suggests that speciation is based on major mutations.[55]

Michael Ruse has identified three historical phases in Gould's thinking about punctuated equilibrium, and I suspect Ruse draws heavily on Mayr's interpretation. The first phase of the theory is represented by the 1972 Eldredge-Gould paper, and according to Ruse offers "a fairly straightforward extension of orthodox Darwinism."[56] Ruse comments, however, that between this first phase and the later second phase (which he dates to Gould's 1980 essay "Is a New and General Theory of Evolution Emerging?") something important happened: Gould was now "downplaying the role of natural selection," and accordingly "the father figure had changed from Charles Darwin to Richard Goldschmidt."[57] Ruse freely acknowledges that at the time his own essay was written (it was published in 1989), Gould "categorically denies that he himself was ever a saltationist in Goldschmidt's or anyone else's sense."[58] Nonetheless, Ruse argues that "one can fairly say . . . Gould (especially) was starting to think of evolution's processes through a lens or filter of discontinuity. . . . In his own mind, he was starting to highlight the essential abruptness of evolution, as opposed to its continuity."[59] Finally, the third phase represents Gould's final position, and dates to about 1982. This, Ruse comments, offers "a pull-back from extremism," though not a "retreat" from hierarchical evolution. The final version of the theory offers a defense of punctuationism, but at the same time disavows Goldschmidtian saltationism.

If one looks at the early history of punctuated equilibrium (as this chapter has done), Ruse's model appears valid. Immediate reactions—both initial reviews and comments in the scientific literature—suggest that the theory was seen as mildly controversial but hardly upsetting by most paleontologists and biologists who encountered it. Intense reaction begins in the early 1980s, shortly after Gould's infamous essay in *Paleobiology*. Where I am inclined to differ with Ruse, however, is in his apparent assumption that these phases reflect actual, substantive modifications to Gould's conception of the theory. This is what I believe he has inherited from Mayr—he has made Mayr's "moderate" and "drastic" readings of punctuated equilibrium into actual stages in the theory's development. Ruse reiterates the position that Gould modified punctuated equilibrium in important ways in his chapter on Gould in *Mystery of Mysteries*, arguing that while Gould's final position is amicable toward natural selection, "one sees that other factors, including brute chance, come increasingly into play."[60]

However, I would point out that brute chance was *always* a central component of the theory. If it was not as explicitly mentioned in the 1972 paper as it was in later publications, it was in the minds of Gould and his collaborators on stochastic modeling as early as December of that year, when the radical fringe group met in Woods Hole. The very first publication of the group (the 1973 paper in *Journal of Geology*) explicitly ties stochastic evolution to punctuated equilibrium:

> In evolutionary terms, this [stochastic model of phylogeny] describes a situation where Phyletic transformation is absent and where new species arise only through speciation. We do not view this as an artificial simplification constructed to ease our calculations; Eldredge and Gould (1972) have argued that it corresponds to biological reality.[61]

In a series of follow-up publications, Gould and his collaborators continued to link punctuated equilibrium with a potentially stochastic and nondirectional view of evolution. For example, a paper coauthored with Raup in 1974 used "a somewhat idealized form of the evolutionary model presented by Eldredge and Gould (1972)" as a basis for its simulation of morphologic change. And lest there be any question concerning the relation of stochastic simulations to traditional Darwinism, Raup and Gould's paper proposes to "discard the questionable model of Phyletic gradualism," concluding that while "this paper is not an attack upon the concept of uni-directional selection. . . . Over long periods of time, undirected selection may be the rule rather than the exception in nature."[62]

CONCLUSION

Why, then, did Schopf fall out with the theory? Schopf's initial support of punctuated equilibrium reflects his general commitment to expanding paleontology into evolutionary theory, and more specifically his conviction that Gould's theory fit within the program of nomothetic, stochastic paleontology. As he described the group's work to an editor at the journal *Science*:

> the important thing about all of this work is that it derives from a quite different conceptual viewpoint than has been used for a century of paleontology. . . . Rather than reading the fossil record 'literally'—and seeking empirical laws derived by summing up individual events[,] our approach more directly utilizes theory and predictions from theory.[63]

However, he differed with Gould on some fundamental points. Most centrally, Schopf was certain that equilibrium models would ultimately best represent evolutionary change over the long haul. In other words, he believed that patterns of evolution and extinction tended to oscillate around a stable mean—the factors that determined the fluctuations might be considered random, but the law of evolution held that the system will always stabilize itself. This notion recalls Charles Lyell's steady state equilibrium model, and in fact the MBL group acknowledged this debt in one of its final publications, 1979's "The Shape of Evolution: A Comparison of Real and Random Clades." The paper concluded by invoking Lyell's "most cherished belief . . . that earth history, like planetary motion, was in a dynamic steady state," but went on to comment that "this belief represents one extreme metaphor in a continuum that places directionalism and notions of inherent progress at the other pole."[64] The authors noted that the tendency of paleontologists was to lean toward the directional pole, and to assert "inherent uniqueness for each period of time and . . . 'directions' in earth history." The paper, however, drew the opposite conclusion, and ended by boldly proclaiming "Lyell's metaphor is due for a renaissance."[65] Schopf's private views on the subject were apparently even more radical, as he expressed to Raup in early 1979:

> in my view, all of paleontology, i.e., all of those fossils, is (are) simply a metaphore [sic] for what is really the statistical mechanics of a series of interacting hollow-curves, each hollow curve being appropriate to a given faunal province, or habitat type, and that because ecological disturbance . . . the forces of physics thru air and water and rock kill off individuals, and because of aging in any organism without 'disturbance,' the hollow curves of a given region rise and fall, sometimes becoming enormous, sometimes evaporating all together.[66]

Note the inversion of the Lyellian metaphor here: as the MBL paper framed it, the geometric regularity of planetary motions was a metaphor for the dynamics of earth history. But according to Schopf's (almost Platonic) formulation, it is the physical processes of natural change themselves that are a metaphor for mechanical geometry.

Despite having the word "equilibrium" in its title, Gould's theory does *not* suggest the same kind of dynamic steady-state model Schopf favored. In Schopf's vision, change is constant and minor. In Gould's, change is virtually nonexistent during lengthy periods of stasis (in fact, "punctuated stasis" might be a better name for the theory), then comes suddenly and with major effect. Gould's paper in 1982, "Darwinism and the Expansion of Evolutionary Theory," frames the matter succinctly. The question, as he saw it, asks "is our world . . . primarily one of constant change (with structure as a mere incarnation of the moment), or is structure primary and constraining, with change as a 'difficult' phenomenon, usually accomplished rapidly when a stable structure is stressed beyond its buffering capacity to resist and absorb?"[67] In Schopf's view, change is easy; in Gould's, it is hard. As Schopf put it in a letter to Gould that same year, explaining (in part) his sudden attacks on punctuated equilibrium:

> I hope you will see it [the letter in *Science* with Hoffman] not as a campaign *against* something (PE), but rather as a campaign *for* something (a view of the world where change is easy—and continuous). I think you hit the nail-on-the-center when you said it is a question of change 'difficult' vs. change 'easy.' My bias, owing to my Woods Hole bryozoan work—I recall those times with great fondness—is for change as 'easy.'[68]

Why should this issue have caused Schopf such concern? Ultimately, because it went to the heart of Schopf's and Gould's disagreement about the substance of the challenge to orthodox Darwinism paleobiology would make. According to Schopf, species are mathematically reducible entities, analogous to particles, and describable (potentially) by a set of paleontological gas laws. As Schopf's letter to Gould continued, he explained "I am as convinced as I can be . . . that with 10^6 to 10^7 living species, and $\approx 10^{10}$ over geologic time, that species are particles in a never-ending biological world." But this mathematical, particulate model had a major liability: geometry and mechanics tend to be paradigmatically *deterministic* sciences, and Schopf wanted to avoid determinism at all costs. So, as he explained to Gould, "in order to *avoid* Raup's determinism . . . I am *forced* to a view that species durations must be quite short ($\approx 2\text{x}10^5$ yrs). If so, change must be easy." According to Schopf, what the steady-state model demonstrates is that change (meaning speciation and extinction) occurs not because certain organisms are more or less fit, better or worse adapted, but rather because according to the predictable rules of equilibrium dynamics there will be a natural ebb and flow. The notion that species persist in stable form for tens of millions of years—as punctuated equilibrium suggests—is tantamount to capitulating to determinism; what, besides better fitness or greater adaptive value, could explain the sudden and rare evolutionary transformations the theory predicts?

Viewed from this perspective, it is clear that Schopf's philosophical objection to punctuated equilibrium is that it is *not radical enough*. In another letter, Schopf spells out his concerns to his friend directly:

> it seems to me, that much of the competition thinking is all wrong for understanding evolution. Sure 'competition' exists. But, every species is in some way a superior competitor *and* an inferior competitor. . . . If all species *through time* are equally 'successful'—and if all species *at any given moment in time* are equally successful—then the notion of 'success' (*sensu latu*) has no place in evolutionary theory.[69]

He continues, "as for punctuated equilibrium—goodness knows I have nothing against the rapidity of speciation. . . . But, the notion of stasis—that the mean duration of a species is millions and millions of years, then that becomes the MAIN SUPPORT for a deterministic view of life!" In this context, Schopf, the opponent of punctuated equilibrium and proponent of gradualism, is chiding Gould for insufficient radical fervor. "Maybe," Schopf concludes, "you can find a middle ground. My sense of our field, such as it is, is that punctuated equilibrium becomes nearly the main argument for biological determinism. It would be ironic indeed, Steve, if you above all, in championing stasis, were seen in retrospect as the chief architect of biological determinism." Whether or not Schopf's own vision of the pattern of life's history was misguided, his irony is a useful caution to the historian: as in the case of Gould's relation to the principles of neo-Darwinism, whether one is "radical" or "conservative" is most definitely a matter of perspective.

NOTES

1. Gould 2002, 55. For reviews of *Structure*, see (among others), Barash 2002; Futuyma 2002; Ghiselin 2002; Hull 2002; Jablonski 2002; Stearns 2002; Wake 2002.

2. Gould 2002, 1005.

3. Schopf to Raup, March 7, 1970. Schopf Papers, Box 3, Folder 30.

4. Gould to Schopf, March 13, 1970. Schopf Papers, Box 5, Folder 14.

5. Eldredge and Gould 1972.

6. A word should be said here about Niles Eldredge's contribution to punctuated equilibrium. As Gould himself acknowledged on many occasions, Eldredge invented the theory. While this is not the place to investigate the subsequent history of the theory's association with Gould (often to the exclusion of Eldredge), it is important to state that this is an important story, and one that ought not be overlooked. For a more detailed examination of the content of the theory (and Eldredge's contribution), see Patricia Princehouse's contribution to this volume.

7. Gould and Eldredge 1977,115.

8. Gould to Schopf, April 14, 1971. Schopf Papers, Box 5, Folder 14.

9. See Eldredge to Schopf, June 21, 1971. Schopf Papers, Box 4, Folder 12.

10. Schopf to Eldredge, June 30, 1971. Schopf Papers, Box 4, Folder 12.

11. Niles Eldredge, interviewed by David Sepkoski, 1/18/2006. During the review process, Gould left the United States for a sabbatical year at Oxford, leaving Eldredge to deal with the revisions. When confronted with Schopf's criticism of the manuscript, Gould counseled Eldredge to not "give an inch," as Eldredge recalls. "So he made me write a tough letter and I wrote a tough letter. And we prevailed, but he [Schopf] really hated it."

12. Richard H. Benson, "Review of 'Speciation and Punctuated Equilibrium: An Alternative to Phyletic Gradualism.'" Schopf Papers, Box 13, Folder 10.

13. Schopf to Raup, Gould, and Simberloff, April 20, 1972. Schopf Papers, Box 3, Folder 30.

14. Raup et al. 1973.

15. Gould To Schopf, December 6, 1976; Gould to Schopf, November 23, 1976. *Paleobiology* Archives, Box 3.

16. Stanley 1975, 648, writes "Macroevolution is decoupled from microevolution, and we must envision the process governing its course as being analogous to natural selection but operating at a higher level of organization." Gould himself admits that Stanley "developed the implications that I had been unable to articulate from our original section on evolutionary trends" (Gould 2002, 980).

17. Gould and Eldredge 1977, 120.

18. Gould and Eldredge 1977, 146. This section, of course, contains the infamous admission that "it may also not be irrelevant to our personal preferences that one of us learned his Marxism, literally at his daddy's knee."

19. Further papers included Schopf et al. 1975, and Gould et al. 1977. See the contribution by John Huss to this volume (ch. 16) for a detailed analysis of the MBL project.

20. Gould to Schopf, December 6, 1976. *Paleobiology* Archives, Box 3.

21. Schopf to Gould, undated (early 1977). *Paleobiology* Archives, Box 3.

22. Gould 2002, 980.

23. Gould 2002, 1014–15.

24. Gingerich, "Review of 'Punctuated Equilibrium,'" December 31, 1976. *Paleobiology* Archives, Box 3. The review does have a critical tone—particularly about the writing and potential biases of the authors—but does not raise substantive objections to the manuscript.

25. Gould 2002, 980. Eldredge recalls that the original 1972 paper "seemed to annoy virtually everyone," but the first instances of criticism he cites do not appear until 1980 (Eldredge 2006, 35).

26. See, for example, Hallam 1978; Stanley 1978; Sepkoski 1978, 1979.

27. Raup, "Review of 'Hierarchical Linear Modeling of the Tempo and Mode of Evolution,'" 1978. *Paleobiology* Archives, Box 4.

28. Bookstein, Gingerich, and Kluge 1978, 133.

29. Gould 1980b, 96.

30. In this sense, the term *ideographic* refers to information that is presented in a pictorial or figurative manner.

31. Raup and Gould 1974.

32. Gould 1980b, 98.

33. Ibid., 1980b, 115.

34. Gould 2002, 1002.

35. Gould 1980a, 120.

36. Gould 2002, 1005.

37. Gould 1980a, 122.

38. Schopf 1981, 1982; Schopf, Hoffman, and Gould 1983.

39. Schopf 1981, 156.

40. Ibid., 158.

41. Ibid., 160.

42. Ibid., 161.

43. Gould to Schopf, April 21, 1981. Schopf Papers, Box 8, Folder 30.

44. Anonymous, "Review of MS by Schopf." Schopf Papers, Box 2, Folder 5.

45. Anonymous, "Comments on 'A Critical Assessment of Punctuated Equilibrium,'" 1982. Schopf Papers, Box 2, Folder 5.

46. Davida E. Kellogg to Douglas Futuyma, March 18, 1982. Schopf Papers, Box 2, Folder 5.

47. Schopf, Hoffman, and Gould 1983, 438.

48. Schopf and Hoffman, "Punctuated Equilibrium and the Fossil Record—Draft A." Schopf Papers, Box 9, Folder 133, p. 1.

49. Schopf and Hoffman, "Punctuated Equilibrium and the Fossil Record—Draft B." Schopf Papers, Box 9, Folder 133, p. 1.

50. Schopf to Horowitz, August 15, 1982, and September 15, 1982. Schopf Papers, Box 9, Folder 106.

51. Schopf to Gould, September 19, 1982. Schopf Papers, Box 9, Folder 106.

52. Indeed, in his 1979 paper "Evolving Paleontological Views on Deterministic and Stochastic Approaches," Schopf argued just this case for the history of the journal *Paleobiology*: "its establishment (if at all!), and subsequent development, were far from inevitable. . . . Yet I wager despite these words, some historian of paleontology a decade or so in the future will somehow say that 'the time had come and the journal was inevitable.' For those who were there, nothing could be further from the truth." (Schopf 1979), 338.

53. Gould to Schopf, November 25, 1977. Schopf Papers, Box 5, Folder 14.

54. Mayr 1982, 1127.

55. In 1982, Mayr wrote to Schopf with his own personal feelings about punctuated equilibrium: "Since the Eldredge and Gould theory was expressly based on my 1954 paper ["Change of Genetic Environment and Speciation"] one might think that I would be completely behind the Gould theory. This, however, is only partially the case. I strongly object to the Goldschmidtian interpretation of rapid speciation in founder populations and I likewise do not agree with the complete stasis of other species." Mayr to Schopf, February 9, 1982. Schopf Papers, Box 8, Folder 32.

56. Ruse 1989, 120.

57. Ibid.,, 122.

58. Ibid., 122.

59. Ibid., 122.

60. Ibid., 138.

61. Raup et al. 1973, 528.

62. Raup and Gould 1974, 307, 314, and 321. See also Gould et al. 1977.

63. Schopf to Gina Bari Kolata, July 2, 1975. Schopf Papers, Box 3, Folder 60.

64. Gould et al. 1977, 39.

65. This line is literally the last sentence of the paper.

66. Schopf to Raup, February 6, 1979. Schopf Papers, Box 3, Folder 30.

67. Gould 1982, 383.

68. Schopf to Gould, November 22, 1982. Schopf Papers, Box 9, Folder 106.

69. Schopf to Gould, December 4, 1981. Schopf Papers, Box 8, Folder 31.

REFERENCES

Archival Sources

Schopf, Thomas J. M. Papers. Smithsonian Institution Archives. RU 007429.
Paleobiology Editorial Papers. University of Illinois Champaign-Urbana Library.

Published Sources

Barash, David P. 2002. Grappling with the ghost of Gould. *Human Nature Review* 2:283–92.

Bookstein, Fred L., Philip D. Gingerich, and Arnold G. Kluge. 1978. Hierarchical linear modeling of the tempo and mode of evolution. *Paleobiology* 4 (2):120–34.

Eldredge, Niles. 2006. Confessions of a Darwinist. *Virginia Quarterly* (Spring): 32–53.

Eldredge, Niles, and Stephen Jay Gould. 1972. Punctuated equilibria: An alternative to phyletic gradualism. In *Models in paleobiology,* ed. T. J. M. Schopf, 82–15. San Francisco: Freeman, Cooper & Co.

Futuyma, Douglas J. 2002. Stephen Jay Gould a la recherche du temps perdu. *Science* 296:661–63.

Ghiselin, Michael T. 2002. An autobiographical anatomy. *History and Philosophy of the Biological Sciences* 24:285–91.

Gould, Stephen Jay. 1980a. Is a new and general theory of evolution emerging? *Paleobiology* 6 (1): 119–30.

———. 1980b. The promise of paleobiology as a nomothetic, evolutionary discipline. *Paleobiology* 6 (1): 96–118.

———. 1982. Darwinism and the expansion of evolutionary theory. *Science* 216:380–87.

———. 2002. *The structure of evolutionary theory.* Cambridge, MA: Belknap Press of Harvard University Press.

Gould, Stephen Jay, and Niles Eldredge. 1977. Punctuated equilibria: The tempo and mode of evolution reconsidered. *Paleobiology* 3 (2): 115–51.

Gould, Stephen Jay, David M. Raup, J. John Sepkoski, Jr., Thomas J. M. Schopf, and Daniel S. Simberloff. 1977. The shape of evolution: A comparison of real and random clades. *Paleobiology* 3 (1): 23–40.

Hallam, Anthony. 1978. How rare is phyletic gradualism and what is its evolutionary significance? Evidence from Jurassic bivalves. *Paleobiology* 4 (1): 16–25.

Hull, David L. 2002. A career in the glare of public acclaim. *Bioscience* 52:837–41.

Jablonski, David. 2002. A more modern synthesis. *American Scientist* 90 (4): 368–71.

Mayr, Ernst. 1982. *The growth of biological thought : Diversity, evolution, and inheritance.* Cambridge, MA: Belknap Press of Harvard University Press.

Raup, David M, Stephen Jay Gould, Thomas J M Schopf, and Daniel S Simberloff. 1973. Stochastic models of phylogeny and the evolution of diversity. *Journal of Geology* 81 (5): 525–42.

Raup, David M., and Stephen Jay Gould. 1974. Stochastic simulation and evolution of morphology—towards a nomothetic paleontology. *Systematic Zoology* 23 (3): 305–22.

Ruse, Michael. 1989. Is the theory of punctuated equilibria a new paradigm? In *The darwinian paradigm,* ed. M. Ruse, 118–45. London: Routledge.

Schopf, Thomas J. M. 1979. Evolving paleontological views on deterministic and stochastic approaches. *Paleobiology* 5 (3): 337–52.

———. 1981. Punctuated equilibrium and evolutionary stasis. *Paleobiology* 7 (2): 156–66.

———. 1982. A critical assessment of punctuated equilibria. I, Duration of taxa. *Evolution* 36 (6): 1144–57.

Schopf, Thomas J. M., Antoni Hoffman, and Stephen Jay Gould. 1983. Punctuated equilibrium and the fossil record; discussion and reply. *Science* 219:438–40.

Schopf, Thomas J. M., David M. Raup, Stephen Jay Gould, and Daniel S. Simberloff. 1975. Genomic versus morphologic rates of evolution: Influence of morphologic complexity. *Paleobiology* 1 (1): 63–70.

Sepkoski, J. John, Jr. 1978. A kinetic model of Phanerozoic taxonomic diversity I. Analysis of marine orders. *Paleobiology* 4 (3): 223–51.

———. 1979. A kinetic model of Phanerozoic taxonomic diversity II. Early Phanerozoic families and multiple equilibria. *Paleobiology* 5 (3): 222–51.

Stanley, Steven M. 1975. A theory of evolution above the species level. *Proceedings of the National Academy of Sciences of the United States of America* 72 (2): 646–50.

———. 1978. Chronospecies' longevities, the origin of genera, and the punctuational model of evolution. *Paleobiology* 4 (1): 26–40.

Stearns, Stephen C. 2002. Less would have been more. *Evolution* 56 (11): 2339–45.

Wake, David B. 2002. A few words about evolution. *Nature* 416: 787–88.

CHAPTER SIXTEEN

The Shape of Evolution:
The MBL Model and Clade Shape

John Huss

INTRODUCTION

In 1971, University of Chicago paleontologist Tom Schopf organized a symposium on models in paleobiology at the Geological Society of America meeting in Atlantic City, New Jersey. In a field populated by specialists on everything from inarticulate brachiopods to pterodactyls, Schopf was skeptical that paleobiology would ever move beyond the description and classification of fossils to the construction and testing of general models and theories unless the field were radically reoriented. The session resulted in the seminal book *Models in Paleobiology* (1972), perhaps best known as the initial vehicle for Niles Eldredge's and Stephen Jay Gould's theory of punctuated equilibrium. Other contributors to the volume included Dave Raup, whose chapter concerned models of morphology, and ecologist Dan Simberloff, who provided a primer on models in biogeography.

Encouraged by the success of that meeting, Schopf cornered three of his fellow symposiasts—Raup, Gould, and Simberloff—for two informal brainstorming sessions in the winter of 1972 at the Marine Biological Laboratory (MBL) in Woods Hole, Massachusetts (Jack Sepkoski, then a geology graduate student at Harvard, also participated). Above all, the group sought to shake up what they perceived to be a moribund field. They began looking for means by which paleontology might move toward becoming a more predictive, general, and theoretical—in a word, *nomothetic*—science, and turned to recent developments in ecology and biogeography for inspiration (Raup et al. 1973). For an idiographic science like paleontology, mired in taxonomic nomenclature and description, the prospect of a predictive model that did not depend on the identity of particular taxa was promising.[1] MacArthur and Wilson's equilibrium model of island biogeography was especially appealing as an exemplar because it predicted the equilibrium number of species on an island as a function of immigration rate of new species and extinction rate of species already present, regardless of taxonomic affinity (MacArthur and Wilson 1963, 1967). It helped that Simberloff had collaborated with E. O. Wilson on experimental tests of island biogeography, which was one of the main reasons Schopf brought him on board. The question was whether an island biogeographic model, with speciation rate substituting for immigration rate, could be ramped up to geologic timescales. Schopf was hopeful.

Schopf arrived at the first MBL meeting with the multivolume, data-rich *Treatise of Invertebrate Paleontology* in tow, and Simberloff was accessorized with one of the first handheld calculators. The group spent long days exchanging ideas and poring over raw data for patterns and trends, but Simberloff, skilled in mathematical modeling and ecological data analysis, found most of the data too fragmentary to analyze. It was against this backdrop of desperation and frustration, on the last day of the first MBL meeting, that the question arose: "what if we were to remove Darwin from evolution?" In other words, what would phylogenetic patterns look like without natural selection as a guiding force? In effect, this question suggested a null hypothesis for evaluating the effects of natural selection on phylogenetic patterns (A couple of years earlier Simberloff had used an analogous approach, generating null communities by random sampling to assess the effects of interspecific competition on the composition of island communities [Simberloff 1970]). The meeting adjourned with the agreement that Raup would write a simulation program that reflected this picture of evolution. Thus was born the MBL model.

THE MBL MODEL

The MBL model began as a computer model in which evolutionary trees grew by a stochastic branching process in discrete time subject to an equilibrium constraint. It "removed Darwin from evolution" in two ways. First, the model was neutral at the lineage level. By assigning each lineage the same probabilities of persistence, branching, and extinction, no lineage had an advantage over any other. Second, natural selection simply went unrepresented in the model. Instead, phylogenetic evolution was modeled as a Monte Carlo process. Each simulation run would begin with a single lineage and a random number "seed." At each time unit, the computer would generate a random (or pseudorandom—the distinction need not concern us here) number which would determine, according to three preset probabilities, the lineage's fate: (1) extinction, (2) persistence to the next time unit and branching to form a sister lineage, or (3) persistence to the next time unit without branching. Each new lineage that evolved followed these same rules with the same probabilities. Initially (in the pre-equilibrial phase) the probability of branching would exceed that of extinction until the total number of species reached a preset equilibrium, at which point the probabilities of branching and extinction would be made equal to one another, and the total number of species would oscillate around the equilibrium value. The oscillations were damped by an algorithm that adjusted extinction and branching probabilities in proportion to the square of the deviation from equilibrium (Simberloff 1974). A simulation run would terminate when the end of the time scale was reached, when all lineages became extinct, or when a total of 500 species had evolved throughout the course of the run, whichever came first (a limitation of computer memory). A taxonomic routine subdivided the resulting phylogeny into clades (sets of lineages sharing a common ancestor) once a certain minimum size (defined as the sum of the durations of all lineages descended from the ancestor) was reached. If a group within an existing clade had grown large enough to meet the minimum size requirement, a new clade would be established (Raup et al. 1973). The model operated at a single taxonomic rank—no subtaxa were recognized. Thus, once a new clade was cropped out of an existing clade, the residual lineages formed a paraphyletic group, which, strictly speaking, is not a clade. The MBL authors did not distinguish between monophyletic and paraphyletic topologies, calling both "clades," and I will follow their usage herein.

The simulation output was a phylogeny, or evolutionary tree. Although in principle the information could simply have been represented numerically, the outputs were instead represented in the familiar iconography of paleontology: phylogenies of clades, and plots of originations, extinctions, and diversity (number of lineages) versus time. In the phylogenies of clades, each clade is represented by a "spindle," varying in width in proportion to the number of lineages in it during any given time interval. The decision to represent simulation results in this way was not arbitrary. By capitalizing on the working paleontologist's familiarity with such diagrams, the paper was able to attract much attention. It is doubtful that presenting the same results as tables of numbers would have had as strong an impact on the paleontological community.

The MBL simulation results were surprising. This relatively simple probabilistic branching model, primed by a random number seed, could simulate a variety of familiar-looking evolutionary patterns: adaptive radiation, competitive replacement, and simultaneous extinction events (Raup et al. 1973). An expanded version of the model incorporating morphologic change at each time interval produced morphologically coherent taxa, morphologic trends, variation in evolutionary rates, and correlated character evolution, despite the absence of natural selection in the model (Raup and Gould 1974). Using this same morphologically explicit version of the MBL model, Schopf, who was increasingly committed to the neutralist view that natural selection intrinsically favored no species over any other, spearheaded an effort to demonstrate that differences in evolutionary rate among taxa as disparate as clams and horses were more apparent than real, our judgments of rate being biased by differences in the morphologic complexity of the taxa involved (Schopf et al. 1975).[2] Once the group devised measures of clade size and shape, the computer-generated clades appeared to be startling simulacra of those compiled from the fossil record, both real and simulated clades exhibiting random variation around a mean shape (Gould et al. 1977).

The question, both then and today, was what to make of these results. Did they suggest we change our views on the fossil record, or rather the methods by which we draw inferences from it? Did they warrant a fundamental rethinking of our picture of evolution or simply offer a novel perspective on it? Did they call into question what paleontologists thought they knew about adaptive radiations, mass extinctions, adaptive trends, and evolutionary rates, or did they simply suggest that more stringent evidentiary standards were called for? Was the capacity of the MBL model to simulate evolutionary patterns in the absence of selection evidence for non-Darwinian evolution? Or were the results simply blown out of proportion? There is evidence of mixed agendas within the group itself, and reactions of the paleontological community were also varied. Modified versions of the MBL model have far outlived the MBL collaboration—the model has truly taken on a life of its own (Sepkoski 1978; Wassersug et al. 1979; Simberloff et al. 1981; Simberloff 1987; Sepkoski and Kendrick 1993; Uhen 1996; Foote et al. 1999a, b; Hubbell 2001). In my discussion here, I shall mainly focus on the arguments concerning clade shape in the 1973 and 1977 papers. I am particularly concerned with the visual argumentation on which these two papers rely in connection with scaling problems pointed out by Steven Stanley (1979) and coworkers (Stanley et al. 1981). Before moving on to these issues, I will comment on some roughly contemporaneous developments in several fields that suggest that the MBL collaboration was part of a larger cultural moment.

DIGRESSION: A CULTURAL MOMENT?

The period from roughly 1968 to 1980 was marked by a critical attitude toward natural selection and competition as explanatory schemata, a widespread interest in the scope of chance effects in pattern formation, and a probing inquiry into the limits of human intuition in reasoning about probability. These cultural elements coalesced in a skeptical stance toward what paleontologists and students of evolution and ecology thought they knew about distributional patterns. A few examples may suffice to illustrate the point.

In molecular biology, Motoo Kimura (1968), assuming a neutral model of molecular evolution, had derived the result that in the absence of selection, the rate of sequence evolution will equal the mutation rate, *regardless of population size*. While the common wisdom had been that genetic drift is only important in small populations, Kimura's result was a bombshell in that it showed that an allele could randomly drift to fixation even in a large population. Of course this did not establish the truth of the neutral theory, but it did remove one obstacle to the theory's acceptance. When coupled with the assumption that most genes are selectively neutral or nearly so, the neutral theory assigned a vastly reduced role to natural selection in molecular evolution. In ecology, Dan Simberloff (1970) took aim at an idea of Darwin's— increased competition among congeneric species—that had all but hardened into dogma. Ecologists had long observed that the species-to-genus ratio is lower on islands than it is on the mainland. This had been taken as evidence that species in the same genus—presumed to be similar in their ecological requirements and habitat preferences—competitively exclude one another on islands due to limitations of space and resources. Yet Simberloff showed via computer simulation that a small random sample from a list of mainland species will have a lower species-to-genus ratio simply because it is unlikely that all of the species in a genus will be sampled (what Simberloff demonstrated in 1970 using a computer C.B. Williams had done as early as 1947 by pulling numbers out of a hat). Thus, simply because islands have fewer species than the mainland, a lower species-to-genus ratio is to be expected, even in the absence of competition.

Paleontologists Niles Eldredge and Stephen Jay Gould explored the implications of Ernst Mayr's model of allopatric speciation for evolutionary patterns in the fossil record, developing a model they called "punctuated equilibria" (Eldredge 1971; Eldredge and Gould 1972). This model predicted that species lineages would exhibit morphologic stasis, with rapid morphologic change concentrated at speciation events. Noteworthy in the 1972 paper was the idea that evolutionary trends might not be driven by selection at the organismal level, but rather by speciation events that are random with respect to selection pressures at the organismal level. Furthermore, the assumption that morphologic change is concentrated at speciation events made for a convenient assumption in computer models of evolution (Michael Foote, personal communication), and indeed, Raup and coworkers invoked punctuated equilibria as a justification for their assumption that new species arise only by branching, and not by intralineage tramsformation (Raup et al. 1973). Yet for all this, the overriding philosophical message of punctuated equilibria was that one's evolutionary model strongly influences the pattern one sees. Those assuming a model of evolution as continuous morphologic change will fail to see evidence of morphologic stasis in the fossil record (Eldredge and Gould 1972).

Determining the evidentiary status of patterns inevitably involves a probabilistic judgment, and in many situations, our probabilistic intuitions lead us astray. Cognitive psychologists Amos Tversky and Daniel Kahneman (1974), in their landmark *Science* paper on judgment under uncertainty, had shown that human beings, when reasoning intuitively about probabilities, rely on heuristics that are generally adaptive but in certain domains are prone to systematic bias. Although there is no evidence that this psychological research directly influenced the MBL group, an important facet of research by the latter was to suggest that paleontologists might be similarly misled in their intuitive judgments of evolutionary patterns. Many randomly generated patterns are virtually indistinguishable from those that are routinely explained in terms of directed evolutionary processes such as adaptation. In fact, Schopf and Raup (1978) would eventually conduct a two-week National Science Foundation (NSF) training workshop, "Species as Particles in Space and Time," which began with exercises in which participants, consisting mainly of younger paleontologists, were asked to offer interpretations of randomly generated patterns—a bit of Rorschach therapy, so to speak—with later sessions devoted to replacing intuitive interpretations with approaches using explicit probabilistic models.[3]

SIMULATION RESULTS: DIVERSITY PATTERNS

Which evolutionary patterns is it possible to generate using a stochastic model of phylogenetic evolution? This was the main question being addressed in the 1973 paper. If most evolutionary patterns traditionally explained in other ways could also be generated stochastically, this would *not* imply that stochastic effects are *actually* responsible for these patterns in nature, nor would it rule out any of the mechanisms typically invoked. What it would imply is that discriminating among the mechanisms potentially responsible for a diversity pattern would require additional evidence *beyond* the pattern itself. In addition, it was hoped that the stochastic branching model would be useful in dividing evolutionary patterns in the fossil record into two components: a baseline component reflecting the dynamic equilibrium between branching and extinction intrinsic to phylogeny in general, and a residual component reflecting departures from the baseline for which specific historical causes might be sought (Raup et al. 1973).

The first MBL paper focused on diversity patterns (Raup et al. 1973). Within the simulation results could be found an array of clade shapes rivaling those of the known fossil record, although the fossil record does have instances of taxa whose diversities undergo wider and more abrupt fluctuations than did the simulated clades. In the MBL model the entirety of the range of simulated clade shapes represents random fluctuation around a mean clade morphology (Raup et al. 1973). What the MBL authors considered to be the most striking result of the simulations was that, in a model where no lineage differs from any other in its evolutionary potential, such a wide variety of clade shapes could be produced (See fig. 16.1, panel B). If these had represented actual fossil taxa, it would have been quite reasonable to expect patterns to be explained in terms of particular extrinsic causes or inherent differences between lineages or taxa. For example, in fig. 16.1, panel B, the four leftmost clades exhibit the kind of pattern that could be taken to indicate three successive competitive replacements. However, there can be nothing the least bit competitive about these replacements, for it is a known feature of the MBL model that all species have equal probabilities of evolution, extinction, and persistence, and that any extrinsic cause (damping of rates near equilibrium) affects all lineages and taxa with equal probability. The waxing (or waning) of a clade is a stochastic effect caused by an accidental excess of originations over extinctions (or extinctions over originations). None of these clades is intrinsically fitter than the other.

One thing the MBL researchers rediscovered and then trumpeted was a thesis known by philosophers as the underdetermination of theory by data. They noted that many fossil patterns, such as adaptive trends and adaptive radiations, are presumed on the basis of the pattern alone to be adaptive, even though the same or similar patterns can be produced (i.e., simulated) in the absence of adaptation. In other words, the pattern itself is insufficient to discriminate among explanatory hypotheses consistent with it. The underdetermination thesis states that data underdetermine theory: any finite set of data is compatible with an infinite number of explanatory hypotheses (just as

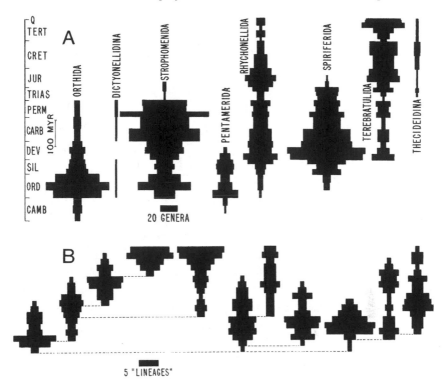

Figure 16.1 Comparison of (A) genera within orders of brachiopods with (B) lineages within clades simulated using the MBL computer program (Stanley et. al. 1981). Diversity is proportional to the width of the spindles. Phylogenetic relationships are indicated by dashed lines. The original figure in Gould et al. (1977) lacked a scale bar for the simulated clades. (From Stanley et al. 1981. Used by permission of The Paleontological Society).

an infinite number of different curves can be fitted to any finite set of points). In scientific practice the way out of such an impasse is to find additional data that eliminate certain of the rival hypotheses. Criteria such as parsimony and coherence with other accepted theories provide practical means by which scientists may confront underdetermination, although exactly how to resolve the problem tidily is still a matter of debate among philosophers of science (see Ladyman 2002, chapter 6 for a primer).

By highlighting the multiple realizability of any given type of pattern, the simulation results opened up the space of possible explanations to include stochastic effects. That the diversity patterns in and of themselves are insufficient to discriminate among stochastic effects and differences rooted in biological properties of the clades leaves two research strategies open. One can either investigate potential biological differences among the clades empirically (if suitable evidence can be obtained), or one can seek a focal level of description of the system that is insensitive to such differences (as in equilibrium modeling of whole biotas).

A second feature became evident upon plotting up the cumulative number of originations, extinctions, and the diversity of coexisting taxa through time. Even though total *lineage diversity* eventually fluctuates around an equilibrium value in the MBL computer simulations, the *taxonomic diversity* (number of taxa in each time unit) consistently exhibits a drop-off near the end of the simulation run (fig. 16.2). This is because the taxonomic algorithm in the MBL computer program requires that a clade within an existing taxon reach a specified minimum size before being recognized as a new taxon of its own. Near the end of a simulation run, time is usually too short for an incipient taxon to evolve a sufficient number of lineages to qualify as a new taxon, resulting in the absence of originations of new taxa at the end of the run. This results in a diversity drop-off, an artifact of the rules by which new taxa are erected in the MBL model. One might expect that artifacts of modeling assumptions are unintended by-products that would be eliminated or explained away if possible, but in fact it often turns out one can learn something from them. That was the case here. Raup and his coauthors (1973) argued that mass extinctions and the present day are the empirical analogues of the "end of the run" in the simulations. Species originating just before the end of the run, a mass extinction, or the present day are unlikely to evolve enough descendant species to ever give rise to a new taxon. Thus, *if* taxonomists tend to erect a new taxon only after a species evolves enough descendants, as in the rules of the model, then in time-series plots of fossil taxa, declines in diversity occurring just before mass extinctions, or just before the present day, might likewise be artifacts. The search for a natural cause for such declines is misguided *if* they are artifacts of taxonomic rules.[4]

The MBL researchers had expected to find that random extinction would lead to mass extinction (Raup, personal communication, 2002). Surprisingly, this did not happen. Although in the simulations a small number of distantly related clades were observed to have their diversities drop to zero during the same time unit or within a short span, large-scale mass extinctions on the scale of the end-Permian or end-Cretaceous extinctions were not observed. What this seemed to imply was that mass extinctions are not likely to occur stochastically, and that searching for common causes is a reasonable research strategy.

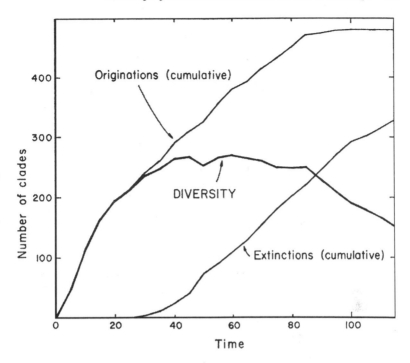

Figure 16.2 Plot of cumulative originations and extinctions, and diversity of coexis-
tent clades through time. Note the decline in clade diversity near the end of the run.
This is an artifact of the taxonomic algorithm. Lineages arising late in the run tend to
have evolved too few descendants to found a new clade. See text for further explanation (From Raup et al. 1973. Used by permission of the University of Chicago Press.
All rights reserved.).

At least in hindsight this epiphany seems relevant to later work by Raup and
Sepkoski on a possible extraterrestrial forcing agent for mass extinctions in
the Phanerozoic (Raup and Sepkoski 1984).

Perhaps the most important result of the first MBL paper was one that
underscored the benefits of a nomothetic approach and the drawbacks of a
narrow inductivism focused on individual clades. Equilibrium behavior in the
MBL simulations was manifest if one tracked the *total diversity* of lineages or
higher taxa at each time unit during the simulation run. No amount of detailed
investigation at the level of individual clades would have revealed this statistical regularity. If such equilibria also obtain in empirical data, it may be that by
focusing on specific events affecting particular taxa during specified intervals,

investigators will fail to observe the lawlike behavior that would be observable at a higher focal level.

THE SHAPE OF EVOLUTION

In 1977, Gould, Raup, Sepkoski, Schopf, and Simberloff published a paper in *Paleobiology* titled "The Shape of Evolution: A Comparison of Real and Random Clades." Therein they described a set of "clade statistics" for measuring the sizes and shapes of clade diversity diagrams. These statistics included duration, maximum diversity, size (number of lineages summed over the duration of the clade), uniformity (an inverse measure of degree of fluctuation in diversity), and center of gravity (the relative temporal position of the clade's median diversity).

From their many and detailed analyses they derived two principal generalizations. The first is that in a diversity-saturated world (whether real or simulated), the average clade has a center of gravity of one half, achieving half of its total diversity about midway through its history before waning to extinction (in other words, the average clade diversity diagram is diamond-shaped). The second (and stronger) claim is that, excluding such pre-equilibrial times as the Cambrian explosion or diversification events following mass extinctions, most of the clade shapes seen in the fossil record—flat tops, adaptive radiations, bottlenecks—lie within the range of stochastic scatter about the average diamond clade shape. In other words, despite their range of shapes, real and random clades look pretty similar.

Yet despite all of the quantitative analysis, the most salient argument remained a visual argument based on an overall gestalt, the likes of which had been doing the heavy lifting since their earliest work.[5] A prominent figure 1 (reproduced herein), occupying three-quarters of the article's facing page presents brachiopod clades and simulated clades for visual comparison (Gould et al. 1977). This comparative figure is curious, for aside from a rather uninformative caption, no further explanation of it is given in the text.[6] The two figures are left for the reader to compare visually. The same range of clade shapes, with margins jagged, bases tapered, and tops either tapered or flat, can be observed in both the simulated clades and the brachiopod clades.[7] In fact, with a few minutes' perusal one can find one-to-one correspondences between brachiopod clades and random clades. The visual argument, which is supplied by the similarity in the overall gestalt between simulated and brachiopod clades, is that stochastic processes can simulate diversity patterns virtually indistinguishable from empirical ones.

SIMULATIONS AND REALITY: STANLEY'S CRITIQUE AND VISUAL BIAS

Yet this visual comparison of clade shapes, striking though it was, seems to have concealed a scaling problem. By 1979, Steven Stanley had formulated a critique of the clade shape results on just these grounds, pointing out that parameters in the MBL simulations had been inappropriately scaled for the visual comparison. Subsequent computer simulations conducted with a research team of his own argued the case even more strongly (Stanley et al. 1981). Before discussing Stanley's critique, it is important to note how the MBL simulations came to be scaled as they were.

In the early phases of the MBL collaboration, there had been considerable discussion over whether their model was supposed to represent a comparative baseline, a possible picture of nature, or sheer fancy (Simberloff, personal communication, 2002). The choice of parameter values for speciation and extinction probabilities, minimum taxon size, and equilibrium diversity therefore reflected a compromise. The extinction and speciation probabilities were set low enough that they were not too unrealistic, high enough that interpretable effects would be detectable in the simulation results, and equal to one another so diversity would not too quickly exceed computer memory. No strong attempt was made to tune the model to empirical values, in part because it was designed for maximum generality. As it was, lineages in the model could be taken to represent any taxonomic level from local populations to higher taxa such as orders or classes. So rather than scaling the simulations to empirical values, it appeared that one could simply choose the taxonomic level that seemed appropriate, given how the simulation had, in fact, been scaled. This presented no real problems so long as the simulations were regarded as a hypothetical exercise, but a valid comparison between simulated clades and empirical clades would require that the model's parameters be appropriately scaled.

The fact that the real and random clades of figure 1 were scaled differently may have been obscured by the fact that visual comparisons of shape per se are scale independent, even though the branching processes that give rise to various clade shapes depend crucially upon scale. Stanley and his coauthors pointed out that the absolute diversities used in the MBL simulations were quite low relative to those of the real brachiopod clades with which they were being compared (Stanley 1979; Stanley et al. 1981). Low diversities in the MBL model increase the volatility of clades—chance increases or decreases in diversity are more probable at the low numbers represented by the simulated clades than at the large numbers represented by the brachiopod clades (Stanley et al. 1981). Subsequent simulations conducted by Stanley's research team argued the case even more strongly (Stanley et al. 1981). To make this more intuitive, we can imagine two clades: clade A with four lineages and clade B with twenty lineages. Ignoring extinction for a moment to make the calculation easier, if each lineage has the same probability of branching (say, 1/2), regardless of the clade it is in, then the probability of clade A doubling in diversity (width) in a single time step is $(1/2)^4$, or 1 in 16. The probability of clade B doubling its diversity in a single time step is $(1/2)^{20}$, or less than one in a million. Thus, the potential for dramatic changes in clade shape simply due to chance is far greater for smaller clades such as those produced by the MBL simulations (where average standing diversity was approximately four lineages) than it is for the brachiopod clades (average standing diversity of twenty genera) with which they were compared in the 1977 paper (Gould et al. 1977; Stanley 1979; Stanley et al. 1981).

Stanley and colleagues (1981) also criticized the MBL simulations for using higher taxa rather than species as units of diversity in the simulations. Simulating the branching and extinction of higher taxa instead of species results in smaller numbers, since there are fewer taxa at higher taxonomic ranks than at lower ranks. This again gives greater sway to chance effects, for the same reason as given previously in the discussion of within-clade average standing diversity. Even if rates are properly scaled to the higher taxonomic level, the small-numbers effect artificially makes chance fluctuations a more important factor than they would be if species were used.

Two lessons emerged from the analyses of Stanley and his coworkers. First, although clade shape per se is independent of scale, the generating processes that produce a given clade shape do depend critically on the scaling of standing diversity, on speciation and extinction rates, and on taxonomic level. The visual comparison of real and random clade shapes obscured this fact, resulting in visual bias. Second, the domain of applicability of the results of the MBL simulations of phylogenetic evolution appears to be restricted to small clades, reinforcing the importance of background conditions in the interpretation of simulation results.

It is important to note the scope of the conclusions of Stanley and colleagues (1981). They did show that sudden extinctions or radiations of large clades are not likely to represent mere stochastic fluctuations; rather, such large changes in diversity are liable to be due to changes in the rates of speciation, extinction, or both. Thus, contra the claims of the MBL group, phylogenetic patterns *are* prima facie evidence of changes in underlying diversification rate.

However, their critique left intact several of the deeper ideas behind the MBL model of phylogenetic diversification. Perhaps foremost is the realization that determining the evidentiary significance of an empirical pattern requires a comparison, whether tacit or explicit, to the pattern expected under a null hypothesis. In fact, it was by means of an appropriately scaled null model that Stanley and colleagues were able to show that the diversity histories of brachiopod clades were unlikely to have occurred stochastically. A second surviving idea of the MBL group is that whatever the causes of individual evolutionary events, it may be possible to find a higher focal level that is governed by its own dynamics.[8] Thirdly, Stanley and his colleagues did not attempt to address the implications of the morphologically explicit version of the MBL model (Raup and Gould 1974). In fact, Stanley's own work on Cope's Rule and species selection (1973, 1975), Raup and Gould's (1974) model of random morphological evolution, and Fisher's (1986) ideas on morphological evolution as diffusion through morphospace eventually coalesced into a new type of evolutionary model known as the *passive diffusion model* (PDM).[9] The analyses of Stanley and colleagues did underscore the importance of scaling in the comparison of the hypothetical clades of the simulations and the empirical clades of the fossil record. Most of the remaining insights stemming from the MBL simulations concerning the importance of focal level, standards of evidence in inferring process from pattern, the possibility of taxonomic artifacts, and the phenomena of morphological evolution that can be obtained using a Monte Carlo model were left untouched by these critiques.

CONCLUSIONS

Historians and philosophers of science frequently treat theoretical models as their focal point in studies of scientific change. Scientists often do as well. When John Maynard Smith summoned paleontology down from the highchair to join the rest of evolutionary biology at the high table, he did so presumably because he saw punctuated equilibrium as a distinctively paleontologic contribution to evolutionary theory. Yet the underlying dynamics of scientific change are far more complex than this. In the present case, two theoretical developments (punctuated equilibria and island biogeography) underwrote the assumptions of a new tool (the MBL simulation model) enabling the interrogation of patterns in the fossil record. Even if punctuated equilibrium falls out of favor as a theory, it certainly did ease the task of modeling evolution (Michael Foote, personal communication, 2000). This raises the interesting idea that the tools of science, simulation included, although relying on theory for their assumptions in their initial incarnation, often take on a life of their own.

Reflection on the history of the MBL model suggests the following:

1. In addition to tracking theoretical models, students of science need to take notice of new tools in studying scientific change (Kaiser 2005). Despite domains in which the MBL model may have been inadequate as a theoretical model of evolution, it represented a new tool in the investigation of evolutionary patterns. The introduction of simulation to paleontology changed paleontologic practice. Rather than relying solely on the interpretation of patterns given by nature, it became possible to generate patterns from known starting assumptions (i.e., a model) and compare them with observed natural patterns, what philosophers of science call the *hypothetico-deductive* method (Hempel 1966).

2. Computer simulations with the MBL model were a tool for thought experiments in evolution. They were useful in discovering potential artifacts of taxonomic procedure (Raup et al. 1973; see figure 16.2 in the present chapter), as well as exploring what types of patterns can and cannot be produced using a stochastic branching model (Stanley et al. 1981). As such, they were of tremendous heuristic value.

3. By helping introduce paleontologists to modeling methods, the MBL group changed the kinds of questions paleontologists are able to ask and answer, and even the ways they are able to resolve disputes. Stanley's argument—that the similarities Gould and his coauthors claimed between simulated and real clades were more apparent than real—is an early illustration of this shift in argumentation (Stanley 1979; Stanley et al. 1981). Rather than appealing solely to their familiarity with the fossil record of some taxonomic group or other, Stanley and his coworkers made their case via a combination of empirical data, computer simulation, and analytic calculation. More recently, when molecular clock estimates placed the origin of eutherian mammals sixty-four million years earlier than their first known fossil appearance (Hedges and Kumar 1998), and Foote et al. (1999a, b) were able, via a combination of simulation and analytic calculation, to ask and answer a tractable question: how low a preservation rate would it take to produce a sixty-four-million-year lag between the true origin of eutherians and their oldest known fossil occurrence? Their conclusion was that a preservation rate sufficiently low to yield even a twenty-eight-million-year gap is extremely unlikely, given our best estimates of mammalian preser-

vation rates. In their model they assumed time-homogeneous stochastic branching, just as in the MBL model (albeit without the equilibrium constraints). It is fair to say that paleontologists were not hashing out conflicts in this way prior to 1973. Formal mathematical models, whether solved analytically or by simulation, have the potential to help the paleontologic community in the adjudication of disputes by localizing points of disagreement to particular assumptions, choice of parameters, or parameter values.

4. The model of non-Darwinian evolution assumed in the MBL model has been modified and pursued as a neutral theory of biodiversity and biogeography, and appears to enjoy some degree of predictive success in the case of tropical forests (Hubbell 2001, 2005; see also Levinton 1979). The crucial macroevolutionary insight is that there may be processes operating at the species level and above that are not simply a scaling up of natural selection at the organismal and population levels (Gould and Vrba 1984). Also, null models of morphological evolution are being developed as a method for recognizing and classifying evolutionary trends (Bookstein 1987; Roopnarine, Byars, and Fitzgerald 1999; Roopnarine 2001).

5. Simulation eventually gave way to analytic calculation in explorations of stochastic branching models (cf. Raup 1985) but it seems reasonable to assume that if the MBL papers had relied solely on analytic calculation rather than computer simulation, and in particular if they had not presented their results in the familiar visual iconography of paleontology, their research would have received far less attention. Paradoxically, it is likely that the decision to couch their argument in visual terms led Gould and coworkers (1977) to overlook the difference in scale between their simulated clades and empirical brachiopod clades. There is a suggestion here for historians and philosophers of science that the distinctive (visual) culture of a scientific discipline may deserve attention in studies of the introduction and proliferation of new ideas and methods.

ACKNOWLEDGMENTS

I thank David Raup for reading and commenting on a very early draft of this paper, and Joanna Trzeciak for her comments on a much later draft. I am also grateful to David Raup and Dan Simberloff for corresponding with me about their research. This essay also benefited from conversations with Pete Wagner and Dave Kaiser.

NOTES

1. The original formulation of the distinction between idiographic and nomothetic science may be found in Windelband (1980 [1894]).

2. By incorporating an equilibrium constraint based on total biomass rather than total number of lineages, the morphologically explicit model provided a plausible stochastic mechanism for the puzzling phenomenon of mammalian dwarfism on islands (Wassersug et al. 1979).

3. Specifically, when processing a pattern intuitively, we tend not to consider the effects of path-dependent processes (Raup 1977). In addition, as G. Udny Yule pointed out, the many "nonsense correlations" between unrelated variables reflect the correlations many variables have with time (Yule 1926).

4. Note that this particular simulation result reveals only a *potential* artifact. The potential seriousness of this as a general problem hinges on the idea that hindsight is involved in the erection of higher taxa. According to the taxonomic algorithm of the MBL computer program, at time t_0, whether a species s_1 in an existing clade c_1 founds a new clade c_2 may depend on states of affairs *subsequent* to time t_0, namely whether s_1 eventually gives rise to a sufficient number of descendant species. An event at some later time $t_0+\Delta t$, such as a mass extinction of multiple species (some of them in the s_1 lineage) may ramify backward to t_0, reducing the number of paraclades at time t_0, as s_1 never ended up giving rise to enough ancestral species to justify placing it at the stem of a new paraclade p_2. This is not an instance of backward causation, but rather an artifact of the rules according to which new paraclades are erected. Mass extinctions occurring in nature may thus have the same effect as the end of the run in the MBL simulations: a drop in taxonomic diversity (number of paraclades) in a given time unit may be an artifact of a mass extinction occurring later. In fact, there are several hypotheses for declines in sampled diversity before mass extinction events: (1) the Signor-Lipps effect, (2) taxonomic artifact, and (3) an actual diversity decline.

5. In the original MBL paper, spindle diagrams for reptile clades were presented for comparison with the simulated clades, and discussion was devoted to both similarities and differences in shape between the two (Raup et al. 1973). Some features of the reptile fossil record (such as the Permo-Triassic extinction, with 50% of extant reptile clades becoming extinct in a single time step) were reflected in patterns of clade shape (simultaneous flat tops) well outside the range obtained in the simulations. Within the MBL research program, these would make excellent candidates for historical explanation. But in the 1977 paper, it was the *similarities* between simulated and empirical clades that were stressed.

6. The caption reads: "Figure 1. Comparison of random with real clades. Top: clades for one run of the MBL program at branching and extinction probabilities of 0.1. Bottom: real clades for genera within orders of brachiopods, from *Treatise on Invertebrate Paleontology*." The figure is cited just once in the text to the article: "The illustrations conventionally used by paleontologists to depict the diversity of life

through time have an undeniable morphology (Figure 1), and paleontologists have spent a century trying to interpret the evident order of these diagrams" (Gould et al. 1977, p. 23).

7. In a paper reviewing the MBL work with stochastic models, Gould (1973) remarks of the comparison between simulated and brachiopod clades, "As an initial gestalt, these random clades look strikingly like real ones."

8. This line has been pursued most vigorously by Schopf (1979; see also Gould 1984), Levinton (1979), and, most recently, Hubbell (2001).

9. For further discussion of PDMs see Slatkin (1981), Grantham (1999), Alroy (2000), and McShea (2000).

REFERENCES

Alroy, J. 2000. Understanding the dynamics of trends within evolving lineages. *Paleobiology* 26:319–29.

Bookstein, F. L. 1987. Random walk and the existence of evolutionary rates. *Paleobiology* 13:446–64.

Eldredge, N. 1971. The allopatric model and phylogeny in Paleozoic invertebrates. *Evolution* 25:156–67.

Eldredge, N., and S. J. Gould 1972. Punctuated equilibria: An alternative to phyletic gradualism. In *Models in Paleobiology*, ed. T. J. M. Schopf, 82–115. San Francisco: Freeman, Cooper and Co.

Fisher, D. C. 1986. Progress in organismal design. In *Patterns and processes in the history of life,* ed. D. M. Raup and D. Jablonski, 99–117. Berlin: Springer.

Foote, M., J. P. Hunter, C. M. Janis, and J. J. Sepkoski, Jr. 1999a. Evolutionary and preservational constraints on origins of biologic groups: Divergence times of eutherian mammals. *Science* 283:1310–14.

———. 1999b. Divergence times of eutherian mammals. *Science* 285:2031a.

Gould, S. J. 1984. The life and work of T. J. M. Schopf (1939–1984). *Paleobiology* 10 (2):280–85.

Gould, S. J., D. M. Raup, J. J. Sepkoski, Jr., T. J. M. Schopf, and D. Simberloff. 1977. The shape of evolution: A comparison of real and random clades. *Paleobiology* 3:23–40.

Grantham, T. A. 1999. Explanatory pluralism in paleobiology. *Philosophy of Science* 66 (supplement): S223–S236.

Hedges, S. B., and S. Kumar. 1999. Divergence times of eutherian mammals. *Science* 285:2031.

Hempel, C. 1966. *Philosophy of natural science.* Englewood Cliffs, NJ: Prentice-Hall.

Hubbell, S. H. 2001. *The unified neutral theory of biodiversity and biogeography.* Princeton, NJ: Princeton University Press.

———. 2005. Large-scale diversity and species-area relationships in tropical tree communities under the neutral theory. In *Tropical rainforests: Past, present, and future*, ed. E. Bermingham, C. W. Dick, and C. Moritz: 41–71. Chicago: University of Chicago Press.

Kaiser, D. 2005. Physics and Feynman's diagrams. *American Scientist* 93:156–65.

Kimura, M. 1968. Evolutionary rate at the molecular level. *Nature* 217:624–26.

Ladyman, J. 2002. *Understanding philosophy of science*. New York: Routledge.

Levinton, J. S. 1979. Theory of diversity equilibrium and morphological evolution. *Science* 204:335–36.

MacArthur, R. H., and E. O. Wilson. 1963. An equilibrium theory of insular zoogeography. *Evolution* 17:373–87.

———. 1967. *The theory of island biogeography*. Princeton, NJ: Princeton University Press.

McShea, D. W. 2000. Trends, tools, and terminology. *Paleobiology* 26:330–33.

Raup, D. M. 1987. Neutral models in paleobiology. In *Neutral models in biology,* ed. M. H. Nitecki and A. Hoffman, 121–32. New York: Oxford University Press.

Raup, D. M., and S. J. Gould. 1974. Stochastic simulation and the evolution of morphology—towards a nomothetic paleontology. *Systematic Zoology* 23:305–22.

Raup, D. M., S. J. Gould, T. J. M. Schopf, and D. Simberloff. 1973. Stochastic models of phylogeny and the evolution of diversity. *Journal of Geology* 81:525–42.

Raup, D. M., and Sepkoski, Jr., J. J. 1984. Periodicity of extinctions in the geologic past. *Proceedings of the National Academy of Sciences* 81:801–5.

Roopnarine, P. D. 2001. The description and classification of evolutionary mode in stratophenetic series: A computational approach. *Paleobiology* 27:446–65.

Roopnarine, P. D., G. Byars, and P. Fitzgerald. 1999. Anagenetic evolution, stratophenetic patterns, and random walk models. *Paleobiology* 25:41–57.

Schopf, T. J. M. 1972. *Models in paleobiology*. San Francisco: W. H. Freeman.

———. 1979. Evolving paleontological views on deterministic and stochastic approaches. *Paleobiology* 5 (3): 337–52.

Schopf, T. J. M. and D. M. Raup. 1978. *Stochastic models in paleontology:A primer*. Notes for a workshop on "Species as Particles in Space and Time," held at the U.S. National Museum, June 5–16, 1978.

Schopf, T. J. M., D. M. Raup, S. J. Gould, and D. Simberloff. 1975. Genomic versus morphologic rates of evolution: Influence of morphologic complexity. *Paleobiology* 1:63–70.

Sepkoski, Jr., J. J. 1978. A kinetic model of Phanerozoic taxonomic diversity I. Analysis of marine orders. *Paleobiology* 4:223–51.

Sepkoski, Jr., J. J., and D. C. Kendrick. 1993. Numerical experiments with model monophyletic and paraphyletic taxa. *Paleobiology* 19 (2):168-184.

Simberloff, D. 1970. Taxonomic diversity of island biotas. *Evolution* 24:23–47.

———. 1974. Permo-Triassic extinctions: Effects of area on biotic equilibrium. *Journal of Geology* 82:267–74.

————. 1987. Calculating probabilities that cladograms match: A method of biogeographical inference. *Systematic Zoology* 36:175–95.

Simberloff, D., K. L. Heck, E. D. McCoy, and E. F. Connor. 1981. There have been no statistical tests of island biogeographical hypotheses. In *Vicariance biogeography: A critique,* ed. G. Nelson and D. E. Rosen, 40–63. New York: Columbia University Press.

Slatkin, M. 1981. A diffusion model of species selection. *Paleobiology* 7:421–25.

Stanley, S. M. 1973. An explanation for Cope's Rule. *Evolution* 27:1–26.

————. 1975. A theory of evolution above the species level. *Proceedings of the National Academy of Sciences USA* 72:646–50.

————. 1979. *Macroevolution.* San Francisco:Freeman.

Stanley, S. M., P. W. Signor, S. Lidgard, and A. F. Karr. 1981. Natural clades differ from 'random' clades: Simulations and analyses. *Paleobiology* 7:115–27.

Tversky, A., and D. Kahneman. 1974. Judgment under uncertainty: Heuristics and biases. *Science* 185:1124–31.

Uhen, M. D. 1996. An evaluation of clade-shape statistics using simulations and extinct families of mammals. *Paleobiology* 22:8–22.

Wassersug, R. J., H. Yang, J. J. Sepkoski, Jr., and D. M. Raup. 1979. Evolution of body size on islands: Computer simulation. *American Naturalist* 114:287–95.

Williams, C. B. 1947. The generic relations of species in small ecological communities. *The Journal of Animal Ecology* 16:11–18.

Windelband, W. 1980 [1894]. Rectorial Address, Strasbourg, 1894. *History and Theory* 19:169–85.

Yule, G. U. 1926. Why do we sometimes get nonsense-correlations between time-series? *Journal of the Royal Statistical Society* 89:1–69.

CHAPTER SEVENTEEN

Ritual Patricide:
Why Stephen Jay Gould Assassinated
George Gaylord Simpson

Joe Cain

INTRODUCTION

The launch and rise of punctuated equilibrium in evolutionary studies has been expertly studied by Ruse (1996) and Sepkoski (this volume; 2005), among others. By all accounts, this discipline building was fast paced, fractious, and contested. It involved internal jockeying and prioritization as much as it involved external struggles for definition and autonomy.

This paper examines connections drawn by discipline builders to their predecessors. Specifically, it focuses on the relationship between two focal points, Stephen Jay Gould (1941–2002) and George Gaylord Simpson (1902–1984). In 1950, Simpson was hailed as paleontology's principal innovator in macroevolutionary theory. Over the 1970s and 1980s, however, Gould led a campaign to systematically deny Simpson any relevance to contemporary developments. This paper examines the rhetorical devices used in that campaign and considers the social function of patricide in the founding rhetoric of new disciplines.

Expressed relations between generations are much studied in sociology and anthropology.[1] Everywhere, generations—itself a label that is part of the fluid negotiation of identities—are crisscrossed by assertions of continuity and break. In the broadest sociological frame, these assertions are functional. They're constructed with purpose and given agency so they may contribute to disciplinary and intellectual ends. In scientific circles, look no further than the *pater familias*, Charles Darwin, and the attribution "Darwinian." It's hard to find a case where paternity is more cherished or more contested.

I argue the legacy Gould attributed to Simpson was a tactical construction. Whatever the actual intellectual and social connections might have been, Gould's rhetorical constructions positioned Simpson in particular, purposeful ways. In short, Gould put Simpson to work. The work accomplished in this case changed over time. I argue that it evolved into a form of patricide, with Gould crafting accounts of the past that eventually made Simpson obsolete. Others around Gould joined in this work. There soon evolved a ritual form of this patricide, creating a routine ceremonial acting out of asserting Simpson's irrelevance.

GEORGE SIMPSON

Simpson dominated paleontology's contribution to evolutionary studies in mid-twentieth-century American biology.[2] Specializing in the study of mammals, his interests were mandarin, with extensive publications in systematics, biogeography, evolution, and morphology . The widest frame to view Simpson involves the notion of "paleozoology," which he contrasted with "neozoology." For Simpson, the goal of paleozoologists was to understand ancient organisms in all the ways his colleagues understood the organisms living around us today. Added to this, Simpson argued, paleozoologists could use geology's panoramic vision to follow patterns and processes over scales simply inaccessible to those who only studied currently living organisms. Simpson sometimes referred to this broad research program as "four-dimensional" or "temporal" biology. It was pandisciplinary in scope. Simpson never abandoned this vision for a fully synthetic biology (Simpson 1983; Cain 1992).

Starting in the late 1930s, Simpson began to produce a series of innovative theoretical works as part of temporal biology. These included attempts to introduce population-level studies into paleontology (Simpson 1937a), the deployment of confidence tests for hypothesis testing (Simpson 1937b; Simpson and Roe 1939), new classification schemes (Simpson 1945), and an attempt to balance transformation and migration in evolutionary narratives (Simpson 1939, 1940). At the same time, Simpson followed the increasing attention cytologists, geneticists, and field naturalists were devoting to the causes of speciation and to the particularly knotty problems related to classification of supra- and super-specific taxa (e.g., Simpson 1937c). As he learned more about technical and theorctical developments in these areas, Simpson saw avenues for new intellectual alliances. Working largely from published literature, he excitedly foraged and consumed, then put these new resources to work. Simpson's efforts found their widest audience in 1944, when Columbia

University Press published *Tempo and Mode in Evolution* (Simpson 1944) as part of its Columbia Biological Series (Cain 2001). Written between 1938 and 1942, this book shows Simpson bubbling over with ideas for how his four-dimensional biology might move evolutionary studies forward.

Tempo and Mode made Simpson a star. Among other virtues, it used paleontology, genetics, and ecology in a joint attack on some longstanding evolutionary problems. It also showed how similar joint attacks elsewhere not only could identify new evolutionary processes but also extend the range of application for familiar explanations. *Tempo and Mode* was read by the growing number of researchers interested in speciation and underlying evolutionary mechanisms. (No surprise, as Simpson aimed this book squarely at them.) Though particular factions in that group stressed different messages from *Tempo and Mode*, all agreed Simpson (1944) deserved standing among other synthetic innovations of the 1930s and 1940s (Cain 2003).

Though Simpson later added to, expanded, and altered the vision he presented in *Tempo and Mode*, there's no doubt this book formed part of the legacy he and his colleagues wanted to pass on to future generations. Together with other texts in the Columbia Biological Series, *Tempo and Mode* became essential reading in graduate training. It served as a benchmark for measuring innovation in new research. It also served as a trading zone (Galison 1999) in which cross-disciplinary and multinational discussion about evolutionary theory and paleontology took place. In such trading, it didn't matter if one was an advocate or critic of Simpson (1944). Like it or loathe it, still, everyone was expected to know it.[3]

SIMPSON AND THE LAUNCH OF PUNCTUATED EQUILIBRIUM

In the 1970s expansion of macroevolution, views on Simpson's legacy came to serve as a positional shorthand.[4] This is especially true with early advocates of punctuated equilibrium (PE). Compare the famous launch paper of Eldredge and Gould (1972), with its successor, Gould (1977), the review paper that claimed victory and converted PE into a coherent research program.[5] These papers invoke Simpson for significantly different purposes.

In 1972, Eldredge and Gould's basic narrative is revolution and radical departure, akin to Kuhn and paradigm shifts. The dominant paradigm, they suggested, was "phyletic gradualism." This had whole populations slowly transforming over time such that one species smoothly grades into its successor. Slow, continuous, and steady. As an explanation for macroevolution,

Eldredge and Gould argued, this paradigm relied on "species extrapolation": whatever explains evolution *within* species also explains evolution *between* species and *between* all larger taxonomic units. For them, PE offered a radical alternative. Most species changed rather little over most of their evolutionary history, they argued. Occasionally, this stasis was punctuated, which occurred when small, peripheral populations became isolated and then rapidly changed, normally as a result of random processes such as genetic drift.

In this 1972 formulation, Simpson is a minor footnote. He's one of a group who crafted the old, extrapolationist paradigm. His major books are cited (Simpson 1953, 1944), but nothing substantive is said about them. The overall narrative thrust is "us versus them." On one side are the orthodox, theory-laden, extrapolating gradualists, Simpson included. On the other side are open-minded discoverers of evolution's true story, reluctantly forced into a fight.

Gould and Eldredge (1977) present a rather different construction for Simpson's value. Instead of rhetoric grounded in paradigm shifts, this paper's narrative is "standing on the shoulders of giants." Simpson, they wrote, towered over the subject of macroevolution. His innovations and wisdom helped the next generation see just a little bit further than their predecessors. Gould and Eldredge (1977) claimed their 1972 paper presented merely a "modest proposal," offered in an effort to "clarify and emphasize" preexisting ideas. "For all the hubbub it engendered," they suggest, "the model of punctuated equilibria is scarcely a revolutionary proposal" (Gould and Eldredge 1977, 117).

Between 1972 and 1977 Gould and Eldredge were sharply criticized not only for the substance of their views but also for their style of argument and claims of radical revolution. One frequent complaint against PE was that the dichotomy of gradualism versus stasis was nothing but a straw man. Another complaint focused on their claims for novelty and innovation. In one way or another, critics said, what was interesting about PE had been said many times before, notably by Simpson.

The "shoulders of a giant" language, I argue, was meant as a peace offering. But it was unsuccessful, satisfying neither PE's opponents nor proponents. On one hand, it gave too much away. Many macroevolutionists of the 1970s tied their identity closely to values of rebel chic: antiestablishment, paradigm breaking, and radical. For them, it just wasn't good enough to present a "modest proposal" about continuity, follow in someone else's footsteps, or stand on someone else's shoulders. That denied the very sense of innovation and break with the past some proponents sought in the first place.

Gould and Eldredge's (1977) concession didn't calm PE's opposition, either. From this perspective, the 1977 version of events still failed to concede sufficiently to precedent and predecessor. It just seemed too easy to spot Simpson's concepts in PE and too easy to see the basic intuitions of macroevolution as something handed down from past generations.[6] From this perspective, Gould and Eldredge's concession seemed nothing but smug, disingenuous, and patronizing.

Gould quickly recognized the failure of their second strategy. He quickly adopted a third rhetorical device—negation—which he embedded in a narrative about the "hardening" of the evolutionary synthesis.

HARDENING AS RHETORICAL DEVICE

Gould was a master of the written word, and his historical works were among his most popular texts. Occasionally Gould selected topics simply for the story told. Much more frequently, however, his topics had tactic value, smartly chosen to accomplish work in a particular moment and cause. He deployed history to expose bias and fraud, to explain the persistence of bad ideas, and to celebrate the work of right-thinking people who struggled against dominant paradigms. For instance, Gould's first historical book, *Ontogeny and Phylogeny*, attacked adaptationism and trumpeted the approach to developmental biology he advocated against genetic reductionism (compare Gould 1977; Gould and Lewontin 1979). *Time's Arrow Time's Cycle* attacked uniformitarianism (Gould 1987). *Wonderful Life* traced a century of research into the Burgess Shale fossils so Gould could further attack ideas of progress and extrapolation, then trumpet alternatives such as chance and contingency (Gould 1989). Regardless of their value as historical scholarship, these works also functioned within Gould's multifaceted defense of his views—history combining with empirical data, theoretical models, and political advocacy.

In the late 1970s, Gould began to focus his historical energies on evolutionary studies in the generation preceding him.[7] He wrote a great deal about the so-called "architects" of the evolutionary synthesis—Theodosius Dobzhansky, Sewall Wright, Ernst Mayr, Julian Huxley, and, of course, Simpson. Gould promoted his views at conferences, in his growing number of public lectures, in the forwards he wrote for other people's books, and in the pages of *Natural History*. Working with Eldredge, he also organized facsimile editions of classic texts from the period, giving themselves the job of writing the historical introductions—telling readers how best to read and appreciate these great books.[8]

In this work, Gould produced a third interpretation regarding Simpson's value. Simply put, it's a clever form of negation, embedded in a thesis about how the evolutionary synthesis hardened into an ideology.

In brief, the hardening thesis constructs a "before, during, and after" sequence for the evolutionary synthesis. *Before*, in the 1930s, was a period of pluralism, tolerance, and diverse thinking about evolutionary mechanisms. It was a Homeric golden age in which discussion was robust and free. *After*, in the 1950s, attitudes hardened like arteries. Diversity has been killed by Hegemony, and the only game in town was adaptation—that "Panglossian paradigm" (as Gould and Lewontin [1979] called it) with its stale focus on natural selection and its bias toward gradual evolutionary change. *In between the 30s and 50s*, a hardening took place that transformed the before into the after. Gould left this middle period largely in a black box, never quite explaining who drove it or why it happened.

Key to Gould's historical analysis of Simpson in the hardening period is Simpson's theory of "quantum evolution." Gould noted it was one of Simpson's "big" ideas: "once his delight and greatest pride" (Gould 1980, 167). Simpson was trying to explain the origin of major taxonomic units and periods of rapid change. He invoked Sewall Wright's shifting balance theory, in which genetic drift has a major impact on small, partially isolated populations. Drift shifts these groups off adaptive peaks and into nonadaptive valleys. In Simpson's thinking, natural selection quickly challenges these groups to scurry up new evolutionary terrain such that, in the blink of a geologist's eye, the quantum of morphological difference is traversed.

That's Simpson's view in 1944. Over the next decade, Gould argued, Simpson lost his excitement for genetic drift and abandoned his bright new idea in favor of knee-jerk extrapolation. This left natural selection to steer all of life's evolutionary change. By the mid-1950s, Gould wrote, Simpson's worldview was entirely taken over by this paradigm: automatically invoked and never questioned. Adaptationism has hardened George Simpson, and along with him, the rest of evolutionary biology.

It's fair to say today that Gould's hardening thesis has been quite successful, becoming conventional wisdom in synthesis historiography. Only a few historians have examined it critically (compare Gerson 1998, and Shanahan 2004, 133). Some of Simpson's scientific colleagues rose to his defense on this point.[9] Simpson himself rejected the suggestion of a hardening.[10] (For an example of Simpson's views circa 1980, see Appendix 1.) Either way, the hardening thesis remains manipulative and tactically valuable. It's another example of Gould putting history to work. If PE was going to have any claim to novelty, Gould needed some way to negate Simpson.

The hardening thesis does precisely this. As a rhetorical device, it diffuses two related pressures. Its "before-during-after" construction allows praise for the so-called architects of synthesis in the 1930s and 1940s, Simpson included. At the same time, it marginalizes their relevance to contemporary debates by separating the peaks of innovation (the 1930s and the 1970s) by a valley of rot (the 1950s and 1960s). At best, Simpson and the synthesis have had their day. But they're showing their age and now desperately need renovation. At worst, the old boys simply have lost the plot, and their dogmatic control of the discipline is now smothering innovation. Note the combination of deference and replacement. Clever.

PATRICIDE

The hardening thesis gave Gould a way to negate Simpson. By marginalizing him, Gould was marginalizing a key problem for PE's claims of novelty and replacement. Curiously, after the hardening thesis was forwarded, Gould's commentaries about Simpson grew increasingly hostile. Taken together, these combine into a rather sharp set of criticisms. Overall, they form an attack on four fronts (table 17.1). During the same time, correspondence between Gould and Simpson show their relationship had completely broken down.

The last of these four fronts is worth noting with a few examples. Simpson died in 1984. Gould's obituary for Simpson, in *Evolution*, certainly is full of praise (Gould 1985). For instance, Gould calls Simpson the "most important paleontologist since Georges Cuvier" (229) Although Gould claims he didn't "wish to dwell" on it, as he closes his obituary, he didn't resist adding some scathing remarks. Simpson, Gould explained, was not an easy man to like. A man who feared for his legacy and who had to be treated gently because he "took offense easily, placing the worst possible interpretation on any event that displeased him" (232).[11]

Gould's anger with Simpson seemed to intensify with time (e.g., Gould 1988). Ten years after Simpson's death, that anger was red hot. One of Simpson's daughters found an unpublished book manuscript of her father's she wanted in circulation (Burns 1996). Gould agreed to write an afterward. He let loose. "I don't want to sound like a two-bit Freudian quack," Gould exclaimed, but Simpson was lonely, dissatisfied, craved recognition, and was incapable of satisfaction. He "wallowed in a miasma of doubt and anger, always fearing that future generations would ignore him and that all his work would ultimately go for naught" (Gould 1996).

Character assassination is common enough. Patricide is more than a single attack on character. In the context of using history to construct heritage, patricide is a systematic attempt to disconnect—to construct not relevance but

TABLE 17.1 Summary of themes used in Stephen Jay Gould's attacks on George Gaylord Simpson

---

1. Simpson's science was wrong.
   a. His paleontology and systematics have been reinterpreted.
   b. He denied the importance of drift and other stochastic processes.
   c. He looked in the wrong place for evolution's key events.
   d. He ignored the importance of hierarchy and cascading systems.
   e. He missed "species selection."

2. Simpson's science was biased and theory-driven.
   a. He was a Panglossian pan-selectionist and a knee-jerk adaptationist.
   b. He assumed extrapolation and reduction could carry the explanatory load.

3. Gould used structural exclusion to remove Simpson from relevance to today's problems, via
   a. The hardening thesis
   b. Suggesting Simpson ultimately left the job of synthesis undone
   c. Suggesting Simpson denied paleontology's virtue and independence by ceding authority to other disciplines, e.g., via extrapolation

4. Gould attacked Simpson's character, representing him as
   a. Hostile, aggressive, mean-spirited
   b. Insecure, pedantic, undermining
   c. Dogmatic, intolerant, unpredictable
   d. A racist

---

irrelevance. It involves crafting narratives in which breaks override continuity and in which the past is not simply a foreign country but a place with no connections whatsoever. While Gould's hardening thesis offers a form of negation, his later representations of Simpson combined to form an exhaustive form of denial. This used every scientific, historical, and personal tool in Gould's formidable arsenal to dethrone Simpson—to dethrone someone his own training had told him to count as a founding father. This is more than negation. It's patricide.

RITUAL PATRICIDE

Patricide is one thing. *Ritual* patricide is quite another. The notion of ritual helps explain the breadth and the persistence of Gould's attacks on Simpson. He wasn't simply angry with Simpson. This isn't simply a case of intellectual rejection combined with a mere dislike for the guy. And Gould didn't simply strike out at Simpson once or twice. He was persistent and systematic, often going well out of his way and carrying on long after the fight needed to be waged. It's hard to find a person Gould demonized more fervently. Even creationists got off lighter.[12]

Using the notion of ritual as ceremonial routine undertaken in the context of a common life, Gould's repeated attacks on Simpson can be understood not as a function of need or vengeance but as a signal. It's an outward manifestation with crucial inner meaning. Twenty years ago, Laporte (1983, 410) suggested efforts to undermine Simpson were part of a bonding process for advocates of PE. The *act of attack* defined affiliation. Gould's persistent attacks on Simpson, then, signaled this bond. As Gould was one of the undisputed leaders in PE circles, the ritualized nature of these attacks served a ceremonial function. Ironically, Gould (1982b) suggested similar ideas about bonding rituals when he wrote about heretics in science. He claimed attacks on unorthodox thinkers, such as the geneticist Richard Goldschmidt, serves as a glue for social groups.[13]

A ritualized form of patricide also explains attacks on Simpson by other PE advocates. The best example is Eldredge. His early scholarship on Simpson offers close exegesis combined with honest disagreement (Eldredge 1985; 1985a, b, c). Later writing seems to go out of its way to negate Simpson, mainly through repeating the claim that *Tempo and Mode* offered nothing more than consistency argument to relate paleontology with population genetics and the new speciation theory. Sometimes swipes are made that seem merely *ad hominem* (consider, e.g., Eldredge 1995, 25–26; 1999, 8, 12, 109, 133–40). I interpret such later writings as acts of ritual patricide.

Patricide is only one kind of ritual. The 1999 *Osiris* volume on commemorative practices illustrates others, following the view of some anthropologists that we should connect repeated actions to rituals, and rituals to social functions (Abir-Am and Elliott 1999). A key idea in the *Osiris* volume is demarcation and boundary work. Rituals serve to separate. They also serve to remove ambiguities in alliance and to license certain forms of identity. Patricide and ritual patricide add two more pieces to this larger repertoire of strategies for managing social connections over time.

CONCLUSION

Simpson's value to PE evolved in five steps. In 1972 he simply plays for the other side. In 1977, he's the giant on whose shoulders PE stands. Thereafter, Gould uses the hardening thesis to simultaneously praise and exclude. In Gould's later writings, exclusion grows into patricide, and that patricide evolves into a ritual. Apparently, it's important we're regularly reminded that Simpson was the *old* guard: stuck in a harmful paradigm and disconnected from the excitement of new developments. Gould could have represented Simpson's legacy in myriad ways, but he chose negation and patricide.

I close with a longstanding concern. There's no question that historians' studies of modern science are read by scientists and other participants in the events described. We know sometimes they use our work for their own purposes. Is there any way we can avoid complicity or conscription into their partisan struggles, one group versus another? I fear that unless we're quite careful, we risk providing ammunition for that combat. Worse, we risk providing ammunition *only to some* of the participants. We must remain ever mindful of the work others might do with the materials we provide.

APPENDIX 1: SIMPSON'S 1980 VIEW ON PE

On 18 July 1980, John Bucher (*Discover* magazine) wrote to Simpson with a request. "*Discover* is doing a story about recent developments in evolutionary theory, particularly the rise of the macroevolution school." He asked Simpson to respond to several questions:

1. How important is Eldredge and Gould's theory?
2. Does it constitute a challenge to the primacy of natural selection?
3. Does it constitute a challenge to the modern synthesis?
4. Do you think they are correct in stating
   a. that evolution proceeds by fits and starts,
   b. that natural selection is not the factor which accounts for the appearance of new species?
5. What does Eldredge and Gould's theory mean for the overall picture of evolution?

On 22 July 1980, Simpson replied

. . . I cannot reply adequately and in full for the same reasons that I have not written a full critique of the views of Eldredge and Gould: to do so would take more time than I can afford to take from teaching, work on three books on other subjects, and research, and such critiques are appearing from other sources.

I think that the views expressed by Eldredge and Gould constitute a potentially important contribution to the growing complex of evolutionary theory that has been called (by me and others) the synthetic theory. On this basis, I appreciate and welcome their views. They are enthusiasts, and they consequently and understandably do tend to overstate both the novelty and the generality of their ideas. In broader and somewhat calmer consideration their main point had long ago been stated in other words as a part of the synthetic theory. The idea that their views approach a general theory of evolution that contradicts and replaces the synthetic theory as of the 1970s and 1980 is not justified in my opinion.

What they call 'punctuation' involves the origin of new species and eventually of higher taxa by changes that are either instantaneous, that is, occurring between one generation and the next, or occur at rapid rates of evolution, followed by either slower rates or no further change ('stasis'). In more general terms it was already stated by Darwin in 1859 that rates of evolution demonstrable from the fossil record vary greatly and may be essentially zero or static or may be relatively very rapid. In *Tempo and Mode in Evolution* (1944) I showed, without claiming particular originality, that although most rates fall into a more or less normal distribution, some are very slow or for long periods nil and others are exceptionally rapid, resulting in seemingly abrupt evolutionary changes in the populations involved. I called the latter 'quantum evolution.' In *Major Features of Evolution* (1953) p. 389, I further generalized this concept:

'Quantum evolution may lead to a new group at any taxonomic level. It is probably that species, either genetic or phyletic, often arise in this way.' [p. 389]

I believe that quantum evolution is essentially the same as the 'punctuation' of Eldredge and Gould. In *Macroevolution* (1979) Steven M. Stanley, who inclines toward the general model of Eldredge and Gould, considers that quantum speciation, which he ascribes to me, is the same as the punctuation of Eldredge and Gould. The difference is that Eldredge and Gould, although not always quite clear on this point, evidently believe that all speciation is quantum speciation. Stanley, incidentally, does not go along with them in that respect.

Eldredge and Gould attack their concept of the synthetic theory—a straw man, as their concept of it is really not that of syntheticists in general—as being 'gradualistic'. This is an ill-defined term. They seem to mean just the idea that successive speciation within a single lineage takes place certainly more slowly or more probably not at all. That it usually is slower than quantum speciation is just what I have said since 1944 (or more exactly 1942). Stanley agrees with me, although he does so with-out clearly indicating that this disagrees with Eldredge and Gould. Successive speciation, or the origin or what are called chronospecies, certainly does occur and may rarely even involve quantum evolution. In this respect the model of Eldredge and Gould is misleading. They also are misleading in the implication that dichotomous speciation by quantum 'punctuation,' does not involve phyletic or lineage continuity. All evolution necessarily and obviously involves a continuity of successive generations of populations. It has long been a part of the synthetic theory that quantum speciation usually involves relatively small populations, often but not necessarily marginal parts of a larger parental specific population. It is possible, but highly improbable and hence rare and hardly provable, that the quantum change may occur through a single individual or pair. Even in such a case there would be phyletic continuity.

I think I have answered your questions:

1: What is the importance of Eldredge and Gould's views? (As those views are neither wholly new nor a complete 'theory' I do not call them a new theory.) They are important adjuncts but not replacements. 3: Is it a challenge to the synthetic theory? 'Challenge' yes on their part, not seen so by me. 4a: Does evolution proceed by 'fits and starts.' It may be said to, but that colloquial expression is likely to be misunderstood. 5: What do Eldredge and Gould's views mean for the overall picture of evolution? They fit in well enough, but may distort it not because of what they include but because of what they omit or try to erase or paint over.

On your other questions:

2: Do those views constitute a challenge to the 'primacy of natural selection'? Eldredge and Gould have not to my knowledge denied that natural selection really occurs. They do tend to downrate and at times to ignore it. They do not clearly face the obvious fact that all organisms not becoming extinct are adapted to their ways of life and ecologies and that this cannot rationally be due solely to although it may include, chance. Positive natural selection is the only demonstrable factor in evolution that is nonchance and usually in the direction of adaptation. Negative natural selection is the obvious general cause of extinction. Eldredge and Gould have not faced these facts. That

natural selection is not the *only* factor or even necessarily the prime
factor in all of evolution was already seen, although less clearly, by
Darwin and is a generally accepted aspect of the synthetic theory.

4b: Are Eldredge and Gould correct in stating that natural selection is
not the factor which accounts for the appearance of new species? I
do not know of anyone who has ever believed that natural selection
alone accounts for the appearance of new species, although perhaps
some late 19th century Neodarwinist did. I think, and I believe most
syntheticists think, that other factors are necessarily involved when
an ancestral species divides into two or more descendant species
but that this process is also usually influenced to some extent by
natural selection. I believe that natural selection often dominates
evolution of successive chronospecies in one nondividing lineage,
and this is widely accepted. Still, it is not quite an orthodox dogma
of synthetic theory, which indeed has no orthodox dogmas. If Eldredge and Gould really said that natural selection is not *a* factor in
the evolution of species (I do not believe they ever have said that),
then, no, they are certainly not correct.

It is another point and perhaps not especially relevant here, that natural
selection also occurs as between different species and not only within
species. This was also known to Darwin but not emphasized by him.
Some present evolutionists (notably Stanley) do strongly emphasize or
indeed, I think, overemphasize it, but this has little real bearing on the
questions you raise.

You may quote this letter if you wish to, but if you quote only parts of
it I want to see a copy for approval before publication.[14]

NOTES

1. Note the explicit distinction between "actual" and "expressed" relations. The
latter are found in language, actions, and artifacts. They are fluid associations that can
be changed easily over time. For an introduction to the sociology of generations and
cohorts, see Ryder (1965), Wyatt (1993, 2–5) and Turner (1999, 246–61).

2. The best overviews of Simpson's biography are Simpson (1976; 1978), Whittington (1986), and Laporte (1983; 2000).

3. Simpson, himself, came to loathe the strong connection biologists made between
his career and *Tempo and Mode*. He often complained of being a *homo unius libri*—a
"one-book man"—and sometimes wondered if others ever bothered to read his later
work. Long lists of revised work became common in his writing. His frustration was

especially strong when people much later linked Simpson only to views he expressed in *Tempo and Mode in Evolution* (1944). Whether he expressed a view in 1944 was one matter, Simpson complained. But so fixed a focus on 1944 forced him to appear as he was in 1944 and prevented him from presenting views based on more recent or more considered material. In short, it forced him to seem old and out of date. Why, he wondered, were his revised views given such lesser weight? (For instance, note the title "Forty Years Later" to the 1984 facsimile (Simpson 1984); also see Simpson to Boucot, 24 December 1979, Simpson Papers, APS Library, series 1, folder "Boucot, Arthur.") As for Darwin's *Origin of Species* (1st edition, 1859; 6th edition, 1872), decisions by others to focus on particular versions of ideas can be tactical choices with strategic consequences.

4. Simpson's own views on PE are variously expressed. Though he frequently turned down requests to speak on the subject, in reply to correspondence he often set out his views (e.g., see Appendix 1; compare correspondence in Simpson Papers, series 1, folder "Coyne, Jerry"). He also actively encouraged opposition to PE (e.g., Simpson Papers, series 1, folder "Corning, Peter A."). Simpson made brief mention of PE in autobiographical statements (Simpson 1976; 1978, 269). In 1980 he spoke on the subject (see lecture notes, Simpson Papers, APS Library, series 5, folder "Punctuated Equilibrium"). These notes are close to Simpson (1983, 171–76).

5. By 1977, other programs within paleobiology and macroevolutionary studies had expanded, too. Gould and Eldredge were working to position themselves as central players in that expanding program.

6. Gould's final comments on the question of originality are given in Gould (2002, 1014–17).

7. Gould's first specific incentive to focus on the synthesis period came from his participation in the evolutionary synthesis conferences organized by Ernst Mayr (Mayr and Provine 1980). Simpson did not attend, and paleontologists of his generation were represented only by E. C. Olson. Gould presented a paper, and certainly contributed to the discussions, as transcripts for the sessions indicate (Gould 1980). Simpson was extremely agitated with Mayr and Gould about this conference, as his correspondence with Mayr indicates (also see Simpson to Verne Grant, 26 May, 1981, Simpson Papers, series 1, folder "Grant, Verne #4"). Transcripts of the conference discussions and related correspondence were deposited in the American Philosophical Society Library.

8. Gould and Eldredge planned to reprint the first edition of Dobzhansky's (1937) *Genetics and the Origin of Species*, Mayr's (1942) *Systematics and the Origin of Species*, and Simpson (1944) as a series titled "Columbia Classics in Evolution." Reprints of the first two appeared, with their introductions serving as "critical evaluations" (Gould 1982a; Eldredge 1982). Later Mayr privately circulated a response and later published on the matter (Mayr 1999). While discussing these forwards with Simpson, Mayr also joked "maybe it will comfort you to know that you are not the only one to be tarred and feathered by the smart Alec's of the AMNH!" (Mayr to Simpson,

12 December, 1982, Simpson Papers, series 1, folder "Mayr, Ernst #4") Eldredge wrote an introduction for the reprint of *Tempo and Mode* in this series, but Simpson exercised his contractual right to refuse the request to reprint (for an explanation, see Simpson to Gould, 26 July, 1980, Simpson Papers, series 1, folder "Gould, Stephen Jay"). Simpson (1984, esp. xxii–xxvi) secured a reprint of his own, complete with his own introduction. Later, Eldredge (1985) published his introduction in another form. (See Simpson Papers, series 1, folder "Raeburn" and folder "Mayr, Ernst #4" for related correspondence.)

9. Verne Grant examined changes to the concept of "quantum evolution" since 1944, both in print (Grant 1985) and in correspondence with Simpson (Simpson Papers, series 1, folder "Grant, Verne #1"), concluding "quantum evolution is the obvious forerunner of punctuated equilibrium" (Grant to Simpson, 28 November, 1980).

10. Simpson frequently pointed to differences in his views over the 1940s and 1950s, frustrated with the fixation on his 1944 book (e.g., Simpson to Levinson, 15 March, 1984, Simpson Papers, series 1, folder "Levinson, Jerrery S.").

11. Compare Gould's (1985) obituary with others, e.g., Olson (1986; 1991) or Whittington (1986).

12. Of course, ritualized killing is not limited to literary forms; see Hsia (1992; 1988) and Forgie (1979).

13. Pope (2005) proposed ritual patricide to explain certain phenomena in American environmentalism.

14. This correspondence is preserved in Simpson Papers, APS Library, series 1, folder "Bucher, John." Minor typographical errors in the original have been corrected here.

REFERENCES

Abir-Am, Pnina G., and Clark A. Elliott, eds. 1999. *Commemorative practices in science: Historical perspectives on the politics of collective memory. Osiris, 2nd series, vol. 14.* Chicago: University of Chicago Press for History of Science Society.

Cain, Joe. 1992. Building a temporal biology: Simpson's program for paleontology during an American expansion of biology. *Earth Sciences History* 11 (1): 30–36.

———. 2001. The Columbia Biological Series, 1894–1974: A bibliographic note. *Archives of Natural History* 28 (3): 353–66.

———. 2003. A matter of perspective: Disparate voices in the evolutionary synthesis. *Archives of Natural History* 30 (1): 28–39.

Eldredge, Niles. 1982. Introduction. In *Systematics and the Origin of Species (facsimile of 1942 1st ed.)*, ed. E. Mayr, xv–xxxvii. New York: Columbia University Press.

———. 1985a. Evolutionary tempo and modes: A paleontological perspective. In *What Darwin began: Modern Darwinian and neo-Darwinian perspectives on evolution*, ed. L. R. Godfrey, 113–37. Boston: Allyn and Bacon.

———. 1985b. *Time frames: The rethinking of Darwinian evolution and the theory of punctuated equilibria.* New York: Simon and Schuster.

———. 1985c. *Unfinished synthesis: Biological hierarchies and modern evolutionary thought.* New York: Oxford University Press.

———. 1995. *Reinventing Darwin. The great debate at the high table of evolutionary theory.* New York: Wiley.

———. 1999. *The pattern of evolution.* New York: W. H. Freeman.

Eldredge, Niles, and Stephen Jay Gould. 1972. Punctuated equilibria: An alternative to phyletic gradualism. In *Models in Paleobiology*, ed. T. J. M. Schopf, 82–115. San Francisco: Freeman, Cooper.

Forgie, George B. 1979. *Patricide in the house divided: A psychological interpretation of Lincoln and his age.* New York: W. W. Norton.

Galison, Peter. 1999. Trading zone: Coordinating action and belief. In *Science studies reader*, ed. M. Biagioli, 137–60. New York: Routledge.

Gerson, Elihu. 1998. The American system of research: Evolutionary biology, 1890–1950. PhD diss., University of Chicago.

Gould, Stephen Jay. 1977. *Ontogeny and phylogeny.* Cambridge, MA: The Belknap Press of Harvard University Press.

———. 1980. G. G. Simpson, paleontology, and the modern synthesis. In *The evolutionary synthesis*, ed. E. Mayr and W. Provine, 153–72. Cambridge, MA: Harvard University Press.

———. 1982a. Introduction. In *Genetics and the Origin of Species (facsimile of 1937 1st edition)*, ed. T. Dobzhansky, xvii–xxxv. New York: Columbia University Press.

———. 1982b. The uses of heresy: An introduction of Richard Goldschmidt's *The material basis of evolution.* In *The material basis of evolution (facsimile)*, edited by R. Goldschmidt, xiii–xlii. New Haven, CT: Yale University Press.

———. 1985. Recording marvels: The life and work of George Gaylord Simpson. *Evolution* 39 (1): 229–32.

———. 1987. *Time's arrow time's cycle: Myth and metaphor in the discovery of geological time.* Cambridge, MA: Harvard University Press.

———. 1988. To him, fossils said the darnedest things. *The New York Times Book Review*, 14 February, 1988, 13–14.

———. 1989. *Wonderful life: The Burgess Shale and the nature of history.* New York: W. W. Norton.

———. 1996. The truth of fiction: An exegesis of G. G. Simpson's dinosaur fantasy. In *The dechronization of Sam Magruder.* Ed. J. Simpson Burns. New York: St Martin's Press.

———. 2002. *The structure of evolutionary theory.* Cambridge, MA: Belknap Press of Harvard University Press.

Gould, Stephen Jay, and Niles Eldredge. 1977. Punctuated equilibria: The tempo and mode of evolution reconsidered. *Paleobiology* 3:115–51.

Gould, Stephen Jay, and Richard Lewontin. 1979. The spandrels of San Marco and the Panglossian paradigm: A critique of the adaptationist programme. *Proceedings of the Royal Society London* B205:581–98.

Grant, Verne. 1985. *The evolutionary process: A critical review of evolutionary theory.* New York: Columbia University Press.

Hsia, R. Po-chia. 1988. *The myth of ritual murder: Jews and magic in reformation Germany.* New Haven, CT: Yale University Press.

———. 1992. *Trent, 1475 : Stories of a ritual murder.* New Haven, CT: Yale University Press.

Laporte, Léo. 1983. Simpson's *Tempo and mode in evolution* revisited. *Proceedings of the American Philosophical Society* 127 (6): 365–417.

———. 2000. *George Gaylord Simpson: Paleontologist and evolutionist.* New York: Columbia University Press.

Mayr, Ernst. 1999. Introduction. In *Systematics and the origin of species, from the viewpoint of a zoologist.* Cambridge, MA: Harvard University Press.

Mayr, Ernst, and William Provine, eds. 1980. *The evolutionary synthesis: Perspectives on the unification of biology.* Cambridge, MA: Harvard University Press.

Olson, Everett C. 1986. Memorial to George Gaylord Simpson, 1902–1984. *GSA Memorials* 16:1–6.

Olson, Everett. 1991. George Gaylord Simpson. *Biographical Memoirs of the National Academy of Science* 60:331–353.

Pope, Carl. 2005. *And now for something completely different. An in-depth response to 'The death of environmentalism.'* Sierra Club, 13 January, 2005 [cited 30 September, 2005]. Available from www.grist.org/news/maindish/2005/01/13/pope-reprint/.

Ruse, Michael. 1996. *Monad to man: The concept of progress in evolutionary biology.* Cambridge, MA: Harvard University Press.

Ryder, Norman. 1965. The cohort as a concept in the study of social change. *American Sociological Review* 30:843–61.

Sepkoski, David. 2005. Stephen Jay Gould, Jack Sepkoski, and the 'Quantitative Revolution' in American Paleobiology. *Journal of the History of Biology* 38 (2):209-237.

———. forthcoming a. Stephen Jay Gould: Iconoclasm and the fossil record. In *Rebels of life: Iconoclastic biologists of the twentieth century,* ed. O. Harman and M. Dietrich. New Haven, CT: Yale University Press.

———. forthcoming b. The 'delayed synthesis': Paleobiology in the 1970s. In *Descended from Darwin: Insights into American evolutionary studies, 1925–1950,* ed. J. Cain and M. Ruse. Philadelphia: American Philosophical Society.

Shanahan, Timothy. 2004. *The evolution of Darwinism: Selection, adaptation, and progress in evolutionary biology.* Cambridge: Cambridge University Press.

Simpson, George Gaylord. 1937a. The Fort Union of the Crazy Mountain Field, Montana, and its mammalian faunas. *Bulletin of the United States National Museum* 169:1–287.

———. 1937b. Patterns of phyletic evolution. *Bulletin of the Geological Society of America* 48:303–14.

———. 1937c. Supra-specific variation in nature and in classification: From the viewpoint of paleontology. *American Naturalist* 71:236–67.

———. 1939. The development of marsupials in South America. *Physis* 14:373–90.

———. 1940. Antarctica as a faunal migration route. *Proceedings of the 6th Pacific Scientific Congress* 1939:755–68.

———. 1944. *Tempo and mode in evolution.* New York: Columbia University Press.

———. 1945. The principles of classification and a classification of mammals. *Bulletin of the American Museum of Natural History* 85:i–xvi, 1–350.

———. 1953. *The major features of evolution.* New York: Columbia University Press.

———. 1976. The compleat paleontologist? *Annual Review of Earth and Planetary Sciences* 4:1–13.

———. 1978. *Concessions to the improbable: An unconventional autobiography.* New Haven, CT: Yale University Press.

———. 1983. *Fossils and the history of life.* New York: Scientific American Books.

———. 1984. Introduction: Forty years later. In *Tempo and mode in evolution (facsimile of 1944 edition)*, ed. G. G. Simpson, xiii–xxx. New York: Columbia University Press.

———. 1996. *The dechronization of Sam Magruder: A novel.* Ed. Joan Simpson Burns. New York: St. Martin's.

Simpson, George Gaylord, and Anne Roe. 1939. *Quantitative zoology: Numerical concepts and methods in the study of recent and fossil animals.* 1st ed. New York: McGraw-Hill.

Turner, Bryan S. 1999. *Classical sociology.* London: Sage.

Whittington, H. B. 1986. George Gaylord Simpson. *Biographical Memoirs of Fellows of the Royal Society* 32:527–539.

Wyatt, David. 1993. *Out of the sixties: Storytelling and the Vietnam generation.* Cambridge: Cambridge University Press.

CHAPTER EIGHTEEN

The Consensus That Changed the Paleobiological World

Arnold I. Miller

INTRODUCTION

Throughout his career, Stephen Jay Gould championed the theme of *contingency*, the concept that unanticipated events serve as nick points that dramatically alter the courses of events that take place in their wakes. While large-body impacts and other mechanisms that induce mass extinctions are often cited as prime examples of contingency in the macroevolutionary arena, most such events have likely been far more subtle than the crash of a ten-kilometer comet or asteroid. Many of the cascading, critical events in the lifetime of a global biota or a species or a person or a scientific discipline can be appreciated only in retrospect. I am sitting at my computer typing this essay because the University of Rochester, which I had never heard of previously, sent me a bulk-mailed pamphlet during the summer after my junior year in high school. And several students on my dormitory corridor during my freshman year at Rochester talked up geology as an enjoyable major, with courses taught by cool professors. And my paleontology instructor during my junior year was Dave Raup, who had taken over the course from another professor that very year. And, during the same semester, I took a course on computers and statistical methods from the recently hired Jack Sepkoski, who would later become my PhD mentor at the University of Chicago, but not until after I had been shipped out to Virginia Tech to work with Richard Bambach on a master's degree. And, during my career at Virginia Tech, Bambach helped me to traverse a minefield of contingency that saw me quit school, move to the Caribbean for awhile, and then return to Blacksburg to finish my degree, before finally going to Chicago. It all seems so inevitable in retrospect (and it almost goes with saying that my parents might never have met. . .).

Of course, the intellectual development of paleobiology as a discipline was similarly contingent on chance meetings and other events that could just as easily not have happened, as evidenced by the essays throughout this volume. One such event, which is the focus of this essay, was the 1981 publication of a three-page paper in *Nature* titled: "Phanerozoic Marine Diversity and the Fossil Record" (Sepkoski et al. 1981). While it might be poetic to suggest that this paper initially escaped the notice of most paleobiologists and its importance was only appreciated in retrospect, the truth is that the publication of the paper had the subtlety of a sledge hammer. Given the paper's authorship— J. John Sepkoski, Jr., Richard K. Bambach, David M. Raup, and James W. Valentine— it was bound to attract the immediate attention of a broad audience, particularly because it represented an unexpected meeting of the minds between two of the coauthors, who had spent much of the previous decade in an intellectual conflict related to the central theme of the paper. For this reason, the paper quickly earned two nicknames in the paleobiological community: the "Kiss and Make Up Paper," and, more famously, the "Consensus Paper."

The Consensus Paper contained but a single figure (fig. 18.1), which compares five, independently compiled (but not quite independent) histograms depicting the trajectory of global marine diversity throughout the Phanerozoic. All five histograms show a similar pattern of initial increase from the Cambrian into the Ordovician, followed by fluctuations around a Paleozoic plateau, a decline into the early Mesozoic associated with a major mass extinction, and then a continuous increase through the Mesozoic and Cenozoic eras. Given the striking similarities among the five histograms, the authors offered two major conclusions. First, they noted that the marine fossil record exhibits a strong underlying signal that transcends the individual peculiarities of the five illustrated datasets. In itself, this was an important conclusion, given the open resistance to taxon counting among many paleontologists at that time, who argued that taxonomic treatments of paleontological specimens remained too uncertain to permit the kind of large-scale compilations at the heart of the Consensus Paper. Clearly, the broad Phanerozoic signal transcended this messiness. Second, the authors argued that, given the variety of methods used to compile the five curves, *the signal was biologically meaningful.* That is, the pattern could be accepted more or less at face value as reflecting, in particular, major episodes of global diversification in the Early Paleozoic and in the Mesozoic and Cenozoic. Importantly, this face-value reading of the pattern also supports the interpretation that Cenozoic diversity far exceeded Paleozoic

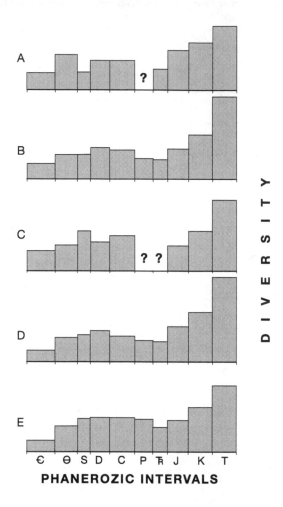

Figure 18.1 Comparison of five different compilations of marine taxonomic diversity through the Phanerozoic (figure 1 of the Consensus Paper; Sepkoski et al., 1981). Because the emphasis here was on the trajectories, rather than the absolute numbers per se, all five histograms were scaled to the same average height. Panel A: median richness of trace fossil assemblages, using data acquired primarily from Seilacher (1974, 1977). Panel B: a slightly modified version of Raup's (1976b) depiction of species diversity. Panel C: median species richness of communities in "open marine environments," from Bambach (1977). Panel D: genus diversity, using data from Raup (1978). Panel E: family diversity, using data compiled by Sepkoski that was also used in the more highly resolved curve that Sepkoski published in the same year Sepkoski (1981).

levels. As will become clear later, it was the second conclusion that was particularly surprising at the time, given the authorship of the paper.

I bothered to summarize my own ontogeny in the opening paragraph not simply to convey my personal set of cascading contingencies, but also to make the point that, as a student, I had the good fortune to spend a fair amount of time with three of the four coauthors of the Consensus Paper. At the time of its publication, I was a first-year PhD student at Chicago, and I cannot claim to have been more than casually aware that the paper was in the works. But, once it was published, I obviously took notice. In the years immediately following, I became fascinated not only with its effects on the science of paleobiology, but also with the thought that there just *had* to have been no small amount of behind-the-scenes intrigue in the forging of the consensus.

This is the story of the run-up to the publication of the Consensus Paper, which actually had a gestation of some four years. But to understand its significance, it is important to first consider the methodological and intellectual foundations involved in mining the paleontological record as a meaningful source of data for investigating the history of biodiversity. Therefore, this essay begins with a brief review of these foundations. Next, I turn to the decade preceding the publication of the Consensus Paper, when two of the authors advocated dramatically different interpretations of Phanerozoic diversification, when another author developed a novel way of investigating the history of diversity that appeared to circumvent the issues being debated by the other two, and when still another author, who would ultimately play *the* crucial role in bringing the consensus to fruition, was first cutting his teeth in the assembly and analysis of global-scale paleontological databases. Then, I discuss the assembly of the consensus during the early days of the next decade, a consensus that first included two authors, then three, then four, and then very nearly, but not quite, five. Finally, I conclude by considering briefly the lasting impact of the Consensus Paper, and its pivotal role in shaping an agenda for paleobiology.

BEGINNINGS

Because soft tissues tend to decay rapidly upon the deaths of organisms, neither they nor their chemical residues are commonly preserved in the fossil record. Furthermore, the preservation of hard parts is contingent on rapid burial by sediments, which is far more likely to occur underwater, where sediments tend to be deposited, than on land, where sediments tend to be eroded. For both of these reasons, the fossil record is biased in favor of the preservation of aquatic—in particular, marine—organisms that possess hard parts. It stands to reason, therefore, that in assessing the history of biodiversity, pale-ontologists focused initially on well-skeletonized marine groups.

Paleontologists have long been documenting the occurrences of taxa pre-served in the fossil record. Although there remain some worries that the *pub-lished* fossil record is biased in favor of data from developed nations in Europe and North America, there is a strong sense that, after more than two centu-ries of intensive investigation, paleontologists have collected a representative cross-section of the taxa actually preserved in the record. In the primary lit-erature, paleontologists have made a habit of noting the localities and strati-graphic intervals in which taxa occur. This information can be assembled to-gether from multiple intervals and localities into databases that record the first (oldest) and last (youngest) known occurrences of a set of taxa in a region of interest or throughout the entire world. Of course, the construction of a meaningful database depends at its heart on the ability of paleontologists to determine the relative ages of strata that are sometimes arrayed among a broad set of localities, but this has been accomplished for much of the geological record using basic principles of biostratigraphy. These data can be used, in turn, to construct graphs that depict changes through time in the number of extant taxa, based on the entirely reasonable assumption that each taxon was extant for the entire interval between its first and last know occurrences in the fossil record.

In 1860, John Phillips used species-level data gleaned from John Mor-ris's (1854) *A Catalogue of British Fossils* to construct an illustration of the Phanerozoic marine diversity trajectory. Although Phillips's data were lim-ited to Great Britain, his curve (figure 4 of Phillips) looks broadly similar to global depictions published more than a century later, and Phillips is there-fore widely credited with having published the first meaningful depiction of Phanerozoic diversification. What is not as widely appreciated is that Phillips also confronted a bias that strikes at the central issue debated in the run-up to the Consensus Paper. Phillips suggested that, all else being equal, if one strati-graphic interval contains a significantly greater thickness or volume of strata than another, it might well contain a large number of fossils and, therefore, greater taxonomic richness. Because of this potential bias, Phillips recognized that interval-to-interval changes in taxonomic richness depicted on a diversity curve might reflect changes in the size of the available sample, rather than a meaningful biological pattern. In the construction of his curve, therefore, Phil-lips recast diversity as the number of species per unit thickness of the interval in question. Had Phillips not applied this correction, the broad Phanerozoic trajectory for Great Britain might well have exhibited peak diversity in the

Paleozoic Era, rather than the Cenozoic, because of the disproportionate representation of Paleozoic rocks in the British stratigraphic record.

As paleontology moved into the twentieth century, there was a growing effort to produce encyclopedic compilations of paleontological information pertaining in particular to systematics, which paleontologists could consult and expand upon as they collected and catalogued fossil organisms worldwide. This process continues even today, and new and revised parts of major paleontological references, such as the *Treatise on Invertebrate Paleontology*, are almost continuously in production. Because these references contain information on the global stratigraphic occurrences of the taxa catalogued therein, usually down to the genus level, paleontologists came to recognize that they could also serve directly as sources of data for studies of Phanerozoic global biodiversity. Among the researchers to take advantage of these data was Jim Valentine, who published a series of global diversity curves at the phylum-through-species levels for an aggregate set of well-skeletonized groups (Valentine 1969, 1970). Valentine recognized a disconnect among Phanerozoic trajectories exhibited in compilations at different taxonomic levels. Perhaps not surprisingly, at least in retrospect, the phylum-level trajectory was nearly flat after the Cambrian, reflecting the early origination and persistence of the vast majority of well-preserved marine phyla. As Valentine moved down the taxonomic hierarchy to the family level, he began to observe the more familiar pattern first illustrated by Phillips, characterized by a significant, nearly continuous increase during the post-Paleozoic. Importantly, however, Valentine's curves were fully global in scope.

Valentine's compilations above the species level were based primarily on data from the *Treatise* and, therefore, directly summarize what was known about the occurrences of fossil taxa at the time that graphs were assembled. No such compilation was possible at the species level because of difficulties developing consistent methods for delineating fossil species, and perhaps also because of the daunting number of fossil species that have been recognized by paleontologists over the years. The *Treatise* does not catalogue the stratigraphic occurrences of species, and neither do other global-scale taxonomic compendia, which have generally been limited to the genus level and above.

If a species-level Phanerozoic trajectory was to be compiled at all, it would have to be based on estimates of what the published fossil record *would* exhibit if species-level data were mined adequately enough to produce a direct assessment. Valentine (1970) presented one such trajectory, based on estimates derived from a combination of three sources: the number of present-day species observed among the nine marine phyla used in several of his compilations at higher taxonomic levels (Valentine 1969), ratios of the number of taxa empirically observed at different taxonomic levels, and changes inferred based on the changing positions of paleocontinents in the number of marine faunal provinces through the Phanerozoic. Strikingly, Valentine's estimated species-level trajectory exhibits an increase of about one order of magnitude during the past 100 million years (i.e., since the mid-Cretaceous; figure 18.2A).

GLOBAL DIVERSITY AND THE SEDIMENTARY RECORD: ENTER DAVE RAUP

As the 1970s unfolded, Dave Raup took up the issue of Phanerozoic diversity, focusing in particular on the veracity of Valentine's estimates and the biological significance—or not—of the Phanerozoic trajectory. At first, Raup (1972) considered the set of family-and genus-level trajectories presented previously by Valentine (1969) and other authors in relation to the estimated Phanerozoic trajectory of global sedimentary rock volume. Like Phillips more than a century earlier, Raup was concerned about the possible biasing effects of secular trends in the availability of fossils, and he drew attention to the clear parallels between the diversity and rock volume trajectories. In particular, both trajectories exhibited a comparable increase through the Mesozoic and Cenozoic eras, raising the possibility that the diversity increase was a preservational mirage: if there are more sedimentary rocks we should expect to find more fossils and, all else being equal, if there are more fossils, there should be more fossil taxa.

Four years later, Raup (1976a, b) continued on this theme, but this time with a twist. In addition to presenting updated estimates of global sedimentary rock volume (figure 18.2B) and outcrop area, Raup also compiled a new estimate of Phanerozoic species diversification. Raup's method for doing so was decidedly different than Valentine's, in that his data came from sampling a well-known bibliographic reference to the systematic literature, the *Zoological Record*, for citations to new fossil species belonging to a large spectrum of higher taxa. By compiling data in this way, Raup was looking to provide a more direct estimate of species-level patterns than that available through Valentine's method. Raup's estimate exhibited two important features (figure 18.2C). First, the Mesozoic-Cenozoic increase was significantly more muted than Valentine's estimate, exhibiting a three-fold increase rather than an order-of-magnitude increase. Second, as in his earlier analyses, Raup continued to observe that the dramatic Cenozoic increase was accompanied by an equally dramatic increase in sedimentary rock volume. For this reason, Raup (1976a) concluded that "there is no compelling evidence for a general increase in the number of invertebrate species from Paleozoic to Recent."

OVERCOMING ROCK VOLUME: RICHARD BAMBACH

By the late 1970s, the field of paleobiology was therefore at an impasse over the question of the long-term Phanerozoic diversity trajectory. While there was growing agreement that the fossil record preserved evidence of a significant diversity increase through the Mesozoic and Cenozoic, the extent of the species-level increase remained contentitious and, more importantly, the biological significance of the increase was being openly questioned because of the rock-volume problem.

As an outgrowth of his interests in the analysis of paleocommunities, Richard Bambach proposed a possible way out of the rock-volume conundrum. Bambach reasoned that changes in global diversity should also be observable at the level of individual communities, given that global patterns are built, in aggregate, of those observed in local collections. While the biological relationship between diversity at different hierarchical levels might not be quite as straightforward as Bambach envisioned, his inference was nevertheless testable, and he set out to assess community-level diversity trends among marine invertebrates throughout the Phanerozoic. Bambach's database included species-level faunal lists for individual paleocommunities compiled from the literature for North America and England and, from these, he assessed period-by-period changes in the median value of paleocommunity species richness. His use of medians, as opposed, say, to the aggregate sum of species contained within a stratigraphic interval, was particularly important, because it appeared to overcome the rock-volume problem. While one might expect to find more fossil assemblages, and therefore more species in aggregate, preserved in younger strata, there is no reason to expect a priori that the richness of the *individual* samples contributing to the calculation of a median should also be inflated in these strata (but see the Aftermath).

In analyzing the large subset of his data that represented paleocommunities from open-marine environments, Bambach (1977) found dramatic Mesozoic-Cenozoic increases in the median species richness of paleocommunities, paralleling those already observed in aggregate, global compilations

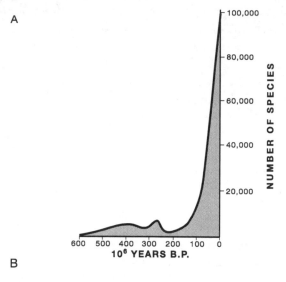

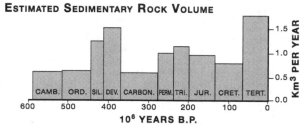

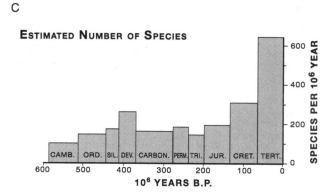

Figure 18.2 Two competing depictions of the Phanerozoic trajectory of marine species diversity, in comparison with sedimentary rock volume. Panel A: Valentine's estimated species trajectory (reprinted from Valentine 1970, figure 2). Panel B: Phanerozoic trends in sedimentary rock volume (reprinted from Raup, 1976a, figure 2). Panel C: Raup's estimated species trajectory (reprinted from Raup, 1976b, figure 2).

for the same interval. This appeared to significantly strengthen the argument that the global signal was biologically meaningful.

During the early 1970s, Jack Sepkoski was a graduate student at Harvard University, where he was recruited by Steve Gould to assemble a database that could be used to assess the diversification of major higher taxa. At the time, Gould was involved in a collaborative project to simulate the diversification of clades using stochastic parameters, and he and his colleagues wanted to know whether the diversity histories of these randomly generated clades bore any resemblance to those observed among actual higher taxa (e.g., Gould et al. 1977). The compilations that Sepkoski undertook were wholly unrelated to his field-based dissertation project on the stratigraphy and paleoecology of Cambrian strata in Montana and adjacent states (Sepkoski 1994; Ruse 1999; Miller 2001), but they quickly captured his imagination, and Sepkoski began to focus increasingly on the question of Phanerozoic diversification.

In 1977, after Sepkoski had joined Raup on the faculty at the University of Rochester, he was in the audience when Richard Bambach presented a colloquium on his within-community research. Bambach's talk, as well as an earlier presentation at the 1976 Geological Society of America meeting in Denver, clearly struck a chord with Sepkoski. In June of 1977, shortly after Bambach's within-community paper was published in *Paleobiology*, Sepkoski wrote a letter to Bambach that was accompanied by two hand-drawn figures. The first was a pair of histograms, compiled at the period level of temporal resolution and drawn in the style of Raup's (1976a) global species-level depiction (figure 18.2C). Sepkoski's new figure compared the Phanerozoic trajectory of global familial diversity, based on his own data compilations, to Bambach's within-community species tabulations; ultimately, these become panels C and E of the Consensus Paper figure (figs. 18.1C and 18.1E, herein). The second figure was a more highly resolved depiction of the global pattern for families that bears a striking resemblance to the overall trajectory in what is arguably the most-reproduced figure in the history of paleontology: Sepkoski's famous rendition of Phanerozoic familial diversity parsed into three evolutionary faunas, published in *Paleobiology* in the same year as the Consensus Paper (figure 5 of Sepkoski 1981).

Given the similarities between the figures that Sepkoski drew up in 1977 and those published in 1981, one cannot help but wonder why there was such a long delay before these or similar figures were submitted to refereed journals. While we can never know for sure, part of the explanation likely relates to Sepkoski's own uncertainty in 1977 about the biological veracity of the pattern. In his letter to Bambach, Sepkoski noted: "I am not yet convinced that the post-Paleozoic rise evident in all our data sets is real. For the familial data, I have discovered it is very easy to simulate nearly identical patterns using very simple (and thus possibly robust) models of sampling. . . .We must get around to doing some rarefaction analyses of Phanerozoic community data!" (Sepkoski letter to Bambach, June 1, 1977).

Rarefaction is a method that can be used for overcoming sampling inhomogeneities from interval to interval, such as that which Sepkoski suspected might be responsible for the Cenozoic diversity increase. Interestingly, rarefaction and related techniques were used much later to assess Phanerozoic diversity trends at both the global and within-community levels (e.g., Miller and Foote 1996; Alroy et al. 2001; Powell and Kowalewski 2002; Bush and Bambach 2004), but Sepkoski himself never followed up on this initial vision, apparently never again conveyed this idea to colleagues or students, and never pursued the development of the kind of database that would have made a global analysis of this kind possible.

It also seems clear that, at least for a time, Sepkoski was strongly influenced by Raup's vision, particularly as they became close colleagues in Rochester and continuing in Chicago, where Raup had also moved, initially to a post at the Field Museum of Natural History, but later to the University. The same issue of *Paleobiology* that contains Raup's species-level compilation and interpretation (Raup 1976a, b) also includes an accompanying paper by Sepkoski (1976), in which he argues that at least a portion of Raup's trajectory is biologically meaningful and relates to species-area effects—a precursor to his series of later papers on logistic modeling of global diversification—but nevertheless accepts the primacy of Raup's interpretation, as evidenced in his introductory paragraph: "These data, as Raup (1976a) has forcefully argued, appear primarily to reflect the vagaries of sampling: the estimated numbers of fossil species in each geologic system correlate strongly with measures of the amount of rock available for study, specifically outcrop area and rock volume" (Sepkoski, 1976, p. 298).

Finally, it must also be kept in mind that the intellectual climate in paleontology during the mid to late 1970s was still fairly conservative. To be sure, there was a growing emphasis on the synthesis and analysis of large paleontological datasets, as evidenced in particular by the birth of *Paleobiology*, which published its first issue in 1975. There remained significant resistance to taxon counting, however. For someone of Sepkoski's relative youth—in June of 1977, he was just short of twenty-nine years old and had only recently completed his PhD—it may be that he did not yet feel secure enough to take a leadership role in this contentious arena.

But by 1980, two years after moving to Chicago, Jack Sepkoski was ready to roll.

SEPKOSKI AND BAMBACH

After a hiatus of some three years following their correspondence of 1977, Sepkoski and Bambach picked up anew on the issue of Phanerozoic diversity during informal discussions in the spring of 1980, while both were attending a conference in Tübingen, Germany. These discussions took place at the home of the German paleontologist Adolf Seilacher, who was hosting the conference (Sepkoski letter to Seilacher, February 10, 1981; Bambach, personal communication 2001). It remains unclear whether Sepkoski or Bambach had determined prior to the Tübingen meeting that the initial comparison of two diversity compilations should be expanded to a five-way comparison. By the time of the meeting, however, the expanded comparison was certainly under active consideration, and one of the five histograms was to be based on the trace fossil record (i.e., preserved trails and tracks of organisms, as opposed to skeletal remains), using data from two publications of Seilacher. One of the more intriguing aspects of the Tübingen living room meeting is that it represented the first of several unsuccessful attempts to convince Dolf Seilacher to become a coauthor of the Consensus Paper (see the following).

In late January 1981, Sepkoski completed a first draft of what was initially a two-author paper, and he sent it out to Bambach for review on February 1. The five-way comparison that Sepkoski and Bambach envisioned was the centerpiece of the paper, and now included, in order (figure 1):

(a) An assessment of trace fossil data from shallow (*"neritic,"* or shelf) and deeper (*"flysch,"* or continental slope and rise) facies averaged together, gleaned primarily from the work of Seilacher (1974, 1977). The style of compilation for the trace fossil assessment was similar to that used by Bambach (1977); taxonomic richness for a given period was based on the median number of taxa per locality.

(b) Raup's (1976b) global species-richness data, with insects removed.

(c) Bambach's (1977) within-community median species richness assessment for open-marine environments.

(d) A modified version of global, genus-level data presented by Raup (1978), which had been collected from the *Treatise on Invertebrate Paleontology.*

(e) The family-level data, used also by Sepkoski in his famous, more highly resolved, Phanerozoic diversity curve (Sepkoski, 1981), and subsequently published as a family-level global compendium (Sepkoski, 1982).

The first draft contained an overview of the relationships among these histograms, including a set of statistical analyses conclusively verifying the obvious visual similarities among them. In most respects, the first draft differed little from the final published version, except for several stylistic modifications, such as the migration of most of the technical discussion about the data to the figure caption. There were, however, a few substantive additions. One was a brief reference to the aforementioned work of Phillips, added at the suggestion of Chicago paleontologist Tom Schopf, who read a draft of the Consensus Paper but certainly was not enamored of it (see the following). More significantly, as we will soon see, two paragraphs were added, one at the suggestion of the fourth author and the other as part of an attempt to sign up a fifth author.

SEPKOSKI, BAMBACH, AND RAUP

Nobody knows for sure how it is that Dave Raup became a coauthor of the Consensus Paper. On February 10, nine days after the two-author draft was sent out to Bambach, copies of the same draft were sent out to Valentine and Seilacher, and it is clear from the accompanying cover letters (Sepkoski letter to Valentine, February 10, 1981; Sepkoski, letter to Seilacher, February 10, 1981) that by that time, Raup had become an author. Because Raup and Sepkoski were close colleagues residing in the same city, it is probable that Sepkoski simply handed a copy of the paper to Raup, and this was likely followed shortly thereafter by conversations between the two; there is, however, no paper trail to confirm this.

While Raup has no recollection of the events leading to his agreement to join the paper (Raup, personal communication 2006), he has offered some broader reflections on his views at the time and the likely reasons *why* he became a coauthor (Raup, personal communication 1996):

I was so delighted to see support for the view that a lot of the increase that Jim Valentine had observed couldn't be substantiated—it was a numerical compromise . . . as I read this second '76 paper [Raup 1976a]— "there is no compelling evidence for the existence of a trend"—that doesn't say that there isn't a trend and, to me, this consensus paper was supportive in that it reduced the increase to something almost down in the noise level of the sample. It was clearly a step in my direction. . . . Also, Jim Valentine had . . . plotted and published the factor of ten [graph]—his certainly represented the very strong conventional wisdom that everybody had—and so the consensus paper—I saw as a severe break with the conventional wisdom—not only Jim Valentine, but everybody else . . . although I happily agreed to be an author on the consensus paper, I don't remember agreeing with it. But it was a lot closer to my position than I thought possible . . . an important element of this is that at no point did I think I had a case for level or declining diversity. The only thing I had a case for was the possibility of level or declining diversity. Therefore, I didn't have an advocacy position to lose." (Miller 2000, 61–62).

Whether the histograms presented in the Consensus Paper reduced the Mesozoic-Cenozoic diversity increase to "something almost down in the noise level of the sample" may be in the eye of the beholder, but most observers have agreed, informally at least, that Dave Raup gave up much more than Jim Valentine when he signed on to the manuscript. Although it is true that, by joining the paper shortly thereafter, Valentine appeared to have at least tacitly accepted that species diversity did not increase by an order of magnitude, there is much more at stake from an evolutionary perspective with respect to the question of whether diversity increased significantly—and many paleontologists would agree that a threefold increase *is* significant—versus the possibility that it did not increase at all, remaining at equilibrium since the Paleozoic Era. That is, there is a much greater theoretical gulf between a zero increase and a threefold increase than there is between a threefold increase and a tenfold increase.

It is also striking that, some twenty-five years after the publication of the Consensus Paper, Raup continued to harbor doubts about its central theme, and these doubts persist even today (Raup, personal communication 2006).

SEPKOSKI, BAMBACH, RAUP, AND VALENTINE

As with Raup, there is no written record of an invitation to Valentine to join the Consensus Paper as an author. In his February 10 letter to Valentine that accompanied a copy of what was by then a three-author manuscript, Sepkoski invited comments from Valentine, "particularly if you feel we have not credited your work sufficiently . . ." (Sepkoski letter to Valentine, February 10, 1981). By the time that Valentine responded, nearly a month later, with comments and suggested revisions, an invitation had been proffered: "If you were actually serious in asking me to join the paper, I would be happy to do so . . ." (Valentine letter to Sepkoski, March 6, 1981). Although Valentine has no recollection of the conversation(s) that resulted in the invitation (Valentine, personal communication 2006), there is little doubt that such a conversation took place during late February or early March.

Valentine's most significant suggestion for revision was the addition of a paragraph emphasizing more explicitly the fact that a remarkably consistent signal emerged from the five data sets, despite their individual inherent biases. A draft of the paragraph, which enumerated the unique peculiarities of each data set, was included in his letter of March 6, and an adapted version of the paragraph was incorporated into the final draft of the paper.

Given that the Consensus Paper includes Raup's estimate of about a three-fold Mesozoic-Cenozoic increase in global species richness, Valentine's willingness to become an author signaled to the profession that he accepted this pattern as being close to reality (Valentine 1970). In reality, however, this was not the case: "as you might guess I didn't (and don't) think three-fold does it, but I expect ten-fold may be too much . . ."(Valentine, personal communication 2006).

SEPKOSKI, BAMBACH, RAUP, AND VALENTINE (BUT NOT SEILACHER)

The published version of the Consensus Paper acknowledges the assistance of two people. The first is Tom Schopf, Sepkoski's colleague at the University of Chicago, who is acknowledged for pointing out Phillips's (1860) work to the authors, and also for providing critical comments on earlier versions of the paper. In fact, Schopf was not happy at all with the paper and sent in a rebuttal to *Nature* (Schopf letter to John Maddox, December 28, 1981). As it happens, after securing a reply from Sepkoski, *Nature* elected to not publish the exchange.

The second acknowledgment went to Dolf Seilacher, also for providing critical comments, but it reflects a more interesting chain of events, because Sepkoski very clearly wanted Seilacher to become a coauthor from the beginning, when the paper was first discussed in Seilacher's living room in Tübingen. Sepkoski did not finally give up on this hope until shortly before the manuscript was submitted to *Nature*.

The main reason to include Seilacher as a coauthor is self-evident: figure 1A in the Consensus Paper was based on his data for trace fossils. Beyond that, there was an additional reason articulated to Seilacher in a letter from Sepkoski: "I view this little paper as a kind of statement of consensus (and the reasons for it) by the compilers of various data sets on diversity as to what they believe is the real pattern of Phanerozoic diversity" (Sepkoski letter to Seilacher, April 8, 1981).

From the very beginning, therefore, Sepkoski understood the political ramifications of the Consensus Paper and was concerned about how it might be perceived if the (passive) contributor of one of the five data sets was not included as an author. Seilacher declined to be an author because he didn't believe that it was appropriate to use his trace fossil data in a paper articulating a broad statement about Phanerozoic marine diversification. In particular, Seilacher was uncomfortable with the averaging together of separate trace fossil data sets from shallow and deep marine settings, noting that the other curves in the Consensus Paper, which were based on body fossils, primarily record biodiversity trends for shallow marine settings: "By lumping the two groups you may get a curve that pleases you, but this might be an accident. The curve you have in mind [based on the record of body fossils] is mainly one of shallow marine diversity. Including the flysch [i.e., the deepwater] counts (which by their high diversity influence the results very much), I am afraid will do no justice to the cause, although the result may seem to fit the general picture" (Seilacher letter to Sepkoski, February 23, 1981).

Given Seilacher's concerns, Sepkoski added several sentences to the manuscript, noting explicitly that, indeed, the trace fossil data set combined information from two different milieus and that the data were therefore more heterogeneous than those used for the body-fossil histograms. But Seilacher remained unmoved, because he felt that the resulting average was primarily a record of the increase in the deep sea, given that the shallow-water data, by themselves, did not exhibit a significant increase. In the end, he even suggested that the paper might be stronger if his data were not included: "Wouldn't it make your conclusions stronger if you leave out the trace fossil data altogether?" (Seilacher letter to Sepkoski, April 28, 1981).

About a month later, the four-author Consensus Paper was submitted to *Nature*; it was accepted in August and published in October with a few revisions, including a change to the title and the addition of a new abstract.

AFTERMATH

To many professionals and graduate students engaged in paleobiological research at the time, there is little doubt that the Consensus Paper was viewed as a big deal, and it was the subject of lively discussion in numerous venues. But what of its scientific impact? While there may be some danger of overstating the paper's importance, it can fairly be noted that, in the immediate aftermath of the Consensus Paper, there was an explosion of research emphasizing the macroevolutionary and paleoecological processes responsible for the major features of Phanerozoic diversification and extinction (e.g., Valentine 1985). The assumption implicit in all of this research was that the patterns that researchers were trying so hard to explain were, after all, *real*, and it was the Consensus Paper that validated this view. The publication in the same year of Sepkoski's highly resolved depiction of Phanerozoic family-level diversity trends (Sepkoski, 1981) was clearly a major one-two punch.

Given the reluctance, in particular, of two of the Consensus Paper's co-authors to fully embrace it, it can also be asked whether the main conclusions of the Consensus Paper are *right*. A quarter century after its publication, a verdict has yet to be reached. Few people would seriously doubt the first conclusion—that the marine fossil record preserves a pattern of significant diversity in the Early Paleozoic, followed by a much larger increase through most of the Mesozoic and Cenozoic at the family level and below. The second conclusion, that the Mesozoic-Cenozoic increase can be accepted at face value as a meaningful biological signal, has been more problematic. A full airing of this topic is beyond the scope of this essay, but it can be said that the biological significance of the Cenozoic increase remains an active subject of discussion and debate among paleontologists. For example, in the view of many paleontologists, Bambach's and (perhaps ironically) Seilacher's data are the linchpins of the Consensus Paper, because their reliance on medians, rather than aggregate sums, overcomes the rock-volume issue, as discussed earlier. That said, those data are not necessarily free of secular biases. Through upper Mesozoic and Cenozoic strata, there is an increase in the percentage of the record preserved in unlithified (i.e., soft), as opposed to lithified, sediments. It has long been suspected that this enhances the extraction and identification of fossils, which, if true, could artificially inflate the number of taxa sampled in an unlithified sample. Several researchers are currently conducting analyses aimed at calibrating the extent of this possible bias.

Regardless of how the story eventually ends, there can be little doubt that, by giving license to a large portion of the macroevolutionary agenda in which the science has been engaged over the last twenty-five years, the Consensus Paper was instrumental in bringing paleobiology to the High Table.

ACKNOWLEDGMENTS

I thank Dave Raup, Richard Bambach, and Jim Valentine for graciously responding to my questions about their recollections and views concerning the Consensus Paper. I also thank Bambach for bringing to my attention the letter and figures that Jack Sepkoski sent to him in June of 1977. Finally, I am grateful to Michael Foote for critically reviewing a draft of this essay.

REFERENCES

Alroy, J., C. R. Marshall, R. K. Bambach, K. Bezusko, M. Foote, F. T. Fürsich, and T. A. Hansen. 2001. Effects of sampling standardization on estimates of Phanerozoic marine diversification. *Proceedings of the National Academy of Sciences* 98:6261–66.

Bambach, R. K. 1977. Species richness in marine benthic habitats through the Phanerozoic. *Paleobiology* 3:152–67.

———. 1985. Classes and adaptive variety: The ecology of diversification in marine faunas through the Phanerozoic. In *Phanerozoic diversity patterns: Profiles in macroevolution,* ed. J. W. Valentine, 191–253. Princeton, NJ: Princeton University Press.

———. 1993. Seafood through time: Changes in biomass, energetics, and productivity in the marine ecosystem. *Paleobiology* 19:372–97.

Bush, A. M., and R. K. Bambach. 2004. Did alpha diversity increase during the Phanerozoic? Lifting the veils of taphonomic, latitudinal, and environmental biases. *Journal of Geology* 112:625–42.

Gould, S. J., D. M. Raup, J. J. Sepkoski, Jr., T. J. M. Schopf, and D. S. Simberloff. 1977. The shape of evolution: A comparison of real and random clades. *Paleobiology* 3:23–40.

Miller, A. I. 2000. Conversations about Phanerozoic global diversity. In *Deep time: Paleobiology's perspective* (special 25th anniversary issue of *Paleobiology*), ed. D. H. Erwin and S. L. Wing, 53–73.

———. 2001. J. J. Sepkoski, Jr. *American National Biography Online:* http://www.anb.org/articles/13/13-02634.htm.

Miller, A. I., and M. Foote. 1996. Calibrating the Ordovician radiation of marine life: Implications for Phanerozoic diversity trends. *Paleobiology* 22:304–9.

Morris, J. 1854. *A catalogue of British fossils.* London: Author.

Phillips, J. 1860. *Life on the Earth.* Cambridge: Macmillan.

Powell, M. G., and M. Kowalewski. 2002. Increase in evenness and sampled alpha diversity through the Phanerozoic: Comparison of early Paleozoic and Cenozoic marine fossil assemblages. *Geology* 30:331–33.

Raup, D. M. 1972. Taxonomic diversity during the Phanerozoic. *Science* 177:1065–71.

———. 1976a. Species diversity in the Phanerozoic: An interpretation. *Paleobiology* 2:289–97.

———. 1976b. Species diversity in the Phanerozoic: A tabulation. *Paleobiology* 2:279–88.

———. 1978. Cohort analysis of generic survivorship. *Paleobiology* 4:1–15.

Ruse, M. 1999. *Mystery of mysteries: Is evolution a social construction?* Cambridge, MA: Harvard University Press.

Seilacher, A. 1974. Fossil-Vergesellschaftungen 20: Flysch trace fossils; evolution of behavioural diversity in the deep sea. *Neues Jarbuch für Geologie und Paläontologie. Monatshefte* 4:233–45.

———. 1977. Evolution of trace fossil communities. In *Patterns of evolution, as illustrated by the fossil record,* ed. A. Hallam, 359–76. Amsterdam: Elsevier.

Sepkoski, J. J., Jr. 1976. Species diversity in the Phanerozoic: Species-area effects. *Paleobiology* 2:298–303.

———. 1981. A factor analytic description of the Phanerozoic marine fossil record. *Paleobiology* 7:36–53.

———. 1982. *A compendium of fossil marine families.* Milwaukee, WI: Milwaukee Public Museum.

———. 1994. What I did with my research career: Or how research on biodiversity yielded data on extinction. In *The mass extinction debates: How science works in a crisis,* ed. W. Glen, 132–44. Stanford, CA: Stanford University Press.

Sepkoski, J. J., Jr., R. K. Bambach, D. M. Raup, and J. W. Valentine. 1981. Phanerozoic marine diversity and the fossil record. *Nature* 293:435–37.

Valentine, J. W. 1969. Patterns of taxonomic and ecological structure of the shelf benthos during Phanerozoic time. *Palaeontology* 12:684–709.

———. 1970. How many marine invertebrate fossil species? A new approximation. *Journal of Paleontology* 44:410–15.

———, (ed.). 1985. *Phanerozoic diversity trends: Profiles in macroevolution.* Princeton, NJ: Princeton University Press and AAAS, Pacific Division, and San Francisco.

* III *

Reflections on Recent Paleobiology

CHAPTER NINETEEN

The Infusion of Biology into Paleontological Research

James W. Valentine

One of the few privileges of age, with the indulgence of the editors, is that these observations can be presented in part as a memoir, as my professional life happens to overlap the rise of paleobiology. I did my graduate training in paleontology during the 1950s, and so was in graduate school (UCLA) when DNA was first identified as genetic material and its structure linked to Mendelian inheritance, and also when evidence supporting the notion of continental drift was accumulating. Thus the foundations for our present understanding of basic features of organic evolutionary change on a molecular level, and of the basic framework of environmental evolution on a global level, were being established just as I was preparing to enter the life of an academic paleontologist. Timing is everything, and interesting times are clearly the best of all—despite the putative Chinese curse.

Many important events during the rise of paleobiology are covered in detail in other chapters; I shall therefore chiefly deal with two areas in which I have worked most and am thus least likely to grievously err, avoiding such obviously important fields as analytical and functional morphology and the study of extinctions, and treating American contributions, while acknowledging that scientists elsewhere have played significant roles as well. One area concerns early attempts to fashion paleobiological hypotheses based on fossil associational and distributional data, chiefly during the 1960s and early 1970s in the United States. I shall argue that these and other studies eventually fueled a subsequent movement of paleobiological findings into biology. The other area concerns the present and hopefully future movement of molecular techniques and findings into paleobiology, and the advantages of cooperative research on mutual problems.

PALEOECOLOGY AND BIOGEOGRAPHY
AND THE REGULATION OF DIVERSITY

At UCLA I worked on assemblages of marine Pleistocene invertebrates preserved in terraces along the coast of California and Baja California, attempting to understand their paleoecological and biogeographic features, and thus describing their community associations and their provincial settings (Valentine 1961). As I was writing my dissertation, the large two-volume *Treatise on Marine Ecology and Paleoecology* was published (Hedgpeth 1957; Ladd 1957). The ecology volume was full of wonderful review chapters covering just the topics highly relevant to a would-be marine paleoecologist. But the paleoecology volume was quite a disappointment. With a few notable exceptions the papers were concerned with what amounted to stratigraphic and other geological problems involving the paleogeography and depositional environments of the rocks, not the biogeography or ecology of the organisms per se. The authors were by and large fine paleontologists, but their research programs were chiefly devoted to systematics and biostratigraphy in aid of geological interpretations, and it showed.

Nevertheless, the treatise provided a most important stimulus for paleoecology and therefore a pathway for the movement of biological concepts into paleontology. During the early 1960s, several texts were published on paleoecological topics, including translations from the Russian of Hecker's (1957) *Introduction to Paleoecology* (into French, 1960; into English, 1965), and Ager's *Principles of Paleoecology* (1963), and a symposium volume also appeared (Imbrie and Newell 1964). These contributions certainly included methods of interpreting the depositional environments and histories associated with the fossils, but also increasingly applied biological ideas to understanding the ancient ecologies, biogeographies, and diversities of the fossils themselves. As for the journals, paleobiological contributions were beginning to appear in the regular paleontology journals of the time; then, in 1965, Elsevier began publishing *Palaeogeograpy, Palaeoclimatology,* and *Palaeoecology,* and in 1968 *Lethaia* made its debut, with a lead paper on Silurian benthic communities by Fred Ziegler, Robin Cocks, and Richard Bambach (1968).

There were several attempts to summarize Phanerozoic diversity during the 1960s, including a terrific paper by Newell (1967), the well-documented but confusingly analysed summary of the fossil record edited by Harland et al.

(1967), and my own review (1969) of marine taxa, chiefly from data in the *Treatise on Invertebrate Paleontology* but also from other sources. Newell's reconstruction used all animal families, binned by Series, and showed a generally increasing diversity trend in the Lower Paleozoic, a mid- to late-Paleozoic plateau, a shallow dip across the Permian-Triassic boundary, and then a rise to about twice the late Paleozoic level, a general pattern that has held through subsequent empirical studies. The Harland et al. data were binned by stages, and invertebrate families showed similar features and had a sharp end-Paleozoic dip. Newell interpreted the post-Paleozoic increase partly to "ecological feedback"—that is, newly evolved organisms provided potential niches for additional originations—an interesting explanation, but not one that could account for both the Paleozoic plateau and the post-Paleozoic rise. My diversity curves were analyzed in Series bins for all major Linnean taxonomic categories of nine phyla with durable skeletons. Marine invertebrate families showed a generally similar pattern to those of Newell and Harland et al., but with more volatility during the Paleozoic plateau and with a deeper end-Paleozoic dip. By comparing the curves of the different taxonomic categories it was possible to show that each category had a separate history, but that the major break among categories was between the ordinal and family levels, and that there were trends among categories that permitted extrapolation to produce an hypothesis of Phanerozoic species diversity, even though the fossil record was not complete enough (or well known enough) at the species level to support a direct empirical evaluation. The trend suggested that species diversity through the Cenozoic rose even more rapidly, and to a far greater extent, than family diversity. These trends invited evolutionary and, in particular, ecological explanations.

I am hardly an expert judge of the progress in ecology during this interval, but it is clear that it was blooming and that a great deal of the work involved both theoretical and data-rich studies that could be applied to paleontology, especially to fossil marine invertebrates. Much of that work dealt with the distribution, abundance, and diversity of species and of biotic associations, and was nicely summarized in monographs and small books such as Slobodkin (1962), Klopfer (1962), MacArthur and Wilson (1963), Levins (1968), Margalef (1968), and many others, much of it stimulated by the work of G. Evelyn Hutchinson (and building, of course, on a rich legacy of earlier studies). These and similar contributions provided conceptual bases for framing the sorts of ecological and biogeographic questions that paleontologists could ask of their fossil materials. And it was not only the ideas themselves, but the frame of mind that they represented—a certain bent toward abstraction and modeling, a willingness to expand on observations and to propose hypotheses that, while seeming plausible, made rather sweeping claims—that was just the attitude paleontologists needed when dealing with 600 million years of faunal evolution set in an environment undergoing incessant change. Or so I thought. After all, data on the geological ranges, assemblages, and geographical distributions of fossil marine invertebrates had been piling up for a century and a half; it was clearly time to examine this trove for its biological implications, using the approaches being successfully employed in those sorts of biological studies of the living biota, suitably modified for application to the fossils.

It was also during this period that plate tectonics became established, and the history of the Earth was radically rewritten. As it happened I'd served under Harry Hess, one of the principal architects of plate tectonics, in the U. S. Navy during World War II (he was executive officer and eventually captain, while I was an enlisted man just out of high school; I can't say that I got to know him personally). Although I did know that he was some sort of earth scientist, having taken depth soundings for him as we crossed various important deep seafloor features (of which I was totally ignorant—and incidentally this sometimes involved steering a course somewhat out of our way), it was startling to find later that he was a leading scientist and had pioneered the notion of sea-floor spreading in the early 1960s. At any rate, the appearance of a global theory that permitted reconstruction of paleogeographies that turned out to be strikingly unlike today's opened a window on past environmental conditions that has still not been fully exploited by paleontologists. Continents and continental fragments had moved across latitudes and hemispheres, sometimes to form supercontinents that then broke up into different configurations and dispersed. Not only were continents drifting into new climatic regions, they must surely have altered the zonal climates there, created new oceanic circulation patterns, and enabled new global patterns of productivity. The environmental mosaic was in fact more like a kaleidoscope of changing environmental conditions, to which there had to have been important biotic responses. Truly the past is another country.

As these advances unfolded I was attempting to integrate the findings on diversity levels with the ecological theorizing of the time (as in Valentine 1969). One idea, borrowed from some niche theory, was that modal niche sizes might shrink or come to overlap at times, permitting a denser packing of species and thus a higher diversity, and that average secular change over Phanerozoic time was in the direction of shrinking niches and higher diversities, absent unusual extinction episodes. Such changes were visualized as occurring in ecospace, a sort of hypervolume whose dimensions were ecological variables, and in which the modes of life of taxa (at various levels) or, alternatively, ecological dimensions of populations, communities, and other ecological units, could in principle be mapped. A second notion was that diversity levels were partly reflecting the biogeographic consequences of climate change. This was especially applied to the history of polar cooling assumed for the Cenozoic, as suggested by early assessments of oceanic paleoclimates. Those estimates were partly based on the earlier appearance of tropical lineages, including many genera, in higher paleolatitudes than they now inhabit. Therefore the opening up of new, cooler high-latitude climate regimes, the thinking went, tended to shrink the latitudinal ranges of the tropical forms, as could be documented, and permitted the rise of new faunal provinces along all continental coasts in both polar hemispheres. And a third feature was that coastlines had been multiplied by continental drift, particularly by the Mesozoic opening and Cenozoic widening of the Atlantic. As each new province would support endemic elements, a global rise in levels of taxonomic diversity would ensue. Thus there could be a global paleobiology. In the early 1970s, Eldredge Moores and I explored further the opportunity represented by the new understanding of past geographies to try to tie Phanerozoic diversity patterns explicitly into the patterns of continental dispersal indicated by global tectonic studies (e.g., Valentine and Moores 1970).

These ideas were expanded in an attempt to evaluate the changing Phanerozoic biosphere in a book, *Evolutionary Paleoecology of the Marine Biosphere* (Valentine 1973). Hypotheses of the effects of secular environmental change were presented for major levels of the ecological hierarchy, from individuals, populations, communities, and provinces to the biosphere, drawing on what seemed to be the more plausible and useful body of ecological theory of the time, and from the then burgeoning but rather descriptive paleoecological literature. The book was quasi-theoretical, using the notion of ecospace and the concepts of density-dependence and density-independence (see Valentine 1972b), hoping to help clarify the sorts of factors that could or could not account for patterns of diversity in space or time. The interpretations in the book have met various fates: some are clearly incorrect, others are likely to be correct, and an amazingly large fraction of them are still under heavy study, remaining quite contentious (see the following). I believe that, if that book has a legacy, it will be simply that it represented a deliberate and concerted attempt to bring a wide range of biological theory into paleontology and to view the fossil record in a global framework—or at least a global context. This was, of course, hardly an individual effort, but was quite in keeping with the goals of a growing community of workers at the time. The book drew on many of these paleobiological contributions—for example, the early comparative community studies of Bretsky (1969) and of Walker and Laporte (1970), and of biodiversity gradients by Stehli, Douglas, and Newell (1969).

As scientists, paleontologists surely have the burden of explaining life's history: this involves establishing the facts about the life of the past, understanding the dynamics associated with temporal changes of past life that thus comprise its history, and explaining the reasons that those changes occurred. This last burden involves biological theory, and since achieving a dynamic picture is hard enough, it is all too easy to stop short of theory, especially as there is a virtual certainty that much of it will prove to be incorrect. I was president elect of the Paleontological Society in 1973, and was understandably enthusiastic about the future of the incorporation of biology into paleontology, and vice versa. Many interesting paleobiological contributions of the time were being scattered through the literature, in biological and general science publications, for there was no specific outlet for them in the United States. So I proposed that the Society create a new journal, to be called *Paleobiology*, as a place where both paleontologists and neontologists could publish on topics of mutual interest, and that would serve to focus the field and establish themes that were accessible to paleontological research and would draw the attention of students. The then president, Porter Kier, gave me a chance to present this proposal to the Council. The idea drew some sharp opposition but was supported by enough council members, including Porter, that it was eventually put to a vote of the general membership and passed the following year. Even before final passage, evolutionary paleobiologist Tom Schopf was informally chosen as coeditor (with paleoecologist Ralph Johnson to help and to keep him under some sort of control) and was using his enormous drive to get the journal off the ground and to ensure its success. *Paleobiology* was a success, with the one reservation that neontologists have not published there as much as hoped, though, judging from citations, they have tended to read it. Taxonomic richness trends were duly discussed in those pages (e.g., Bambach 1977, followed up elsewhere in 1983, who first documented and evaluated the expansion of benthic ecospace as new life modes were evolved, adding a significant biological dimension to diversity increases), and much else paleobiological. It is indeed a pleasure to find today any number of biological theories vying vigorously in the literature of paleontology, and of paleobiological hypotheses being tested in the living biota.

Beginning in the late 1970s, Jack Sepkoski produced a series of marvelous studies (e.g., Sepkoski 1981), based on statistical approaches to evaluating stage-level bins of data from the treatise and subsequent publications, which established a family diversity curve that proved to roughly parallel other biodiversity measures and thus was to be celebrated as the consensus curve (Sepkoski et al. 1981) and which continues to provoke and inspire much research (see Miller, this volume, and Benton, this volume). I have been told (though not by Jack) that at least one of the reasons that Jack had taken up Phanerozoic diversity problems was that he thought my curves (and presumably the previous ones) were likely to be incorrect. Dave Raup, a major figure in the rise of paleobiology, had noted that the fossil record contained numbers of biases, such as variations in the amount of preserved and exposed marine sediment through time, and had been subjected to variable collection intensities and entailed other sources of error (Raup 1972; rejoined in Valentine 1972). Raup therefore suggested that temporal trends in species diversity, especially the post-Paleozoic rise, might be artifacts of fossil preservation (see also Raup 1976), and Jack proceeded to investigate this notion, with a massive effort to more accurately compile the geologic ranges of fossil taxa, and to develop of a set of models that borrowed partly from the ecological theory of the day. It appears to me now, from the quarter century or so of subsequent and very fruitful work by many investigators, that although the record is certainly biased and that some aspects of the diversity curves are affected, marine species diversity has nevertheless truly increased dramatically since the Paleozoic, especially through the Cenozoic. It is in fact possible that the incompleteness of the fossil record has tended to bias our data toward a smoothing of a significant volatility in species diversity patterns, and we are underestimating the highs and lows. These questions remain under vigorous study.

PALEOBIOLOGY AND MODES OF SPECIATION

Genetic change was understood to be a major component of the engine of evolution, and with the forging of the modern synthesis some microevolutionists lay claim to be the main arbiters of the evolutionary process. To the extent that this claim was true, paleontologists would essentially be reduced to figuring out whether the evolutionary potentials and constraints revealed by population genetics were compatible with fossil evidence—whether any particular constraint, for example, had indeed played a significant role in life's history—because statistical evidence concerning the transmission of traits, so important to population genetics, was essentially impossible to recover from fossils in any regular fashion.

However, much of the tempo and many of the modes of speciation can be indicated by morphological patterns preserved in the fossil record, and the wonderful paper on punctuated equilibrium by Eldredge and Gould (1972) exploited this circumstance. This paper gave rise to an incredibly rich literature, partly on the significance of microevolutionary change in secular evolutionary directions (see chapters by Hallam, Princehouse, and Sepkoski, this volume). A critical observation in that paper is that species sorting can override microevolution, being on a higher level in a hierarchy of evolutionary processes (in this case a macrocvolutionary level). The disposition of the products of microevolution depends on the differential fate of the lineages that are produced, a principle that arises from hierarchy theory and one that is true even if speciation were entirely gradual and changes were gradational. This point got everyone's attention, led to a vigorous dialogue between neontologists and paleontologists, and stimulated a lot of paleontological research. It also made comparative diversity curves among clades and functional types, depending as they do on origination and extinction dynamics, a more interesting topic to neontologists (and see Stanley 1979). There is no question but that the Eldredge and Gould paper was an outstanding success in bringing paleobiology to the attention of the neontological community, from among whom there was strong criticism as well as support, forcing attention on the fossil record.

PALEOBIOLOGY AND THE EVOLUTION OF MAJOR MORPHOLOGICAL NOVELTIES

Most closely related living species display differences in morphological features explicable by microevolutionary processes that have long been amenable to experimental approaches involving the tracing of the transmission patterns of genetic traits. However, the genetic origins of the disparate body architectures of higher taxa—the sorts of morphologies that paleontologists routinely confront—could not be explored extensively until the genes involved in the development of these major novelties could be studied directly, thanks to the rise of molecular biology.

I was at UC Davis from the mid-1960s through the mid-1970s, associated with some outstanding evolutionary population geneticists (such as Theodosius Dobzhansky and Francisco Ayala). Both Doby and Francisco were quite open minded about the explanatory reach of population genetics, even though they themselves worked at the species level and below. I couldn't understand how their models would provide the genetic explanation for the pattern of appearance of novelties above the species level in the fossil record. One of the features of the record that bothered me was the early and abrupt appearance of high-level taxa , and equally bothersome, the tailing off of the appearance of new phyla, classes, and orders through time, in that order; why was evolution running out of new ideas? I proposed a model (Valentine 1981) that in essence involved the invasion of empty ecospaces; large ones were required for the origin of the highest taxa, medium-sized ones for taxa of intermediate rank, and so on. The ecospace mosaic, composed of tessera (niches), was invaded by species-level lineages, but as it filled up, the opportunity to produce novelties was progressively reduced. Thus diversity among higher taxa was regulated ecologically. This is the model I still prefer, though for a while it seemed that producing a genome for a distinctive body plan might somehow constrain further genomic change.

There had been some theorizing about the nature of eukaryotic gene regulation in development, as by Britten and Davidson (1969), and in 1976 Cathryn Campbell and I applied such ideas to the Cambrian explosion, though the molecular mechanics of gene regulation were not known at the time. The evo-devo revolution began soon after, though; among the many landmarks in this work were the discovery of transcription factors with distinctive binding motifs such as homeobox genes (e.g., McGinnis et al. 1984), and comparative studies of the functions of such regulatory genes among different phyla (e.g., Patel et al. 1989). A large fraction of the basic toolkit of regulatory genes found among metazoan phyla has been in place since before the origin of crown sponge clades (Nichols et al. 2006), though certainly some important genes are known only in eumetazoans, most regulatory gene families had originated before the last common ancestor of radiates and bilaterians appeared (Putnam et al. 2007). But it turns out that the evolution of the metazoan genome is in great part owing to changes in *cis* regulation, in alternative exon splicings, and in the regulation of gene expression by swarms of the various types of RNAs that tend to the DNA molecules and their products. It is evolution between factors controlling the expression of regulatory genes and the pattern of their effects on target genes that seems chiefly responsible for morphological evolution, though those factors must have to be scrutinized by selection at the population level . Here, then, is a major area of biology with relevance to paleontology that is unfolding as I write.

The significance of these findings for paleontology is enormous, and in my opinion foretells an interesting future in which molecular developmental biology will (or should) infuse some major areas of paleontological theory and practice, and as this happens paleontology will inform hypotheses of genomic evolution. After all, evolution of the developmental genome during divergences and radiations is chiefly responsible at the genetic level for the wonderful diversity of forms found among our fossils (Valentine, Jablonski, and Erwin 1999). And on the other hand, the fossil record holds major clues—indeed, produces major findings—that can suggest the nature of genomic evolution. For example, an unparalleled period of body plan evolution occurred just before and during the Cambrian explosion, and it must have been during those times that the distinctive genomes of the crown phyla were evolved (Valentine and Jablonski 2003). The fossil record holds the potential for understanding the pattern of accumulation of morphological differences (implying corresponding genomic changes) among those Cambrian body plans, though it is proving frustratingly difficult to work them out. Nevertheless it does appear that the explosion taxa evolved from significantly smaller, largely vermiform progenitors adapted to the various major marine environments, and from the diversity curves it appears that the ratio of origin of major novelties to speciation rates was much higher then than at any time later in the Phanerozoic. Further, it is likely to have been the increasing occupation of adaptive zones—that is, expansion into ecospace, and corresponding shrinkage of adaptive opportunities as in the tessera model, that has slowed the evolution of novel body types, for the opportunities for change within the incredibly flexible metazoan genomes seem just as great as ever (see Valentine 2004 for references). Origins of some novelties also occur during the postexplosion Phanerozoic, of course, many among crown taxa, and in those cases joint studies of morphological change in the fossil record and of genomic differences among corresponding living groups are possible now, have the potential for increasing our understanding of the great radiations among stem groups, and are highly desirable (Valentine and Jablonski 2003).

PATHS TO RECIPROCATION

In summary, I believe that the route of paleobiology to a significant position within biology has been a two-way street. It has clearly involved the importation of significant amounts of biology into paleontology, at which point paleontological findings become directly relevant to biologists. There have certainly been lots of advances within paleontology that are very important to the field, but which to biologists are sort of inside baseball—not things that they follow. But as paleobiology produces findings that are increasingly relevant to understanding how the biological world works, the findings then become appreciated within the appropriate neontological communities, and paleontology is accordingly imported into biology. Because of the distinctive nature of the materials available in these two disciplines, neontological studies may be pursued in reductionist programs involving living organisms, while paleobiological studies must be more integrative. That sounds to me like the basis for a fruitful collaboration.

ACKNOWLEDGMENTS

Dedicated to Jack Sepkoski, Steve Gould, and Tom Schopf, absent friends. I thank Dave Jablonski and Richard Bambach for reviewing this chapter.

REFERENCES

Ager, D. V. 1963. *Principles of paleoecology*. New York: McGraw-Hill.

Bambach, R. K. 1977. Species richness in marine benthic habitats through the Phanerozoic. *Paleobiology* 3:152–67.

———. 1983. Ecospace utilization and guilds in marine communities through the Phanerozoic. In *Biotic interactions in recent and fossil benthic communities*, ed. M. J. S. Tevesz and P. L. McCall, 719–46. New York: Plenum.

Bretsky, P. W. J. 1969. Evolution of Paleozoic benthic marine invertebrate communities. *Palaeogeography, Palaeoclimatology, Palaeoecology* 6:45–59.

Britten, R. J., and E. H. Davidson. 1969. Gene regulation for higher cells: A theory. *Science* 165:349–57.

Eldredge, N., and S. J. Gould. 1972. Punctuated equilibria: An alternative to phyletic gradualism. In *Models in paleobiology*, ed. T. J. M. Schopf, 82–115. San Francisco: Freeman, Cooper.

Harland, W. B., C. H. Holland, M. R. Housse, N. F. Hughes, A. B. Reynolds, M. J. S. Rudwick, E. Satterwhite, L. B. H. Tarlo, and E. C. Willey. (Eds.) 1967. *The fossil record*. London: Geological Society of London.

Hecker, R. F. 1957. *Introduction to paleoecology*. Moscow: Academy of Sciences (in Russian; French translation 1960, Éditions Technip, Paris; English translation 1965, American Elsevier, New York).

Hedgpeth, J. W., (ed.) 1957. Treatise on marine ecology and paleoecology. 1. Ecology. *Geological Society of America Memoir* 67 (1):1–1296.

Imbrie, J, and N. Newell. 1964. *Approaches to paleoecology*. New York: Wiley.

Klopfer, P. M. 1962. *Behavioral aspects of ecology*. Englewood Cliffs, NJ: Prentice-Hall.

Ladd, H. S. (ed.) 1957. Treatise on marine ecology and paleoecology. 2. Paleoecology. *Geological Society of America Memoir* 67 (2):1–1077.

Levins, R. 1968. *Evolution in changing environments*. Princeton, NJ: Princeton University Press.

MacArthur, R. H., and E. O. Wilson. 1967. *The theory of island biogeography*. Princeton, NJ: Princeton University Press.

Margalef, R. 1968. *Perspectives in ecological theory.* Chicago: University of Chicago Press.

McGinnis, W., M. Levine, E. Hafen, A. Kuroiwa, and W. J. Gehring. 1984. A conserved DNA sequence in homeotic genes of the *Drosophila Antennapedia* and *bithorax* complexes. *Nature* 308:428–33.

Newell, N. D. 1967. Revolutions in the history of life. *Geological Society America Special Papers* 89:63–91.

Nichols, S. A., W. Dirks, J. S. Pearse, and N. King. 2006. Early evolution of animal cell signaling and adhesion genes. *Proceedings of the National Academy of Science USA* 103:12451–56.

Patel, N. H., E. Martin-Blanco, K. G. Coleman, M. C. Poole, M. C. Ellis, T. B. Kornberg, and C. S. Goodman. 1989. Expression of engrailed proteins in arthropods, annelids, and chordates. *Cell* 58:955–68.

Putnam N., M. Srivastava, U. Hellston, B. Dirks, J. Chapman, A. Salamov, and A. Terry. 2007. Sea anemone genome reveals ancestral eumetazoan gene repertoire and genomic organization. *Science* 317:86–94.

Raup, D. M. 1972. Taxonomic diversity during the Phanerozoic. *Science* 177:1065–71.

———. 1976. Species diversity in the Phanerozoic: An interpretation. *Paleobiology* 2:289–97.

Sepkoski, J. J., Jr. 1981. A factor analytic description of the Phanerozoic marine fossil record. *Paleobiology* 7:36–53.

Sepkoski, J. J., Jr., R. K. Bambach, D. M. Raup, and J. W. Valentine. 1981. Phanerozoic marine diversity and the fossil record. *Nature* 293:435–37.

Slobodkin, L. 1962. *Growth and regulation of animal populations.* New York: Holt, Rinehart and Winston.

Stanley, S. M. 1979. *Macroevolution.* San Francisco: Freeman and Co.

Stehli, F. G., R. G. Douglas, and N. D. Newell. 1969. Generation and maintenance of gradients in taxonomic diversity. *Science* 164:947–49.

Valentine, J. W. 1961. Paleoecologic molluscan geography of the Californian Pleistocene. *University of California Publications in Geologic Science* 34:309–442.

———. 1969. Patterns of taxonomic and ecological structure of the shelf benthos during Phanerozoic time. *Palaeontology* 12:684–709.

———. 1972a. Phanerozoic taxonomic diversity: A test of alternate models. *Science* 180:1078–79.

———. 1972b. Conceptual models of ecosystem evolution. In *Models in paleobiology,* ed. T. J. M. Schopf, 192–215. San Francisco: Freeman, Cooper.

———. 1973. *Evolutionary paleoecology of the marine biosphere.* Englewood Cliffs, NJ: Prentice-Hall.

———. 1981. Emergence and radiation of multicellular organisms. In *Life in the universe,* ed. J. Billingham, 229–57. Cambridge, MA: The MIT Press.

———. 2004. *On the origin of phyla.* Chicago: University of Chicago Press.

Valentine, J. W., and C. A. Campbell. 1976. Genetic regulation and the fossil record. *American Scientist* 63:673–80.

Valentine, J. W., and D. Jablonski. 2003. Morphological and developmental macro-evolution: A paleontological perspective. *International Journal of Developmental Biology* 47:517–22.

Valentine, J. W., D. Jablonski, and D. H. Erwin. 1999. Fossils, molecules, and embryos: New perspectives on the Cambrian explosion. *Development* 126:851–59.

Valentine, J. W., and E. M. Moores. 1970. Plate-tectonic regulation of faunal diversity and sea level: A model. *Nature* 228:657–59.

Walker, K. R., and L. F. Laporte. 1970. Congruent fossil communities from Ordovician and Devonian carbonates of New York. *Journal of Paleontology* 44:928–44.

Ziegler, A. M., L. R. M. Cocks, and R. K. Bambach. 1968. The composition and structure of Lower Silurian marine communities. *Lethaia* 1:1–27.

CHAPTER TWENTY

From Empirical Paleoecology to Evolutionary Paleobiology: A Personal Journey

Richard Bambach

INTRODUCTION

I am a paleoecologist whose career paralleled the evolution of evolutionary paleobiology. While I was a college undergraduate in the 1950s the transition from descriptive paleontology to evolutionary paleobiology began. In the 1960s paleoecology emerged as a biologically oriented subdiscipline, but its theory was predominantly borrowed, sometimes inappropriately, from neontology. Evolutionary paleobiology developed during the 1970s, driven by new theory (especially punctuated equilibrium and species selection/sorting). In the 1980s and 1990s evolutionary paleobiology became the unifying theme in paleontology, with comprehensive databases for the whole Phanerozoic making large-scale comparisons over time feasible and the excitement about mass extinctions generating multidisciplinary research efforts. Events in my career exemplify the transformation of traditional paleontology into evolutionary paleobiology. I hope this personal memoir from the perspective of an empirical paleoecologist illuminates some of the human aspects of doing science and helps reveal how some of the change in paleontology came about.

My driving interests are discovery and learning history, not the quest for underlying theory. This does not mean, however, that I am uninterested in theory. I do want to understand and explain the patterns I discover. However, rather than starting my research projects by developing conceptual models with predicted consequences to test, I usually choose a general relationship (such as the number of species in different habitats), compile data relevant to it, and discover what pattern emerges before I seek an explanation. This style may come from my weak math history (which I won't bore you with, and would rather not revisit, myself), or perhaps it is because I am visually oriented rather than responsive to symbolic logic. I use graphs much more than equations, although the data in a graph often are simply a special case of a generality expressed by an equation.

My first meeting with Dave Raup illustrates the difference between the empirically and theoretically inclined. On a field trip to the Rochester area in late October of 1970, Fred Ziegler and I were invited to Dave's house for dinner. During the evening conversation Dave suddenly turned to me and said, "Since the statistical likelihood of preserving any individual is so small, how can you believe you can find communities in the fossil record?" I replied that the likelihood of preservation may be low, but preservational events, like storms that deposit sediment quickly, generally affect locations, not isolated individuals. Successful preservational events may be infrequent, but when one occurs almost everything in a local setting gets buried together. Therefore we do see community associations in the fossil record. Dave was motivated by theory and had asked an interesting theoretical question, testing an interpretation, but he had not done fieldwork on fossil assemblages or sedimentology. He had omitted a step in the common sequence of preservation. Fortunately, Fred and I had been careful to figure out how our assemblages formed, so I could answer Dave and convince him, I hope, that we weren't just assuming we could see community associations. To this day Dave's question reminds me that I must always evaluate and test the criteria I use to reach conclusions. This minor incident reveals that we need both theory and observation to get a full answer to a question. Theory is necessary to test ideas and to explain, but observation of the appropriate phenomenon is also necessary to ensure that the questions and theories relate properly to the situation.

BACKGROUND — GETTING STARTED

I first became aware of paleontology by looking at books in the study of the father of Anne Cooper, a girl I had a crush on in junior high school. Sadly, we never dated; but her father, G. Arthur Cooper, was the great brachiopod paleontologist. When I started college I intended to be a marine biologist, so I majored in biological sciences at Johns Hopkins. I also took courses in geology and paleontology and ended up doing undergraduate research for Harold Vokes, who had been president of the Paleontological Society in 1951. My task was to identify the mollusks Vokes had collected from a shell-bed in the Brightseat Formation, the first unit recognized as Paleocene in age found to outcrop in the Atlantic Coastal Plain. The novelty of discovering new things from the past, coupled to the challenge of working on the jigsaw puzzle of the history of life, got to me, and I chose paleontology over neontology.

My first jobs were temporary positions in the geology department of the Smithsonian Institution. I worked in the mineralogy division between my two senior years in college (don't ask; I was a chronic class cutter) and, after finishing college, worked in mineralogy and paleontology for a year before joining the U.S. Navy to fulfill my military obligation (the draft was still active in those days), going to officer candidate school and serving three years on active duty. I completed my naval service too late to enter a graduate program in 1961, so I went back to the Smithsonian. G. A. Cooper was chief curator (chair) of the department of geology and my boss.

One day in the late fall of 1961, Dr. Cooper inquired, "I understand you are thinking of going to graduate school." I said yes, I was interested in Paleozoic bivalves. He responded, "Have you thought of Yale?" and pointed out that A. Lee McAlester, who studied Devonian bivalves, had recently been appointed to the Yale faculty. With my checkered undergraduate record I had not dreamed of a top-flight program, but flattered by Cooper's suggestion of his graduate alma mater, I applied. Apparently my work record in the Navy and at the Smithsonian, plus kind letters of recommendation from the right people, made up for my past deficiencies, and I started graduate school at Yale in the fall of 1962.

Cooper was a very traditional systematic paleontologist. Two incidents illustrate his focused approach and, although both incidents occurred after I left the museum, show the intellectual orientation of the program in which I got my first professional experience. Once, when I was visiting his lab while in graduate school, he commented, "I could never have discovered evolution working on brachiopods." He was observing the pattern of punctuated equilibrium with brachiopod fossils eight years before it was named and he felt he couldn't have discovered evolution because he didn't see the transitions occur as new forms originated. Apparently he thought he wouldn't have deduced evolution if he couldn't directly observe gradual evolution of form. The second incident happened after I gave him a reprint of one of my early papers. He came up to me at a meeting and actually said, "Dick, that was an interesting paper, but you know we should describe all the fossils before we start interpreting them." He was always concerned that there weren't enough systematists (he may not have been wrong).

The transition from traditional paleontology to modern paleobiology was well underway by the early 1960s. Norman Newell, who spearheaded the development of modern paleoecology and focused attention on mass extinctions as "crises in the history of life," and John Imbrie, a pioneer in quantitative analyses of fossils, led an innovative graduate program at Columbia University that would produce Leo Laporte, Steve Gould, and Niles Eldredge. Ralph Gordon Johnson was developing taphonomic studies and quantitative evaluation of fossil assemblages at the University of Chicago, Jim Valentine was studying Pleistocene molluskan distributions on the West coast at the University of California at Davis, Dave Raup was already working on early computer models of coiling morphology at Johns Hopkins, and in 1960 Dick Beerbower published a textbook on paleontology in which half the book was conceptually based and only half devoted to systematics.

Although several schools were involved in the early paleobiology revolution, the Yale graduate program was the first to blossom, with numerous paleobiologically oriented students. Lee McAlester, who finished his dissertation in 1959, catalyzed the success of the program. Among other things, he had picked up a sense of marketing from business courses he took as an undergraduate at S.M.U., and he produced brochures advertising the paleobiological emphasis of the Yale program. That, plus the long-established reputation of the geology program, attracted a group of students who became highly successful paleoecologists. When I arrived as a beginning graduate student in 1962, Marty Buzas and Ian Speden were still working with Karl Waage, who was the senior invertebrate paleontologist on the faculty, and by the time I defended my dissertation in 1968 Peter Bretsky, Ken Walker, Steve Stanley, Dave Meyer, Sara Bretsky, Charley Thayer, Jeff Levinton, Jeremey Jackson, and Geerat Vermeij were student colleagues.

Three things were particularly stimulating for the paleontology students at Yale. First was the interaction within the student group itself. Students always learn as much from each other as they do from formal coursework, and we kept each other thinking. Second, a fortunate selection of visitors and new faculty, especially Leo Laporte and Don Rhoads, brought in ideas that further broadened our perspective. Laporte, then a rising leader in carbonate sedimentology and paleoecology, spent a year at Yale as a visiting professor. Don Rhoads, who did his doctorate with Ralph Gordon Johnson at Chicago, brought a fresh approach for looking at animal-sediment relationships when he joined the Yale faculty; he also established a research base at Woods Hole that benefited us. The third thing was the outstanding faculty in other disciplines. John Rodgers, a true polymath, expert in stratigraphy as well as tectonics and structure, was a great inspiration for all things geological. Most influential, however, was the great ecologist, G. Evelyn Hutchinson.

The paper that Jeff Levinton and I did on analyzing bivalve mortality patterns from size-frequency distributions (Levinton and Bambach 1970)—one of only three citations to research by the Yale student paleoecology group in the influential symposium volume *Models in Paleobiology* (Schopf 1972)—began as a term paper by Levinton in Hutchinson's course on ecological principles. Some of us also took Hutchinson's course on major features of evolution. It was "evolutionary paleobiology" before anyone had thought of the term. I still have my course notes; in spring of 1963 we covered the diversity of all major forms of life, pre-Cambrian life, origin of metazoa, origin of protostomes and deuterostomes, and relationships among the animal phyla. To be sure, the phylogenies are out of date, because the course was given before molecular phylogenies were available, but Hutchinson asked all the right questions. Studying with Hutchinson is still one of my inspirations.

COMMUNITY PALEOECOLOGY — THE FIRST FULLY PALEOBIOLOGICAL SUBDISCIPLINE

My dissertation topic was the remarkable Silurian bivalve fauna from Arisaig, Nova Scotia. Art Boucot offered me the Arisaig bivalves from his collections to add to those from my two summers collecting. When I visited Cal Tech to get them in May of 1965 Art had me stay with his post-doc, Fred Ziegler. Fred had just finished his dissertation at Oxford and had already published his seminal paper on Silurian marine communities and their environmental significance (Ziegler 1965). We found a lot of common interests, and this began our long association. As my visit ended, Fred gave me a suite of bivalve specimens from collections he had made in England and asked if I would identify them and describe their life habits. The specimens were from the collections Fred thought best represented the communities he had studied in Britain, and he was checking to see if his interpretations of environment for the assemblages would be reflected in my interpretations of the bivalves. Happily, they were, and Fred cut up the four-page letter I sent him and used it as the basis for my contribution to what became Ziegler, Cocks, and Bambach (1968), the paper that initiated the journal *Lethaia*. That paper, which describes the spectrum of five community types across marine shelf environments of Early Silurian times in detail, was my first publication, and one that became a frequently cited paper, aided by the beautiful reconstructions of communities Fred commissioned and by its place in a new journal (*Lethaia*, volume 1, page 1).

Although my dissertation only dealt with bivalves in the classic specialize-on-a-group-during-a-time mode, I focused much of the text on ecological aspects of the fauna. For instance, I argued that my bivalves were from "essentially untransported, undisturbed fossil assemblages which accumulated directly from local living assemblages" (Bambach 1969, 43), using concepts established by Ralph Gordon Johnson from the Recent to help justify this conclusion, and I related my assemblages to the community spectrum Ziegler had initially described in Great Britain.

Paleocommunity work became fashionable in the late 1960s and early 1970s, but general interest faded in the later 1970s, primarily because most community studies were simply descriptive. Community paleoecologists, myself included, had not developed any compelling general theory about ecological data in paleontology. This failing was one of the problems at the early high point of community paleoecology—the Penrose Conference on benthic marine community ecology and paleoecology, organized by Leo Laporte, Jim Valentine, and Peter Bretsky, held in December of 1970. In my opinion, the failure of the 1970 Penrose Conference, in which a group of leading marine ecologists were brought together with a group of paleoecologists, related to two problems: (1) inadequate development of theory on the part of paleontologists, and (2) lack of appreciation by the biologists for the biological information contained in the fossil record. In general, the paleontologists were simply applying biological concepts to the fossil record rather than bringing new theory to the table or presenting observations that would stimulate the development of new theory. We were eager but we weren't ready, and we had no theory to offer to the neontologists. On the other hand, the neontologists were distracted by the complexity of the fluctuations in standing-crop communities (this predated metapopulation biology and macroecology). They not only weren't thinking about pattern and process on geological timescales, they couldn't see it would be of any value for them to do so. For instance, after a presentation by a paleontologist on a pioneering effort to use computer simulation to extract information about population dynamics from the fossil record, I overheard one prominent ecologist say to another, "Who would even want to do that?" They weren't ready, either.

The climate was sufficiently bleak at the 1970 Penrose Conference that Ken Walker and I called for a discussion of the problem during one of the free-time periods. Most of the paleontologists came, but only two of the thirty biologists showed up. During that discussion Ken and I realized that one issue dividing the conference was the difference between standing crop assemblages, which were all the ecologists had ever studied, and fossil assemblages. The problem to the ecologists was a lack of exact contemporaneity among components of fossil assemblages, which incorporate live and accumulated dead specimens together. Ken and I had already been thinking about "time-averaging" and had mentioned it in an abstract we had already submitted for an upcoming Southeastern Section GSA meeting (Bambach and Walker 1971). Realizing the concept was important, we wrote up an extended abstract on "time-averaging" for a discussion paper (precursor of poster sessions) at the GSA annual meeting (Walker and Bambach 1971). Time-averaging has become an important research topic in taphonomy, so something did come from the 1970 Penrose Conference. Ironically, at a second Penrose Conference on ecology and paleoecology twenty-eight years later, I spoke with some trepidation about the differences in samples between paleoecology and neontology and, to my surprise, the neontologists insisted that time-averaged assemblages were an advantage when looking at general patterns because they integrated the variability that characterizes the standing-crop systems they deal with. In this and other ways the paleoecologists had developed a theoretical base by 1998 from which to present their ideas, and the ecologists had expanded their theoretical horizons, as well.

Prior to 1975 all my publications were empirical and related to observing or categorizing paleoecologically interesting phenomena. None were on evolutionary topics. From my initial specialization came papers interpreting original shape in deformed bivalves, interpreting the functional morphology of an odd Silurian bivalve species, and my collaboration with Levinton, interpreting survivorship in bivalves from size-frequency distributions of preserved shells. My interest in community paleoecology resulted in my collaboration with Ziegler, describing Silurian communities, collaboration with Ron Kreisa on describing and interpreting depositional environments in Early Carboniferous deltaic deposits, and collaborations with Ken Walker on categorizing and interpreting modes of feeding in invertebrates and on aspects of community structure. However, seeds of future work related to evolutionary paleobiology are found in some of my abstracts for presentations at Geological Society of America meetings.

Ideas that would be more fully developed later, when I introduced the guild concept into marine paleoecology (Bambach 1983), began in a study I did in the spring of 1970 with Jane Fisher, an honors student at Smith College. In one of her collections from the Middle Devonian we found that there were three groups of four species in which the species within each group had about the same abundance, but each member of the group had a very different life habit. Furthermore, whenever several species were about equally abundant in the other collections those species also represented different modes of life. This got me interested in controls on community structure. Ken Walker had made similar observations for Ordovician assemblages, and we did an abstract for the Southeastern Section of the GSA (Bambach and Walker 1971) in which we observed, "community dominance seems to be more a function of the pattern of niche subdivision than an inherent property of the community as a whole." In 1974 I gave a presentation on "Resource Partitioning in Paleozoic Benthic Communities" (Bambach 1974) in a symposium on the structure and classification of ancient communities at the annual meeting of the GSA. That abstract and presentation featured the abundance relations within *niche groups* (sets of autoecologically similar species). When I met Jack Sepkoski in 1975 he asked me if a masters student of his could test my ideas about the abundance patterns I had mentioned in my abstract. I loaned them the data, and Dale Schwartz (now Springer) showed that my naive views on abundance relations were incorrect. So I did not write up my original ideas, but continued to contemplate community structure, until niche groups became guilds in my 1983 paper (and I didn't overinterpret the data). Further benefits for me were that one of my happiest professional associations, my friendship with Jack Sepkoski, and the acquisition of a good graduate student, Dale Springer, came from that work.

The second example of early work presaging later, more sophisticated studies also appeared initially only in a GSA abstract. In preparing to look for a job back home, Charles Calef, an American who did his doctoral work on Middle Silurian brachiopods with Stuart McKerrow at Oxford, spent a postdoctoral year with me at Virginia Tech. Both of us were interested in the contrast between the restricted environmental distribution of abundant bivalves and the widespread distribution of articulated brachiopods in early Paleozoic facies. Our suggested explanation, "Low Nutrient Levels in Lower Paleozoic (Cambrian-Silurian) Oceans," became the title of our abstract for the 1973 annual meeting of the GSA (Calef and Bambach 1973). We speculated that in the early Paleozoic the land lacked plant cover, which would limit nutrient supply to epeiric seas, especially to areas distant from the shoreline. This might explain the abundance of low metabolic rate animals like brachiopods offshore and the restriction of abundant, high metabolic rate bivalves only to inshore settings where more nutrients would have been available. We didn't write this work up as a paper because, given the vagaries of sedimentation rates, we were stumped about methods of comparing rates and abundances quantitatively from place to place. But that abstract began a long development of ideas that eventually resulted in three papers, the most well known of which is "Seafood through Time" (Bambach 1993). These papers argue that the success of different groups with different metabolic properties and energetics at different times implies that there have been global changes in nutrient concentrations and productivity over time.

TRANSITION TO EVOLUTIONARY PALEOBIOLOGY

By 1974 I was editing the Paleontological Society newsletter and attending Paleontological Society Council meetings. That fall the proposed new journal *Paleobiology* received final approval at the Council meeting. The editors of the *Journal of Paleontology* were concerned that their journal's circulation would drop if general interest articles were siphoned off into the new journal. *Paleobiology* was finally approved with a compromise on funding; subscription to the *Journal of Paleontology* was to remain automatic with payment of Society dues, but subscribing to *Paleobiology* required an additional fee. This has meant that *Paleobiology* has never had a circulation equal to the *Journal of Paleontology*, but, thanks to careful work starting with the initial editors, Tom Schopf and Ralph Johnson, *Paleobiology* has been a success and has always paid its own way.

The *Models in Paleobiology* symposium and the series of papers Raup, Gould, Schopf, and Simberloff began at the Marine Biological Laboratory in the early 1970s are the two projects commonly regarded as marking the beginning of evolutionary paleobiology. No Yale paleoecology students were included as participants in either project. Likewise, only a few of the participants in the *Models* symposium had been at the ecological 1970 Penrose Conference. But we all began to unify under the term *paleobiology* as time went on. For instance, the initial editorial board for *Paleobiology* in 1975, selected by Tom Schopf (who had organized the *Models* symposium) and Ralph Gordon Johnson (who had participated in the *Models* symposium) represented the full spectrum of interests, including five Yale paleoecology students and, counting the two editors, seven of the participants in the *Models* symposium among the thirty-eight members.

Empiricists need data and empirical evolutionary study requires comparison of data across time. It was not possible to reach broad evolutionary conclusions until adequate data from different times became available. The first study with significant evolutionary implications, in which I collaborated (Levinton and Bambach 1975), appeared in the first issue of *Paleobiology*. Jeff and I mined our dissertations to compare deposit-feeding bivalve communities in the Recent with several from the Silurian. Niche subdivision turned out to be similar in both intervals for this community type, implying that the efficiency of feeding utilized by modern deposit feeders had already evolved by the Silurian.

The disparity between estimates of marine diversity through time stimulated my first comparative project encompassing the whole Phanerozoic (Bambach 1977). In 1970 Valentine suggested that diversity had increased tenfold (an order of magnitude) between the mid-Paleozoic and today (Valentine 1970). But in 1972 Raup speculated that diversity might have been constant since the Cambrian radiation. He reasoned that the apparent increase in diversity could be an artifact of differential quality of the geologic record, with apparent low diversity early, when there is less rock preserved, and more apparent diversity later, when the record is less degraded (Raup 1972). I thought there had to be some way to improve on this order of magnitude uncertainty. Paleontology would be a pretty weak science if it couldn't even determine whether diversity had changed over time, much less by any particular amount. I realized comparing within-assemblage diversity avoided the record-failure issue. If you had a Cambrian assemblage, it had survived the vicissitudes of geologic time, as was true of assemblages of any age. If within-habitat richness (alpha diversity) had changed, then global diversity should have done so as well, unless there had been a coincidental balancing change in the number of communities. If the number of communities had remained about constant—not an unreasonable assumption—then global diversity change should parallel within-assemblage (alpha diversity) change. If there had been no within-habitat change (no change in alpha diversity), then perhaps apparent global diversity change was an artifact. I collected data on 386 paleocommunities and found that alpha diversity had increased about two and a half times between the mid-Paleozoic and later Cenozoic. Although the project was structured as an empirical discovery project, with the intention of finding some way of differentiating between Valentine's and Raup's positions, the results, which included documenting that the increase in alpha diversity was concentrated in relatively stable, open marine assemblages, with no diversity increase in high-stress habitats, needed explanation, and I attempted some speculations (Bambach 1977, 161–63). I have worked on how ecosystem structure might influence diversity and diversification ever since.

One small incident in getting this paper into the press shows the new attitude of rigor that developed with the growth of paleobiological thinking. I gave the manuscript to Tom Schopf, the editor of *Paleobiology*, while we were having dinner together at the start of the 1976 GSA meeting in Denver. On the last day of the meeting he came up to me and said, "I read your manuscript and I don't believe it." He was not convinced by the difference in modes of diversity, despite the large amount of data and my use of medians as a very conservative metric of central tendency. He wanted a statistical test and I had not done one (nor, with my background, did I know how to do one on those data). That evening I sat in the lobby of the Brown Palace Hotel with Jack Sepkoski and told him of Tom's doubts. Jack thought my results were obviously significant on their face and offered to find a method to test them (see acknowledgments, Bambach 1977, 163), which he did, and then Tom was satisfied.

The plate tectonic revolution affected all geology, even paleobiology. As a paleoecologist I needed to know how changing plate positions, by shifting continental blocks into different environmental settings, had determined the habitats of the fossil assemblages I studied. During the 1970s Fred Ziegler's interests shifted from mapping environments from community distributions to working on global paleogeographic reconstructions. I collaborated with Fred on a preliminary reconstruction of Silurian paleogeography in 1977, and in 1978–1979 I spent the academic year on research leave from Virginia Tech, working in Fred Ziegler's lab at the University of Chicago. Chris Scotese was a graduate student in the lab and Judy Parrish a post-doc. With Chris programming the computer and compiling paleomagnetic information and the rest of us gathering lithologic data to evaluate paleoclimate and environments, we produced a set of seven global reconstructions that became widely known, one for a short interval in each time period of the Paleozoic.

The year in Chicago also deepened my acquaintance with Jack Sepkoski and Dave Raup. Jack had joined the Chicago faculty and Dave had moved from Rochester to the Field Museum. Jack and I began to collaborate on several projects, one of which generated our personal favorite epithet, "leading casual theorists in paleontology." We contemplated having tee shirts made with a "casual theorist" logo, but never did. I am not citing this reference because there are names attached. They are not important, only the humor matters; that and the demonstration that the new approach of paleobiology was not always welcomed.

During a visit to Cambridge, Massachusetts, in early May of 1979, I was sitting in Steve Gould's office at Harvard when the telephone rang. Steve interrupted his caller to ask me to list what I thought had been five successful and five unsuccessful topics that had been prominent in paleontology in the last twenty years. I did so as he continued with the phone call. Thus I became a part of his "informal survey" of paleontologists reported in "The Promise of Paleobiology as a Nomothetic, Evolutionary Discipline" (Gould 1980), a piece that will always be a good read. After finishing the phone call Steve looked at my list and we began to discuss the topics I had designated. The discussion continued as we walked to his apartment and settled into his living room. We disagreed about functional morphology, a topic I favored and Steve felt was unimportant. I persisted in my advocacy because of the understanding functional morphology gives us about how animals lived in the past. Steve eventually erupted with an impassioned, "But Dick, it doesn't tell us anything about evolution!" At the time I thought, "Doesn't Steve care about ecology? Is evolution all he cares about?" It was a shock that my brilliant friend might have any limitations to his interests. But as I thought about his remark, it dawned on me that there was another side to the issue. Maybe I didn't think enough about evolution. This was my wake-up call (although I still disagree with Steve's opinion), and I have been thoughtful about including an evolutionary perspective, when it is appropriate, in all my later ecological work.

The consensus paper on global marine diversity (Sepkoski et al. 1981) marks the end of my transition to full-scale evolutionary paleobiology. We concluded that the diversity pattern observed in a variety of data sets reflects the actual course of diversity through time. This established the diversity curve seen in both Sepkoski's family and genus compilations as an icon of marine macroevolution. Arnie Miller discusses the consensus paper in another contribution to this volume, so I will just comment that the similarity of diversity patterns observed in studies at different taxonomic scales and susceptibilities to loss of geologic record suggested that the fossil record contains a reliable signal of diversity. Several compilations on species and genus diversity, which were global in scope, were subject to record failure. However, my examination of alpha diversity avoided the issue Raup had raised about record failure, but was not itself a measure of global (gamma) diversity. Sepkoski's compilation of family diversity was a measure of gamma diversity, but at a high level. However, the range-through technique he used, which records the presence of a taxon in all intervals, from its first to last appearance, whether the taxon is observed in each interval or not, and the generally wide geographic distribution of members of many families together, also reduced the danger of artifact from record failure.

ECOLOGICAL ASPECTS OF MATURING EVOLUTIONARY PALEOBIOLOGY

Paleobiology was finally established as the theoretical heart of the "new" paleontology in the early 1980s, and evolutionary paleobiology became the overarching approach. In the late 1970s Steve Gould became the most visible spokesman for the field, and Steve Stanley had published a textbook on macroevolution, the first book to have that word in its title. As the 1980s be-

gan, Gould's exhortations to develop theory at an appropriate paleontological scale, coupled to the dramatic discovery of the iridium anomaly at the
Cretaceous/Tertiary boundary and its implications for the end-Cretaceous
extinction, made paleontological perspectives relevant to macrobiology. Jack
Sepkoski's compilation of stratigraphic ranges of marine families at the stage
level (Sepkoski 1982) gave us a comprehensive database at a new level of temporal resolution, permitting a range of comparative analyses and quantitative
evaluation. Data on the distribution and structure of marine communities developed by me, Sepkoski, and others, plus the paleogeographic reconstructions that the Ziegler lab produced, opened the way to studying ecological
patterns in regional as well as global context. Both paleoecologists and those
interested in evolution now could take serious comparative looks at the fossil
record, and paleontological data were being recognized as carrying biological
information beyond the scope of neontology.

Four group meetings and resulting publications illustrate the way in which
ecological views were integrated with evolutionary viewpoints in this first,
mature phase of evolutionary paleobiology. One emphasized ecology, another
focused on diversity patterns, and two incorporated all aspects of evolutionary
paleobiology in large-scale overviews.

The most ecological was the Paleontological Society symposium on biotic
interactions in recent and fossil benthic communities held at the 1981 GSA
meeting and published in 1983. Evolutionary aspects of biotic interactions received attention. Two of the four sections (and half the pages) of the proceedings' volume (Tevesz and McCall 1983) dealt with biotic interactions through
time and the effects of interactions on community evolution. My initial study
of guilds (Bambach 1983), in which I showed that within-community diversity was related to the number of guilds represented in the assemblage, was
part of the section on community evolution.

Jim Valentine organized a symposium on diversity patterns for the Pacific
Division, American Association for the Advancement of Science meeting in
June of 1982. In his preface to the published proceedings (Valentine 1985, ix)
Jim noted, "Considering the plethora of symposia in paleontology, it is remarkable that this is the first treatment of Phanerozoic diversity as such." He went
on to comment, "most of the present papers are routinely involved or explicitly
concerned with theory or at least with explanation, and should be of interest
to biologists as well as paleobiologists." The purely descriptive days were in
the past. I had a long paper in this volume in which I argued that diversity
expanded within classes only with the evolution of modes of life or invasion of
ecospace previously unutilized by members of that class (Bambach 1985).

NASA sponsored a workshop on the evolution of complex and higher organisms, organized by Dave Raup, that met in July 1981 and January and July 1982. It was the epitome of interdisciplinary interaction and dealt with evolutionary paleobiology in its most comprehensive form, considering all aspects that influenced the evolution of complex life on our planet. The workshop participants included a variety of paleontologists, botanists, zoologists, geochemists, and extraterrestrial experts. I think I was included because I had worked on paleogeography and plate tectonics as well as ecosystem paleontology. The group report (Milne et al. 1985) was one of the foundations for a grant program at NASA that included paleontology, and evolved into the astrobiology program.

The Dahlem Conference on patterns and processes in the history of life held in Berlin in June 1985 was also comprehensive evolutionary paleobiology. It focused on four topics (Raup and Jablonski 1986): (1) directions in the history of life, (2) organismic evolution: the interaction of microevolutionary and macroevolutionary processes, (3) causes and consequences of extinction, and (4) the neontologico-paleontological interface of community evolution: how do the pieces in the kaleidoscopic biosphere move? Participants included both neontologists and paleontologists. It still took some effort to get the neontologists on the same page with the paleontologists, but this time, unlike the Penrose Conference fifteen years before, we got together and agreed there were interesting biological patterns expressed in the fossil record not directly accessible in the modern world. Dave Raup, Dave Jablonski, Dan Fisher, Steve Gould, Jeff Levinton, Jim Valentine, Adolf Seilacher, Tony Hallam, Jack Sepkoski, Geerat Vermeij, Dan Simberloff, Karl Niklas, and I participated, a real cross-section of paleontological disciplinary interests, now all interested in and contributing to evolutionary paleobiology.

During the last twenty years my views on ecology and evolution have continued to mature. I finally wrote up "Seafood through Time" (Bambach 1993) for a symposium Geerat Vermeij and Mellissa Rhodes organized for the Fifth North American Paleontological Convention. I followed that up with two related papers, one on the possible connection between evolution of the biota on land and changing energetics in the marine fauna and the other on the diversity history of predators. I collaborated with Norman Gilinsky on several purely evolutionary studies in the late 1980s, published on community evolution and testing the theory of coordinated stasis in collaboration with J. Bret Bennington in the mid-1990s, and wrote on the meaning of evolution and faunal succession in short-course volumes in the late 1990s. I have also worked on the Permian mass extinction in collaboration with Andy Knoll and others and recently I published a review paper on mass extinctions. Finally, my old interest in guilds and ecospace has now become a stimulating collaboration with Andy Bush at the University of Connecticut. We are exploring much more rigorously than I did in the 1980s how the evolution of new modes of life and consequent changes in the use of ecospace have influenced change in alpha diversity and ecosystem structure over time.

POSTSCRIPT

Steve Gould was right when he criticized the simple transfer of biological concepts to the fossil record, which characterized much early work in paleobiology. He said (Gould 1980, 102–103), "Our primary stance towards evolutionary biology has been fundamentally uncreative: we have used our data to exemplify the principles of neontology, not to search for new principles or to think seriously about how existing theory might work itself out in uncharacteristic ways through the vast times at our disposal. We seem satisfied as long as we can show that fossil organisms and communities worked pretty much as their modern counterparts do." I think that paleobiology matured when we finally moved beyond noting that animals in the past functioned like animals and realized that the context in which they functioned was different than today. It took the refinements of collecting data on fossils in their ecological context, plus work on the implications of the patterns of evolution of organisms over geological time, before we could really claim to be adding to conceptual biology.

Gould was the principal advocate for developing theory in paleontology, but I do have a criticism of his views. Despite his extensive empirical experience with his beloved Bermudan and Caribbean land snails, Steve never seriously tried to bring ecological conditions and concepts into his thinking on evolution, evolutionary processes, and evolutionary theory. They all got lumped under "contingency." His focus on exaptation and his desire to limit the term *adaptation* only to features evolved de novo for their function is a case in point. In a seminar discussion in 1995 he and I tangled over the importance of selective modification. I pointed out that it was important because it was nearly (although not completely) ubiquitous. He argued that the initial origin of features is more significant than the modification of extant features. This idea was also behind his focus on the disparity of the Cambrian fauna compared to later diversification. I do not deny that the initial origin of a feature is fundamental in a way that the modification of the feature, once it is extant, is not, but I do think turning a four-legged land mammal into a whale is about as interesting as turning a tunicate into a shark, or even a sponge into a tunicate. The early evolution of most higher taxa is interesting, but that some groups, once established, are so functionally plastic that they can evolve to fill many roles is equally so. Understanding the ecological theater is necessary for understanding the constraints that determine the plot of the evolutionary play.

A lot of directional or constraining factors derive from ecological relationships—the effects both of organisms on each other and of the environment on organisms. The presumed unpredictability of contingent history may be a lot more regular and channeled than Steve Gould claimed. I may never get beyond several very general arguments along these lines (Knoll and Bambach 2000; Bambach, Knoll, and Sepkoski 2002), but I do think that, given this particular planet with its location relative to its parent star and its internal workings (like plate tectonics)—yes, all contingent, but only as to location in the universe and initial conditions—that the course of evolution, while unquestionably unique in detail, would probably run along a generally parallel path if we "ran the tape over again." We will only have a fully mature profession when we understand not only how evolution operates, but which evolutionary patterns are deterministic and which are chance, and when we also understand the context in which evolutionary change has occurred and can explain the successes and failures along the way.

ACKNOWLEDGMENTS

My deepest thanks go to all the people named in this paper. They provided stimulation and leadership for a whole profession, and many also became good friends, making my career in paleobiology a pleasure. I also want to thank David Sepkoski for inviting me to participate in this project, Steve Stanley and Doug Erwin for discussions that reminded me of events in the dim past, and Susan Bambach for suggesting many ways to improve my early drafts.

REFERENCES

Bambach, R. K. 1969. *Bivalvia of the Siluro-Devonian Arisaig Group, Nova Scotia.* Unpublished PhD diss., Yale University, New Haven, CT.

———. 1974. Resource partitioning in Paleozoic benthic communities. *Geological Society of America Abstracts with Programs* 6:643.

———. 1977. Species richness in marine benthic habitats through the Phanerozoic. *Paleobiology,* 3:152–67.

———. 1983. Ecospace utilization and guilds in marine communities through the Phanerozoic. In *Biotic interactions in recent and fossil benthic communities,* ed. M. Tevesz and P. McCall, 719–46. New York: Plenum.

———. 1985. Classes and adaptive variety: The ecology of diversification in marine faunas through the Phanerozoic. In *Phanerozoic diversity patterns: Profiles in macroevolution,* ed. J. W. Valentine, 191–253. Princeton, NJ: Princeton University Press.

———. 1993. Seafood through time: Changes in biomass, energetics and productivity in the marine ecosystem. *Paleobiology* 19:372–97.

Bambach, R. K., A. H. Knoll, and J. J. Sepkoski, Jr. 2002. Anatomical and ecological constraints on Phanerozoic animal diversity in the marine realm. *Proceedings of the National Academy of Science (USA),* 99:6854–59.

Bambach, R. K., and Walker, K. R. 1971. Trophic relationships: An approach for analyzing the structure of fossil benthonic communities. *Geological Society of America Abstracts with Programs* 3:292–93.

Calef, C. E., and Bambach, R. K. 1973. Low nutrient levels in lower Paleozoic oceans. *Geological Society of America Abstracts with Programs* 5:565.

Gould, S. J. 1980. The promise of paleobiology as a nomothetic, evolutionary discipline. *Paleobiology* 6:96–118.

Knoll, A. H., and R. K. Bambach. 2000. Directionality in the history of life: Diffusion from the left wall or repeated scaling of the right? In *Deep time: Paleobiology's perspective.* Supplement to *Paleobiology* 26:4, ed. D. H. Erwin and S. L. Wing, 1–14.

Levinton, J. S., and R. K. Bambach. 1970. Some ecological aspects of bivalve mortality patterns. *American Journal of Science* 268:97–112.

———. 1975. A comparative study of Silurian and Recent deposit-feeding bivalve communities. *Paleobiology* 1:97–124.

Milne, D., D. M. Raup, J. Billingham, K. Niklas, and K. Padian, eds. 1985. *The evolution of complex and higher organisms.* Washington, DC: NASA SP-476.

Raup, D. M. 1972. Taxonomic diversity during the Phanerozoic. *Science* 177:1065–77.

Raup, D. M., and D. Jablonski, eds. 1986. *Patterns and processes in the history of life.* Dahlem Konferenzen Life Sciences Research Report 36, Berlin: Springer-Verlag.

Schopf, T. J. M., ed. 1972. *Models in paleobiology.* San Francisco: Freeman, Cooper.

Sepkoski, J. J., Jr. 1982. *A compendium of fossil marine families. Milwaukee Public Museum Contributions in Biology and Geology 51.*

Sepkoski, J. J., Jr., R. K. Bambach, D. M. Raup, and J. W. Valentine. 1981. Phanerozoic marine diversity: A strong signal from the fossil record. *Nature* 293:435–37.

Stanley, S. M. 1979. *Macroevolution: Pattern and process.* San Francisco: W. H. Freeman.

Tevesz, M., and P. McCall. 1983. *Biotic interactions in Recent and fossil communities.* New York: Plenum.

Valentine, J. W. 1970. How many marine invertebrate fossil species? A new approach. *Journal of Paleontology* 44:410–15.

————, ed. 1985. *Phanerozoic diversity patterns: profiles in macroevolution.* Princeton University Press, Princeton, New Jersey, ix + 441 pages.

Walker, K. R., and R. K. Bambach. 1971. The significance of fossil assemblages from fine-grained sediments: Time-averaged communities. *Geological Society of America Abstracts with Programs* 3:783–84.

Ziegler, A. M. 1965. Silurian marine communities and their environmental significance. *Nature* 207:270–72.

Ziegler, A. M., L. R. M. Cocks, and R. K. Bambach. 1968. The composition and structure of Lower Silurian marine communities. *Letahia* 1:1–27.

CHAPTER TWENTY-ONE

Intellectual Evolution across an Academic Landscape

Rebecca Z. German

INTRODUCTION

The scientists who fermented the field of paleobiology were committed to education. Either through gradual transformations of existing courses or more marked leaps in the development of new programs, Tom Schopf, Dave Raup, and Steve Gould carried their enthusiasm for science into the formal classroom and informal seminars. They were immersed in both the scientific and philosophic changes to the standard canon of paleontology. Yet the differences among them were evident not only in their science and publications. Their teaching was distinctive, both in the small focus of what happened in a classroom, and in the larger scope of their visions of graduate training.

An analysis of the teaching at Chicago, Rochester, and Harvard, from roughly 1973 through 1983, would be easy if there had been a grand ceteris paribis experiment. But there were significant differences among the three institutions in terms of their history, the geology and biology departments in which paleobiology found a home, and the resources available at each. A historian or philosopher would assemble more data, track down details, and weave these specific histories into a more cohesive picture. I am a scientist, with a predilection for controlled variables, replicated trials, and testable hypotheses that are not available. What I can do is tell the story from my perspective.

I had the good fortune—or luck, as Tom Schopf would say, to learn evolution and paleobiology from Schopf at the University of Chicago as an undergraduate in mathematics (1973–1977), and then, as a graduate student, first with Raup in geology at the University of Rochester (1977–1979) , and finally in biology with Gould at Harvard University (1979–83). Mathematics was one of the easiest majors at Chicago, as it had the fewest requirements. I had already become interested in evolutionary biology, as a way of doing applied modeling, but I needed physical science credits. I came to paleontology to fill a requirement and stayed for a PhD thesis. Although I ultimately left paleontology for neontology (with a brief reprise for the 1991 PS short course "Analytical Paleobiology," organized by Norm Gilinsky and Phil Signor), the education in paleobiology from Schopf, Raup, and Gould has stood me in good stead.

Today's emphasis on teaching in science may be a relatively recent innovation, but good teachers existed even before Plato. If there is value in examining the history of the development of the field of paleobiology, then of some interest is how the founders of that field engaged their students and taught their subject.

My perceptions are just those, perceptions, filtered through the sieve of time, and refracted by the lens of a student's eye. Nor was I a fixed, impartial external observer; I grew over this time, and changed not only my views, but my ability to learn from others. This essay is not the carved-in-stone views of a historian, but the scribbled-in-sand memories of a student. If there is a reason for telling this story it is because that time was exciting and challenging, and all students have a debt to their teachers that can never be fully repaid.

THOMAS SCHOPF AND THE UNIVERSITY OF CHICAGO

In 1975, Tom Schopf's course, "Paleontology" (Geophysci 221) was arguably one of the most exciting courses in evolution at the University of Chicago, but also one of the most exacting. It habitually dropped from nearly twenty students to five or six by the end of the ten-week term. The course was taught by Schopf's implementation of the Socratic method, which to students well versed in the Aristotelian legacy of Chicago, felt more like trial by humiliation. Yet if one could endure the methods ("you are the most stupid mammal of them all"), the content of the course was exhilarating. The class was analogous to Simpson's *Tempo and Mode in Evolution,* paleontology without pictures of fossils. This was a class of ideas, of theories, of testable hypotheses. Many natural science classes were either a recitation of the history of the field, or worse, an exercise in stamp collecting. These are the features that characterize group A, and these characterize group B. Here is a list of Latin binomials. Schopf worked his way through the major groups, but each was only an example of biological theory or process or larger nongroup specific pattern. The class on mammals was about higher level diversity. In most biology classes, the teacher's taxonomic group is clear from day one. What group Schopf had worked on was irrelevant; the ideas of evolution and ecology as applied to the fossil record were everything.

Schopf's exams were backbreaking. He expected students to have the same command of detail that he did, and to muster that detail to defend the ideas. One left convinced of failure, a frequently honest and accurate assessment, but nevertheless knowing more than at the start of the exam.

No one ever had the illusion that Schopf cared about students. What students respected though, was his clear commitment to the field. He thought that paleobiology as an endeavor was clearly *the* most exciting thing to study. In class, he challenged the received wisdom of paleontology and the standard bearers of evolutionary biology. It didn't matter what the students thought of him; what mattered was what they thought about the discipline. If students were willing to learn, he was, in his own way, willing to teach. The students who took his classes all left with not only an understanding of paleontology, but what makes good science.

DAVID RAUP AND THE UNIVERSITY OF ROCHESTER

In the mid-1970s Dave Raup started building the Center for Evolution and Paleobiology at the University of Rochester. He moved the Rochester paleontology group, which included Jack Sepkoski, Dan Fisher, and Curt Teichert, to space in the middle of the biology department. He enticed Bob Selander and Conrad Istock from the biology department into participating. Most of all, he started recruiting graduate students to a department that had never had the reputation in paleontology that places like Chicago and Harvard had.

What Raup did for bringing disparate faculty together, he did double for bringing students together. He promoted an atmosphere where students talked, debated, and collaborated. He organized evening discussion groups but assigned students to lead them. This seems ordinary now, but it was novel at that time. He would start a discussion, and then step back and watch it run away.

One of the most intellectually invigorating outgrowths of the collaborations in the Rochester program was interactions between students and the younger faculty, Dan Fisher and Jack Sepkoski. Relatively new PhDs they were the young turk's young turks. The students looked up to them, and in return they helped build an intellectual give-and-take that characterized the student's perceptions of faculty interactions. They, too, had very different styles. Jack, who helped bring a quantitative and statistical rigor to paleobiology, believed in problem solving and computer programming and data sets. Dan polished ideas as if they were a fine-carved wooden spoon, combining esthetics and utility. But both challenged the students: what do you think, what does your data say, how does it fit in the bigger picture?

The students responded to the collaboration promoted by the faculty by forming an easy community that talked together, worked together, and grew together. Graduate school is often a time of meaningful friendships, but seldom was it as comfortable and intellectually stimulating as it was at Rochester.

As a teacher Dave Raup did not have the religious fervor of Tom Schopf. He used to lecture holding a cup of coffee, a cigarette, and a piece of chalk. Students had a betting pool, from time to time: would he drink out of his coffee cup after knocking cigarette ash into it or would he try and smoke the piece of chalk? But his lectures were adventures in subtle probing. Whether he was teaching introductory historical geology or graduate paleobiology, the material was never distilled, never simplified. Science was complex, and Raup taught it in its fully glorious complexity.

The watchword of the Rochester program was *nomothetic*. Whoever in the larger collaboration originally coined the idea as applied to paleobiology did not matter to the graduate students. It was perceived as a gift from Dave Raup to the students. One could study an organism, a group, a set of taxa, but it was only in the context of a larger question. The student's tribute to this idea was a short-lived, then mimeographed, publication called "The Nomothetic News" (formerly the "Idiographic Ideal").

The message to students at Rochester was that they were free to think about what interested them. They didn't have to have a group, a geologic time, a horizon. Raup, and by extension Sepkoski and Fisher, thought that paleobiology was changing the intellectual landscape, and that students were entirely capable of participating as partners.

STEPHEN JAY GOULD AND HARVARD UNIVERSITY

Steve Gould and the Museum of Comparative Zoology had always been able to attract the best students in paleontology. At the very start of the paleobiology revolution, before the journal was founded, as the MBL group was beginning its work, Dan Fisher and Jack Sepkoski actively chose to work with Steve. He continued to train some of the most interesting paleontologists until his death.

Gould never had to build a graduate program the way that Dave Raup did. Harvard had one of the top-ranked programs in organismic and evolutionary biology. One could be Steve Gould's student and work in Dick Lewontin's lab, or in Wally Gilbert's lab. E. E. Williams or A. W. Crompton could serve on a paleontology thesis committee. The student cohort was large and diverse and full of potential friends.

The Harvard graduate experience was less organized and orchestrated than that at Rochester. There were few required courses, occasional seminars. Lunchtime lab meetings across the entire department, which were far and away the most interesting scientific discussions with faculty, were on a come-if-you-wish basis. Gould extended his ferocious independence to his students. For someone who did not blindly accept natural selection as the be all and end all of evolution, his students faced a stiff selection gradient. The good ones would survive.

One of Gould's greatest gifts to his graduate students was his weekly meetings. To call it a journal club is a misnomer: the papers read were seldom published, but manuscripts people had sent to him for comment. Gould used these discussions to teach not only about the results, but also how to do science. The discussion of Sepkoski's 1981 paper on the three great faunas of the Phanerozoic was memorable. Gould described how this paper was the logical outgrowth of the work on diversity that had come out of the MBL collaborations. But equal parts of the discussion were about what models could do, and why testing them was critical. Gould, who was never the mathematician that Raup was, nor the strong believer in laws of science that Schopf was, said that this paper represented exactly what good science should be, not just integrating across data, but also across ideas. It started with a model, generated by theory, with roots in history. But testing of that model, using data and statistics was critical. Finally, the interpretation of that data, and finding of specific, unexpected results was ultimately one of the most satisfying aspects of doing science.

The same kind of intellectual multitasking, teaching different objectives in one lesson, was true of the graduate teaching experience with Gould. Through the early 80s, as his commitment to the then current incarnation of the fight against creationism grew, along with his outside speaking and writing commitments, Gould taught fewer formal graduate seminars. One teaching commitment he never relinquished was his undergraduate course "The History of Earth and Life." This course, for science and nonscience majors, had a significant lab, which most of his graduate students taught. The graduate students, as usual, had much independence in designing and organizing the labs. The competition among graduate students made this an occasionally uncomfortable proposition, but as a learning process it was invaluable. Gould had specific ideas on what topics needed to be covered, but the students had to figure out ways, on a fiscally and temporally limited budget, to convey the ideas. His input into the design of labs, both in terms of conveying concept and practical advice, was incredible. For a lab on allometry of cranial volume, he lent the original seeds and bearings used in early anthropological studies, acquired for historical value at a point when no one but he cared.

Gould never seemed to be teaching professionalism. Yet the lessons of professionalism were there for students who cared to learn them. He never asked what a student was doing, but he was always willing to read an abstract or discuss a paper before submitting. He listened to countless student rehearsals for talks at meetings, and inevitably pointed out improvements. But he used to shake his head and say that he wasn't very good at it. The lessons were in watching his interactions, and his teaching. He defended teaching large classes of nonscientists on the basis that reaching out to these students was an invaluable chance to educate. The graduate students used to attend the lectures, whether required to or not, to watch him teach. He had a way of speaking to students that reached out them. It wasn't formal, and it didn't feel much like a lecture. From the graduate student perspective it was the same tone and intent that he used when talking to a group of four or five students.

Gould never expected his students to have *his* varied interests, he just expected them to have some varied interests. His open door policy, both in time and in ideas, encouraged the students to develop their own research programs. Gould's students learned by example, and his example was one of both horizontal integration, using taxa as model systems for understanding larger ideas, and vertical integration from history and theory, through data and analysis to interpretation.

LESSONS IN PALEOBIOLOGY

Despite the differences in personality, achievement, and philosophy of evolution, Tom Schopf, Dave Raup, and Steve Gould had many messages in common for their students. First and foremost was that paleobiology was hands down the single most exciting intellectual innovation to come along in evolutionary biology. A graduate student discussion about what constituted a scientific revolution, and whether gel electrophoresis was merely a method to fuel a revolution that had yet to happen, brought a bemused smile to Dave Raup. Another message was that a broad-based approach held endless promise. Tom Schopf taught that the value of paleobiology lay in the scope of geologic time and the freedom from taxa for understanding evolution. Dave Raup pushed students to consider quantitative methods beyond standard statistics and to look for creative ways of expressing and testing evolutionary pattern with large-scale data sets. Steve Gould wanted students to appreciate not only the future, but the historical roots, the theoretical models, and the necessity of bringing the results back to the model. Yet the intellectual interactions among these colleagues led them all to consider alternatives. Dave Raup used G. Udny Yule's work on modeling and stochastic processes in graduate seminars. Schopf understood the need for data to test the models. Gould, despite his protests that he was the historian of the group, offered an occasionally frustrating (because it was so tough) course in quantitative methods and statistics.

Schopf, Raup, and Gould taught a great deal of evolutionary paleobiology to their students. One came away from their classes understanding the scope of the history of life, and the mechanisms that generated that history. Timescales and taxa are hierarchical, as are questions about pattern and process. But beyond these lessons in paleontology were larger messages about science and teaching.

Most teachers maintain that leadership by example is an excellent didactic method. The feelings that students had at Chicago, Rochester, and Harvard was that faculty were too excited and involved with the ideas to worry about examples. Study what you love, they seemed to say (though it is doubtful any of these would have used the word "love" in any didactic situation). Schopf, Raup, and Gould were all passionate about their work, and with or despite different personality quirks they conveyed that passion to their students. Do not be content with button collecting, look for the big idea or the uniting concept. Find a model and develop it. Measure your data, test your model. If the question requires novel methods, invent them. If the problem looks like something in another field, find the people in that field. Collaboration is good, and listen to what the collaborators have to say. Don't be afraid to challenge the paradigm, stepping on toes is good for the toes.

The students trained in these programs absorbed these lessons to varying degrees. They went on to a variety of careers and subdisciplines of science, and carried the lessons of Tom, Dave, and Steve with them. Organized pedagogy and formal mentoring were not crucial during the growth of paleobiology. If Schopf, Raup, or Gould thought about these words, they certainly didn't use them with students. The causality ran in the other direction: the growth and ideas of paleobiology turned the founders of the field into great teachers. The ideographic knowledge about paleobiology and evolution changed a generation of students who became teachers of the same material. The nomothetic reasoning and analysis had an impact outside the bounds of paleobiology.

CHAPTER TWENTY-TWO

The Problem of Punctuational Speciation and Trends in the Fossil Record

Anthony Hallam

As a frequent English visitor to the United States during the 1960 and 1970s, I was privileged to witness at first hand the emergence of paleobiology there, and got to know more or less well most of the leading protagonists. One of the consequences is the book I edited on patterns of evolution as illustrated by the fossil record (Hallam 1977).

My experience as a student at Cambridge in the mid-1950s, reading for a degree in geology, was of a rather dreary, intellectually unstimulating atmosphere. Although the geology department at Cambridge was generally considered to be one of the strongest centers in the country for invertebrate paleontology, the subject was still treated as essentially the handmaiden of stratigraphy. This was epitomized by about 400 stratigraphically arranged fossils kept in a few drawers that undergraduates were expected to learn. The only rationale we could discern was that we were being trained to be nineteenth-century-style geological survey geologists who could go out into the field and distinguish there Silurian from Jurassic. In practice, the only way we could distinguish the stratigraphically long-ranging inarticulate brachiopod *Lingula* from the Cambrian *Lingulella* was that the latter occurred in a black, slaty matrix!

There was, however, one notable strength in the otherwise pedestrian teaching. We were encouraged to undertake detailed analyses of morphology and, where appropriate, a rigorous and careful evaluation of functional morphology. This was especially important in vertebrate paleontology, whose practitioners were trained in zoology and worked in the neighboring Museum of Zoology. A background in vertebrate anatomy is obviously vital in making sense of a great diversity of disarticulated fossil teeth and bone fragments. This very respectable tradition came to the fore during the paleobiological revolution of the 1970s, with the seminal reanalysis of fossils from the Burgess Shale by Harry Whittington and his students Derek Briggs and Simon Conway Morris, and later in the century by research on the Upper Devonian tetrapods by Jenny Clack and colleagues in the University Museum of Zoology, and on the recognition elsewhere in the United Kingdom of the identity of the conodont animal.

This work epitomized the so-called idiographic approach at its best. In the United Kingdom there was a general skepticism toward the alternative nomothetic approach involving conceptual models, and when I was a student and for a very long time afterward, virtually no interest in evolution. I was, however, inspired by George Gaylord Simpson's *The Major Features of Evolution* (1953), an expansion of his ground-breaking *Tempo and Mode in Evolution*, published a decade earlier.

What did begin to emerge in the United Kingdom in the late 1950s and 1960s was an increased interest in fossils as environmental indicators rather than biostratigraphic markers, including study of trace fossils, which no longer could be dismissed as fucoids. This involved the discovery of German research earlier in the century, such as that of Othenio Abel, and particularly the contemporary work of Dolf Seilacher. This newly emerging interest characterized a parallel change in the United States. On my first extended visit in the mid-1960s, on a Harkness Fellowship, the general atmosphere in paleontology that I encountered in university departments of geology was as sleepy as in the United Kingdom, but Yale provided a shining exception. I encountered there for the first time an intellectually stimulating atmosphere, which I was told was generated initially by the biologist G. Evelyn Hutchinson. Most of the drive of the young faculty members Lee McAlester and Don Rhoads, and an exceptional bunch of graduate students including Dick Bambach, Pete Bretsky, Jeff Levinton, and Steve Stanley, was in paleoecology, though one should note Stanley's outstanding monograph on bivalve functional morphology published a few years later (Stanley 1970). This work was especially interesting to me, as it was the macroinvertebrate group I was most interested in for my own Jurassic research as being both the most abundant and diverse, with many modern relatives whose life habits were well known. Hitherto the only Jurassic macroinvertebrates that were paid any serious attention to were the ammonites, because of their biostratigraphic value.

Beside the Yale group there were, of course, exceptional individuals scattered across the country, the most notable being Jim Valentine in California, whose initial interest in Pleistocene mollusks was soon to expand to embrace the general pattern of metazoan evolution. He was also a pioneer in exploring the influence of plate tectonics on fossil distributions (Valentine and Moores 1970). By the end of the 1960s, a general acceptance of continental drift had taken place in the western world, though by the beginning of the 1970s continental drift had been transformed into the more sophisticated and comprehensive plate tectonics. Much of the paleobiological drive in the newly burgeoning field of paleobiogeography came from the United States, especially concerning vicariance biogeography, some of which involved intensive polemical battles, as I experienced at first hand at a meeting at the American Museum of Natural History to which I had been invited (Nelson and Rosen 1981). I was also flattered to be invited to speak at what came to be described as the official centenary conference of Darwin's death, appropriately organized at Darwin College, Cambridge, where I first met Michael Ruse. The title of my lecture (specified by the inviting committee) and subsequently published article (Hallam 1983) was plate tectonics and evolution, an overly ambitious subject, it might be thought, but one that evidently intrigued the molecular biologists.

Following the publication of the seminal paper by Alvarez et al. (1980), a huge interest in the subject of mass extinctions developed, again mainly in the United States, and the controversies surrounding this subject have been a major preoccupation of mine ever since (Hallam 1989b, 2004). What I wish to consider in the remainder of this article, however, concerns an earlier controversy, on modes of speciation as discernible from the fossil record.

THE SPECIATION MODE CONTROVERSY

My first encounter with Steve Gould was in rather unusual circumstances. I had been greatly impressed by a sophisticated review on allometry and size in ontogeny and phylogeny (Gould 1966) and was amazed to learn that it had been written by a mere graduate student at Columbia University, who had just been appointed assistant professor at Harvard, to fill a vacancy created by the removal of Harry Whittington from Cambridge, Massachusetts, to Cambridge, England, to take up the Woodwardian Chair of Geology. In late 1967 I was on my way to a UNESCO conference in Montevideo and chose to stop over at Boston, expressly to visit Steve. I thought that there would be no difficulty in obtaining a hotel room in Boston for an overnight stay, but in my innocence I had descended on the city at the time of the World Series, and there was absolutely no prospect of success. Having explained my predicament to Steve on my first visit to the Museum of Comparative Zoology, he kindly offered me his sofa in the sitting room of the modest apartment that he and his wife had recently rented. Thus began a close friendship that continued until his tragically premature death.

We had a lot to discuss, including our shared admiration for out paleobiological hero, George Gaylord Simpson. Steve expressed his mild frustration that he had gone to Columbia to experience at least close proximity, only to find that Simpson had decamped himself—Simpson had promptly removed himself to Arizona! Our main talking point, however, was our common interest in the Lower Jurassic oyster *Gryphaea*, which had been the subject of evolutionary speculation since quite early in the century. Both Steve and I had done some biometric work, and were skeptical, for a variety of reasons, of Trueman's celebrated hypothesis.

Several years later I was able to invite Steve to spend a few months in Oxford, where I found him an office in the University Museum of Natural History, a location that intrigued him as the site of the celebrated—if now notorious— debate between Huxley and Wilberforce over a century earlier. While in Oxford we investigated another evolutionary succession of Middle and Upper Jurassic *Gryphaea* species (Hallam and Gould 1975). The statistical work was done, incidentally, by one of Steve's students at Harvard, Jack Sepkoski, who of course subsequently became a leading light in the Chicago school of paleobiology.

As regards punctuated equilibria, I was fortunate to be present at a Geological Society of America symposium in 1971 organized by Tom Schopf entitled "Models in Paleobiology" (one of the first uses of the term) to which I had been invited by Tom to speak on the topic of Models in Population Dynamics. By far the most famous paper emerging from the symposium (Schopf 1972) was the one by Eldredge and Gould (1972) on punctuated equilibria. Though it did not create much of a stir at the meeting, their provocative work subsequently aroused a lively debate that only began to fade in the mid-1980s, as other controversial topics, such as mass extinctions, came to the fore.

With respect to my own research, I was stimulated by the punctuated equilibrium hypothesis to undertake a survey of Jurassic bivalve species in Europe, based on a combination of examination of museum specimens in several countries, study of monographs, and fieldwork. Because of the parochialism and narrow typological concepts adopted by nineteenth-century paleontologists, there was a vast excess of species names, but I was able to reduce the number to several hundred taxa as they would be recognized today, using modern concepts allowing for a range of variation. The results were unequivocal: the overwhelming speciation mode was punctuated equilibria (Hallam 1978) with the only convincing case for phyletic gradualism being the early Jurassic *Gryphaea*. Subsequent work on this genus indicated a good case of a paedomorphocline (Hallam 1982; Gould 2002). The extent to which the change within the three successive species *arcuata*, *mccullochi*, and *gigantea* is punctuational or gradualistic has been the subject of an argument between myself and one of my former students (Johnson 1994; Hallam 1998). The results I obtained were of course, *Gryphaea* apart, based only on a broad survey, but the detailed taxonomic analyses of two families with distinctive morphologies—the Pectinidae and Trigoniidae—by two of my students came to similar results. In particular, my estimated species durations proved in some cases to be underestimates rather than the overestimates that could have been discerned from older literature.

The stimulation provoked by the Eldredge and Gould hypothesis came to a head at a symposium on macroevolution held in 1980 at the Field Museum of Natural History in Chicago, attended by leading lights in the field of evolutionary biology as well as a number of paleobiologists; this was the first such meeting ever held. Needless to say, there were some lively interchanges, especially those involving population geneticists, who were ill at ease with the proposed decoupling of macro- from microevolution. I was personally gratified by the response to my intervention from the floor to outline the essence of my Jurassic bivalve results using chalk and blackboard (something now impossible to do in these days of exclusive PowerPoint projection). There was a stirring from among the largely biological audience of a sort that I have never experienced before or since, which was essentially their response to a recognition of a whole field of research relevant to evolutionary studies of which they had hitherto been unaware.

This was my modest contribution to paleobiologists demanding a seat at the High Table of evolutionary studies. What was evidently most disturbing to the population geneticists, who were the dominant evolutionary biologists at that time (though since eclipsed by the developmental geneticists) was the concept of evolutionary stasis, that so many species could have persisted unchanged in their morphology for millions of years. The relatively sharp punctuational change from an ancestral to a descendent species was not seen as a problem, because the inferred geological instant could have extended for thousands or even tens of thousands of years, which was more than enough time for the most dramatic morphological changes readily accountable by conventional neo-Darwinian theory.

SPECIES SELECTION

With the advent of punctuated equilibrium theory, following the seminal article by Eldredge and Gould (1972), the need arose to account for phyletic trends in the fossil record without recourse to anagenesis, which implies gradual change in species lineages through time. This has been attempted by punctuationists promoting the theory of species selection. This is especially true of Gould, who persisted until his magnum opus, published shortly before his death (Gould 2002) in promoting species selection as a key level of selection above that of individuals, as part of his hierarchical model of evolution. Whereas I am a strong supporter of punctuated equilibria, I differ strongly on the subject of species selection (Hallam 1998). Recent revival of interest in heterochrony (McKinney and McNamara 1991), which was defined by Gould (1977, 2) as the "changes in relative time of appearance and rate of development of characters already present in ancestors" has demonstrated that many phyletic trends can be accounted for without invoking cladogenesis (Hallam 1989a; McNamara 1990). Lower Jurassic *Gryphaea* provides a good example (Hallam 1982).

As Erwin and Anstey (1995) point out, speciation is one of the most intractable problems of evolutionary biology. The process of speciation generally requires too long a duration to be directly observable by biologists, who can only extrapolate from the populational events they observe, whereas for paleobiologists the fossil record has too poor a temporal resolution to allow study of the detailed history of speciation events. Furthermore, they suffer from lack of access to many of the characters that define species. In effect, both biologists and paleobiologists are obliged to rely on indirect evidence. There remains among biologists today neither a unified definition of species nor a unified explanation of mechanisms of speciation (Williams 1992; Dennett 1995). The debate among paleobiologists about speciation in the fossil record is of necessity about the evolution of morphospecies. Although problems undoubtedly persist, for instance the phenomenon of sibling species, it appears from a number of biological studies that the general paleobiological assumption—that morphological transitions observed in the fossil record are a valid proxy for speciation—is a reasonable one (Gould and Eldredge 1993; Erwin and Anstey 1995).

Both gradualistic and punctuational models of speciation pose problems for understanding the underlying evolutionary mechanism. Calculated selection coefficients seem to be too weak to justify conventional natural selection as the cause for long-term patterns, including sustained gradualism (Erwin and

Anstey 1995), but an alternative, such as genetic drift, does not seem plausible because it fails to explain why a trend should continue in one direction for a long period of time. The most promising explanation involves heterochrony (McNamara 1990). The most interesting aspect of punctuated equilibrium is the widespread acceptance by paleobiologists today of stasis in the evolutionary history of species, ranging up to millions of years, a fact that was not anticipated by population geneticists. This stasis is not seen as reflecting complete immobility but instead as signifying fluctuations of little or no accumulated consequence, and temporal spread within the range of geographic variability among contemporary populations (Gould and Eldredge 1993).

An apparent paradox emerges. Species lineages may be very stable, but adaptive diversification in evolution is rampant, so how can adaptive change be concentrated phylogenetically around episodes of development of reproductive discontinuity (Williams 1992)? A possible explanation, attributed to Douglas Futuyma, proposes that morphological change may accumulate within species, but unless it can be "locked up" by reproductive isolation (= speciation) it will be eliminated during subsequent interbreeding (Gould and Eldredge 1993; Gould 2002). Alternative explanations that have been offered to account for stasis involve stabilizing selection and developmental constraints. If stabilizing selection is to be a prominent mechanism for mediating long-term stasis, either communities must be at equilibrium and/or the physical environment must be invariant over long periods of time. However, most evidence from both fossil and extant communities suggests that neither assumption holds true. Instead, communities are not at equilibrium and selection pressures vary in strength and constancy. The most likely outcome of environmental change is biogeographic change—in other words, environmental tracking. Thus habitat tracking is an important ingredient of stasis (Eldredge 1995). An outstandingly well-documented example concerns Quaternary beetles, researched by a former colleague of mine, Russell Coope. It might reasonably be expected that such a speciose group as beetles would speciate in response to the slightest environmental fluctuations, as it were, at the drop of a hat. Instead, beetle species have remained stable through a time of strongly fluctuating climates, with a succession of latitudinal migrations corresponding to alternating episodes of warming and cooling (Coope 1994).

RIVAL EXPLANATIONS FOR EVOLUTIONARY TRENDS

The traditional explanation for phyletic trends in the fossil record has been in terms of lineage anagenesis, which may be either gradualistic or punctuational. The alternative, within the framework of punctuated equilibrium theory, considers trends as the products of differential speciation, with speciation not a process of transformation but of replication of lineages. In the words of Gould (1982, 85) this involves repeated cladogenesis (branching) filtered through the net of differential success at the species level. Gould actively promoted what he called *species selection* as a major challenge to conventional Darwinian evolutionary theory, in a series of publications culminating in his swan song (Gould 2002). He argued that if trends primarily occur within a pattern of punctuated equilibrium, and if the differential success of species that must power such trends arises from truly emergent, special-level selection, then the Darwinian model of macroevolution as extrapolated from selection among organisms must fail and trends have legitimate authority at what he called "the second tier," implying a hierarchical model of evolution.

There is, however, an alternative punctuated equilibrium explanation, now generally interpreted as species sorting, which offers a less radical challenge to Darwinian theory (Vrba 1989; Eldredge 1995). Sorting is a simple description of differential birth-and-death processes among reproducing entities at whatever level they may be situated, and is not a causal explanation. For a process of species sorting to qualify as species selection, it must have a measure of independence from selection among organisms, and only this requires emergent structures and dynamics. As Dawkins (1982) states, species selection cannot shape an organismal adaptation, and hence tends not to be treated by evolutionary biologists as a major factor in evolution. Selection at the level of the individual is bound to remain immensely more important because of the vastly greater frequency of opportunity for selection. Even more dismissive is another distinguished neo-Darwinian biologist, who refer to species selection as a "species fallacy" (Williams 1992).

A reasonable question that can be posed by paleobiologists concerns the supporting evidence. Unlike the widely discussed and supported theory of punctuated equilibrium, which has generated a huge literature, species selection has been almost totally ignored by the paleobiological community, and even the paradigm cases have been challenged (Hallam 1998). Whether one prefers species selection or the less evolutionarily controversial species sorting, both imply a multiplicity of species of the clade under consideration at given stratigraphic horizons but, as I have observed (Hallam 1989a) the supporting evidence is commonly lacking. Whereas some larger-scale trends clearly involve cladogenesis, with a species bush rather than a simple lineage (a good example is provided by the modern interpretation of horse evolution) a large proportion of what have been described as trends appear to involve anagenesis, commonly in conjunction with heterochrony (McNamara 1990; McKinney and McNamara 1991). It can indeed by argued that heterochrony is a crucial factor in the generation of evolutionary trends, with the focus of attention shifting to intrinsic factors related to ontogeny.

This last subject is, of course, one of burgeoning interest, with the emergence within the last couple of decades of developmental genetics. The subject of evolution and development (evo-devo for short) has indeed become one of the new fashions. It is somewhat ironic that what was probably Steve Gould's greatest scholarly contribution to the study of evolution, his book on ontogeny and phylogeny (1977), was essentially sidetracked by him as he pursued his hierarchical model of evolution involving species selection. However, in his defense, he pointed out to me on more than one occasion in the last decade of his life, the scientific tools to take the matter further—in other words those of developmental genetics—were not available earlier.

REFERENCES

Alvarez, L. W., W. Alvarez, F. Asaro, and H. V. Michel. Extraterrestrial cause for the Cretaceous—Tertiary extinction: Experimental results and theoretical interpretation. 1980. *Science* 208:1095–1108.

Coope, G. R. 1994. The response of insect faunas to glacial-interglacial climatic fluctuations. *Philosophical Transactions of the Royal Society of London* B344:19–26.

Dawkins, R. 1982. *The extended phenotype.* Oxford: Freeman.

Dennett, D. C. 1995. *Darwin's dangerous idea.* London: Penguin.

Eldredge, N. 1995. Species, speciation, and the context of adaptive change in evolution. In *New approaches to speciation in the fossil record,* ed. D. H. Erwin and R. L. Anstey, 39–63. New York: Columbia Univ. Press.

Eldredge, N. and S. J. Gould. 1972. Punctuated equilibria : An alternative to phyletic gradualism. In *Models in paleobiology,* ed. T. J. M. Schopf, 82–115. San Francisco: Freeman Cooper.

Erwin, D. H., and R. L. Anstey. 1995. Speciation in the fossil record. In *New approaches to speciation in the fossil record,* ed. D. H. Erwin and R. L. Anstey, 11–38. New York: Columbia University Press.

Gould, S. J. 1966. Allometry and size in ontogeny and phylogeny. *Biological Review* 41:587–640.

———. 1977. *Ontogeny and phylogeny.* Cambridge, MA: Harvard University Press.

———. 1982. Darwinism and the expansion of evolutionary theory. *Science* 216:380–87.

———. 2002. *The structure of evolutionary theory.* Cambridge, MA: Harvard University Press.

Gould, S. J., and N. Eldredge. 1993. Punctuated equilibrium comes of age. *Nature* 366:223–27.

Hallam, A., ed. 1977. *Patterns of evolution as illustrated by the fossil record*. Amsterdam: Elsevier.

———. 1978. How rare is phyletic gradualism? Evidence from Jurassic bivalves. *Paleobiology* 4:16–25.

———. 1982. Patterns of speciation in Jurassic *Gryphaea*. *Paleobiology* 8:354–66.

———. 1983. Plate tectonics and evolution. In *Evolution from molecules to man*, ed. D. S. Bendall, 367–86. Cambridge: Cambridge University Press.

———. 1989a. Heterochrony as an alternative to species selection in the generation of phyletic trends. *Geobios*, Mem. Sp. 12:193–98.

———. 1989b. *Great geological controversies*, 2nd ed. Oxford: Oxford University Press.

———. 1998. Speciation patterns and trends in the fossil record. *Geobios* 30:921–30.

———. 2004. *Catastrophes and lesser calamities: The causes of mass extinctions*. Oxford: Oxford University Press.

Hallam, A., and S. J. Gould. 1975. The evolution of British and American Middle and Upper Jurassic *Gryphaea:* A biometric study. *Proceeding of the Royal Society of London* B189:511–42.

Johnson, A. L. A. 1994. Evolution of European Lower Jurassic *Gryphaea (Gryphaea)* and contemporaneous bivalves. *Historical Biology* 7:167–86.

McKinney, M. L., and K. J. McNamara. 1991. *Heterochrony: The evolution of ontogeny*. New York: Plenum.

McNamara, K. J. 1990. The role of heterochrony in evolutionary trends. In *Evolutionary trends*, ed. K. J. McNamara, 59–74. London: Belhaven.

Nelson, G., and D. E. Rosen, eds. 1981. *Vicariance biogeography: A critique*. New York: Columbia University Press.

Schopf, T. J. M., ed. 1972. *Models in paleobiology*. San Francisco: Freeman Cooper.

Simpson, G. G. 1953. *The major features of evolution*. New York: Columbia University Press.

Stanley, S. M. 1970. Shell form and life habits in the Bivalvia (Mollusca). *Geoogicacal Society of America Memoir* 125.

Valentine, J. W., and E. M. Moores. 1970. Plate-tectonic regulation of faunal diversity and sea level: A model. *Nature* 228:657–59.

Vrba, E. S. 1989. Levels of selection and sorting with special reference to the special level. *Oxford Surveys in Evolutionary Biology* 6:111–68.

Williams, G. C. 1992. *Natural selection: Domains, levels and challenges*. New York: Oxford University Press.

CHAPTER TWENTY-THREE

Punctuated Equilibrium versus Community Evolution

Arthur J. Boucot

INTRODUCTION

Many problems in science are related to sampling. The concept of punctuated equilibrium is a prime example. It suggests that for most species of the past, including those from which the modern biota evolved, the actual change from one species of a genus to another species of that genus was a geologically abrupt, allopatric event. Eldredge and Gould's (1972) concept postulates that most species-level evolution takes place in very small populations geographically situated at or near the margins of the parent species' distribution (outer limits of species A in fig. 23.1). The concept is nonecological, and unsupported by ecological evidence from the present or the past. It disregards gradual evolutionary change as a significant process in speciation and assumes there is no evidence for moderate rates of evolution, only extremely rapid rates or stasis.

The concept of community evolution also involves sampling questions. The concept is diagrammed (Boucot 1978, fig. 1; 1982, fig.1; and 1983, fig. 10), concluding that within an evolving community group, over time, there is an inverse relation between rate of morphological change in species and relative abundance; that is, abundant genera show little evidence through the duration of the community group for lineal speciation, anagenesis, whereas uncommon-to-rare-genera do show such evidence, manifested by nonreversible morphological changes. Community evolution supports the concept of gradual morphological change where ancestral and descendent species within

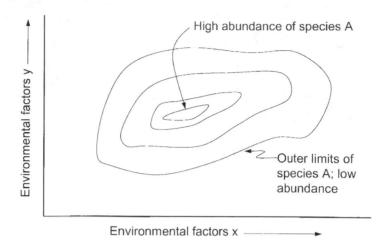

Figure 23.1 Diagram indicating that any species will environmentally be characterized by an area, or areas, of maximum abundance from which abundance will decline peripherally to zero. Near the outer limits of any species' distribution the occurrences will be in environments distinct from those present in the more central regions of high abundance.

the uncommon-to-rare-genera occur at rates inversely proportional to relative abundance within a community group.

The two opposing concepts stand or fall on the validity of the sampling that allegedly supports one while refuting the other.

PUNCTUATED EQUILIBRIUM

It is appropriate to consider the concept of punctuated equilibrium in a volume devoted to how the fossil record provides a better understanding of the evolutionary process. The concept has captured the imagination of students down to the middle school textbook level as the modal form of species-level evolution as demonstrated by the marine, benthic megafossil record. The concept of punctuated equilibrium is appealing, particularly to biostratigraphers, since it implies that there is a morphological time gap between most species of a genus. Common presence of such morphological gaps would make life easier for the biostratigrapher than would the more difficult task of dealing with gradational relations between species. If correct, the concept would make it easier to date rocks with fossils, since there would be no morphological overlap between most species of a genus. This is not the case when one consults practicing biostratigraphers or neontologists. Biostratigraphers are well aware of the difficulty in assigning many collections of a form to a species— that is, does it belong to a previously described species, is it a new species, or is it intermediate morphologically between two previously described species, or does it merely represent variation within a previously described species? The uncommon-to-rare-genera and their species, the bulk of the taxa in the fossil record, generally belong to lineages with many known gaps between their species. These are generated by the frequent environmental changes that have occurred over time at any one locality; stenotopic taxa of one community group being replaced by stenotopic taxa of another community group, and from the habitual failure to provide large enough collections to obtain the uncommon-to-rare-genera. These gaps, most often caused by environmental changes between communities, can be interpreted in a variety of ways. However, when morphological gaps between species are filled now and then, it is almost invariably the result of gradational, phyletic/anagenetic relations between the species.

Allopatry

The concept of allopatric speciation was emphasized by the late Ernst Mayr in a lengthy series of papers and books. Mayr's taxonomic work in New Guinea and adjacent southwestern Pacific islands on the bird faunas led him to conclude that most species originated allopatrically. His work on these islands was then extended to major portions of the diverse modern nonmarine fauna, and by implication, to marine faunas as well. His conclusion that allopatry is the most reasonable explanation for the origins of most of the modern biota makes good sense. However, as far as I know, Mayr never addressed the question of when did this allopatry originate for individual species or faunas. Is it relatively ancient in origin, following which there was extended phyletic evolution, or did it occur very recently?

When one considers the fossil record it appears that it is consistent with the overall importance of allopatry. The pattern appears to be that following an extinction event there is shortly thereafter a reorganization of the survivors into many new community types that incorporates those taxa produced at the same time by radiations.[1]

This gives rise to new ecological-evolutionary units and subunits. These taxa, survivors and new ones, then evolve vicariantly, phyletically within their respective community groups. However, if, later in time, barriers to reproductive communication are introduced, there then is diacladogenetic splitting of many taxa into descendant taxa (Boucot 1978, fig. 3) that will also continue to evolve phyletically within their respective biogeographic units. In order to follow the evolutionary path one needs to consider the effects of extinction and subsequent radiation, with attention being given to both community evolution and possible biogeographic changes.

Eldredge and Gould's "genius" was that they took the data of the fossil record ranging from abundant-to-rare genera and their species, then assumed that the extensive gaps in that sampling record for the more numerous genera (the uncommon-to-rare-genera) represented punctuated events rather than gradualism. While the negative evidence was used in the punctuated interpretation, the far less abundant evidence for gradualism provided by some of the more common genera and their time sequence of species was ignored as minor rather than as showing overall gradualism. One needs to keep in mind that when additional, new samples of rare genera are provided between the gaps they are usually described as new species based on minor but consistent morphological features, thus narrowing the gaps between the alleged parent and descendant species of that genus. Fortey (1985) very cleverly made it clear how unreasonable the sampling "requirements" posed by the punctuated equilibrium concept were to arrive at a gradualistic interpretation of samples from the fossil record. I have provided tabular compilations (1978, 1990) of the relatively minor number of taxa for which there is good evidence for gradualism. Simpson (1943; and personal discussion in the early 1980s) adhered to the concept of mammalian evolution being characterized by "chronoclines" featuring gradual, species-level evolution rather than a punctuated mode.

The Argument

The punctuated equilibrium concept is based on the assumption that speciation within a genus most often takes place at the very outer geographic limits of the parent species—that is, it is an allopatric model. It assumes implicitly that the allopatrically evolved, geographically peripheral species have the inherent capability of invading the more central, higher-abundance area (fig. 23.1) occupied by the parent species and replacing the parent population of the species. There is no evidence from the present to support this assumption. If the concept was valid, one should find evidence of geographically marginal, small populations of various species invading and replacing the parent populations. These small populations of allopatrically derived species originated in a somewhat different environment from the more central and denser parent species. If such replacement were the case we would have numerous examples of later

Quaternary, nearshore island-endemic species reinvading the mainland and replacing the parent species. Pygmy proboscidians, rhinos, deer, and so on, and giant rodents and lagomorphs should have reinvaded the mainland and replaced the parent populations. (One can now include the recently described [Brown et al. 2004] small-bodied hominins from the island of Flores, Indonesia!) However, there is no evidence of this occurring. Neither is there such evidence from the curious endemics on some isolated mountaintops nor for the subterranean obligate troglobites. The endemic taxa found in many ancient lakes, modern and from the past, show no evidence that they provided taxa that reinvaded the rivers from which they were derived. The concept is based on negative evidence. Also exceptional to their conclusions are examples of heterochrony (see Raff et al. 1990, for a useful discussion), where rapid morphological changes have taken place without evidence of intermediate forms (no discussion of this possibility is ever made by the advocates of the punctuated concept, nor the evidence favoring heterochrony provided by paleontological evidence).

Additionally, it has been the habit of taxonomists when describing a species to prepare a diagnosis that details the characters that separate the new species from other species within the genus; most editors are reluctant to accept descriptions indicating a level of uncertainty; that is, is there a reasonable possibility that the species being described is really not very distinct or gradational with previously described species, or is the result of intraspecific variation? It is usually easier to separate species based on published diagnoses than it is with actual samples—that is, easier with the literature than with the study of specimens.

There is also the problem (Boucot 1978, 566) that some biostratigraphers have the unfortunate habit of uncritically assigning specific names to material without really doing a thorough investigation and critical comparison of the material. This tends to give a false sense of species-level continuity through time when there may be none. This problem is more prominent in faunal lists unaccompanied by careful taxonomic descriptions.

In selective terms, the concept of punctuated equilibrium ignores the fact that selection pressures in the area of a species' greatest abundance will normally be very different from those at its low-abundance periphery. The communities present in these two regions will be distinct, reflecting different environments and selective pressures. None of these evolutionary factors have been discussed or considered by proponents of the punctuated equilibrium concept.

Turning to the problem of rates of evolution among species, Simpson (1953, 360) strongly advocated a gradual change and varied rates of evolution. The punctuated equilibrium concept only recognizes lengthy intervals of stasis, broken by brief intervals of very rapid evolution . . . no intermediate rates, which doesn't agree with the fossil record in the experience of most practiced taxonomists (Gingerich 1976, 1977, 1984, 1987).

Alleged support for the concept of punctuated equilibrium is provided by the fact that most described genera and their species differ in one or more morphological characters from species alleged to have evolutionarily descended from them. From a sampling perspective, this reflects the basic fact that most genera, past and present, are monotypic or represented by only a very few species, that this restricted number of species is frequently very stenotopic and biogeographically provincial—that is, members of a relatively small population. Because of these small populations it is critical that very large fossil collections be made in order to capture a sample of these essentially uncommon-to-rare species. This is seldom done. The geologists, who have usually acted as the fossil collectors, have seldom made large fossil collections. For example, Whittard (1966, 298), a distinguished, hard-working trilobite worker commented "[G]enerally it is only by extensive collecting that the rarities are found. . . ." Because of this sampling failure it is no wonder that a more complete fossil record of the rapidly evolving, uncommon-to-rare-genera and their species is lacking. All we ordinarily have are some examples of gradual, species-level change among the less common genera and their species. There is little or no evidence for such changes for most genera, those that are uncommon-to-rare normally represented through time by a single species. These less common genera and their species usually form a very small proportion of the genera and species present during any time interval. It is this sampling artifact that supports the theory of punctuated equilibria and misdirects its followers.

In view of the above, if one takes a narrow, legalistic viewpoint, while it is true that most species of most genera are morphologically distinct from one another, such acceptance ignores basic sampling problems! In actuality, all samples supporting gradualistic changes are restricted to the small percentage of the less common genera.

Eldredge's doctoral thesis on late Middle Devonian phacopid trilobites (1972) from the central Appalachians and adjacent regions in the midcontinent involve two time-successive forms of the trilobite *Phacops*, and has been allegedly the prime example of punctuated equilibrium. Eldredge studied a number of trilobite assemblages in New York and adjacent states that were isolated from each other in time by meters of trilobite-barren beds. Eldredge and Gould (1972) appealed to one geographically isolated, poorly correlated

Phacops collection containing the younger species (the seventeen file subspecies of *Phacops rana*) as being from beds contemporaneous with the ancestral species; however, no solid evidence favoring this conclusion was provided. They then viewed the poorly correlated collection as representing an allopatrically derived form of the ancestral species that later reinvaded and replaced the descendent species in its area of greater abundance. They concluded that most species present in the fossil record were morphologically distinct from other cogeneric species, suggesting that one could infer on purely negative evidence that the transition had occurred allopatrically somewhere else. As pointed out earlier, most species occur in uncommon-to-rare-genera where the nature of sampling results in abundance gaps in the record of many potential lineages. The negative evidence provided by incomplete sampling shows that these gaps can have many interpretations.

In 1971 Eldredge and Gould presented their concept orally at the annual meeting of the Geological Society of America. My own initial reaction to the concept was negative (I recall getting strong support from the audience!). Why? Because I had been working for over twenty years for the U.S. Geological Survey's Paleontology and Stratigraphy Branch, where my duties involved age-dating Silurian and Devonian brachiopod collections sent in for study by field geologists, chiefly from U.S. locations. While doing this work I slowly realized that within the time interval, which turned out to be mostly late Lower Silurian through Devonian, I was repeatedly recognizing the same faunas during geologically different time intervals. I had difficulty being certain about specific identities of some forms; had they been previously described or new species . . . the decision was complicated due to inadequate older specific descriptions and the possibility that specifically intermediate forms were involved. Eldredge and Gould published their concept in 1972, virtually assuring that most collections of the same genus would be represented by species characterized by distinct morphological gaps between ancestor-descendent species. In 1977 Gould and Eldredge tried to dismiss all of the previously published records of gradualistic change, although in the same year Gould published a volume (1977) devoted to significant morphological changes involving heterochrony; that is, another route that does not call for an allopatric explanation at all. I wrote up my objections (1978). Not everyone agreed with Eldredge and Gould, as indicated by varied papers opposing the concept for one reason or another. I was disappointed when eventually it became obvious that the objections were ignored by the paleobiologists (as contrasted with the old-fashioned paleontologists who actually worked with fossils and biostratigraphy). I consider Gould, who never actually published any taxonomic study nor mastered any group of fossils or stratigraphic interval, to belong to the former group. The paleobiologist's concept, while being very logical, often ignores facts in the fossil record that are not in accord with what we know about the ecological tolerances of organisms and their evolution. Discussions during the early 1980s with G. G. Simpson made it clear that he felt the punctuated equilibrium concept made little sense, based on his own knowledge of vertebrate fossils, as indicated in his 1943 paper on the implications of a mammalian chronocline.

The paleobiologists had, and still have, little tolerance for alternate views of the fossil record. My experience with a short manuscript submitted in 1977 to *Paleobiology* is typical. The manuscript, which criticized an earlier paper by the editor (Schopf et al. 1975) spent a year in review, following which I received a rejection letter. I then submitted the manuscript to the *Journal of Paleontology*. The editor of that journal asked G. G. Simpson to review the manuscript, which he did. He saw nothing wrong with the manuscript, but indicated that it should be published in *Paleobiology,* the journal in which the paper being criticized was published, rather than in the *Journal of Paleontology*. The editor of the *Journal of Paleontology* then answered that the manuscript had been rejected earlier by *Paleobiology*. The manuscript was then published in the *Journal of Paleontology*. My experience is far from unique.

Furthermore, there has been some confusion between Simpson's (1944, 1953) Quantum Evolution and so-called punctuated equilibrium, probably engendered in no small part by Gould and Eldredge's (1977) paper. These are *not* synonyms! Simpson, the leading evolutionary paleontologist of the twentieth century, recognized the need for an evolutionary explanation dealing with the virtually complete absence of genera and species intermediate between families and higher taxa thought to enjoy an ancestor-descendant relation. Simpson's *apologia* was a suggestion, consistent with the precepts of microevolution, that absence from the fossil record or the present of such intermediate genera and species could be explained as follows: (1) very small populations; (2) very rapid evolutionary changes; (3) conditions of high endemism. When these three conditions are satisfied the chances of finding intermediate genera and species in the fossil record become vanishingly small, as is also true for the present. So-called punctuated equilibrium, on the other hand, is an attempt to allopatrically explain the absence of transitional species belonging to a genus with transitional populations present in the fossil record. As previously detailed, this frequent absence of intermediate, morphologically transitional populations of uncommon-to-rare species is mostly a sampling problem.

HOW ARE STRATIGRAPHIC RANGES
OF FOSSILS DETERMINED?

Before beginning any discussions on community evolution, the evolutionary aspects of associations of fossils, extinctions, and radiations, it is helpful to know just how the stratigraphic ranges of fossils are determined. For almost two hundred years many geologists—those now considered biostratigraphers—have concerned themselves with working out the stratigraphic ranges of fossils. Today, and in the past, this is mostly done by reference to well known, rapidly evolving groups with a wide geographic distribution and relatively short stratigraphic ranges. For example, in the Paleozoic today reliance in the Cambrian is largely on offshore, deeper water agnostid and olenid trilobites, in the later Cambrian through the earlier Devonian on graptolites and conodonts, beginning in the later Early Devonian through the Permian on goniatites and also on conodonts. Chitinozoans are today assuming greater importance in the Ordovician through much of the Devonian. The major groups relied on have changed through time as knowledge about them has improved; in d'Orbigny's day the Paleozoic relied heavily on rugose corals and brachiopods, supplemented by trilobites. The groups being relied on are engaged in a kind of horse race, with prominence given to groups changing over time. For example, until intensive work on Foraminifera began in the 1920s, due to impetus from the petroleum industry, they were not of much biostratigraphic value in the Cenozoic or later Mesozoic. We still rely largely on fossils for biostratigraphy, although in principle the future should bring heavier reliance on both radiogenic and stable isotope standards, as well as on magnetic reversal time scales.

COMMUNITY EVOLUTION

Introduction

Introductory discussions of evolution have often been accompanied by a tree of life diagram featuring an amoeba-like, protozoan blob at its base giving rise upward to varied branches culminating in the highest vertebrates, ourselves on side branches, with varied invertebrates such as the insects and other morphologically complex animals on other side branches. This view of evolution emphasizes changes in various groups of animals over time. It suggests that such changes are transitional, although a time ordinate is not provided for the diagram. This overly simplistic view of evolution has been refuted since the mid-nineteenth century by the fossil record. The fossil record very positively indicates that there are numerous breaks in the transitions from one group of animals, or plants, for that matter, to another. Georges Cuvier, at the beginning of the nineteenth century, employing vertebrate evidence chiefly from the Mesozoic and Cenozoic, with which he was familiar, was the first to recognize the presence of significant extinctions and what we now consider subsequent radiations among some of the survivors. His younger contemporary, Alcide d'Orbigny, in the mid-nineteenth century extended Cuvier's concepts to the marine invertebrate fossil record.

My purpose here is to consider the evolutionary significance of fossils between the many extinctions and subsequent radiations recognized in the fossil record, while incorporating data on community reorganization that follow extinctions, together with subsequent biogeographic changes involved with the cosmopolitan to provincial spectrum. However, before discussing their evolutionary implications we need to consider the nature of the fossil record, frequently considered to be a record of the history of life.

The Fossil Record

Unfortunately, the fossil record definitely does *not* record the history of life! It is important to consider what it records and what it does not record before arriving at any evolutionary implications regarding successive extinctions and subsequent radiations.

First, the record of the nonmarine—both terrestrial and aquatic—organisms is far less complete than that of marine organisms, owing to the physical effects of both erosion and oxidation in particular, and the biological action of bacteria, fungi, and scavengers. In the nonmarine realm there is essentially no fossil record for montane organisms, since erosion quickly destroys their remains. The few bits that survive become submerged in the overwhelming biomass of lowland and valley organisms, with most valleys being at lower elevations. Cave organisms are essentially unknown prior to the later Pleistocene, owing to the effects of erosion in quickly removing the relatively near surface cave environments. Desert environments are subject to rapid, destructive erosion accompanied by strong oxidation. Lake deposits are far better environments for nonmarine preservation, but they mostly involve lacustrine and nearshore biota, plus that swept in from nearby riverine sources.

Turning to the far better record of life preserved in the marine environment we find, first, that soft-bodied organisms have a very poor fossil record, owing to the activity of scavengers, bacteria, and fungi, as well as sediment feeders. A few unusual deposits, so-called Lagerstätte, actually provide a good record of co-occurring soft-bodied and well-skeletonized organisms (see Whittington 1985, fig. 5.2, for the famous Burgess Shale Middle Cambrian biota, and Chen Junyuan et al. 1991 for the more recently discovered but equally useful Lower Cambrian Chengjiang biota). But overall our knowledge of fossilized, soft-bodied marine megabiota is very poor, although they form a majority of the living marine taxa, and presumably did so in the past, if the two cited Cambrian examples and a few others in younger strata are any guide. Our knowledge of abyssal benthos is abysmal, owing both to the poorly skeletonized nature of many taxa and the vigorous activities of deposit feeders in recycling nutrients of dead organisms; the only exceptions being a few ostracods and arenaceous Foraminifera. Oceanic, skeletonized plankton, such as Foraminifera, Radiolaria, diatoms, coccoliths, pteropods, and the like, are abundant in the Cretaceous-to-present abyssal sediments, but poorly mineralized organisms such as copepods are totally absent. Pre-Cretaceous abyssal benthic and planktonic animal remains were presumably removed by seafloor spreading and subsequent intense metamorphism of the subducted strata. For this reason geologists have difficulty recognizing pre-Cretaceous oceanic deposits, except for a few Jurassic remnants. We have no fossil record for very small, interstitial organisms living in the upper sediment layers. However, there is an entire class consisting of microscopic remains with organic-walled exoskeletons, such as acritarchs and chitinozoans, plus less well known groups, that have only recently begun to be studied by palynologists. So . . . our rich marine record is essentially restricted to well-skeletonized, largely megascopic organisms abundantly preserved in relatively shallow water, continental shelf depth environments of the past, especially including the vast epicontinental seas that intermittently covered large parts of the continents.

All that now follows is based on the assumption that the fossil record of the well-skeletonized continental shelf depth organism from the Cambrian to the present is a sample representative of the majority of past organisms; if not, what follows has very limited significance. Also to be considered is the fact that the younger strata, in the Cambrian-to-present sequence, have received almost exponentially more study than the older; that is, we know far, far more about the Miocene biota than that of the Cambrian, which is reflected in the number of specialists working in each time interval, Cambrian through Cenozoic.

After having considered the overall nature of the sample we next need to examine just how it is distributed through evolutionary time. In particular, we need to analyze the many extinctions and subsequent radiations that have molded the fossil record into a time sequence of basic subdivisions, the ecological-evolutionary units and subunits.

Ecological-Evolutionary Units and Subunits

D'Orbignyan étages. Alcide d'Orbigny (1849–1852, 1850–1852) made it clear by means of extensive documentation of varied fossils, chiefly marine mega-fossils, that the fossil record is characterized by a number of specific time intervals, each characterized by a largely unique set of taxa, chiefly genera in the mid-nineteenth-century sense. Many of these genera have since been split into additional genera, but the overall nature of d'Orbigny's intervals remains the same. D'Orbigny visualized these intervals, his *étages* and *sous-étages*, as bounded above by Cuverian-type extinctions and below by appearances of new taxa that we would now consider as in radiations. Today we have an appreciation of the absolute time involved in d'Orbigny's units and recognize that they are of an episodic, not periodic nature (Boucot 1994). Within the Phanerozoic there are a large number of his units; I estimated (1996, table 2) that from the Cambrian to the present there are at the moment almost forty, including the recently recognized Late Permian, Lopingian Stage.

D'Orbigny paid attention not only to the stratigraphic ranges of taxa, but also to their relative abundances while defining his units. In this connection today it is vital to recognize that extinctions and radiations are recognized not only by the stratigraphic ranges of taxa but also by significant appearances and disappearances of community groups (defined in the following), and also involve significant abundance changes for those taxa persisting across extinction boundaries.

Figure 23.2 diagrammatically indicates the nature of an evolving community group, one of many within a d'Orbignyan unit, an ecological-evolutionary unit or subunit.

Oppelian subdivision of d'Orbignyan units. Oppel (1856–1858) made it clear, chiefly by means of the then-known stratigraphic ranges of varied Jurassic ammonoid cephalopods, that the Jurassic could be finely subdivided into a number of time intervals by employing the overlapping time ranges of the varied ammonoid taxa. We now recognize that what Oppel was considering was the fact that within any evolving community there are a number of lineages of the more rapidly evolving taxa that enable one to subdivide geological time very

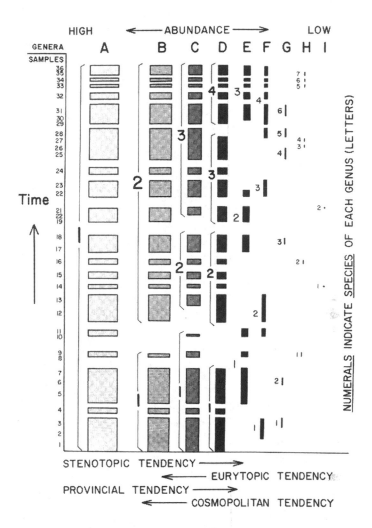

Figure 23.2 Diagram showing the characteristics of an evolving community group. The numerals in the left-hand column represent samples—note the numerous unsampled intervals (unsampled due to lack of exposure or other reasons, such as too small sample size). The letters represent unrelated genera associated together in the evolving community group. The numerals below each letter represent the species of each genus that have been recognized in the sampled intervals—oldest species below; youngest species above. Line widths in each column are proportional to the abundance of each genus. Note that the species of each genus tend to maintain similar relative abundance levels; that is, the species of rare genera tend to remain rare, and vice versa. The patterning is for artistic effect only. Note that the species of the rare genera are the most rapidly evolving and vice versa (modified from Boucot, 1990, fig. 409).

intimately when one overlaps the ranges of the various evolving taxa (see fig. 23.2 for a diagrammatic illustration).

Extinction. Since Cuvier's early nineteenth-century demonstration of extinctions affecting vertebrates, it has become increasingly clear that the Phanerozoic fossil record also includes an extensive set of marine invertebrate extinctions (see Boucot 1996, table 2 for a listing). John Phillips (1840) recognized the presence of the major marine invertebrate extinctions when defining Paleozoic, Mesozoic, and Cenozoic for the first time. D'Orbigny's units are largely terminated through time by such extinctions. We now recognize that there are a large number of such Phanerozoic extinctions, major to minor. I have emphasized (1996) that rare genera and their species tend to be more extinction prone as well as being more stenotopic; that is, more specialized. The literature on these types of extinctions is voluminous. However, when it comes to extinction causation there is little understanding or agreement (see Kaiser and Boucot 1996 for discussion). Most of the attempts at explaining causation have focused on physical possibilities, such as bolide impacts (the popular end-Cretaceous item), major episodes of volcanism, major climatic changes, regression-transgression, and so forth, but for none of these is the evidence overwhelming or conclusive (Boucot 1996). There has been little attempt to suggest biological causation, which may reflect the nonbiological backgrounds of most of those involved in explaining causation.

It is important that extinction events not be confused with merely local environmental changes that represent local ecological changes in biota. These are sampling problems that may be resolved by employing as large a number of well sampled, well dated, geographically widespread stratigraphic sections as possible. Normally, in the careful measurement of a stratigraphic section, accompanied by careful, bed-by-bed fossil collecting, many more than one faunal type (community group) are observed. These faunal changes include the disappearance of one set of genera and species and their replacement by another set. These changes, if global or even if affecting only several major biogeographic units probably represent extinction events, but if merely restricted to one section or several adjacent sections only represent local environmental changes and have nothing to do with extinction.

An additional caveat is how the uppermost, preextinction stratigraphic ranges of genera and their species should be recognized. This, again, is a sampling problem. In this case it involves the simple fact that within any one ecological-evolutionary unit or subunit it is necessary to find localities, which correspond to stratigraphic sections, where each community group present during the unit or subunit occurs immediately prior to the extinction horizon. Failure to do this will inevitably result in an apparent earlier extinction of the more stenotopic taxa far before their true extinction. This is when their community group appears to disappear earlier in one section than would often be the case if more ecologically varied stratigraphic sections were sampled. The moral here is that ecological factors must be taken into consideration in order to better understand many paleontological phenomena.

Radiation. It has been recognized that most ecological-evolutionary units and subunits begin with a radiation. The few exceptions involve low diversity intervals after some of the major extinction events (Sheehan 1985, 1996), such as the Permo-Triassic, end-Ordovician and mid-Late Devonian, with the possibility that many more such low diversity units will be recognized when more detailed biostratigraphic work has been carried out in strata immediately following other extinctions. I have pointed out (1990, 587) that there is currently no understanding of why level-bottom environment radiations tend to occur chiefly following extinctions. An explanation for why some taxa, at varied taxonomic levels, radiate whereas others do not is lacking. An appeal to the "empty niches" left by an extinction event is not an adequate explanation (Boucot 1996). Until we understand the environmental characteristics of the taxa that became extinct and those that radiated later to fill the empty niches the puzzle will remain unsolved. Why radiations in the level-bottom environment take place only near or at the beginning of ecological-evolutionary units and subunits, whereas those of nonlevel bottom complexes (reef complexes of communities, bryozoan thickets, pelmatozoan thickets, etc.) generally take place significantly later is not understood. Charlesworth (1990) has reviewed the genetic basis for radiations which, although useful, does not help in understanding the post-extinction synchrony of the level-bottom extinctions, the bulk of the marine taxa in the fossil record.

It is worth noting (Kaiser and Boucot 1996) that the initial taxa taking part in a radiation tend to be morphologically average in size, followed later by marked morphological divergence—that is, increase in disparity. This fact has not yet been satisfactorily explained evolutionarily.

Finally, in order to recognize ecological-evolutionary units and subunits one needs to have a global grasp of the marine benthic biostratigraphy, especially of the time intervals concerned.

WHAT IS COMMUNITY EVOLUTION?

Introduction

The overall nature of the sample, the continental shelf equivalent, and well-skeletonized organisms have been considered previously. The next challenge is to summarize their characteristics through geological time, Cambrian to present. Does the abundant fossil record of the well-skeletonized, relatively shallow-water organisms consist of a globally, largely homogeneous mixture of taxa, changing gradually through time? We are all well aware that organisms on land and sea today are not uniformly distributed, pole to pole, and are not surprised that this has also been the case in the past, although the levels of provincialism to cosmopolitanism have varied back and forth through time; this is biogeography in the past and present. We are also well aware that organisms on land and sea within any small region are not uniformly distributed in a homogeneous manner—that is, they are characterized by clumps dominated here and there by one or another organism or a small group of organisms mixed together with less common, varied organisms. These clumps, on land and sea, are usually referred to as *communities*, although the term *guild* is almost always more appropriate.

What Is a Community, and How Is It Defined?

If one defines a community as a group of interacting living plants and animals—everything from the viral to the top predator levels—then using the term *community* would be singularly inappropriate for paleontologists. But, judging from many of their publications, biologists indulging in community studies very often use the term community when the term guild might be more appropriate. This being the case, paleontologists have followed in the same path.

It is important to understand what the term community means paleontologically. The term community has been defined in many ways by various authors (Boucot 1981, 177–81; 1999, 3–6, 13–17). Paleontologically, I define the term as a regularly recurring set of taxa, mostly genera and their species, for which the relative abundances, common to very rare, remain approximately fixed. "Approximately fixed" does not mean to the nth decimal point. It does mean that abundant taxa will be characterized by the same order of magnitude as will rare taxa. For example, an abundant taxon may range from 35% to 50% or even 60%, but will not sink to 5% or rise to 90% within the same community. Figure 23.1 makes it clear, however, that in other communities, such a taxon may not be abundant or may ultimately sink almost to zero in abundance before disappearing completely. Bennington and Bambach (1996) provide a good example of what happens when an overly restrictive, nonbiological community definition is used, when they analyzed samples from the Pennsylvanian of the Central Appalachians. Their approach should be avoided because it considers almost every sample as a distinct community, which makes no sense either biologically or paleontologically.

Once the concept of biotic community has been defined, based on samples from the present, it is reasonable to ask what happens as one reaches back in time to these co-occurring taxa. Attention to this problem extends back into the 1930s, but has attracted far more attention beginning with the 1970s. The concept has been labeled *community evolution*.

Community Evolution's Characteristics

Ecologists are unaware of the characteristics of community evolution, since they deal with too brief a time interval for recognizable evolution to have occurred, and almost invariably are unaware of the fossil record of the ancestral community, or even the fossil record of many species within the community. The only exception I can think of is A. W. B. Powell's (1937, 1939, 1950) perceptive work on the New Zealand Neogene and present fauna that is apparently unknown to ecologists elsewhere. Some ecologists confuse community evolution with community succession, but these are unrelated phenomena. Community succession is characterized by a time succession of distinct community types replacing one another; each type modifies the environment in such a way as to make it suitable for the following community—that is, a community of fireweeds ultimately followed by a climax forest does not involve community evolution since few if any taxa are shared or related systematically.

Anyone who routinely prepares, sorts, and identifies fossils from many localities within a relatively narrow stratigraphic interval of a particular region soon becomes aware that the presence of certain combinations of genera and their species, with relatively similar abundances, are characteristic of the varied faunas. The concept of faunal facies is an old one among biostratigraphers and geologists working with fossiliferous rocks. Does this fact have any evolutionary significance? Keep in mind that the term "faunal facies" refers, when more narrowly defined (not "shelly facies" or "warm-water facies," as contrasted with "brackish-water oyster facies, pentamerid facies, stringocephalid facies," and so forth) corresponds to what the biologist and paleoecologist refers to as a community, although the term guild might be more appropriate for both. I have suggested (1975) that each of these more narrowly defined faunal facies, and community types, that occur within individual ecological-evolutionary units and subunits be referred to as Community Groups.

What are the characteristics of community evolution within ecological-evolutionary units and subunits? Figure 23.2 outlines in general terms just what is involved. Note that numerically abundant genera, indicated by the widest lines, are usually represented by only a single species from beginning to end of the ecological-evolutionary unit or subunit, whereas the uncommon-to-rare-genera, indicated by the progressively narrower lines, are represented by more than one species. I have listed (Boucot 2006) some examples of articulate brachiopod genera, within which there are gradational relations at the species level between one taxon or another, chiefly from time intervals familiar to me.

In order to more reliably recognize community groups, one needs sufficient experience with the biotas comprising a particular ecological-evolutionary unit to be able to recognize the units. Boucot and Lawson (1999) provide a Silurian sample that illustrates what I mean.

In order to recognize communities through evolutionary time it is necessary to prepare, sort, attempt to identify and then describe a large number, in my experience, at least, a few thousand fossil collections of marine invertebrates from a significant time interval. It is this type of sampling experience that provides conclusions on community (community evolution as I use the term is essentially the same as the older, more narrowly construed *biofacies* or *faunas* of previous generations). The paleobiologists, as far as I know, have not gone through this type of disciplined exercise, thus depriving themselves of any personal understanding of the problems involved in recognizing community evolution.

Cooordinated Stasis

Brett and Baird (1992, 1995) introduced the term *coordinated stasis*. They used material chiefly from New York, Pennsylvania, and eastern Ontario, and considered it to be "a regional phenomenon." They (1995) presented extensive data on the later Middle Devonian Hamilton Group faunas through a significant interval of time, some millions of years. They made a reliable case for the abundant genera maintaining their specific identity through the entire interval—that is, no evidence for phyletic/anagenetic evolution. They did not deal with the very uncommon-to-rare-genera and their species, implying that these too would show no evidence for phyletic/anagenetic evolution during the later Middle Devonian. However, in their Silurian examples, which were listed but not considered in any detail, there are some good evolutionary sequences of species belonging to some genera, such as the articulate brachiopod *Eocoelia* as well as some of the ostracod genera. They divided the Silurian and earlier Devonian into a series of faunal-lithologic units representing locally specialized environments, but not covering enough area regionally to get at the full spectrum of available communities and their evolving and non-evolving taxa. Their Early Devonian units (Helderberg through Schoharie and earliest Onondaga) are characterized by some phyletically evolving articulate brachiopod groups, such as the *Howellella* to *Acrospirifer murchisoni*, *Dalejina* to *Discomyorthis*, and small *Nanothyris* through larger *Amphigenia*. Their Early Devonian units involve too few communities to make the conclusion that species of less common-to-very-rare genera are subject to more rapid phyletic/anagenetic evolution. But, they do make the important point that the more abundant-to-common genera show no evidence for phyletic/anagenetic evolution through significant time intervals, as was also concluded earlier when considering the characteristics of community evolution. Consideration of a much wider geographic area during the Silurian-earlier Middle Devonian would show that the situation corresponds evolutionarily to our conclusions about community evolution. In 1996 Brett, Ivany, and Schopf reviewed the evolutionary implications of their earlier work.

Chronofauna

Olson (1952, 1977; see also 1983) introduced the term Chronofauna to describe a time sequence of Late Carboniferous-Early Permian nonmarine vertebrate faunas involving a high level of phyletic/anagenetic evolution. Unfortunately, vertebrate paleontologists have paid little attention to this pioneering work. Olson's term is very close to what I have called Community Group. I discussed the overall similarities between the term chronofauna and community group (1990, 530), including comments by Olson. Webb's exceptional work on Cenozoic North American Mammal chronofaunas (1983, 1989) made the point that a number of them represent significant time intervals of probable climatic stability of one kind or another associated with relatively fixed assemblages of mammals that underwent species-level evolution through time; in many cases, evidence of significant extra-North American immigration was largely restricted to the initial stage of each mammalian chronofauna, and he found causation other than that of climatic cum vegetational control elusive. Webb and Opdyke (1995) have explored the possibility that the North American land mammal chronofaunas may correspond to similar entities in other parts of the world, although the evidence is still too fragmentary to be conclusive. It is of interest here that diverse non-North American carnivores, aeluroids, only appear very late in North America (Hunt and Tedford 1993) with a few felids; North American Cenozoic carnivores feature such forms as canids and ursids, among others.

Simpson strongly adhered to Olson's view. In the early 1980s, while discussing the concept of community evolution with him, I recall his strong objections to the notion that vertebrate paleontologists had not been aware of the concept for some time. He pointed out forcefully that the term "fauna" (see Simpson 1969 for his community-ecological use of the word; also see quote from Simpson 1977, in Boucot 1978, 556), when used for time-successive vertebrate faunas characterized by evolutionarily closely related species and genera, amounted to the same thing. He was obviously correct. His firsthand knowledge of vertebrate faunas, based on his own taxonomic and biostratigraphic studies and supplemented by the work of his colleagues, strongly supports his position.

Thoughts about Community Evolution

There is more than one way to consider community evolution. In a series of publications I have emphasized the overall fixity, monotony in the generic content present in the community groups *within* individual ecological-evolutionary units and subunits, and the inverse relationship between the relative abundances of the genera and their species—that is, species of the uncommon-to-rare-genera evolve rapidly over time, whereas the abundant-to-common genera normally show little or no evidence for evolution through time, phyletic/anagenetic evolution.

Other ways of thinking about group evolution are to focus on assemblages of genera with similar ecological and behavioral attributes. Bambach (1983) introduced the idea of trying to trace ecological and behavioral guilds through time in an effort to better understand potential biotic interactions. Unfortunately, his work depends on understanding the behavior and ecology of the well-skeletonized organisms he discussed, while ignoring the basic fact that soft-bodied organisms are virtually absent, as well as ignoring the virtual absence of such things as nonmineralized plankton from the fossil record. One cannot assume, for example, that in the Cambrian predators were relatively insignificant, because our poor knowledge of the soft-bodied taxa would overlook predators that were certainly present. When trying to interpret the feeding habits of extinct organisms one faces many insuperable obstacles, as is also the case with their reproductive and physiological characteristics. For example, in the Paleozoic there are many presumed suspension and filter feeders, yet very little is known of what they actually fed on, or whether they were very selective or not. We are unable to reliably reconstruct food webs of the past until we approach strata closely allied to the present, such as in the Miocene and Pliocene, where we can extrapolate back from the present. Trying to behaviorally interpret the functional morphology of wholly extinct groups is highly speculative in most cases, although there are some exceptions. In any event, the application of their guild concept to reconstructing community evolution has not yet proved very helpful.

In 1978 I coined the terms *metacladogenesis* and *diacladogenesis* to distinguish between major radiations (metacladogenesis), involving new families and higher taxa that probably involve movement from one adaptive peak to another, as contrasted with relatively trivial changes (diacladogenesis) at the species level within a genus, such as the presence of modern sister species of some genera on either side of the Isthmus of Panama. This line of thinking might be expanded by emphasizing that the phyletic/anagenetic evolution within the uncommon-to-rare-genera of an evolving community group represent another form, similar in many regards to diacladogenesis in that there is no significant ecological change or abundance change indicated, whereas the reverse, of course, is true for metacladogenesis. One might almost conclude that the term *evolution* is confusing, since there are these two major rate, morphological, behavioral, physiological, and so on modes. For most people the word evolution means the major, metacladogenetic mode that involves the change from terrestrial tetrapods to wholly aquatic whales, from lizards to snakes, from primitive archosaurs to birds, or from insectivores to bats. Yet, both modes presumably involve microevolutionary changes.

THE BIOGEOGRAPHIC COMPLICATION AND DIVERSITY

Diversity through time at the familial, generic, and species levels has interested paleontologists for some time as a potential body of evidence that may tell us something about evolution. It has normally been compiled at the generic and familial levels, since at the species level there is too much uncertainty about the definition of species—that is, which described species are potential synonyms, which workers tend to be lumpers, and which splitters.

Many compilers of generic and familial diversity have scoured the literature for higher taxa and presented numbers of taxa/geological time intervals, sometimes series, sometimes stages, sometimes normalized for the absolute time duration of the various series and stages.

In none of these compilations has there been any attempt to normalize for varying numbers of biogeographic units in the individual time units, whether relative or absolute. This is an error. Why? It is obvious that during highly provincial time intervals there will be far more genera, as a result of evolution under conditions of reproductive isolation, than during more cosmopolitan time intervals (Boucot 1978, fig. 3). I labored this obvious result or conclusion with examples taken from articulate brachiopods in the mid-Paleozoic (1975).

Family diversity is far less affected by changing levels of provincialism through time. In any event, it is obvious that presently published accounts of changing generic diversity through time do not give a true picture of changing diversity because they ignore the considerable effect caused by changing levels of provincialism.

CONCLUSIONS

This treatment began by questioning whether the concepts of punctuated equilibrium or community evolution are based on reliable, adequate sampling. After examining the evidence it becomes clear that punctuated equilibrium fails the sampling test. It assumes, without considering past evidence, that newly evolved allopatric taxa have the inherent capability of reinvading the region of the parent species' environment, an environment distinct from that of the allopatric. Second, it assumes that there are only two rate modes characterizing species-level evolution—static and very rapid—whereas the fossil record clearly shows that there is a rate continuum from the very rapid, tachytelic, moderate, horotelic, to very slow, bradytelic, to use Simpson's terms. Third, it assumes, again without considering the evidence, that the abundance spectra of all genera and their species, from abundant to rare, has no influence on sampling that determines the presence or absence of transitional forms between the species of the uncommon-to-rare-genera, which are the majority. Fourth, it assumes that the published descriptions of old and new species, which often suggest a distinct morphological break between most species of a genus, are reliable.

The concept of community evolution, on the contrary, is based on a full consideration of the sampling possibilities inherent in the abundance spectrum, from abundant to rare. It recognizes that uncommon-to-rare taxa will most often have many sampling gaps between species of their genera. It recognizes that there is no evidence favoring the random movement in evolutionary time of stenotopic species from one environment to another, from membership in one community to another.

In view of the evidence provided here it is time to carefully reconsider the problems associated with the punctuated equilibrium concept.

ACKNOWLEDGMENTS

I am grateful to Christine Janis, Brown University, for guidance in the matter of Cenozoic chronofauna papers, and to George Poinar, Oregon State University, for carefully advising me about many aspects of the manuscript.

NOTE

1. Here and elsewhere the term *radiation* is used for what paleontologists commonly term "adaptive radiation," although there is almost never any evidence demonstrating adaptations except that the abrupt appearance of new families and higher taxa, with significantly distinctive, novel morphologies may have involved adaptations.

REFERENCES

Bambach, R. K. 1983. Ecospace utilization and guilds in marine communities through the Phanerozoic. In *Biotic interactions and fossil benthic communities,* ed. M. J. S. Tevesz and P. L. McCall, 719–46. New York: Plenum Press.

Bennington, J. B., and R. K. Bambach. 1996. Statistical testing for paleocommunity recurrence: Are similar fossil assemblages ever the same? *Palaeogeography, Palaeoclimatology, Palaeoecology* 127:107–33.

Boucot, A. J. 1975. *Evolution and extinction rate controls.* Amsterdam: Elsevier.

———. 1977. Rates of evolution and number of descriptive terms. *Journal of Paleontology* 51:414–15.

———. 1978. Community evolution and rates of cladogenesis. *Evolutionary Biology* 11:545–655.

———. 1981. *Principles of benthic marine paleoecology.* New York: Academic Press.

———. 1982. Ecostratigraphic framework for the Lower Devonian of the North American Appohimchi Subprovince. *Neues Jahrbuch für Geologie und Paläontologie, Abhandlungen* 163:81–121.

———. 1983. Area-dependent-richness hypotheses and rates of parasite/pest evolution. *The American Naturalist* 121:294–300.

———. 1990. *Evolutionary paleobiology of behavior and coevolution.* Amsterdam: Elsevier.

———. 1994. The episodic, rather than periodic nature of extinction events. *Revista de la Sociedad Mexicana de Paleontología* 7:15-35.

———. 1996. Epilogue. *Palaeogeography, Palaeoclimatology, Palaeoecology* 127:339-59.

———. 1999. Introduction: Community evolution. In *Paleocommunities: A case study from the Silurian and Lower Devonian,* ed. A. J. Boucot and J. D. Lawson, 3-6. Cambridge: Cambridge University Press.

———. 2006. So-called background extinction rate is a sampling artifact. *Palaeoworld.* 15:127-34.

Boucot, A. J., and J. D. Lawson, eds. 1999. *Paleocommunities – a case study from the Silurian and Lower Devonian.* Cambridge: Cambridge University Press.

Brett, C. G., and G. C. Baird. 1992. Coordinated stasis and evolutionary ecology of Silurian-Devonian marine biotas in the Appalachian basin. *Geological Society of America Abstracts with Programs* 24(7) : 139.

———. 1995. Coordinated stasis and evolutionary ecology of Silurian to Middle Devonian faunas in the Appalachian Basin. In *New approaches to speciation in the fossil record,* ed. D. H. Erwin and R. L. Anstey, 285-315. New York: Columbia University Press.

Brett, C. G., L. C. Ivany, and K. M. Schopf. 1996. Coordinated stasis: An overview. *Palaeogeography, Palaeoclimatology, Palaeoecology* 127:1-20.

Brown, P., T. Sutikna, M. J. Norwood, R. P. Soejono, Jatmiko, E. Wayhu Saptomo, and R. Awe Rue. 2004. A new small-bodied hominin from the Late Pleistocene of Flores, Indonesia. *Nature* 431:1055-61.

Charlesworth, B. 1990. The evolutionary genetics of adaptations. In *Evolutionary innovations,* ed. M. H. Nitecki, 47-70. Chicago: University of Chicago Press.

Chen Junyuan, J. Bergström, M. Lindström, and Hou Xianguang. 1991. The Chengjiang fauna—Oldest soft-bodied fauna on Earth. *Research & Exploration, National Geographic Society* 6:8-19.

D'Orbigny, A. 1849-1852. *Cours élémentaire de paléontologie et de géologie stratigraphiques.* Premier Volume 1-299, Tome Second, Fascicule I, 1-382. Tome Second, Fascicule II, 383-847, Tableaux. Paris: Masson.

———. 1850-1852. *Prodrome de paléontologie universelle.* Premier Volume, IX-LX, 394. Deuxième Volume, 1-427, Troisième Volume, 1-196, Table Alphabétique et Synonymique des Genres et des Espèces, 1-189. Masson: Paris.

Eldredge, N. 1972. Systematics and evolution of *Phacops rana* (Green, 1832) and *Phacops iowensis* Delo, 1935 (Trilobita) from the Middle Devonian of North America. *American Museum of Natural History Bulletin 147* 2:47-133.

Eldredge, N., and S. J. Gould. 1972. Punctuated equilibria: An alternative to phyletic gradualism. In *Models in Paleobiology,* ed. T. J. M. Schopf, 82-115. San Francisco: Freeman, Cooper.

Fortey, R. A. 1985. Gradualism and punctuated equilibria as competing and complementary theories. *Special Papers in Palaeontology* 33:17-28.

Gingerich, P. D. 1976. Paleontology and phylogeny patterns of evolution at the species level in Early Tertiary Mammals. *American Journal of Science* 276:1–28.

———. 1977. Patterns of evolution in the mammalian fossil record. In *Patterns of evolution,* ed. A. Hallam, 469–500. Amsterdam: Elsevier.

———. 1984. Punctuated equilibria: Where is the evidence? *Systematic Zoology* 33:335–38.

———. 1987. Evolution and the fossil record: Patterns, rates, and processes. *Canadian Journal of Zoology* 65:1053–60.

Gould, S. J. 1977. *Ontogeny and phylogeny.* Cambridge, MA: Harvard University Press.

Gould, S. J., and N. Eldredge. 1977. Punctuated equilibria: The tempo and mode of evolution reconsidered. *Paleobiology* 3:115–51.

Hunt, R. M., and R. H. Tedford. 1993. Phylogenetic relationships within the aeluroid Carnivora and implications of their temporal and geographic distribution. In *Mammal phylogeny, placentals,* ed. F. S. Szalay, M. J. Novacek, and M. C. McKenna, 53–73. Berlin: Springer Verlag.

Kaiser, H. E., and A. J. Boucot. 1996. Specialisation and extinction: Cope's Law revisited. *Historical Biology* 11:247–65.

Olson, E.C. 1952. The evolution of a Permian vertebrate chronofauna. *Evolution* 6:181–196.

———. 1977. Permian lake faunas: A study in community evolution. *Journal of the Palaeontological Society of India* 20:146–63.

———. 1983. Coevolution or coadaptation? Permo-Carboniferous vertebrate chronofauna. In *Coevolution,* ed. M. H. Nitecki, 307–38. Chicago: University of Chicago Press.

Oppel, A. 1856–1858. *Die Juraformation Englands, Frankreichs und des sudwest Deutschlands.,* Wurtemb. Naturwiss. Verein Jahresber., vols. 12–14 (1856, 1–438; 1858, 439–694, 695–857). Stuttgart: Elmer und Seubert.

Phillips, J. 1840. Palaeozoic series. In *The penny cyclopaedia of the society for the diffusion of useful knowledge,* vol. 17, ed. G. Long, 153–54. London: Charles Knight.

Powell, A. W. B. 1937. Animal communities of the sea-bottom in Auckland and Manukau Harbours. *Royal Society of New Zealand Bulletin* 66:354–401.

———. 1939. Note of the importance of recent animal ecology as a basis of Paleoecology. *Pacific Science Congress Proceedings* 6:607–17.

———. 1950. Mollusca from the Continental Shelf, Eastern Otago. *Records of the Auckland Institute and Museum* 4 (1): 73–81.

Raff, R., B. Parr, A. Parks, and G. Wray. 1990. Heterochrony and other mechanisms of radical evolutionary change in early development. In *Evolutionary innovations,* ed. M. H. Nitecki, 71–98. Chicago: University of Chicago Press.

Schopf, T. J. M., D. M. Raup, S. J. Gould, and D. S. Simberloff. 1975. Genomic versus morphologic rates of evolution: Influence of morphologic complexity. *Paleobiology* 1:63–70.

Sheehan, P. M. 1985. Reefs are not so different: They follow the evolutionary pattern of level-bottom communities. *Geology* 13:46–49.

———. 1996. A new look at ecological evolutionary units (EEUs). *Palaeogeography, Palaeoclimatology, Palaeoecology* 127:21–31.

Simpson, G. G. 1943. Criteria for genera, species, and subspecies in zoology and paleozoology. *Annals of the New York Academy of Science* 44 (2): 145–78.

———. 1944. *Tempo and mode in evolution.* New York: Columbia University Press.

———. 1953. *The major features of evolution.* New York: Columbia University Press.

———. 1969. The first three billion years of community evolution. In *Diversity and stability in ecological systems. Brookhaven Symposium in Biology No. 22,* 162–77. Upton, NY: Brookhaven National Laboratory.

Webb, S. D. 1983. The rise and fall of the Late Miocene ungulate fauna in North America. In *Coevolution,* ed. M. H. Nitecki, 267–306. Chicago: University of Chicago Press.

———. 1989. The fourth dimension in North American terrestrial mammalian communities. In *Patterns in the structure of mammalian communities,* ed. D. W. Morris, Z. Abramsky, B. J. Fox, and M. R. Willig, 181–203. Special Publications No. 28. Lubbock: Museum of Texas Tech University.

Webb. S. D., and N. S. Opdyke. 1995. Global influences on Cenozoic land mammal forms. In *Effects of past global change on life,* 184–208. Studies in Geophysics, Board of Earth Sciences and Resources Commission on Geosciences, Environment, and Resources, National Research Council. Wasington, DC: National Academy Press.

Whittard, W. F. 1966, The Ordovician trilobites of the Shelve Inlier. *Palaeontographical Society Monograph* Pt. 8:265–306.

Whittington, H. B. 1985. *The Burgess Shale.* New Haven, CT: Yale University Press.

CHAPTER TWENTY-FOUR

An Interview with David M. Raup

Edited by David Sepkoski and David M. Raup

Throughout his career, David Raup has played a central role in the renaissance of analytical, or quantitative paleobiology. He was one of the first paleontologists to seriously apply computers to paleontological problems, and was a key innovator in two major areas of quantitative analysis: (1) understanding the geometry of morphological constraints by modeling mollusk shells using computer simulations, and (2) examining the fossil record using advanced statistical techniques. In the latter area, Raup collaborated on several pioneering projects during the 1970s and 1980s, including simulations of clade diversity with Stephen Jay Gould, Tom Schopf, and Daniel Simberloff, and the now-infamous proposal of periodicity in mass extinctions, with Jack Sepkoski. Raup's studies of morphology and biocrystallography earned him a substantial reputation while still a young paleontologist in the 1960s, but later in his career he established himself as one of the leading theorists of the effect of extinction—particularly mass extinction—on macroevolutionary dynamics. Additionally, he was the coauthor (with Steven M. Stanley) of the highly influential textbook *Principles of Paleontology*, and has published two highly-regarded popular books on the subject of extinction: *The Nemesis Affair*, and *Extinction: Bad Genes or Bad Luck?*[1]

What follows are excerpts from a conversation with Raup about his life in paleontology. Here he discusses his background, interests, and influences, and describes some of the major events in his career.

Raup grew up in a scientific family—his father, Hugh, was a noted Harvard botanist—and he began his scientific education at Colby College and the University of Chicago. Having begun study of geology at Colby, Raup transferred to Chicago to complete the BA. The timing of this move was significant. As Raup explains, study at Chicago introduced him to an exciting new world of biological study in paleontology:

> This was a transition time at Chicago from the old Hutchins program to what came after.[2] I went as a simple undergraduate transfer with no particular ambitions and found myself a graduate student because of a peculiar mismatch of academic programs and calendars. I had only two years of college but was put into what were then called the divisions, and was immersed in graduate study with very little to work with. In time, I developed a strong interest in paleontology and had some good mentors.

At Chicago, Raup's primary mentor was Everett (Ole) Olson, an eminent vertebrate paleontologist with significant interests in quantitative and biological study in paleontology.[3] Surprisingly, however, Raup reports being unenthusiastic about the novel approach being taken at the time at Chicago:

> I developed this very strong interest in paleontology and wanted to continue, but was completely disgusted by the looseness of the program there and wanted something a little bit more classical. It was all about numbers and models, which I had no use for. I wanted very traditional paleo and biostratigraphy: collecting, describing, and classifying fossils. And so I set my sights on graduate school at the University of Michigan in Ann Arbor. I visited Ann Arbor, liked the museum, liked the atmosphere, liked the people, applied, and was turned down by return mail because Michigan did not honor my degree—I was at that time working for an SB, a terminal Bachelors of Science degree that Chicago had instituted for people who were never to return. (I did return to Chicago many years later but as department chairman.) And so I got the degree and went home not knowing what I was going to do. I ended up going to the hometown school.

The hometown school was, of course, Harvard, where Raup's father had recently met a young paleontologist named Bernhard Kummel. After introducing himself to Kummel, Raup was soon enrolled as a graduate student in paleontology. Ironically, up to this point Raup had little interest in mathematics or quantitative paleontology, which he saw as "shallow" and "sloppy," and "too theoretical for use." Otherwise,

I had very good high school math and a little college math served to make that operational. Beyond that I didn't have any formal training that I can remember, certainly no math courses. Not even a statistics course. I took a population genetics course at Harvard which gave me some applied statistics.

At Harvard, Raup found an inspiring teacher in Kummel, who, despite being fairly traditional, was "a tremendous cheerleader, inspirer."

He didn't know much but that didn't make any difference. He put faith in his students and pushed them. If you came up with an idea, he wanted it published. He got us to go to meetings, he got us to write papers, and if we got crazy ideas that he did not understand, he was still totally supportive. So, in terms of straight book learning, I didn't get that much. Bernie was a disciple of Marshall Kay and was pushing the models of tectonics and sedimentation and geosynclines and so forth, which were exciting and very new. It was a way of synthesizing a whole lot of geologic and paleontologic data and this was exciting. This got me back to models.

In addition to working with the fairly traditionally minded Kummel, Raup also began studying with Ernst Mayr:

In terms of education, a much greater influence was Ernst Mayr. I worked as closely as I could with Ernst. He was always very busy. He interacted a lot with Norman Newell, his former colleague at the American Museum. I had a special respect for Norman. This was his heyday: he was working on the great Permian reefs in Texas and on the modern reefs in the Bahamas with a bunch of good students, including John Imbrie, who was then just completing his graduate work. I remember visiting John in New York to learn his biometrical methods. This was when John was just getting going. So in terms of biometrics and applied statistics, I really got started with John.

Raup recalls that Ernst Mayr "indirectly suggested" his dissertation topic:

I was interested in fossil echinoderms (that's what I'd been doing at Chicago), and Ernst had just completed a paper on geographic distributions and speciation in living sea urchins around the world.[4] He was anxious to have somebody do the same sort of thing with the fossil record. Then Bernie, on the basis of that, linked me up with Wyatt Durham at Berkeley, who knew certain groups of echinoids—fossil sand dollars—backwards and forwards, and had big collections. So I went to Berkeley and spent a few days with Wyatt and he got me started on the dissertation, which in effect Ernst had suggested. It involved looking for phyletic change and geographic speciation in fossil sequences. This brought in Imbrie's biometrics and applied statistics and it sort of became a whole.

While still finishing his PhD at Harvard, Raup's first job was a faculty appointment at Caltech, but he was quickly hired away by Johns Hopkins. It was at this point that he was introduced to computers, almost by accident:

I had a small research grant from the National Science Foundation (1961) to study variation in intertidal arthropods (sand fleas) on some beaches in the southeastern United States. This was at the height of public concern about radioactive fallout and everybody was anxious about mutagenic effects of high-energy radiation. I had come across information on the radioactive black sands that are scattered along the South Carolina and Georgia coasts. I had in mind studying natural populations of sand fleas by making large collections and doing biometrics to see whether those that lived on the radioactive sands had noticeably more variation than those on normal beaches. I got a Geiger counter and a station wagon full of alchohol and spent several weeks picking up sand fleas on beaches from Myrtle Beach, South Carolina, to Fernandina Beach, Florida.

I then spent months with an assistant measuring them, and then, in the mode that I was accustomed to, made various scatter plots of the half dozen dimensions that I had measured on the poor beasts. I was looking for outliers in the plots, but it soon became clear that the plotting was too laborious and prone to error. I needed computer plotting and so I learned enough FORTRAN to program the IBM 1620 they had at Johns Hopkins. The automatic plotting was successful but the overall project a complete bust: morphologic variation was not associated with level of radioactivity in the environment. But the experience got me started with computer programming.

Throughout the 1960s, computers were anathema to most paleontologists. I remember the first computer paper I published in *Science*. It was on simulating molluscan shell form and the manuscript went to two peer reviewers.[5] One turned out to be John Imbrie, who gave it high marks, and the other was Ellis Yochelson, who recommended rejection on the grounds that the work was not science. Thanks to John, the paper was accepted. I was gratified, many years later, when Ellis asked me to collaborate with him on a similar computer simulation study.

In the early 1970s, Raup became involved with a group of similarly minded colleagues who were interested in applying analytical models to paleontology. This group included Stephen Jay Gould and Thomas J. M. Schopf, the latter of whom organized the seminar (and resulting book) *Models in Paleobiology* that included Niles Eldredge's and Gould's famous first paper on punctuated equilibrium. Raup contributed to the volume as well, and also participated in a series of informal meetings with Gould, Schopf, and Daniel Simberloff that took place at the Marine Biological Laboratory (MBL) at Woods Hole beginning in late 1972. The outcome of the first MBL meeting was the groundbreaking joint-authored paper "Stochastic Models of Phylogeny and the Evolution of Diversity," published in the *Journal of Geology* in 1973, which became an icon of the new theoretical approach to macroevolutionary modeling.[6] This is what Raup recalls about those meetings:

> Tom had brought the multivolume *Treatise on Invertebrate Paleontology* in a box or boxes and we put it on the table. The question was: "What can we do that's different?" Dan Simberloff, a mathematical ecologist, was there to see how his work on island biogeography could be transferred to the fossil record. Also, Dan was recognized for his computational expertise. My recollection is that it was a three-day meeting. We got nowhere. Dead zero. We were pretty frustrated by this and I think it was toward the end of the afternoon of the final day that I suggested random simulation models. What would evolutionary trees look like if they were truly random—that is, if extinction or survival of lineages was merely chance? How would patterns of biodiversity differ from those seen in the real world of fossils? Could mass extinctions be simulated by a random model?
>
> I'm sure Dan must have been well aware of what this implied because he had done similar things with bird distributions in the tropics. The idea was not meant to suggest that things like the extinction of species occur without cause. Rather, that there are so many different causes of extinction operating in any complex ecosystem that ensembles of extinctions may behave *as if* governed by chance alone. In the extinction case, this approach would predict that species extinctions would occa-

sionally cluster in time merely by coincidence—not because there was some common cause for the clustering. I don't remember what reaction the others had to my proposal but, lacking any better idea, I said I'd go home and program it. That was the end of the meeting.

In an earlier paper (1969), Raup had developed the "grazing track model," which describes the grazing patterns of individual animals as "partially stochastic." Although the paper did not consider stochastic factors in much detail, it was Raup's first use of the concept, and probably gave him the idea to apply a random model to phylogenetic trees.

By then, I was far enough along with computers that I could handle random number generators. During the MBL meeting, I guess, the basic question that I must have asked was, what would evolution look like without natural selection? At that time, however, and throughout the first MBL meeting, we were talking not about natural selection of form or function but about clades and cladogenesis. That is, only the pattern of branchings and terminations in a simulated evolutionary tree. The question was whether patterns of radiations and extinctions of typical groups of lineages (clades) could be simulated by a random birth-death model of branching and termination of lineages.

The main problem in programming was the graphics: to get the trees displayed on a line printer designed only for text. I think Jack Sepkoski, working as Steve's graduate assistant, spent the best part of six months trying to figure out my algorithm. I write very, very dirty code.

The simulations described in our first paper were quite simple. A starting lineage had fixed probabilities of surviving, dying, or surviving and splitting into two lineages. If that lineage died out before branching, the simulation was automatically restarted. As the number of coexisting lineages (biodiversity) increased toward an arbitrarily specified carrying capacity, branching probability was lowered or extinction probability raised so that biodiversity fluctuated about the designated carrying capacity. Specifying a carrying capacity for the system was necessary to keep the computer's memory from saturating—and it also made sense ecologically.

We then just tweaked the probabilities until we found a range that looked like the real world. At that time, I don't think we had the data to make good estimates of what those probabilities actually were. And we were a little vague on whether our "lineages" were species or groups of species (higher taxa such as genera or families). In other words, we had made little effort to use realistic scaling in the simulation experiments.

The simulation program also included an algorithm to assemble lineages into groups analogous to the families and orders used by taxonomists. We could then look for patterns of origination and extinction of major groups of our imaginary organisms.

Our conclusion was that the random model did indeed come close to mimicking some real-world patterns—including some mass extinctions. This produced a lot of debate and controversy. For extinctions, it implied that collections of near-simultaneous extinctions need not have a common cause. Chance and coincidence would do the job. Some critics found our results impossible and claimed our random number generator must be faulty. In fact, the major flaw was our failure to use proper scaling, and we were obliged to back down on some of our stronger claims—especially those about mass extinctions. The near-simultaneous extinction of so many species at the end of the Cretaceous, for example, simply could not be explained by a model in which all species were dying for different reasons.

After the first MBL paper was published in 1973, I coauthored a paper with Steve Gould that applied the simulation idea to evolving morphology.[7] Using the original MBL program for branching lineages, and explicitly calling each lineage a species, morphological characters were allowed to "mutate" (by chance) at each lineage branch point. Thus, we could observe patterns of morphologic change in hypothetical animals in the absence of natural selection. This was a much larger break with convention than the original MBL model. We expected to find chaos but instead found orderly patterns strikingly similar to those paleontologists have described and interpreted for years. Many people found it impossible to have produced such natural looking patterns in a random simulation.

By this point, Raup was becoming somewhat concerned about the possibly exaggerated claims for random factors in evolution—a concern his friend and collaborator Tom Schopf did not share.

Although the 1974 paper on random change in morphology was holding up and making a significant contribution to the discipline, Steve and I were still smarting from our failure to scale the original MBL work properly. Tom, on the other hand, saw the MBL work as predicting a wonderful new world wherein evolution was predictable from natural laws.

Tom became more and more fanatical in his thinking of species as particles. For him, the Holy Grail was a set of "gas laws" for paleontology. We were all enthusiastic, of course, and there were several subsequent multiauthored papers (some just in response to criticism). But over the next few years the original MBL group dispersed. Tom became more fixed in his "gas law" ambitions and the rest of us moved on to other things.

In retrospect, I am reasonably sure that the MBL work raised the consciousness of a lot of people about the potential for random processes and the necessity of always considering them as a null expectation in any natural phenomenon for which causation was being sought. And, of course, lots of other people have beaten the same drum. At Chicago and elsewhere, it became standard to insist that the graduate students test whatever idea they had against a set of null expectations.

As time went on, Raup's interest was drawn more and more to the problem of extinction, largely as an outgrowth of the MBL collaboration. He also began to question some of the inferences of that earlier work, especially the idea that all species participating in a mass extinction died out for different reasons.

Tom remained wedded to the first MBL paper, where mass extinctions were events without identifiable causes. I was by that time beginning to think that occasional big rare events of environmental stress might be causing the big mass extinctions. My approach was thus becoming more deterministic, while Tom's remained stochastic. We diverged as I became more and more convinced that the externalities were important and he became more and more convinced that they weren't.

In 1981, at a meeting in Barcelona, Spain, on "Concept and Method in Paleontology," Raup presented a paper on extinction (subsequently published in *Acta Geologica Hispanica*[8]) with a very provocative title: *Extinction: Bad Genes or Bad Luck?*

It dealt with the history of trilobites, a group that dominated the marine world in the Cambrian but slowly dwindled to extinction 300 million years later at the end of the Permian. I posed a very simple, rather narrow question: if trilobites had the same extinction and speciation rates as other marine invertebrates of the Paleozoic, could their decline over the

300 million years have been a statistical accident? That is, an accidental excess of extinctions over speciation events cumulated over a long time? The answer was a resounding NO.

The phrase "bad genes or bad luck" is kind of flip, although I got a lot of mileage out of it. It was much too simplistic—it's bad luck to have bad genes.

Despite the results of the trilobite exercise, I continued to probe the possibility that extinction patterns could be modeled as random processes. The approach had been successful in other fields – such as analyses of stream drainage patterns in geomorphology. And, of course, the gas laws themselves! Motion of individual molecules could not be predicted, but that of ensembles of molecules in a gas could.

Following Leigh Van Valen's lead, I published several papers on the shapes of survivorship curves for species and higher taxa. Leigh had shown in 1973 that taxa do not age in the sense of increasing mortality rate with their time in existence. It was analogous to atoms of a radioactive element not becoming more likely to decay with age. This was a stunning example of how we could treat species as particles and get away with it. The fate of any one species could not be predicted but that of an *ensemble* of species could be known with remarkable accuracy simply by assuming that species had a constant probability of extinction throughout their existence.

In a series of papers, I then explored survivorship patterns in large datasets of fossil invertebrates, working at several taxonomic levels. Although the generality of random models remained a central question, my thinking converged on the idea that these models were legitimately applicable at some scales but not others. In particular, the rather orderly decay of cohorts of species and genera was interrupted occasionally by pulses of extinction far outside the expected limits of the models. And it is these pulses that invite a search for single causes such as global catastrophes of some sort—meteorite impact, global effects of volcanism, and so on.

Although mass extinctions triggered by major catastrophes certainly suggest bad luck for the victims, these events were clearly selective—as, for example, the extinction of every single dinosaur species at the end of the Cretaceous despite ample survival of other groups in similar environments. In this sense, the dinosaurs qualify for the bad genes label. Unquestionably, groups such as the dinosaurs all went extinct for the same reason. We don't know for sure what the reason was, but it was not a statistical fluke. Dinosaurs must have shared some quality, perhaps having nothing to do with their fitness in normal times, which made them vulnerable to the rare circumstances of the global catastrophe.

This led me to an exciting idea about the evolutionary process, namely that the normal processes of natural selection and adaptation cannot cope with rare events. The best analogy I know is hypothetical but effective. Consider the following scenario. Suppose the Earth is blanketed by high-energy radiation at the levels postulated in the 1960s for nuclear winter. Organisms vary enormously in their somatic reaction to radiation with, for example, mammals being far more severely affected than, say, insects. One can easily imagine a radiation level that would kill all exposed, terrestrial mammals yet leave insects unscathed. This would produce a highly selective extinction, yet one in which the victims were well adapted to the range of conditions normally encountered. The bottom line is that natural selection cannot contribute to species survival under conditions that don't happen very often.

As Raup's views about the nature of extinction—and its role in evolution—have changed over the years, he has become somewhat skeptical of some aspects of traditional Darwinism and the Modern Synthesis:

As I got more deeply into extinction I ransacked Darwin for his views. What little Darwin said about extinction was dead wrong, absolutely dead wrong. It was all competitive replacement. And the modern synthesis largely carried that over, although it didn't really even talk about it much. There was a tacit acceptance of the Darwinian view that species die out because better ones come along. My principal gripe with the Synthesis—and it is really minor—is that it virtually ignored extinction. Rather like a demographer ignoring death.

Of course, I have other gripes with the synthesis. In this, I was much influenced by Steve Gould in trying to separate macro from micro-evolution—the basic question being whether the extrapolation from population-level processes to all of evolution is valid. This is one area where I see the Modern Synthesis as making a colossal leap of faith. Will Provine loves to talk about the synthesis as being a major constraint in that an awful lot of evolutionary thinking was culled out and thrown away in that period. And I'm not enough of a historian of biology to know how significant that is, but the brainwashing we all had was profound. I have great hopes for some of the molecular biologists who are working in evolution now because they are free of some of the brain-washing in evolutionary theory the rest of us got. My hope is that they'll stumble on something very, very different.

Finally, we'll close the interview with a few short answers to questions:

How comfortable are you, thinking of yourself as a catastrophist?
Oh, very comfortable—I get it directly from my father.

To what extent do you think that the work that you did with the MBL simulations and elsewhere was genuinely experimental?
Oh, I think it was. Now, the word "experiment" has been distorted. If somebody makes a measurement (perhaps using a telescope or microscope or even calipers), we call it an experiment. It's not the experiment I grew up with where you put A into B to see what happens. So, I mean experiment in its original sense. If done properly, I think simulation models can be applied to processes you haven't actually seen. Yet we can observe what happens—sort of—under the conditions we specify, if the underlying process is understood.

Simulation modeling is wonderful and has contributed mightily in many fields but I sense its use has gotten a bit out of hand. I've noticed in reading reports in *Science* and *Nature*, for example, that you have to read pretty far into an article before you are sure whether the work is based on observation or on simulation. The two modes, observation and simulation modeling, are merging rather dangerously. You find major stories in the *New York Times* that are not careful to make the distinction. I fear sometimes the researchers themselves aren't making the distinction.

Do you think what happened in paleobiology during the 1970s might qualify as a Kuhnian revolution?
I think so—although incomplete.

Finally, in the introduction to your book on extinction, Stephen Jay Gould describes you in the following way: "If Dave has any motto, it can only be: think the unthinkable."[9] Is this accurate?
I think that's probably accurate. The more people agree, the more I'm likely to disagree, often to my detriment.

NOTES

1. Raup, David. 1986 *The nemesis affair: A story of the death of dinosaurs and the ways of science.* New York: W. W. Norton, and Raup, David. 1991. *Extinction: Bad genes or bad luck?* New York: W. W. Norton.

2. Robert Maynard Hutchins was president of the University of Chicago from 1929 to 1951, and is generally credited with drastically changing the character of the university during his tenure. In addition to abolishing the school's vaunted Big Ten football program, he also strengthened the university's famous "great books" core program, divided graduate faculty into four main divisions, and attracted major scientific endeavors to the university—such as the Manhattan Project's Metallurgical Laboratory, where the first self-sustaining nuclear chain reaction took place under the bleachers of the old football stadium.

3. Rainger, Ronald. 1997. Everett C. Olson and the development of vertebrate paleoecology and taphonomy. *Archives of Natural History* 24:373–96.

4. Mayr, Ernst. 1954. A monograph of the Echinoidea. *Evolution* 8:1–18.

5. Raup, David M. 1962. *Computer as aid in describing form in gastropod shells.* *Science* 138:150–52. This is the first paper to describe what has become known as the "Raup model" of gastropod shell coiling.

6. Raup, David M., Stephen Jay Gould, Thomas J. M. Schopf, and Daniel S. Simberloff. 1973. Stochastic models of phylogeny and the evolution of diversity. *Journal of Geology* 81:525–42.

7. Raup, David M., and Stephen Jay Gould. 1974. Stochastic simulation and evolution of morphology: Towards a nomothetic paleontology. *Systematic Zoology* 23:305–22.

8. Raup, David M. 1981.Extinction: Bad genes or bad luck? *Acta Geologica Hispanica* 16 (1-2): 25–33.

9. Gould, Stephen Jay, in Raup, 1991, xvi.

CHAPTER TWENTY-FIVE

Paleontology in the Twenty-first Century

David Jablonski

Paleontology is enjoying a time of intellectually rich and diverse activity. The rhetoric is not pitched as high (for the most part) as in the late 60s, 70s, and 80s, when paleontology was fighting its way back to the high table of biology, but this is for the best of reasons: it's there. Differential speciation and extinction rates among clades and times, variation in the frequency of phenotypic stasis and change, phylogenetic effects and other historical factors, intrinsic constraints and other developmental factors, among many other elements, are now standard considerations in evolutionary analyses (see discussions in Jablonski 2000, 2005b, 2007, and many of the chapters herein). Paleontologists cannot claim all of the credit for bringing these issues forward, but their conceptual and empirical contributions have enriched evolutionary biology in many ways.

The development of its own core set of methods, phenomenologies, and scientific issues was essential for the health of paleontology as a field, but many of the most vital research areas have lain, and will lie, at its intersection with other disciplines in the biological and physical sciences. To be sure, the foundations of paleontology—the systematic and phylogenetic treatment of specimens collected in the field and curated in museums and other repositories for future analyses, which are as crucial to the discipline as systematics is fundamental to so many aspects of neontology—must remain strong for this enterprise to continue. In fact, these foundations and their accompanying infrastructure, such as the Paleobiology Database and other large, interactive compendia, must be strengthened to support the broader and deeper range of questions that are driving the field. However, paleontology has always thrived when and where its connections to the other disciplines within the earth and life sciences were operating at full tilt, and the value of intensifying those connections will be the primary thesis of this chapter. I will address six major areas where especially active, interdisciplinary work can be expected in coming years: the separation of paleontological signal from noise and artifact; evolutionary developmental paleobiology; the relative role of external drivers and intrinsic factors; evolutionary tempo and mode at the species level and other hierarchical levels; the nonlinear effects of extinction and the nature of post-extinction recoveries; and spatial dynamics.

SEPARATING PALEOBIOLOGICAL SIGNAL FROM NOISE AND ARTIFACT

Paleontologists have always been aware of the incompleteness of the fossil record (see Darwin 1859; Simpson 1960). Everyone agrees that the record fails to capture most soft-bodied species (the paleontological history of sea anemones and tunicates will never match those of bivalves and mammals), but even for groups with robust skeletons, sampling and preservation can create two very different kinds of problems: they can generate noise, so that analyses will fail to detect biological pattern when it is there (essentially the statistician's Type II error—incorrectly accepting a null hypothesis), or they can generate artificial pattern that will be mistaken for true biological signal (a Type I error). Any scientific field must critically evaluate its data, and this must occur constantly as new questions gain prominence and new demands arise. As Grantham notes (this volume), methodological issues have been pivotal for consensus-building within paleontology and for productive interactions with other fields. Taphonomy, the study of the fossilization process and the quality of the fossil record, exploded as a research area in recent decades for just this reason (see, for example, Kidwell and Flessa 1995; Martin 1999; Behrensmeyer, Kidwell, and Gastaldo 2000; Kidwell and Holland 2002; and Benton, this volume).

Noise may simply involve the random failure of sampling and preservation, for example in the documentation of stratigraphic ranges of species. Most dinosaur species are known only from the sedimentary deposit—the geologic formation—of their original discovery, so that those species effectively have no measurable duration (Dodson 1990). On a larger scale, incomplete sampling of stratigraphic ranges can blur sharp extinction events because range terminations tend to be smeared back from the true event (reviews in Kidwell and Holland 2002; Fastovsky et al. 2004).

Noise is usually not considered to be as serious a paleontological problem as artificial pattern—after all, results that rise above noise into statistical significance are likely to be especially robust. Artifacts created by imperfections of the fossil record must be actively filtered out, and this is a more difficult task. Consider mass extinctions again. If the sedimentary record declines in volume, outcrop area, or fossiliferousness at particular times, for example because sea-level drops so that erosion and nondeposition undermine the marine record, then stratigraphic ranges will be artificially truncated, potentially creating an apparent extinction event where none had occurred (e.g., Foote 2003, 2005).

Taphonomic problems have seen intensive research on scales from single outcrops to global compendia, and many concerns have been allayed or met with increasingly robust and sophisticated methods for assessing, and sometimes reducing, the risk of Type II errors (see table 25.1 for some of the major concerns regarding the nature of the fossil record that have been addressed most intensively, and their current status). Implications of these successes are rippling through the discipline, and beyond. For example, Kidwell's finding that rank-order abundance data are robust to time averaging and other biases in molluscan communities has prompted a new wave of quantitative paleoecological research, and implies that neoecologists can use dead shells to characterize modern communities (Kidwell 2001; Zuschin and Oliver 2005; NRC 2005). Dead shells are often orders of magnitude more abundant in present-day marine samples than live individuals, and the finding that those dead shells provide reliable information on at least some aspects of the composition and ecological structure of their source community will permit much more efficient and statistically robust analysis of molluscan communities. In some instances, dead shells can be used to reconstruct community states for the decades immediately preceding human disturbance, a valuable and elusive benchmark for conservation biology (Kowalewski et al. 2000; Edgar and Sampson 2004; Ferguson and Miller 2007; Kidwell 2007, 2008).

Taphonomy will continue to be a vibrant subdiscipline of paleontology, developing new ways to test and compensate for the incompleteness of the record as new biological and geological questions arise. At all temporal and spatial scales, the modeling approaches explored by workers such as Foote (2003), Hunt (2004a), and Peters (2005), and especially those treating the interaction of several factors (e.g., Foote 2003, 2005, and in a far cruder approach, Valentine et al. 2006) show considerable promise. At the same time, there will always be room for brute-force empirics like the Pliocene-Pleistocene inventory undertaken by Jablonski et al. (2003) in their evaluation of the "Pull

TABLE 25.1 Potential biases in the fossil record that have recently received extensive attention. Note that references provided are representative and not exhaustive.

Problem	Expected to	Current Status	Representative References
Time averaging	Inflate local diversity, distort relative abundances (by compressing multiple generations of short-lived species into the same sedimentary deposit as a single generation of long-living species), conceal phenotypic change by collapsing a directional trend into a single highly variable pooled sample	Many ecological and evolutionary patterns are robust in many macrofossil groups; very rapid spatial shifts, as in high-amplitude Pleistocene climate flickers, may create problems.	Behrensmeyer et al. 2000 (review); Kidwell 2001, 2002; Bush et al. 2002; Hunt 2004a,b; Olszeweski and Kidwell 2007. Problems: Roy et al. 1996; Lyons 2003; Jernvall and Fortelius 2004
Pull of the Recent	Inflate global diversity of young time intervals because rich sampling of modern biota can extend stratigraphic ranges of geologically young taxa across intervals where fossils of those taxa are lacking, but cannot affect the ranges of extinct taxa	Negligible for marine bivalves, untested for most other groups	Jablonski et al. 2003a
Range truncation by gaps in fossil record	Artificially emphasize extinction and origination bursts	When preservation, origination, and extinction are modeled through the Phanerozoic to simultaneously maximize the statistical fit of these parameters to the observed data, major extinction and origination pulses emerge as robust features; smaller ones are difficult to distinguish from artifacts of record	Foote 2003, 2005

Backward and forward smearing (Signor-Lipps effect and its relatives)	Produce artificially gradual origination or extinction patterns, simply owing to sampling of typically patchy stratigraphic ranges near an origination or extinction pulse	Can be evaluated using increasingly robust methods for setting statistical confidence limits on range endpoints, which have generally confirmed sharpness of extinction events	Marshall 2001; Holland 2003; Wang and Marshall 2004; Wang and Everson 2007
Fossil preservation bias	Taxa building shells of less stable minerals (aragonite, as in most bivalves and almost all snails) artificially underrepresented in the older fossil record relative to taxa using the more stable mineral (calcite, as in brachiopods, scallops, oysters, and sea urchins)	Important for individual communities, but not an overriding factor for diversity trends in marine bivalves and brachiopods (although body size appears to play a significant role)	Bush and Bambach 2004; Behrensmeyer et al. 2005; Kidwell 2005; Ros and De Renzi 2005; Cooper et al. 2006; Valentine et al. 2006
Tropical under-sampling	Tropical regions undersampled relative to temperate regions during the Cenozoic (perhaps artificially depressing evolutionary rate differential and standing diversities between the two regions); opposite is true for the Paleozoic (perhaps inflating tropical diversity in the Paleozoic tropics relative to both Paleozoic temperate zones and the Cenozoic tropics)	Significant bias, requiring statistical approaches to factor this effect out of analyses that encompass latitudinal diversity gradients	Allison and Briggs 1993; Jablonski 1993; Bush and Bambach 2004; Kiessling 2005; Valentine et al. 2006; Jablonski, Roy, and Valentine 2006a
Temporal variation in outcrop area or volume of fossiliferous rocks	Global and regional diversity will track sampling bias imposed by availability of fossiliferous sediments	Active area of research and controversy. Difficult to account for first-order global diversity patterns strictly in terms of sampling biases, but may be important on regional scale; "common cause" linking true diversity and outcrop area is again being considered	Alroy et al. 2001; Crampton et al. 2003; Jablonski et al. 2003; Bush et al. 2004; Bush and Bambach 2004; Barnosky et al. 2005; Peters 2005; Smith 2007; Smith and McGowan 2007

of the Recent" in marine bivalves. Clearly, close interaction with the earth sciences will be essential, in terms of understanding how sea level fluctuations and other aspects of the fabric of the sedimentary record impinge on biological signals (for reviews see Kidwell and Holland 2002; Hannisdal 2006), and improvements in high-resolution dating of the sedimentary records may have a major impact here. This flow of information is—or should be—very much a two-way street, because the methods of taphonomy are essential to the earth sciences whenever sedimentary (often biogenic) particles are used to determine temperature, salinity, aridity, and absolute ages of paleoenvironments (e.g., Martin 1999; Kowalewski and Bambach 2003; Kowalewski and Rimstidt 2003); the concentration, transport, temporal averaging, and loss of those particles are just as capable of setting up Type I and Type II errors in time series of physical parameters as they are for biological data.

Museum collections, which capture centuries of paleontological work but are rarely unbiased samples of a time, place, or clade, present difficult challenges, but could be especially rewarding as sources of insights into sampling intensities, rank abundances, morphological variance, spatial distribution, and many other variables of keen interest (e.g., Teichert et al. 1987; NRC 2002; Barbour Wood, Kowalewski, and Ward 2004; Allmon 2005; Harnik 2009; Jablonski, Roy, and Valentine 2006; see also Graham et al. 2004; Suarez and Tsutsui 2004; Guralnik and Van Cleve 2005; Lutolf, Kienast, and Guisan 2006; Solow and Roberts 2006; Stuart et al. 2006). A well-developed set of protocols and statistical approaches enabling an expanded scientific role of these collections, which contain species and morphologies from the far reaches of the rarity spectrum and often represent localities long since paved over or mined out, would be a major step forward for the historical sciences.

EVOLUTIONARY DEVELOPMENTAL PALEOBIOLOGY

The renaissance of evolutionary developmental biology (evo-devo) is one of the great success stories of the past decade, as attested by the steady stream of books, symposia, and new or reoriented journals in this area. The molecular and tissue-level processes that generate complex biological forms are becoming increasingly well understood, launching an exciting burst of evolutionary thinking and analysis, although the layers of complexity to be revealed in molecular developmental processes—at once elegant orchestrations and Rube Goldberg contraptions—constitute in themselves a research program that dwarfs the volume of work having an evolutionary focus.

The analysis of developmental sequences of extant species in a phylogenetic context can provide great evolutionary insights, but data restricted to highly derived living forms have their limitations. For example, the extinction of basal or intermediate forms can obscure the actual sequence of developmental transformations, as attested by the wide range of views on the nature of early bilaterians in the late Precambrian (see reviews by Erwin and Davidson 2002; Valentine 2004, 2006). In addition, the fossil record documents not only the nature of past phenotypes—both in extant clades and in extinct lines—but also their environmental context, which allows framing and testing novel evo-devo hypotheses. To take one example among dozens, combined fossil and developmental data provide a much fuller picture of the evolutionary changes in both fore- and hindlimb during the evolution of whales (e.g., Thewissen and Williams 2002; Richardson, Jeffery, and Tabin 2004; Thewissen et al. 2006), and suggest that the transition from terrestrial to marine settings (a) occurred in the tropics despite the focus of modern cetacean diversity at higher latitudes, and (b) included at least some early aquatic forms that were freshwater rather than marine (Clementz et al. 2006). A similar profitable interplay between paleontology and evo-devo has brought a deeper understanding of the origin of tetrapods (e.g., Coates, Jeffery, and Ruta 2002; Shubin 2002; Shubin, Daeschler, and Jenkins 2006; Ruta, Wagner, and Coates 2006; Wagner, Ruta, and Coates 2006; Davis, Dahn, and Shubin 2007; see also Raff 2007) and of many other groups, ranging from barnacles to birds.

If paleontology only provided insights into the morphology, development, and ecological context of transitional forms in the origin of novel phenotypes, this would guarantee it a key role in evolutionary developmental biology. But it can provide a much wider range of macroevolutionary insights, although this potential has yet to be fully tapped. First, the fossil record should be used more extensively to test hypotheses on the macroevolutionary consequences of the architecture of developmental systems. Whenever phylogenetic analysis can be combined with developmental data to characterize major developmental differences among clades (usually by bracketing deep phylogenetic nodes with extant species), paleontologists can assess the macroevolutionary role of those differences. We would like to know, for example, whether the tempo and mode of large-scale phenotypic evolution varies with such developmental factors as: genome organization dominated by multiple, slightly divergent copies of genes versus single-copy genes with large batteries of regulatory binding sites versus genes generating many isoforms via alternative splicing or translation initiation (all of these being ways to expand the effective genome size; not mutually exclusive, but apparently varying in importance among clades); the degree of modularity versus developmental integration within the body; the relative timing of induction events or morphogenetic processes; or whether intraspecific variation derives mainly from developmental plasticity versus genetic polymorphism. We don't really know how stable such developmental properties have been within clades over geologic timescales. Some paleontogically informed work on these and other questions in comparative evo-devo has begun (e.g., Valentine 2000, 2004; Salazar-Ciudad and Jernvall 2004; Goswami 2006, 2007; Kavanagh et al. 2007), but basic information is still sparse. This will be a very interesting area for research.

Second, paleontology can add an ecological dimension to the origin of evolutionary novelty. A few neontologists have promoted an "eco-evo-devo," but this approach has largely involved the microevolutionary causes and consequences of plasticity—that is, the transduction of environmental signals into developmental ones (e.g., Gilbert 2001; Sultan 2003). This more tightly focused work certainly touches on interesting issues, but the macroevolutionary questions are even more fascinating. Work in this area could follow many directions (see, for example, Gould's [1977b] early attempt to link modes of heterochrony to life-history strategies), but an obvious avenue involves one of the most striking facts about the fossil record: that evolutionary novelties do not arise randomly in time and space. Temporally, the first occurrences of higher taxa and divergent morphologies are concentrated in the early Paleozoic, as the Cambrian explosion and the Ordovician radiations, with secondary bursts following the major mass extinctions, particularly the huge end-Permian event (see reviews by Jablonski 2000, 2005b, 2007; Erwin 2005, 2006a). Spatially, post-Paleozoic marine invertebrate orders tend to originate in the tropics, and in shallow, variable habitats, and some evidence suggests consistent patterns in Paleozoic invertebrates and terrestrial plants (reviewed in Jablonski 2005a). These patterns, a macroevolutionary manifestation of Van Valen's (1973) famous dictum that "Evolution is the control of development by ecology," remain poorly understood. Are they strictly ecological in nature, or do genomic or developmental mechanisms more directly promote these large-scale inhomogeneities in the origin of novelty? For temporal patterns, at least, current thinking favors a strictly ecological view—open ecospace plus ecological feedbacks promoting evolutionary inventiveness—but we are still too ignorant of the macroevolution of developmental processes (including constraints and trade-offs, as appear to operate between developmental flexibility and developmental precision) to rule out a hybrid view involving both external and internal factors (see discussions in Marshall 2006; Jablonski 2007).

One especially useful analytical and conceptual tool has been the tracking of clade morphology within a multivariate space defined by the statistical treatment of a large array of morphological measurements, or morphospace, in a taxon-free approach. This approach was made famous by Gould's (1989) contention that Cambrian life, typified by the exquisite fossils of the Burgess Shale, encompassed a greater range or variety of biological form than seen in the present-day biota. We now have a much better theoretical and empirical grasp of morphospace analysis, including the strengths and weaknesses of different quantitative approaches—for example, those involving discrete versus continuous characters—and the variety of implicit assumptions attached to different types of morphospaces (see Foote 1996a, 1997; Roy and Foote 1997; McGhee 1998, 2007; Eble 2000; Wagner 2000; Ciampaglio, Kemp and D. W. McShea 2001; Stockmeyer Lofgren, Plotnick, and Wagner 2003; Villier and Eble 2004; Erwin 2007; Gerber, Neige, and Eble 2007). Although Gould's contention of *greater* disparity in Cambrian seas may be difficult to support (definitive tests are still lacking), a wide variety of taxa analyzed by a wide range of methods converge on a pervasive pattern, as previously noted, with rapid deployment in morphospace early in a clade's history, particularly during the early Paleozoic, after mass extinctions, and in the course of major environmental transitions (as in the invasion of land by plants and tetrapods). Two active research directions that seem particularly promising are, first, comparative analyses among clades that simultaneously use different metrics of diversity, and, second, incorporation of phylogenetic data into morphospace analysis.

First, discordant behavior among diversity metrics has sometimes been viewed as an inconvenience or worse, but could actually be a source of major insights. To take the classic example, quantitative morphospace analysis has largely validated the biological interpretation of discordant temporal patterns at different levels in the Linnean taxonomic hierarchy (an approach spearheaded by Valentine 1969, 1973, and heavily employed ever since); the rapid production of evolutionary innovations, as loosely reflected by the occurrence of higher taxa, relative to species-level diversification, is clearly a real phenomenon and not simply an artifact of sampling or the nested structure of Linnean taxa (as had been suggested by, for example, Smith and Patterson 1988; Smith 1994; Forey et al. 2004; see Foote 1996b; Jablonski 2000, 2005a, 2007; Erwin 2007). The proliferation of reproductively isolated units can be significantly decoupled from the net gain of large-scale evolutionary novelties or expansion in morphospace (fig. 25.1), but only under certain circumstances, whose limits remain undetermined. The full set of concordances and discordances

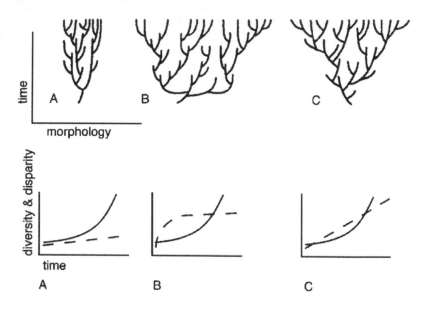

Figure 25.1 Discordances between morphological disparity and taxonomic diversity under different evolutionary scenarios. Top row: idealized diversity patterns. Bottom row: predicted diversity and disparity curves over time (solid line = taxonomic diversity; dashed line = morphological disparity). A: Morphological evolution is constrained though not fully inhibited; taxonomic diversification is not accompanied by morphological diversification. B: Morphological steps are large early in the clade's history; morphologic diversification outstrips taxonomic diversification. C: No constraint on morphological evolution, no trend in morphological step size; morphologic diversification is concordant with taxonomic diversification. From Wesley-Hunt (2005).

among taxonomic, genetic, functional, and morphological diversity have yet to be explored systematically (for discussions or examples of work at these interfaces see the reviews cited previously and Wagner 1995, 1997; Roy, Jablonski, and Valentine 2004; Ricklefs 2005; Wainwright et al. 2005; Wesley-Hunt 2005; Petchey and Gaston 2006; Vellend and Geber 2005; Brakefield 2006; Jablonski, Finarelli, and Roy 2006; Whibley et al. 2006; Wainwright 2007). Further work would be valuable on pinpointing significant evolutionary novelty amid the welter of quantitative phenotypic change that occurs over the history of any major group (e.g., from very different perspectives but emphasizing the functional roles of discrete—as opposed to continuous—characters, see Jablonski and Bottjer 1990; Vermeij 2006; Bambach, Bush, and Erwin 2007; the extensive ecomorphological literature on vertebrates has largely been disconnected to this macroevolutionary effort but could profitably be brought to bear here; see, for example, Wainwright and Reilly 1994; Wainwright 2007).

Second, clades have only rarely been mapped directly into morphospaces, although this phylogenetically based approach holds great potential for clarifying evolutionary dynamics (see Wagner 1995, 1997; Alroy 2000; Eble 2000; Stone 2003; Boyce 2005; Stayton 2006, for different approaches). For example, the proximity of two taxa in a morphospace could reflect recency of common ancestry, the final steps of a clade's trend toward convergence toward another, or mutual protracted converging trends, among other possibilities; regions of morphospace could be filled piecemeal from the edges by a number of clades or by a single radiating clade; evolutionary step sizes could vary through the history of a clade or a biota, or remain stochastically constant. And clades might interfere with one another's trajectories through morphospace, so that patterns of morphospace occupation might be constrained and then released at times of dense occupation followed by extinction (e.g., Ciampaglio 2004; McGowan 2004a, 2004b, 2005); even less explored is the potential for shifts in morphospace occupation to promote diversification in other clades, as must have occurred, for example, when vascular plants attained the tree growth habit. Work in the direction of uniting morphospace approaches with model-based comparative methods would promote rigorous comparisons of differential diversification in this context (as potential starting points, see, for example, Pagel 2002; Freckleton, Harvey, and Pagel 2002; Butler and King 2004; and Sidlauskas 2007). However, application of comparative approaches to arrays of living taxa is severely hampered by the need to assign ancestral morphologies—that is, quantitative characters, as used in defining most morphospaces—for nodes within the phylogeny in some of the most heavily used methods (see Webster and Purvis 2002; Erwin 2007; Wiens et al. 2007, and for one alternative see Harmon et al. 2003); in densely sampled, well-preserved clades, paleontological data can at least come close to estimating ancestral character states without the methodological constraints that impose either prohibitively broad confidence limits or simplistic evolutionary models.

The quest to analyze clades within true developmental morphospaces (Eble 2003) has gone more slowly, because realistic generative models are difficult to construct for complex forms. Empirical morphospaces generally involve simple linear measurements only indirectly related to developmental processes; even theoretical morphospaces (most famously Raup's seminal molluscan forms) tend to be gross abstractions of development and growth of real organisms. The potential for testing patterns of morphospace occupation against expectations from developmental processes is spectacular, and our burgeoning understanding of development at the molecular, cellular, and tissue level, laying a foundation for such work (for exciting steps along this path using mammalian teeth as a case study, see Jernvall 2000; Jernvall et al. 2000; Salazar-Ciudad and Jernvall 2002; Kavanagh et al. 2007).

EXTERNAL DRIVERS AND INTRINSIC FACTORS

The relative role of intrinsic and extrinsic factors in shaping macroevolutionary dynamics was one of Gould's (1977a) "eternal metaphors of paleontology." This issue involves much more than the selection-versus-constraints controversy that dominated twentieth-century discussions of punctuated equilibria and other modes of species-level change. The coming decades may see research activities that focus along three major lines, focusing on the physical environment, the external biotic environment, and the intrinsic properties of organisms and taxa.

Empirical Links to Physical Environmental Drivers

Few workers doubt that climate changes and other large-scale perturbations have had significant biotic effects: asteroid impacts, glacial cycles, and other intense changes in the global environment have surely left their mark on the biota (e.g., Jansson and Dynesius 2002; Janis 2003; Jackson and Erwin 2006). Major strides in the precision of a variety of geochemical methods (e.g., NRC 2005; West et al. 2006) will lead paleontology to strengthen and broaden interactions not only with the life sciences but with the physical sciences. The earth sciences have been the traditional home for most subdisciplines of paleontology (vertebrate paleontology and Quaternary studies being the most obvious exceptions), and this partnership should intensify as paleontological material serves as more than simply the bearer of geochemical ratios, but as a rich phenomenology of biological events and trends embedded in a dynamic environment. In fact, a key contribution of paleontology to our understanding of the earth-life system is a varied and detailed set of environmental parameters lacking present-day analogs, providing a broader envelope of boundary conditions against which to develop and test general models (see NRC 2005).

The wealth of long-term and high-resolution geochemical time-series actually creates a new problem: the ease of finding an environmental event to coincide with virtually every biological one. The challenge is to separate causation from an increasingly dense set of correlations by adding information beyond simple timing. For example, the nature of a given event held to have biological import—say, a pulse of global warming—carries subsidiary predictions on the nature of biotic responses, such as an expansion of thermophilic organisms, and a retreat of cryophilic ones, into higher latitudes. Advances in radiometric dating and temporal correlation techniques will permit high-resolution testing of the synchrony or displacement of events among regions, response rates and lags among clades, and other lines of evidence that will help to tease out causal linkages (Sadler 2004; Erwin 2006a).

The most intriguing findings on the role of physical drivers come when biotic responses confound expectations, an outcome likely to increase in coming years. Most famously, the extreme glacial-interglacial cycles of the past two million years brought unexceptional speciation and extinction rates (e.g., Willis et al. 2004; Barnosky 2005; NRC 2005). Instead, marine and terrestrial species generally shifted their spatial positions to track favorable climates, and they largely did so individualistically rather than as cohesive communities (Willis et al. 2004; Jackson and Williams 2004; and Jablonski 2007, who discusses the lingering controversy over this interpretation, which seems to derive at least in part from the use of unrealistic null models). This behavior, which gives rise to a series of transient communities, raises a host of interesting issues, regarding, for example, the relatively low extinction intensities as species associations are broken and realigned, and the web of apparently co-evolved relationships between, for example, plants and pollinators that seems to require more continuous associations than implied by the paleontological data; new models and theory have resulted from a deeper appreciation of these apparent paradoxes (e.g., Roy et al. 1996; Jackson and Overpeck 2000; Hewitt and Nichols 2005; Huntley 2005; Roy and Pandolfi 2005; Thompson 2005; Rowe, Heske, and Paige 2006; Jablonski 2008a.).

The combination of improved dating methods with the ongoing refinement and discovery of geochemical proxies for environmental factors is likely to usher in a significant increase in the breadth and rigor of work on the role of extrinsic drivers in the fossil record. Analyses can go far beyond the search for impact-associated Iridium spikes and the familiar curves tracing global temperature and sea level, to factors such as nutrient input (e.g., Bambach 1993, 1999; Payne and Finnegan 2006) and changes in atmospheric and oceanic composition, and the feedbacks on many of these variables that biotic system impose in turn (for a sampling of provocative hypotheses, see Harper, Palmer, and Alphey 1997; Knoll 2003; Katz et al. 2004; Beerling and Berner 2005; Falkowski et al. 2005). Potential environmental triggers for the Cambrian explosion (Peterson et al. 2008; Marshall 2006), and environmental, rather than intrinsic biotic, causes for the evolutionary lags immediately after the end-

Permian and end-Cretaceous mass extinctions (Payne et al. 2004; D'Hondt 2005; Knoll et al. 2007) are just a few of the major issues being addressed in new ways with this improved and more egalitarian partnership. Combined paleontological and geochemical research will also permit the tracking of biotic responses to more subtle or more localized perturbations, such as the uplift of the Isthmus of Panama, which not only unleashed the Great American Interchange of terrestrial plants and animals (MacFadden 2006 and references therein) but had profound oceanographic effects that drove selective extinctions in the marine faunas (Todd et al. 2002); and the tectonic assembly of the West Pacific archipelagos, which also permitted biotic interchanges of terrestrial clades but may have promoted explosive diversification among marine forms (Crame and Rosen 2002).

The Role of the Biotic Environment

The extrinsic biotic environment—the ecological interactions that every organism, population, species, and clade is subject to—is also likely to influence large-scale evolutionary dynamics. Perhaps the most familiar large-scale examples are hypotheses of escalation, where predatory clades drive changes in the diversity and morphology of their shelled prey (Vermeij 1987), and niche construction or ecosystem engineering, where the activity of one lineage or group of lineages creates ecological opportunities for others (Odling-Smee, Laland, and Feldman 2003, less explored paleontologically but presumably exemplified by the origin of arborescent plant growth forms on land [Bateman et al. 1998] and the alteration of sediment properties by the deposition of dead shells in marine settings [Kidwell and Jablonski 1983; Kidwell 1986, and in a nice instance of conceptual flow from paleo- to neoecology, Gutiérrez et al. 2003]). Testing hypotheses for the macroevolutionary role of such interactions is difficult, for several reasons: (a) direct evidence of the intensity or continuity of specific interactions is often lacking in the fossil record (although a few modes of predation and parasitism leave unambiguous traces); (b) many of the interactions of interest do not correspond to the classical pairwise species interactions that dominate ecological theory and experiment. Instead, they involve more diffuse, multispecies effects over broad swaths of time and space; (c) the geologic record is so rich in biotic events hypothesized to reflect the onset or consequences of such interactions (extinctions and originations, radiations and declines) that rigorously pinpointing cause and effect is a major challenge (see Jablonski 2008a for an overview). At least three approaches have been profitable in moving beyond temporal correlations by incorporating subsidiary information, and these are likely to continue as active research areas.

First, *spatial structure*: clades should not show the expected, putatively interactive dynamics until their ranges overlap not only temporally, but spatially (e.g., Roy 1994; Miller 1998; Aberhan, Kiessling, and Fürsich 2006). For example, if prey clades decline at high latitudes while their putative predators are still restricted to the tropics, we could reject a biotic-interaction hypothesis for the declines in favor of, say, one of climatic deterioration.

Second, *biomechanical capability*: clades should not show the expected, putatively interactive dynamics until the appropriate morphologies or behaviors are in place. For example, some predatory clades arose tens of millions of years before the durophagous weaponry that allowed them to penetrate shelled prey (Vermeij 1987; Walker and Brett 2002). Unless the derived characters that mark the origination of a target clade are necessary and sufficient to promote the hypothesized interaction (as is demonstrably not the case for a number of predatory fishes and arthropod lineages, for example), taxonomic data will be actively misleading.

Third, *more realistic models of interaction dynamics*: macroevolutionary thinking has been dominated by simplistic expectations, such as synchronous diversifications in sets of positively interacting clades (or for sets of clades where one exploits the other, as in phytophagous insects and their hosts), and the "double-wedge" model for negative interactions, in which two negatively interacting clades (via competition or predation, for example) show reciprocal diversity trends such that one clade dwindles to extinction or marginal status as the superior competitor expands (e.g., Benton 1996). This approach is flawed in at least two ways. First, these models are not the only expectation for clade interactions. For example, net diversification might continue in negatively interacting clades but at a damped rate relative to unimpeded processes (e.g., Sepkoski 1996). In some situations, negative interactions might actually promote diversification, as when interclade competition or predation promotes population fragmentation, divergence, and, ultimately, adaptive radiation (Vamosi 2005; Nosil and Crespi 2006). On the other hand, congruent phylogenies of plants and their insect pests or other indicators of apparent coevolution can arise through strongly asynchronous diversifications, and this asynchrony may even be the rule (Labandeira 2002; Lopez-Vaamonde et al. 2006; for further examples of macroevolutionary lags in clade dynamics, see Jablonski 2007). Second, and more generally, long-term dynamics may often be dominated by diffuse interactions, where the selective milieu or biotic background exerts more persistent and deterministic effects than pairwise species sets. Some neontologists also increasingly emphasize diffuse interactions in evolutionary ecology (e.g., Janzen 1980, who forcefully drew the line between pairwise and diffuse coevolution; see Maurer 1999; Strauss, Sahli, and Conner 2005; Thompson 2005; McGill et al. 2006 for a variety of perspectives). The rich array of biotic interactions seen over the short run in living communities is not necessarily reducible to pairwise interactions nor readily scaled up to clade-level dynamics in any simple way. New approaches are needed to incorporate this more complex ecological dimensionality into macroevolutionary theory. Frequency distributions or multivariate contours in morphospaces through a series of time-space coordinates for traits thought to be important to an interaction would be valuable, for example.

Intrinsic Biotic Factors

The properties of organisms and clades condition their evolutionary behavior: elephants and mice respond differently to the shift of a stream channel through an old field; widespread and restricted species differ in vulnerability to hurricanes, volcanic eruptions and other localized perturbations; species with highly subdivided genetic population structure are expected to (and evidently do) have greater propensities to speciate than species having more extensive gene flow. Many workers have evaluated the role of intrinsic factors in determining the dynamics of clades in the fossil record (for reviews see Jablonski 1995, 2000, 2005a, 2007, 2008b; McKinney 1997; Sepkoski 1998; Jablonski, Roy, and Valentine 2003), and an even richer literature assessing extinction risk and (via the comparative biology literature) speciation or net diversification tendency exists for present-day organisms. I will therefore make only two points here. First, more interaction between paleontology and neontological areas such as ecology and conservation biology is sorely needed, not only because the science will benefit on both sides, but because we need all the insights we can get into the potential biotic responses to present-day stresses. Undue emphasis has probably been placed on supposed parallels between the Big Five mass extinctions of the fossil record and modern extinctions: thankfully we have not reached the point where relatively widespread, abundant marine *genera* are suffering double-digit extinction intensities. However, plenty of other paleontological insights into the links of extinction selectivities and biotic recoveries to intrinsic and extrinsic factors are applicable to modern biodiversity (see Jablonski 1995, 2001, 2005a; Erwin 2001, 2006b; NRC 2005, and below; and for a discussion of widespread and abundant species that *have* suffered severe losses, see Gaston and Fuller 2007).

Second, studies of the evolutionary roles of intrinsic biotic factors need to go beyond analyses of single traits. The factors that influence, say, extinction risk or speciation probability are not independent and features must interact, so that different combinations can heighten or damp effects in complex ways. For example, in present-day mammals, carnivorous habit, long generation time and narrow geographic range in combination typify species most vulnerable to extinction (with geographic range the most important single variable); species having the opposite traits are less vulnerable, and species with mixtures of character states are intermediate in vulnerability (Purvis, Jones, and Mace 2000, 2005; see also Cardillo et al. 2005, who recognize threshold effects related to body size). Paleontological analyses that rank intrinsic biotic traits in terms of their relative influence on speciation, extinction, and other aspects of evolutionary dynamics, or quantify interaction effects among traits, are still scarce. However, in keeping with the theme running through this chapter, a wide range of potential methods are available in other fields, such as path analysis and generalized linear modeling, and application of these approaches to paleontological questions will bring novel insights and even lead to novel questions (see, for example, Simpson and Harnik 2009; Jablonski and Hunt 2006).

TEMPO AND MODE OF EVOLUTION AT THE SPECIES LEVEL AND ABOVE

One of the great successes of twentieth-century paleontology was to establish that the tempo and mode of evolution at the species level and above often does not correspond with simple expectations from short-term observations on living populations (for reviews, see Jablonski 2000, 2007; Gould 2002). To take just two obvious examples, (1) the great evolutionary lability of experimental, domesticated, and wild populations over short evolutionary time scales (e.g., Hendry and Kinnison 1999) translates into surprising evolutionary stasis for many species in the fossil record (e.g., Gould 1982, 2002; Eldredge et al. 2005; see also Gingerich 2001; Hunt, 2007), and (2) the seemingly pervasive short-term selection for large body size in present-day populations (Kingsolver and Pfennig 2004) translates into a surprisingly heterogeneous pattern of body-size evolution over longer timescales, even in analyses that map modern populations onto phylogenies (Jablonski 1996, 1997; Moen 2006). There is great scope for conceptual and empirical work in this general area, and again, recent strides in dating and correlation methods offer enormous potential for quantifying taxonomic and morphologic rates of evolution with unprecedented precision (Sadler 2004; Erwin 2006a), but I think that the first order of business falls under two headings.

First, *determinants of evolutionary tempo and mode at the species level.* Although theoretical discussions on evolutionary tempo and mode tend to focus on end members (pure anagenesis, pure morphological stasis, etc.; see fig. 25.2), in practice all potential combinations, along with intermediates, occur in the fossil record. These variations are evidently not attributable sampling artifacts (see Jackson and Cheetham 1999; Jablonski 2000 for discussion), and so the challenge is to make biological sense of the observed distribution of evolutionary patterns at the species level. Early suggestions that stasis and punctuation might be more common in benthic (not pelagic), multicellular (not protistan), sexual (not clonal) organisms, or in unstable environments, no longer appear to be tenable as definitive generalizations (see Erwin and Anstey 1995; Jackson and Cheetham 1999; Jablonski 2000, 2007). Further, combinations of different tempos and modes within a single lineage are not uncommon: indeed, some well-studied lineages, including the famous coiled oyster *Gryphaea* and the Eurasian mammoth, are considerably more punctuational than previously thought, but apparent gradualistic segments remain (Johnson 1993; Jones and Gould 1999; Lister et al. 2005; and see Knappertsbusch's [2000, 2001] exhaustive analysis of a microplanktonic alga).

General protocols have been proposed for species-level studies of tempo and mode, and new analytical methods have emerged that take much better account of sampling, preservation, and other challenges (see Jablonski 2000; Kidwell and Holland 2002; Bush et al. 2002; Roopnarine 2003, 2005; Hunt 2004a, 2006, 2007; Hannisdal 2006). The field is thus poised for a new and more rigorous attack on this issue. Understanding the controls on species-level phenotypic change over geologic timescales would be a major step in understanding the linkages and discontinuities between micro- and macroevolution. The observed patterns of phenotypic change must be consistent with mechanisms operating at lower levels, of course, but we still need to understand how and why short-term evolutionary change manages to persist over geologic timescales in some lineages but not (most) others. Scales of gene flow and geographic range (and, from a sampling standpoint, the spatial scale of the analysis, this being the Achilles' heel of most paleontological studies of species-level change) may prove to be important elements of a general explanation (Jablonski 2000; Eldredge et al. 2005), but the work needed to answer this question has barely begun.

Second, *the hierarchical dissection of forces driving evolutionary change.* Another major contribution of twentieth-century paleontology was to provide

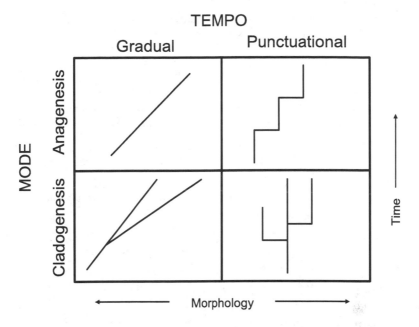

Figure 25.2 All possible combinations of evolutionary tempo and mode have been recorded from the fossil record. Research is sorely needed on whether clades vary in their frequencies of the different combinations, and if so, why. The upper left quadrant is classic phyletic gradualism, and the lower right is punctuated equilibrium.

theoretical and empirical support for a hierarchical view of evolution (e.g., Valentine 1973; Stanley 1979; Eldredge 1985; Gould 2002; and reviews by Jablonski 2000, 2007). Although many of the details remain controversial, most workers, including some who had been most adamant in their opposition, now accept that species sorting—differential speciation and extinction, sometimes termed *emergent fitness*—can be a potent evolutionary force in addition to, and sometimes in opposition to, natural selection at the level of bodies within populations (e.g., Dawkins 1989; Williams 1992; Maynard Smith 1989, 1998). Although rarely couched in these terms, the extensive neontological literature on biotic properties that govern differences in net diversification between sister clades, and the comparative methods developed to support these analyses, provides rich corroboration to what had previously been an almost exclusively paleontological enterprise (Jablonski 2000, 2007, 2008b; Coyne and Orr 2004). However, the raw origination and extinction rates that underlie the waxing and waning of clades, or that might fuel directional trends, are difficult to extract from neontological data on net diversification; paleontological analyses remain essential. Most urgently needed are comparative analyses among clades that not only identify intrinsic biotic factors that govern differential speciation and extinction rates, as previously discussed, but that pinpoint the hierarchical level at which those factors are effective, preferably in a well-resolved phylogenetic framework.

The role of strict-sense species selection, where speciation or extinction differentials are shaped by species-level characters such as geographic range or genetic population structure, in particular, needs further exploration. Some authors emphasize Stanley's (1979) broad-sense definition of species selection, where (as we now say) the key criterion is emergent fitness at the species level—that is, differential birth and death [replication]—rather than emergent properties [interaction] at the species level (see Grantham 1995, 2001, 2002, 2007; Jablonski 2000, 2007, 2008b; Gould 2002; and Lieberman and Vrba 2005 for discussions). This argument merges clade-level effects driven by organismic traits such as body size into a single category with those driven by species-level traits such as geographic range. To some extent this is a semantic argument: any sorting process irreducible to selection on bodies validates a hierarchical evolutionary theory. However, broad definitions can obscure important distinctions, and even Gould (2002) conceded that strict-sense species selection represents the "best cases" of hierarchical factors in evolution. The first challenge is to identify emergent properties in a consistent way and to test their roles empirically at different hierarchical levels.

Several authors have generated tentative lists of species-level traits, from geographic range size and several aspects of geographic range *shape* to genetic population structure and sex ratio (see Jablonski 2007). One statistical approach to operationalize emergent properties in the fossil record reinforces the view that geographic range is a species-level property that influences both speciation and extinction probabilities independent of the underlying organismic basis of the range (see Jablonski and Hunt 2006). Because every species has a geographic range, this opens a large domain for the impact of emergent properties on clade dynamics; the macroecological literature on the causes and consequence of geographic range size (e.g., Gaston 2003) provides a vehicle for extensive integration of paleontological and neontological work on this problem. But much more work is needed at this basic level.

The second challenge is to go beyond situations where processes at one level overwhelm those at other levels, to quantify the relative contribution of processes at different levels to shaping a given large-scale pattern. Theoretical attempts to address multilevel processes by expanding Price's covariance selection equations (Arnold and Fristrup 1982; Damuth and Heisler 1988; Okasha 2004, 2007; Rice 2004) have yet to find much empirical application, but this is clearly an area needing further development. Further, differential proliferation of a trait or a lineage at a given hierarchical level cannot simply be assumed to be driven by selection at that level. Effects might propagate upward, as when selfish genetic elements decrease the fitness of their organismic hosts, or downward, as when low extinction rates owing to broad geographic ranges result in a net shift of coloration frequencies among bird clades. Such hitchhiking effects are likely to be common (see Jablonski 2000, 2005a, 2007, 2008b; Levinton 2001; Gould 2002), but have rarely been tested paleontologically (see Wagner 1996 and Simpson 2009 for pioneering work). We thus need approaches that can evaluate not only the relative impact of organismic and species-level traits on evolutionary rates and patterns, but the potential for upward and downward causation to drive evolutionary trends and other macroevolutionary patterns. Again, the intersection with neontological work is more extensive then generally appreciated, as much of the comparative biology literature is an attempt to take into account hitchhiking of phenotypic traits along evolutionary lines, or "phylogenetic effects."

NONLINEAR EFFECTS OF EXTINCTION AND ORIGINATION

If nothing else, the fossil record is a rich record of evolutionary failure: perhaps 95% of all species that have ever lived are extinct. The paleontological literature on extinction is enormous, especially for the Big Five mass extinctions, but much remains to be done in documenting extinction intensities and selectivities. Especially deserving more investigation is the potential for nonlinear effects of extinction—that is, that the nature and consequences of extinction are more complex than implied by simple extrapolation from effects seen at low intensities. The same might be said for origination, as also discussed briefly in the following. Evidence is accumulating that the victims of at least some of the Big Five events differ qualitatively as well as quantitatively from those lost during quieter times (for review see Jablonski 2005a), but a more comparative approach is sorely needed, not only among the major extinction events (and among the clades that encounter them), but more importantly among intervals of differing extinction intensities. Two research areas stand out.

First, *analyses of extinction intensity versus selectivity*. The contrast in selectivity between some of the Big Five extinctions and their immediately adjacent time bins, with many factors important during background times losing efficacy during mass extinctions, helps to explain why these rare events can play such a major role in rechannelling evolution while accounting for just a small fraction of the total losses over the past half-billion years (Jablonski 2005a). However, we cannot understand these processes without comparative analyses of selectivity among events of different intensities (see McKinney 1995 for an early attempt at exactly this). These shifts in selectivity could represent gradational transitions or may involve threshold effects at a particular extinction intensity; one way to read the evidence is that more intense extinctions tend to be less selective (Jablonski 2005a). Alternatively, selectivity could depend on the nature of the perturbation itself; each mass extinction does seem to have unique properties, although selectivities converge in many aspects despite different drivers (Jablonski 2005a). Thresholds, if they exist, may be clade-specific, so that volatile groups such as ammonoids can be pushed into a mass extinction regime even as more phlegmatic groups such as bivalves are relatively unfazed. If so, then smaller extinction events, such the Eocene-Oligocene, may prove to be a predictable, quantitative mix of background and mass extinction regimes.

As already noted, conservation biology and paleontology have much to gain from interactions in terms of clarifying general principles and empirical rules related to extinction processes, so long as comparisons are mindful of the very real differences between ancient and modern situations in scale and driving mechanisms. Even now, Big Five mass extinctions dwarf present-day losses, although more extreme paleontological data and theory will become increasingly relevant as the modern biota buckles under human pressures. For example, some of the ideas on selectivity developed from work on ancient extinctions may explain the failure of intrinsic factors to predict extinction risk in the most heavily stressed elements of the modern biota, such as freshwater fishes and Australian marsupials (Duncan and Lockwood 2001; Fisher, Blomberg, and Owens 2003; and see Reynolds, Webb, and Hawkins 2005 on apparent body-size selectivity hitch-hiking on geographic range size in fishes); the loss of selectivity may become more pervasive among living species whenever habitat destruction is the dominant extinction mechanism (Russell et al. 1998). At the same time, the phylogenetic approaches applied to many conservation questions could profitably be applied to paleontological data. This could provide not only more rigorous approaches to hitch-hiking and other phylogenetic effects, but insights into the role of evolutionary tree topology in determining the impact of a given extinction intensity (for example, random extinction can remove entire species-poor subclades while merely thinning the more profuse ones; see Heard and Mooers 2000, 2002; Purvis et al. 2000).

Second, *extinction cascades and supercharged recoveries.* Given that species are embedded in webs of ecological interactions, should we expect that every species loss or, in the aftermath of an extinction, the origin of every species, will be an independent event? Instead, the expectation from a large body of ecological theory and modeling is that secondary extinctions can cascade through food webs (e.g., Koh et al. 2004; Ebenman and Jonsson 2005; Eklöf and Ebenman 2006), and evidence is accumulating for such extinction cascades in some modern communities (review by Ebenman and Jonsson 2005). The question is whether such effects are significant at macroevolutionary scales. It is striking that the Pleistocene shows so little evidence of cascading extinction in the face of kaleidoscopic shifts in community composition, with the possible exception of the cluster of megafaunal extinctions apparently initiated by human hunting at the very end of the last glacial maximum. Does the overall mildness of Pleistocene extinction imply that all of those community changes over twenty glacial-interglacial cycles involved precise relays among functionally equivalent species, as membership in local communities shifted with climatic cycles? This seems highly unlikely and thus raises an intriguing set of questions. As already discussed with respect to the evolutionary role of the biotic environment, evaluation of extinction cascades will require data beyond simple tallies of taxonomic losses. The same broad categories of subsidiary data are relevant here, and may extend to quantifying linkages within ancient food webs, no small task given that many of the participating species are unfossilized and direct evidence for interactions is scarce. Some modeling approaches appear to be promising (e.g., Roopnarine 2006; Roopnarine et al. 2007), although as with so many large-scale phenomena, the ultimate solution will almost certainly require more than simply scaling up the current generation of ecological models. The potential role of extinction cascades in determining both the selectivity and intensity of losses, and more generally, the imperfect correlations between losses of taxonomic and functional diversity, is too important and too poorly understood to be ignored any longer (see Jernvall and Wright 1998 for an underappreciated, paleontologically informed analysis of present-day primates from this perspective).

Just as extinctions might propagate through cascading coextinctions, originations might promote positive feedbacks that spur further diversifications. Organisms modify their environments to such an extent (e.g., Lewontin 1983; Jones et al. 1997; Bruno, Stachowicz, and M. D. Bertness 2003; Odling-Smee,

Laland, and Feldman 2003) that the origin of evolutionary novelties and major clades must affect the dynamics of other lineages at a variety of scales. As in ecology, most paleontological work on clade interactions has focused on negative feedbacks such as competition and predation, so that the potential for positive effects has been seriously neglected. Unfortunately, such positive evolutionary interactions can rarely be confirmed simply by detecting elevated diversification. Even without the envelope of statistical error around diversification rates that might disallow such tests, the many alternative explanations for pulses or phases of exceptional diversification, such as the extinction of competitors or enemies and the capture of key innovations, will be difficult to exclude. Intrinsic diversification rates will be difficult to measure independent of biotic interactions (although the maximal rates seen for many clades in the wake of major extinctions may provide the closest estimate [see Miller and Sepkoski 1988], so that rates boosted by positive interactions should be intermediate between the values for unfettered diversification and those seen when mired in predominantly negative interactions). Such positive evolutionary interactions are at least as difficult to detect in modern assemblages, where confounding variables are just as plentiful and the time dimension must be inferred. For example, Emerson and Kolm (2005) found a positive relation between the numbers of endemic species and widespread species on five of the seven Canary Islands (a more appropriate comparison than, say, numbers of endemic species versus total species richness), but the number of insects and plants on Hawaiian islands were equally or more strongly accounted for by physical factors such as island elevation and isolation (see also Cadena et al. 2005 on further problems when islands are not in endemic/nonendemic equilibrium; and also Kiflawi et al. 2007, Whittaker et al. 2007, Birand and Howard 2008, Gruner et al. 2008, and accompanying replies).

All of these difficulties notwithstanding, diversifications must often have a component of positive evolutionary feedback, particularly after mass extinctions that have removed important components of the biotic environment. Consider the lower taxonomic and morphological diversity of clades inhabiting a temperate soil lacking earthworms, or a tropical marine shelf lacking coral reefs, for example. And, as mentioned before, the waxing and waning of marine taxa that contribute skeletal material to benthic sediments might well have promoted the diversification of taxa taking advantage of those changes in seafloor properties (as suggested for echinoderms during the Ordovician diversification; see Sprinkle and Guensberg 1995; Rozhnov 2001). The question is how pervasive such effects really have been and whether they are separable from simpler models of independent clade dynamics. We simply do not know, because the question has not been tackled systematically. Once again, attempts drawing solely on the dynamics of taxonomic richness among clades are unlikely to be sufficient, and interchange with neontological attempts to come to grips with these processes will be valuable in both directions.

SPACE: THE FINAL FRONTIER

Paleontology has tended to focus on two highly disparate spatial scales: local stratigraphic sections on one hand, and global compendia on the other. However, analyses at intermediate, regional scales have much to offer. Spatially structured dynamics have been recorded for many macroevolutionary events, including the Ordovician radiations, the marine Mesozoic revolution, recoveries from the end-Ordovician, end-Permian, and end-Cretaceous mass extinctions, and the demise of the Pleistocene megafauna (for reviews see Miller 1998; Jablonski 2005a, b, 2007, 2008a; also Zhan and Harper 2006; Krug and Patzkowsky 2007). In each of these instances, the availability of spatially explicit data engenders new hypotheses: that the Ordovician radiations may have been promoted by tectonic activity; that the origin of major marine groups is driven by processes focused in onshore, heterogeneous environments; that the Mesozoic marine revolution may have been mediated by regional changes in climate and nutrient inputs; and that recoveries from mass extinctions involve not only in situ evolution but biotic interchanges, with invasion intensity varying among regions. Even the latitudinal diversity gradient of increasing taxonomic, functional, and morphological variety from poles to tropics, the most pervasive biological pattern in the global biota, appears to be shaped significantly by interregional dispersal rather than by the in situ diversification generally assumed to predominate (see Jablonski 1993; Wiens and Donoghue 2004; Jablonski, Roy, and Valentine 2006; Roy and Goldberg 2007). Spatially explicit paleontological data appears to have great potential to inform, and perhaps fundamentally change, our views on large-scale evolutionary processes.

In addition to tracking the spatial dynamics of clades and adaptations, it will be important to gain a fuller picture of how the spatial structure of the physical and biological world affects large-scale patterns. For example, even though global geography and the number and discreteness of biological provinces changes dramatically over geologic time (e.g., Briggs 1996), we still have much to learn on the role this spatial template has played in setting extinction rates, origination rates, interregional differentiation, and other dynamical aspects of biotas at this scale (and some have even argued that provinciality has a trivial effect on global diversity, see Benton and Emerson 2007). We do not even know for sure whether global diversity is higher today, with a relatively narrow tropics but with latitudinal gradients featuring a series of biogeographic provinces, each containing endemic species, or in the mid-Miocene, when the taxon-rich tropics were at their maximum areal extent for the past 15 million years. The research agenda framed by Valentine (1969, 1973 and subsequent papers) should be addressed with newly available data—for example, more rigorous and plentiful stratigraphic, morphologic, and ecological data, including relative and absolute abundances—and new analytical and modeling methods.

Spatially explicit paleontological data will also be valuable in exploring large-scale ecological questions, and the potential feedbacks between paleontology and the burgeoning field of macroecology are rich and varied. Many of the intrinsic biotic properties discussed before, such as abundance, body size, geographic range size, and life history traits (all key variables to macroecology, according to Gaston and Blackburn 1999, 2000), can themselves vary spatially within and among clades. With the added time dimension of paleontology, the prospects are very bright for of an evolutionary macroecology operating at the interface of paleontology and neontological macroecology (review by Jablonski et al. 2003b; see also Jernvall and Fortelius 2004; Lyons et al. 2004; Smith et al. 2004; Raia et al. 2006; Liow et al. 2008); and as already noted, if geographic range or rarity are species-level traits or can confer emergent fitness on species, then the macroecology literature is rich in potential cases of species selection and thus ripe for collaborative work. As noted earlier, simultaneous treatment of multiple variables will be key to a deeper understanding of controls on spatial and temporal distributions in the fossil record.

As with other research areas discussed previously, paleontological analyses of spatial dynamics should be extended beyond taxonomic patterns to incorporate functional and morphological data. Most of the work done so far in this area is strictly neontological (e.g., Shepherd 1998; Roy et al. 2001, 2004; Neige 2003; McClain 2005; Stevens, Willig, and Strauss 2006), and has detected intriguing discordances among the different aspects of biodiversity that will be important in understanding the origin and maintenance of large-scale biotic patterns (e.g., Valentine et al. 2002; Roy, Jablonski, and Valentine 2004; Stevens, Willig, and Strauss 2006). A few studies have profitably ventured into the fossil record, but this work has barely scratched the surface of what should develop into a major research area (e.g., Dommergues, Laurin, and Meister 2001; Navarro, Neige, and Marchand 2005). In one particularly strong instance of multidisciplinary analysis, Hellberg, Balch, and Roy (2001) found that the morphological divergence of living marine snail populations from fossil Pleistocene ones was (unexpectedly) greater in areas they invaded during post-glacial warming, whereas genetic variation in these newly occupied areas was, as expected, much lower than in continuously occupied warm-water refugia. Even more powerful (although limited to the youngest part of the geologic record) would be analyses incorporating ancient DNA from multiple paleopopulations, but relatively few studies have taken full advantage of the spatial and morphological data of the fossil record (for some early steps, see reviews by Gugerli, Parducci, and Petit 2005; Willerslev and Cooper 2005; also Weinstock et al. 2005). Comparative analyses that draw on the rich Pleistocene phenotypic record of many marine and terrestrial groups, coupled with ancient or modern DNA in a spatially explicit framework, would provide an unparalleled look at the relation between genetic and morphological variety among groups having differing biologies, starting locations, and evolutionary histories.

CONCLUSIONS AND SUMMARY

Although paleontology has its own research agenda, central questions, and specialized methodologies, many of its most productive avenues will remain at its intersection with other subdisciplines within the biological and physical sciences. Some of the research areas likely to yield exciting results in the coming years include:

1. Separating paleobiological signal from noise and artifact. This work is essential for rigorous analysis of paleontological patterns, particularly as questions demand increasingly higher-resolution data, which are inevitably more prone to distortion by sampling and other artifacts. The aim is always to strike a balance between naive optimism and undue pessimism, and we have seen notable successes in this regard.
2. Evolutionary developmental paleobiology. The fossil record is nothing if not rich in morphological data, and these data in their temporal, environmental, and phylogenetic context provide a powerful basis for a partnership with evolutionary developmental biology. Paleontology can do far more than document the timing of, and in many cases the steps leading up to, particular novelties, although of course such information is essential for understanding the evolution of developmental systems. Preserved ontogenies, and detailed phylogenetic estimates around key evolutionary

transitions, permit the testing of more general hypotheses on the nature of the developmental changes that have shaped and constrained the evolutionary trajectories of clades over time. Comparative studies permit statistical approaches to the relation between development and evolution; for example, the macroevolutionary consequences of contrasting developmental architectures (and how evolutionary changes in those architectures—the number and structure of regulatory networks, modules, and alternative pathways—have shaped phenotypic evolution), the environmental factors that promote or damp the origin of evolutionary novelties, and the relation between developmental properties and the dynamics of morphospace occupation.

3. External drivers and intrinsic factors. This is one of the longest-standing areas of paleontological research, encompassing issues that were addressed even before Darwin by Cuvier, Lamarck, Lyell, and others. Nevertheless, the field is poised to make major strides in understanding the links of biotic events to physical environmental drivers, the role of the biotic environment, and the role of intrinsic biotic factors in setting large-scale evolutionary and ecological patterns. Technological advances are enabling unprecedented resolution in absolute dating and inferences about atmospheric and ocean composition. The analysis of spatial structure, biomechanical properties, and more realistic models of interaction dynamics on macroevolutionary scales, will take paleontology beyond weak inferences based solely on matching or reciprocal taxonomic diversity trends in potentially interacting clades. Development of protocols for partitioning evolutionary effects among organismic and species-level traits, and for ranking the effects of multiple factors singly and in combination, will open new avenues of research that will also be important for present-day ecology and conservation biology.

4. Tempo and mode of evolution at the species level and above. All possible combinations of species-level evolutionary change occur in the fossil record (punctuation versus gradualism, cladogenesis versus anagenesis), and so the crucial question is how and why clades differ in tempo and mode. Similarly, most workers now agree that evolution can operate simultaneously at multiple hierarchical levels, and the crucial question is how to partition the relative effects of organismic selection within populations, and the components of emergent fitness, that is, differential speciation and extinction owing to (a) organismic traits and to (b) species-level traits. These processes can reinforce or oppose one another, or can operate on such different features and over such different timescales as to be, in Gould's

term (2002), "orthogonal" to one another. Neither theory nor data have sufficiently explored these alternatives.

5. Nonlinear effects of extinction and origination. The evolutionary effects of extinction cannot simply be understood solely in terms of intensity—that is, the number or proportion of taxonomic losses. Extinction selectivity appears to change as intensity increases, but this shift needs be studied across a wide spectrum of intensities rather than the end-member approach used so far. We know even less about the potential for biotic interactions to magnify the effects of a given extinction intensity, for example by promoting coextinctions as food webs and other biotic relationships break down, or to produce positive feedbacks on diversification by promoting heightened origination as one clade creates opportunities for one or more other clades. Implications for conservation biology and related fields are potentially great, but should not be cast in terms of oversimplified analogies between the present-day, undeniably severe, situation and the Big Five mass extinctions of the geologic record.

6. The spatial fabric of large-scale patterns in the fossil record has been surprisingly neglected after very promising work by Valentine (1969, 1973). On the one hand we need to know more about the geographical and environmental dimension in large-scale evolutionary patterns. On the other, the changing geography of the planet has almost certainly played a major role in origination, extinction, and biotic interchange on a variety of scales, but this role has probably varied among clades and time intervals. Research that adds a time dimension to macroecological variables seems especially promising.

The research areas noted here, which are only a subset of the opportunities on the horizon, underscore the kinds of data and conceptual insights that paleontology can bring to what Van Valen has called the evolutionary half of biology. These insights will come most readily if paleontology continues to expand and strengthen its interactions with the earth and life sciences; the blending of fields across scales is always a great challenge but can yield enormous scientific benefits.

ACKNOWLEDGMENTS

I thank Douglas H. Erwin, Gene Hunt, David Sepkoski, James W. Valentine, and especially Susan M. Kidwell for valuable reviews. Supported by a grant from NASA, and by the John Simon Guggenheim Foundation.

REFERENCES

Aberhan, M., W. Kiessling, and F. T. Fürsich. 2006. Testing the role of biological interactions in the evolution of mid-Mesozoic marine benthic ecosystems. *Paleobiology* 32:259–77.

Allmon, W. D. 2005. The importance of museum collections in paleontology. *Paleobiology* 31:1–5.

Alfaro, M. E., D. I. Bolnick, and P. C. Wainwright. 2005. Evolutionary consequences of many-to-one mapping of jaw morphology to mechanics in labrid fishes. *American Naturalist* 165:E140–E154.

Alroy, J. 2000. Understanding the dynamics of trends within evolving lineages. *Paleobiology* 26:319–29.

Alroy, J., C. R. Marshall, R. K. Bambach, K. Bezusko, M. Foote, F. T. Fürsich, T. A. Hansen, et al. 2001. Effects of sampling standardization on estimates of Phanerozoic marine diversification. *Proceedings of the National Academy of Sciences, USA* 98:6261–66.

Arnold, A. J., and K. Fristrup. 1982. The theory of evolution by natural selection: A hierarchical expansion. *Paleobiology* 8:113–29.

Bambach, R. K. 1993. Seafood through time: Changes in biomass, energetics, and productivity in the marine ecosystem. *Paleobiology* 19:372–97.

———. 1999. Energetics in the global marine fauna: A connection between terrestrial diversification and change in the marine biosphere. *Geobios* 32:131–44.

Bambach, R. K., A. M. Bush, and D. H. Erwin. 2007. Autecology and the filling of ecospace: Key metazoan radiations. *Palaeontology* 50:1–22.

Barbour Wood, S. L., M. Kowalewski, and L. W. Ward. 2004. Quantifying collection biases in bulk, museum and literature-based molluscan sample data. *Geological Society of America Abstracts with Programs* 36 (5): 456.

Barnes, I., P. Matheus, B. Shapiro, D. Jensen, and A. Cooper. 2002. Dynamics of Pleistocene population extinctions in Beringian brown bears. *Science* 295:2267–70.

Barnosky, A. D. 2005. Effects of Quaternary climate change on speciation in mammals. *Journal of Mammalian Evolution* 12:247–64.

Barnosky, A. D., M. A. Carrasco, and E. B. Davis. 2005. The impact of the species-area relationship on estimates of paleodiversity. *PLoS Biology* 3 (8): 1356–61.

Bateman, R. M., P. R. Crane, W. A. DiMichele, P. R. Kenrick, N. P. Rowe, T. Speck, and W. E. Stein. 1998. Early evolution of land plants: Phylogeny, physiology, and ecology of the primary terrestrial radiation. *Annual Review of Ecology and Systematics* 29:263–92.

Beerling, D. J., and R. A. Berner. 2005. Feedbacks and the coevolution of plants and atmospheric CO_2. *Proceedings of the National Academy of Sciences, USA* 102:1302–5.

Behrensmeyer, A. K., F. T. Fürsich, R. A. Gastaldo, S. M. Kidwell, M. A. Kosnik,

M. Kowalewski, R. E. Plotnick, R. R. Rogers, and J. Alroy. 2005. Durability bias in the Phanerozoic marine invertebrate fossil record. *Paleobiology* 31:607–23.

Behrensmeyer, A. K., S. M. Kidwell, and R. Gastaldo. 2000. Taphonomy and paleobiology. *Paleobiology* 26 (Supplement to No. 4):103–47.

Benton, M. J. 1996. On the nonprevalence of competitive replacement in the evolution of tetrapods. In *Evolutionary paleobiology*, ed. D. Jablonski, D. H. Erwin, and J. H. Lipps, 185–210. Chicago: University of Chicago Press.

Benton, M. J., and B. C. Emerson. 2007. How did life become so diverse? The dynamics of diversification according to the fossil record and molecular phylogenetics. *Palaeontology* 50:23–40.

Birand, A., and D. J. Howard. 2008. The relationship between proportion of endemics and species diversity on islands: Expectations from a null model. *Ecography* 31:286–88.

Boyce, C. K. 2005. Patterns of segregation and convergence in the evolution of fern and seed plant leaf morphologies. *Paleobiology* 31:117–40.

Brakefield, P. M. 2006. Evo-devo and constraints on selection. *Trends in Ecology and Evolution* 21:362–68.

Briggs, J. C. 1996. *Global biogeography*. Amsterdam: Elsevier.

Bruno, J. F., J. J. Stachowicz, and M. D. Bertness. 2003. Inclusion of facilitation into ecological theory. *Trends in Ecology and Evolution* 18:119–25.

Bush A. M., and R. K. Bambach. 2004. Did alpha diversity increase through the Phanerozoic? Lifting the veils of taphonomic, latitudinal, and environmental biases. *Journal of Geology* 112:625–42.

Bush, A. M., M. J. Markey, and C. R. Marshall. 2004. Removing bias from diversity curves: The effects of spatially organized biodiversity on sampling-standardization. *Paleobiology* 30: 666–86.

Bush, A. M., M. G. Powell, W. S. Arnold, T. M. Bert, and G. M. Daley, 2002. Time-averaging, evolution, and morphologic variation. *Paleobiology* 28:9–25.

Butler M. A., and A. A. King. 2004. Phylogenetic comparative analysis: A modeling approach for adaptive evolution. *American Naturalist* 164:683–95.

Cadena, C. D., R. E. Ricklefs, I. Jiminez, and E. Bermingham. 2005. Is speciation driven by diversity? *Nature* 438:E1–E2.

Cardillo, M., G. M. Mace, K. E. Jones, J. Bielby, O. R. P. Bininda-Emonds, W. Sechrest, C. D. L. Orme, and A. Purvis. 2005. Multiple causes of high extinction risk in large mammal species *Science* 309:1239–41.

Ciampaglio, C. N. 2004. Measuring changes in articulate brachiopod morphology before and after the Permian mass extinction event: Do developmental constraints limit morphological innovation? *Evolution and Development* 6:260–74.

Ciampaglio, C. N., M. Kemp, and D. W. McShea. 2001. Detecting changes in morphospace occupation patterns in the fossil record: Characterization and analysis of measures of disparity. *Paleobiology* 27:695–715.

Clementz, M. T., A. Goswami, P. D. Gingerich, and P. L. Koch. 2006. Isotopic records from early whales and sea cows: Contrasting patterns of ecological transition. *Journal of Vertebrate Paleontology* 26:355–70.

Coates, M. I., J. E. Jeffery, and M. Ruta. 2002. Fins to limbs: What the fossils say. *Evolution and Development* 4:390–401.

Cooper, R. A., P. A. Maxwell, J. S. Crampton, A. G. Beu, C. M. Jones, and B. A. Marshall. 2006. Completeness of the fossil record: Estimating losses due to small body size. *Geology* 34:241–44.

Coyne, J. A., and H. A. Orr. 2004. *Speciation*. Sunderland, MA: Sinauer.

Crame, J. A., and B. R. Rosen. 2002. Cenozoic palaeogeography and the rise of modern biodiversity patterns. In: *Palaeobiogeography and biodiversity change: The Ordovician and Mesozoic-Cenozoic radiations*, ed. J. A. Crame and A. W. Owen, 153–68. London: Geological Society of London Special Publications 194.

Crampton, J. S., A. G. Beu, R. A. Cooper, C. M. Jones, B. Marshall, and P. A. Maxwell. 2003. Estimating the rock volume bias in paleobiodiversity studies. *Science* 301:358–60.

Damuth, J., and I. L. Heisler. 1988. Alternative formulations of multilevel selection. *Biology and Philosophy* 3:407–30.

Darwin, C. 1859. *On the origin of species*. London: J. Murray.

Davis, M. C., R. D. Dahn, and N. H. Shubin. 2007. An autopodial-like pattern of Hox expression in the fins of a basal actinopterygian fish. *Nature* 447:473–76.

Dawkins, R. 1989. The evolution of evolvability. In *Artificial life*, ed. C. G. Langton, 201–220. Redwood City, CA: Addison-Wesley.

D'Hondt, S. 2005. Consequences of the Cretaceous/Paleogene mass extinction for marine ecosystems. *Annual Review of Ecology, Evolution and Systematics* 36:295–317.

Dodson, P. 1990. Counting dinosaurs: How many kinds were there? *Proceedings of the National Academy of Sciences, USA* 87:7608–12.

Dommergues, J.-L., B. Laurin, and C. Meister. 2001. The recovery and radiation of Early Jurassic ammonoids: Morphologic versus palaeobiogeographical patterns. *Palaeogeography, Palaeoclimatology, Palaeoecology* 165:195–213.

Duncan, J. R., and J. L. Lockwood. 2001. Extinction in a field of bullets: A search for causes in the decline of the world's freshwater fishes. *Biological Conservation* 102:97–105.

Ebenman, B., and T. Jonsson. 2005. Using community viability analysis to identify fragile systems and keystone species. *Trends in Ecology and Evolution* 10:568–75.

Eble, G. J. 2000. Theoretical morphology: The state of the art. *Paleobiology* 26:520–28.

———. 2003. Developmental morphospaces and evolution. In *Evolutionary dynamics*, ed. J. P. Crutchfield and P. Schuster, 33–63. Oxford: Oxford University Press.

Edgar, G. J., and C. R. Sampson. 2004. Catastrophic decline in mollusc diversity in eastern Tasmania and its concurrence with shellfish fisheries. *Conservation Biology* 18:1579–88.

Eklöf, A., and B. Ebenman. 2006. Species loss and secondary extinctions in simple and complex model communities. *Journal of Animal Ecology* 75:239–46.

Eldredge, N. 1985. *Unfinished synthesis*. New York: Oxford University Press.

Eldredge, N., J. N. Thompson, P. M. Brakefield, S. Gavrilets, D. Jablonski, R. E. Lenski, B. S. Lieberman, M. A. McPeek, and W. Miller III. 2005. The dynamics of evolutionary stasis. *Paleobiology* 31:133–45.

Emerson, B. C., and N. Kolm. 2005. Species diversity can drive speciation. *Nature* 434:1015–17.

Erwin, D. H. 1996. The geological history of diversity. In *Biodiversity in managed landscapes*, ed. R. C. Szaro and D. W. Johnson, 3–16. New York: Oxford University Press.

Erwin, D. H. 2001. Lessons from the past: Biotic recoveries from mass extinctions. *Proceedings of the National Academy of Sciences, USA* 98:5399–5403.

———. 2005. The origin of animal body plans. In *Form and function: Essays in honor of Adolf Seilacher*, ed. D. E. G. Briggs, 67–80. New Haven, CT: Peabody Museum, Yale University.

———. 2006a. Dates and rates: Temporal resolution in the deep time stratigraphic record. *Annual Review of Earth and Planetary Sciences* 34:569–90.

———. 2006b. *Extinction: How life on Earth nearly ended 250 million years ago*. Princeton, NJ: Princeton University Press.

———. 2007. Disparity: Morphologic pattern and developmental context. *Palaeontology* 50:57–74.

Erwin, D. H., and R. L. Anstey. 1995. Speciation in the fossil record. In *New approaches to speciation in the fossil record*, ed. D. H. Erwin and R. L. Anstey, 11–38. New York: Columbia University Press.

Erwin, D. H., and E. H. Davidson. 2002. The last common bilaterian ancestor. *Development* 129:3021–32

Falkowski, P. G., M. E. Katz, A. J. Milligan, K. Fennel, B. S. Cramer, M.-P. Aubry, R. A. Berner, M. J. Novacek, and W. M. Zapol. 2005. The rise of oxygen over the past 205 million years and the evolution of large placental mammals. *Science* 309:2202–4.

Fastovsky, D. E., Y. Huang, J. Hsu, J. Martin-McNaughton, P. M. Sheehan, and D. B. Weishampel. 2004. Shape of Mesozoic dinosaur richness. *Geology* 32:877–80.

Ferguson, C. A., and A. I. Miller. 2007. A sea change in Smuggler's Cove? Detection of decadal scale Compositional transitions in the subfossil record. *Palaeogeography, Palaeoclimatology, Palaeoecology* 254:418–29.

Fisher, D. O., S. P. Blomberg, and I. P. F. Owens. 2003. Extrinsic versus intrinsic factors in the decline and extinction of Australian marsupials. *Proceedings of the Royal Society of London* B270:1801–8.

Foote, M. 1996a. Models of morphological diversification. In *Evolutionary paleobiology*, ed. D. Jablonski, D. H. Erwin, and J. H. Lipps, 62–86. Chicago: University of Chicago Press.

———. 1996b. Perspective: Evolutionary patterns in the fossil record. *Evolution* 50:1–11.

———. 1997. The evolution of morphological diversity. *Annual Review of Ecology and Systematics* 28:129–52.

———. 2003. Origination and extinction through the Phanerozoic: A new approach. *Journal of Geology* 111:125–48.

———. 2005. Pulsed origination and extinction in the marine realm. *Paleobiology* 31:6–20.

Forey, P. L., R. A. Fortey, P. Kenrick, and A. B. Smith. 2004. Taxonomy and fossils: A critical appraisal. *Philosophical Transactions of the Royal Society of London* B359:639–53.

Freckleton, R., P. H. Harvey, and M. Pagel. 2002. Phylogenetic analysis and ecological data: a review of the evidence. *American Naturalist* 160:712–26.

Gaston, K. J. 2003. *The structure and dynamics of geographic ranges.* Oxford: Oxford University Press.

Gaston, K. J., and T. M. Blackburn. 1999. A critique for macroecology. *Oikos* 84:353–68.

———. 2000. *Pattern and process in macroecology.* Oxford: Blackwell Science.

Gaston, K. J., and R. A. Fuller. 2007. Biodiversity and extinction: Losing the common and the widespread. *Progress in Physical Geography* 31:213–25.

Gerber, S., P. Neige, and G. J. Eble. 2007. Combining ontogenetic and evolutionary scales of morphological disparity: A study of early Jurassic ammonites. *Development and Evolution* 9:472–85.

Gilbert, S. F. 2001. Ecological developmental biology: Developmental biology meets the real world. *Developmental Biology* 233:1–12.

Gingerich, P. D. 2001. Rates of evolution on the time scale of the evolutionary process. *Genetica* 112–13:127–44.

Goswami, A. 2006. Cranial modularity shifts during mammalian evolution. *American Naturalist* 168:270–80.

———. 2007. Cranial modularity and sequence heterochrony in mammals. *Evolution and Development* 9:290–98.

Gould, S. J. 1977a. Eternal metaphors of paleontology. In *Patterns of evolution,* ed. A. Hallam, 1–26. Amsterdam: Elsevier.

———. 1977b. *Ontogeny and phylogeny.* Cambridge, MA: Harvard University Press.

———. 1982. Darwinism and the expansion of evolutionary theory. *Science* 216:380–87.

———. 1989. *Wonderful life.* New York: Norton.

———. 2002. *The structure of evolutionary theory.* Cambridge, MA: Harvard University Press.

Graham, C. H., S. Ferrier, F. Huettman, C. Moritz, and A. T. Peterson. 2004. New developments in museum-based informatics and applications in biodiversity analysis. *Trends in Ecology and Evolution* 19:497–503.

Grantham, T. A. 1995. Hierarchical approaches to macroevolution: Recent work on species selection and the "effect hypothesis." *Annual Review of Ecology and Systematics* 26:301–22.

———. 2001. Hierarchies in evolution. In *Palaeobiology II,* ed. D. E. G. Briggs and P. R. Crowther, 188–92. Oxford: Blackwell Science.

———. 2002. Species selection. In *Encyclopedia of evolution,* ed. M. Pagel, 1086–87. Oxford: Oxford University Press.

———. 2007. Do emergent properties block attempts to reduce macroevolution? *Palaeontology* 50:75–86.

Gruner, D. S., N. J. Gotelli, J. P. Price, and R. H. Cowie. 2008. Does species richness drive speciation? A reassessment with the Hawaiian biota. *Ecography* 31:279–85.

Gugerli, F., L. Parducci, and R. J. Petit. 2005. Ancient plant DNA: Review and prospects. *New Phytologist* 166:409–18.

Gutiérrez, J. L., C. G. Jones, D. L. Strayer, and O. O. Iribarne. 2003. Mollusks as ecosystem engineers: The role of shell production in aquatic habitats. *Oikos* 101:79–90.

Guralnick, R., and J. Van Cleve. 2005. Strengths and weaknesses of museum and national survey data sets for predicting regional species richness: Comparative and combined approaches. *Diversity and Distributions* 11:349–59.

Hannisdal, B. 2006. Phenotypic evolution in the fossil record: Numerical experiments. *Journal of Geology* 114:133–53.

Harmon, L. J., J. A. Schulte II, A. Larson, and J. B. Losos. 2003. Tempo and mode of evolutionary radiation in iguanian lizards. *Science* 301:961–64.

Harnik, P. G. 2009. Unveiling rare diversity by integrating museum, literature, and field data. *Paleobiology* 35, in press.

Harper, E. M. 1998. The fossil record of bivalve molluscs In *The adequacy of the fossil record,* ed. C. R. C. Paul and S. K. Donovan, 243–67. Chichester: Wiley.

Harper, E. M., T. J. Palmer, and J. R. Alphey. 1997. Evolutionary response of bivalves to changing Phanerozoic sea-water chemistry. *Geological Magazine* 134:403–7.

Heard, S. B., and A. Ø. Mooers. 2000. Phylogenetically patterned speciation rates and extinction risks change the loss of evolutionary history during extinctions. *Proceedings of the Royal Society of London* B267:613–20.

———. 2002. Signatures of random and selective extinctions in phylogenetic tree balance. *Systematic Biology* 51:889–98.

Hellberg, M. E., D. P. Balch, and K. Roy. 2001. Climate-driven range expansion and morphological evolution in a marine gastropod. *Science* 292:1707–10.

Hendry, A. P., and M. T. Kinnison. 1999. The pace of modern life: Measuring rates of contemporary microevolution. *Evolution* 53:1637–53.

Hewitt, G. M., and R. A. Nichols. 2005. Genetic and evolutionary impacts of climate change. In *Climate change and biodiversity,* ed. T. E. Lovejoy and L. Hannah, 176–92. New Haven, CT: Yale University Press.

Holland, S. M. 2003. Confidence limits on fossil ranges that account for facies changes. *Paleobiology* 29:468–79.

Hunt, G. 2004a. Phenotypic variation in fossil samples: Modeling the consequences of time-averaging. *Paleobiology* 30:426–43.

———. 2004b. Phenotypic variance inflation in fossil samples: An empirical assessment. *Paleobiology* 30:487–506.

———. 2006. Fitting and comparing models of phyletic evolution: Random walks and beyond. *Paleobiology* 32: 578–601.

———. 2007. The relative importance of directional change, random walks, and stasis in the evolution of fossil lineages. *Proceedings of the National Academy of Sciences, USA* 104:18404–408.

Huntley, B. 2005. North temperate responses. In *Climate change and biodiversity*, ed. T. E. Lovejoy and L. Hannah, 109–24. New Haven, CT: Yale University Press.

Jablonski, D. 1993. The tropics as a source of evolutionary novelty: The post-Palaeozoic fossil record of marine invertebrates. *Nature* 364:142–44.

———. 1995. Extinction in the fossil record. In *Extinction rates,* ed. R. M. May and J. H. Lawton, 25–44. Oxford: Oxford University Press.

———. 1996. Body size and macroevolution. In *Evolutionary paleobiology,* ed. D. Jablonski, D. H. Erwin, and J. H. Lipps, 256–89. Chicago: University of Chicago Press.

———. 1997. Body-size evolution in Cretaceous molluscs and the status of Cope's rule. *Nature* 385:250–52.

———. 2000. Micro- and macroevolution: scale and hierarchy in evolutionary biology and paleobiology. *Paleobiology* 26 (Suppl. to No. 4): 15—52.

———. 2001. Lessons from the past: Evolutionary impacts of mass extinctions. *Proceedings of the National Academy of Sciences, USA* 98:5393–98.

———. 2005a. Evolutionary innovations in the fossil record: The intersection of ecology, development and macroevolution. *Journal of Experimental Zoology* 304B:504–19.

———. 2005b. Mass extinctions and macroevolution. *Paleobiology* 31 (Suppl. to No. 2): 192–210.

———. 2007. Scale and hierarchy in macroevolution. *Palaeontology* 50:87–110.

———. 2008a. Biotic interactions and macroevolution: Extensions and mismatches across scales and levels. *Evolution* 62:715–39.

———. 2008b. Species selection: Theory and data. *Annu. Rev. Ecol. Evol.. Syst.* 39:20–42.

Jablonski, D., and D.J. Bottjer. 1990. The origin and diversification of major groups: Environmental patterns and macroevolutionary lags. In *Major evolutionary radiations,* ed. P. D. Taylor and G. P. Larwood, 17-57. Oxford: Clarendon Press.

Jablonski, D., J. A. Finarelli, and K. Roy. 2006. What, if anything, is a genus? Testing the analytical units of paleobiology against molecular data. *Geological Society of America Abstracts with Programs* 38 (7): 169.

Jablonski, D., and G. Hunt. 2006. Larval ecology, geographic range, and species survivorship in Cretaceous mollusks: Organismic vs. species-level explanations. *American Naturalist* 168: 556–64.

Jablonski, D., K. Roy, J. W. Valentine, R. M. Price, and P. S. Anderson. 2003. The impact of the Pull of the Recent on the history of bivalve diversity. *Science* 300:1133–35.

Jablonski, D., K. Roy, and J. W. Valentine. 2003. Evolutionary macroecology and the fossil record. In *Macroecology: Concepts and consequences*, ed. T. M. Blackburn and K. J. Gaston, 368–90. Oxford: Blackwell Science.

Jablonski, D., K. Roy, and J. W. Valentine. 2006. Out of the tropics: Evolutionary dynamics of the latitudinal diversity gradient. *Science* 314:102–6.

Jackson, J. B. C., and A. H. Cheetham. 1999. Tempo and mode of speciation in the sea. *Trends in Ecology and Evolution* 14:72–77.

Jackson, J. B. C., and Erwin, D. H. 2006. What can we learn about ecology and evolution from the fossil record? *Trends in Ecology and Evolution* 21:322–28.

Jackson, S. T. 2004. Quaternary biogeography: Linking biotic responses to environmental variability across timescales. In *Frontiers of biogeography*, ed. M. V. Lomolino and L. R. Heaney, 47–65. Sunderland, MA: Sinauer.

Jackson, S. T., and J. T. Overpeck. 2000. Responses of plant populations and communities to environmental changes of the Late Quaternary. *Paleobiology* 26 (Suppl. to No. 4):194–220.

Jackson, S. T., and J. W. Williams. 2004. Modern analogs in Quaternary paleocology: Here today, gone yesterday, gone tomorrow? *Annual Review of Earth and Planetary Sciences* 32:495–537.

Janis, C. M. 2003. Tectonics, climate change, and the evolution of mammalian ecosystems. In *Evolution on planet Earth: The impact of the physical environment*, ed. A. Lister and L. Rothschild, 319–38. New York: Academic Press.

Jansson, R., and M. Dynesius. 2002. The fate of clades in a world of recurrent climatic change: Milankovitch oscillations and evolution. *Annual Review of Ecology and Systematics* 33:741–77.

Janzen, D. H. 1980. When is it coevolution? *Evolution* 34:611–12.

Jernvall, J. 2000. Linking development with generation of novelty in mammalian teeth. *Proceedings of the National Academy of Sciences, USA* 97:2541–2645.

Jernvall, J., and M. Fortelius. 2004. Maintenance of trophic structure in fossil mammal communities: Site occupancy and taxon resilience. *American Naturalist* 164:614–24.

Jernvall, J., S. V. E. Keränen, and I. Thesleff. 2000. Evolutionary modification of development in mammalian teeth: Quantifying gene expression patterns and topography. *Proceedings of the National Academy of Sciences, USA* 97:14444–48.

Jernvall, J. and P. Wright. 1998. Diversity components of impending primate extinctions. *Proceedings of the National Academy of Sciences, USA* 95:11279–83.

Johnson, A. L. A. 1993. Punctuated equilibria vs. phyletic gradualism in European Jurassic *Gryphaea* evolution. *Proceedings of the Geologists' Association* 104:209–22.

Jones, C. G., J. H. Lawton, and M Shachak. 1997. Positive and negative effects of organisms as physical ecosystem engineers. *Ecology* 78:1946–57.

Jones, D. S., and S. J. Gould. 1999. Direct measurement of age in fossil *Gryphaea*: The solution to a classic problem in heterochrony. *Paleobiology* 25:158–87.

Katz, M. E., Z. V. Finkel, D. Grzebyk, A. H. Knoll, and P. G. Falkowski. 2004. Evolutionary trajectories and biogeochemical impacts of marine eukaryotic phytoplankton. *Annual Review of Ecology, Evolution, and Systematics* 35:523–56.

Kavanagh, K. D., A. R. Evans, and J. Jernvall. 2007. Predicting evolutionary patterns of mammalian teeth from development. *Nature* 449:427–32.

Kidwell, S. M. 1986. Taphonomic feedback in Miocene assemblages: Testing the role of dead hardparts in benthic communities. *Palaios* 1:239–55.

———. 2001. Preservation of species abundance in marine death assemblages. *Science* 294:1091–94.

———. 2002. Time-averaged molluscan death assemblages: Palimpsests of richness, snapshots of abundance. *Geology* 30:803–6.

———. 2005. Shell composition has no net impact on large-scale evolutionary patterns in mollusks. *Science* 207:914–17.

———. 2007. Discordance between living and death assemblages as evidence for anthropogenic ecological change. *Proceedings of the National Academy of Sciences, USA*, 104: 17701–706.

———. 2008. Ecological fidelity of open-marine molluscan death assemblages: Effects of post-mortem transportation, shelf health, and taphonomic inertia. *Lethaia* 41:199–217.

Kidwell, S. M., and K. W. Flessa. 1995. The quality of the fossil record: Populations, species, and communities. *Annual Review of Ecology and Systematics* 26:269–99.

Kidwell, S. M., and S. M. Holland. 2002. The quality of the fossil record: Implications for evolutionary analysis. *Annual Review of Ecology and Systematics* 33:561–88.

Kidwell, S. M., and D. Jablonski. 1983. Taphonomic feedback: Ecological consequences of shell accumulation. In *Biotic interactions in recent and fossil benthic communities*, ed. M. J. S. Tevesz and P. L. McCall, 195–248. New York: Plenum.

Kiessling, W. 2005. Habitat effects and sampling bias on Phanerozoic reef distribution. *Facies* 51:27–35.

Kiflawi, M., J. Belmaker, E. Brokovich, S. Embinder, and R. Holzman. 2007. Species diversity can drive speciation: Comment. *Ecology* 88:2132–35.

Kingsolver, J. G., and D. W. Pfennig. 2004. Individual-level selection as a cause of Cope's rule of phyletic size increase. *Evolution* 58:1608–12.

Knappertsbusch, M. W. 2000. Morphologic evolution of the coccolithophorid *C. leptoporus* from the Early Miocene to Recent. *Journal of Paleontology* 74:712–30.

——. 2001. A method of illustrating the morphological evolution of coccoliths with 3D animations applied to *Calcidiscus leptoporus*. *Palaeontologia Electronica* 4 (1) : 12.

Knoll, A. H. 2003. The geological consequences of evolution. *Geobiology* 1:3–14.

Knoll, A. H., R. K. Bambach, L. P. C. Jonathan, S. Pruss, and W. W. Fischer. 2007. Paleophysiology and end-Permian mass extinction. *Earth and Planetary Science Letters* 256:295–313.

Koh, L. P., R. R. Dunn, N. S. Sodhi, R. K. Colwell, H. C. Proctor, and V. S. Smith. 2004. Species co-extinctions and the biodiversity crisis. *Science* 305:1632–34.

Kowalewski, M., G. E. Avila Serrano, K. W. Flessa, and G. A. Goodfriend. 2000. Dead delta's former productivity: Two trillion shells at the mouth of the Colorado River. *Geology* 28:1059–62.

Kowalewski, M., and R. K. Bambach. 2003. The limits of paleontological resolution. In *High resolution approaches in stratigraphic paleontology*, ed. P. J. Harries, 1–48. New York: Plenum/Kluwer.

Kowalewski, M., and J. D. Rimstidt. 2003. Lifetime and age spectra of detrital grains: Toward a unifying theory of sedimentary particles. *Journal of Geology* 111:427–40.

Krug, A. Z., and M. E. Patzkowsky. 2007. Geographic variation in turnover and recovery form the late Ordovician mass extinction. *Paleobiology* 33:435–54.

Labandeira, C. 2002. The history of associations between plants and animals. In *Plant-animal interactions: An evolutionary approach,* ed. C. M. Herrera and O. Pellmyr, 26–74. Oxford: Blackwell Science.

Levinton, J. S. 2001. *Genetics, paleontology, and macroevolution,* 2nd ed. New York: Cambridge University Press.

Lewontin, R. C. 1983. Gene, organism, and environment. In *Evolution from molecules to men,* ed. D. S. Bendall, 273–85. Cambridge: Cambridge University Press.

Lieberman, B. S., and E. S. Vrba. 2005. Stephen Jay Gould on species selection: 30 years of insight. *Paleobiology* 31 (Suppl. to No. 2): 113–21.

Liow, L. H., M. Fortelius, E. Bingham, K. Lintulaakso, H. Mannila, L. Flynn, and N. C. Stenseth. 2008. Higher origination and extinction rates in larger mammals. *Proceedings of the National Academy of Sciences, USA* 195:6097–6102.

Lister, A. M., A. V. Sher, H. van Essen, and G. Wei. 2005. The pattern and process of mammoth evolution in Eurasia. *Quaternary International* 126–28: 49–64.

Lopez-Vaamonde, C., N. Wikström, C. Labandeira, H. C. J. Godfray, S. J. Goodman, and J. M. Cook. 2006. Fossil-calibrated molecular phylogenies reveal that leaf-mining moths radiated millions of years after their host plants. *Journal of Evolutionary Biology* 19:1314–26.

Lutolf, M., F. Kienast, and A. Guisan. 2006. The ghost of past species occurrence: Improving species distribution models for presence-only data. *Journal of Applied Ecology* 43:802–15. Lyons, K. L. 2003. A quantitative assessment of the range shifts of Pleistocene mammals. *Journal of Mammalogy* 84:385–402.

Lyons, S. K., F. A. Smith, and J. H. Brown. 2004. Of mice, mastodons and men: Human-mediated extinction on four continents. *Evolutionary Ecology Research* 6:339–58.

MacFadden, B. J. 2006. Extinct mammalian biodiversity of the ancient New World tropics. *Trends in Ecology and Evolution* 21:157–65.

Marshall, C. R. 2001. Confidence limits in stratigraphy. In *Palaeobiology II,* ed. D. E. G. Briggs and J. R. Crowther, 542–45. Oxford: Blackwell Science.

———. 2006. Explaining the Cambrian "explosion" of animals. *Annual Review of Earth and Planetary Sciences* 34:355–84.

Martin, R. E. 1999. *Taphonomy: A process approach.* New York: Cambridge University Press.

Maurer, B. A. 1999. *Untangling ecological complexity.* Chicago: University of Chicago Press.

Maynard Smith, J. 1989. The causes of extinction. *Philosophical Transactions of the Royal Society of London* B325:241–52.

———. 1998. The units of selection. In *The limits of reductionism in biology,* ed. G. R. Bock and J. A. Goode, 203–17. Chichester: Wiley.

McClain, C. R. 2005. Bathymetric patterns of morphological disparity in deep-sea gastropods from the western North Atlantic basin. *Evolution* 59:1492–99.

McGhee, G. R., Jr. 1998. *Theoretical morphology.* New York: Columbia University Press.

———. 2007. *The geometry of evolution.* New York: Cambridge University Press.

McGill, B., B. J. Enquist, M. Westoby, and E. Weiher. 2006. Rebuilding community ecology from functional traits. *Trends in Ecology and Evolution* 21:178–84.

McGowan, A. J. 2004a. Ammonoid taxonomic and morphologic recovery patterns after the Permian-Triassic. *Geology* 32:665–68.

———. 2004b. The effect of the Permo-Triassic bottleneck on ammonoid morphological evolution. *Paleobiology* 30:369–95.

———. 2005. Ammonoid recovery from the Late Permian mass extinction event. *Comptes Rendus Palevol* 4:449–62.

McKinney, M. L. 1995. Extinction selectivity among lower taxa: Gradational patterns and rarefaction error in extinction estimates. *Paleobiology* 21:300–313.

———. 1997. Extinction vulnerability and selectivity: Combining ecological and paleontological views. *Annual Review of Ecology and Systematics* 28:495–516.

Miller, A. I. 1998. Biotic transitions in global marine diversity. *Science* 281:1157–60.

Miller, A. I., and J. J. Sepkoski, Jr. 1988. Modeling bivalve diversification: The effect of interaction on a macroevolutionary system. *Paleobiology* 14:364–69.

Moen, D. S. 2006. Cope's rule in cryptodiran turtles: Do the body sizes of extant species reflect a trend in phyletic size increase? *Journal of Evolutionary Biology* 19:1210–21.

National Research Council (NRC). 2002. *Geoscience data and collections: National resources in peril.* Washington, DC: National Academies Press.

————. 2005. *The geological record of ecological dynamics: Understanding the biotic effects of future environmental change.* Washington, DC: National Academies Press.

Navarro, N., P. Neige, and D. Marchand. 2005. Faunal invasions as a source of morphological constraints and innovations? The diversification of the early Cardioceratidae (Ammonoidea; Middle Jurassic). *Paleobiology* 31:98–116.

Niege, P. 2003. Spatial patterns of disparity and diversity of the Recent cuttlefishes (Cephalopoda) across the Old World. *Journal of Biogeography* 30:1125–37.

Nosil, P., and B. J. Crespi. 2006. Experimental evidence that predation promotes divergence in adaptive radiation. *Proceedings of the National Academy of Sciences, USA* 103:9090–95.

Odling-Smee, F. J., K. N. Laland, and M. W. Feldman. 2003. *Niche construction.* Princeton, NJ: Princeton University Press.

Okasha, S. 2004. Multilevel selection and the partitioning of covariance: A comparison of three approaches. *Evolution* 58:486–94.

————. 2007. *Evolution and the levels of selection.* New York: Oxford University Press.

Olszewski, T. D., and S. M. Kidwell. 2007. The preservational fidelity of evenness in molluscan death assemblages. *Paleobiology* 33:1–23.

Pagel, M. 2002. Modelling the evolution of continuously varying characters on phylogenetic trees: the case of hominid cranial capacity. In *Morphology, shape and phylogenetics,* ed. N. MacLeod and P. Forey, 269–86. London: Taylor and Francis.

Payne, J. L., and S. Finnegan. 2006. Controls on marine animal biomass through geological time. *Geobiology* 4:1–10.

Payne, J. L., D. J. Lehrmann, J. Wei, M. J. Orchard, D. P. Schrag, and A. H. Knoll. 2004. Large perturbations of the carbon cycle during recovery from the end-Permian extinction. *Science* 305:506–9.

Petchey, O. L., and K. J. Gaston, 2006. Functional diversity: Back to basics and looking forward. *Ecology Letters* 9:741–56.

Peters, S. E. 2005. Geologic constraints on the macroevolutionary history of marine animals. *Proceedings of the National Academy of Sciences, USA* 102: 12326–331.

Peterson, K. J., J. A. Cotton, J. G. Gehling, and D. Pisani. 2008.The Ediacaran emergence of bilaterians: congruence between the genetic and the geological fossil records. *Philosophical Transactions of the Royal Society of London* B363: 1435–43.

Purvis, A., K. E. Jones, and G. M. Mace. 2000. Extinction. *BioEssays* 22:1123–33.

Purvis, A., P.-M. Agapow, J. L. Gittleman, and G. M. Mace. 2000. Nonrandom extinction and the loss of evolutionary history. *Science* 288:328–30.

Purvis, A., M. Cardillo, R. Grenyer, and B. Collen. 2005. Correlates of extinction risk: Phylogeny, biology, threat and scale. In *Phylogeny and conservation,* ed. A. Purvis, J. L. Gittleman, and T. Brooks, 295–316. Cambridge: Cambridge University Press.

Raff, R. A. 2007. Written in stone: fossils, genes and evo-devo. *Nature Reviews Genetics* 8:911–20.

Raia, P., C. Meloro, A. Loy, and C. Barbera. 2006. Species occupancy and its course in the past: Macroecological patterns in extinct communities. *Evolutionary Ecology Research* 8:181–94.

Reynolds, J. D., T. J. Webb, and L. A. Hawkins. 2005. Life history and ecological correlates of extinction risk in European freshwater fishes. *Canadian Journal of Fisheries and Aquatic Sciences* 62:854–62.

Rice, S. H. 2004. *Evolutionary theory: Mathematical and conceptual foundations.* Sunderland, MA: Sinauer.

Richardson, M. K., J. E. Jeffery, and C. J. Tabin. 2004. Proximodistal patterning of the limb: Insights from evolutionary morphology. *Evolution and Development* 6:1–5.

Ricklefs, R. E. 2005. Small clades at the periphery of passerine morphological space. *American Naturalist* 165:651–59.

Roopnarine, P. D. 2003. Analysis of rates of morphologic evolution. *Annual Review of Ecology, Evolution and Systematics* 34:605–32.

———. 2005. The likelihood of stratophenetic-based hypotheses of genealogical succession. *Special Papers in Palaeontology* 73:143–57.

———. 2006 Extinction cascades and catastrophe in ancient food webs. *Paleobiology* 32:1–19.

Roopnarine, P. D., K. D. Angielczyk, S. C. Wang, and R. Hertog. 2007. Trophic network models explain instability of Early Triassic terrestrial communities. *Proceedings of the Royal Society of London* B274:2077–86.

Ros, S., and M. De Renzi. 2005. Preservation biases, rates of evolution and coherence of databases: Bivalvia as a case study. *Ameghiniana* 42: 549–58.

Rowe, K. C., E. J. Heske, and K. N. Paige. 2006. Comparative phylogeography of chipmunks and white-footed mice in relation to the individualistic nature of species. *Molecular Ecology* 15:4003–20.

Roy, K. 1994. Effects of the Mesozoic Marine Revolution on the taxonomic, morphologic and biogeographic evolution of a group: Aporrhaid gastropods during the Mesozoic. *Paleobiology* 20:274–96.

Roy, K., D. P. Balch, and M. E. Hellberg. 2001. Spatial patterns of morphological diversity across the Indo-Pacific: Analyses using strombid gastropods. *Proceedings of the Royal Society of London* B268:2503–8.

Roy, K., and M. Foote. 1997. Morphological approaches to measuring biodiversity. *Trends in Ecology and Evolution* 12:277–81.

Roy, K., and E. E. Goldberg. 2007. Origination, extinction, and dispersal: Integrative models for understanding present-day diversity gradients. *American Naturalist* 170 (Supplement to No. 2):S71–S85.

Roy, K., D. Jablonski, and J. W. Valentine. 2004. Beyond species richness: Biogeographic patterns and biodiversity dynamics using other metrics of diversity. In

Frontiers of biogeography, ed. M. V. Lomolino and L. R. Heaney 151–70. Sunderland, MA: Sinauer.

Roy, K., and J. M. Pandolfi. 2005. Responses of marine species and ecosystems to past climate changes. In *Climate change and biodiversity*, ed. T. E. Lovejoy and L. Hannah, 160–75. New Haven, CT: Yale University Press.

Roy, K., J. W. Valentine, D. Jablonski, and S. M. Kidwell. 1996. Scales of climatic variability and time averaging in Pleistocene biotas: Implications for ecology and evolution. *Trends in Ecology and Evolution* 11:458–63.

Rozhnov, S. V. 2001. Evolution of the hardground community. In *Ecology of the Cambrian radiation*, ed. A. Yu. Zhuravlev and R. Riding, 238–53. New York: Columbia University Press.

Russell, G. J., T. M. Brooks, M. L. McKinney, and C. G. Anderson. 1998. Present and future taxonomic selectivity in bird and mammal extinctions. *Conservation Biology* 12:1365–76.

Ruta, M., P. J. Wagner, and M. I. Coates. 2006. Evolutionary patterns in early tetrapods. I. Rapid initial diversification followed by decrease in rates of character change. *Proceedings of the Royal Society of London* B273:2107–11.

Sadler, P. M. 2004. Quantitative biostratigraphy: Achieving finer resolution in global correlation. *Annual Review of Earth and Planetary Sciences* 32: 187–213.

Salazar-Ciudad, I., and J. Jernvall. 2002. A gene network model accounting for development and evolution of mammalian teeth. *Proceedings of the National Academy of Sciences, USA* 99:8116–20.

———. 2004. How different types of pattern formation mechanisms affect the evolution of form and development. *Evolution and Development* 6:6–16.

Sepkoski, J. J., Jr. 1996. Competition in macroevolution: The double wedge revisited. In *Evolutionary paleobiology*, ed. D. Jablonski, D. H. Erwin, and J. H. Lipps, 211–55. Chicago: University of Chicago Press.

———. 1998. Rates of speciation in the fossil record. *Philosophical Transactions of the Royal Society of London* B353:315–26.

Shepherd, U. L. 1998. A comparison of species diversity and morphological diversity across the North American latitudinal gradient. *Journal of Biogeography* 25:19–29.

Shubin, N. H. 2002. Origin of evolutionary novelty: Examples from limbs. *Journal of Morphology* 252:15–28.

Shubin, N. H., E. B. Daeschler, and F. A. Jenkins, Jr. 2006. The pectoral fin of *Tiktaalik roseae* and the origin of the tetrapod limb. *Nature* 440:764–71.

Sidlauskas, B. 2007. Testing for unequal rates of morphological diversification in the absence of a detailed phylogeny: A case study from characiform fishes. *Evolution* 61:299–316..

Simpson, C. 2008. Species selection and driven mechanisms jointly generate a large-scale morphological trend in monobathrid crinoids. *Paleobiology* 34, in press.

Simpson, C., and P. G. Harnik, 2009. Assessing the role of abundance in marine bivalve extinctions over the post-Paleozoic. *Paleobiology* 35, in press.

Simpson, G. G. 1960. The history of life. In *Evolution after Darwin. Volume 1. The evolution of life,* ed. S. Tax, 117–80. Chicago: University of Chicago Press.

Smith, A. B. 1994. *Systematics and the fossil record.* Oxford: Blackwell.

———. 2007. Marine diversity through the Phanerozoic: Problems and prospects. *Journal of the Geological Society, London* 164:731–45.

Smith, A. B., and A. J. McGowan. 2007. The shape of the Phanerozoic marine palaeodiversity curve: How much can be predicted from the sedimentary rock record of western Europe? *Palaeontology* 50:765–74.

Smith, A. B., and B. Patterson. 1988. The influence of taxonomic method on the perception of patterns of evolution. *Evolutionary Biology* 23:127–216.

Smith, F. A., J. H. Brown, J. P. Haskell, S. K. Lyons, J. Alroy, E. L. Charnov, T. Dayan, et al. 2004. Similarity of mammalian body size across the taxonomic hierarchy and across space and time. *American Naturalist* 163:672–91.

Solow, A. R., and D. L. Roberts. 2006. Museum collections, species distributions, and rarefaction. *Diversity and Distributions* 12:423–24.

Sprinkle, J., and T. E. Guensburg. 1995. Origin of echinoderms in the Paleozoic Evolutionary Fauna: The role of substrates. *Palaios* 10:437–53.

Stanley, S. M. 1979. *Macroevolution.* San Francisco: W. H. Freeman.

Stayton, C. T. 2006. Testing hypotheses of convergence with multivariate data: Morphological and functional convergence among herbivorous lizards. *Evolution* 60:824–41.

Stevens, R. D., M. R. Willig, and R. E. Strauss. 2006. Latitudinal gradients in the phenetic diversity of New World bat communities. *Oikos* 112:41–50.

Stockmeyer Lofgren, A., Plotnick, R. E., and Wagner, P. J. 2003. Morphological diversity of Carboniferous arthropods and insights on disparity patterns through the Phanerozoic. *Paleobiology* 29:349–68.

Stone, J. R. 2003. Mapping cladograms into morphospaces. *Acta Zoologica* 84:63–68.

Strauss, S. Y., H. Sahli, and J. K. Conner. 2005. Toward a more trait-centered approach to diffuse (co)evolution. *New Phytologist* 165:81–90.

Stuart, B. L., K. A. Dugan, M. W. Allard, and M. L. Kearney. 2006. Extraction of nuclear DNA from bone of skeletonized and fluid-preserved museum specimens. *Systematics and Biodiversity* 4:133–36.

Suarez, A. V., and N. D. Tsutsui. 2004. The value of museum collections for research and society. *BioScience* 54:66–74.

Sultan, S. E. 2003. Commentary: The promise of ecological developmental biology. *Journal of Experimental Zoology (Mol. Dev. Evol.)* 296B:1–7.

Teichert, C., W. C. Sweet, and A. J. Boucot. 1987. The unpublished fossil record. *Senckenbergiana Lethaea* 68:1–19.

Thewissen, J. G., Cohn, M. J., Stevens, L. S., Bajpai, S., Heyning, J., Horton, W. E. Jr. 2006. Developmental basis for hind-limb loss in dolphins and origin of the cetacean bodyplan. *Proceedings of the National Academy of Sciences USA* 103:8414–18.

Thewissen J. G., and E. M. Williams, 2002. The early radiations of Cetacea (Mammalia): Evolutionary pattern and developmental correlations. *Annual Review of Ecology and Systematics* 33:73–90.

Thompson, J. N. 2005. *The geographic mosaic of coevolution.* Chicago: University of Chicago Press.

Todd, J. A., J. B. C. Jackson, K. G. Johnson, H. M. Fortunato, A. Heitz, M. Alvarez, and P. Jung. 2002. The ecology of extinction: Molluscan feeding and faunal turnover in the Caribbean Neogene. *Proceedings of the Royal Society of London* B269:571–77.

Valentine, J. W. 1969. Taxonomic and ecological structure of the shelf benthos during Phanerozoic time. *Palaeontology* 12:684–709.

———. 1973. *Evolutionary paleoecology of the marine biosphere.* Englewood Cliffs, NJ: Prentice-Hall.

———. 2000. Two genomic paths to complexity in metazoan evolution. *Paleobiology* 26:513–19.

———. 2004. *On the origin of phyla.* Chicago: University of Chicago Press.

———. 2006. Ancestors and urbilateria. *Evolution and Development* 8:391–93.

Valentine, J. W., D. Jablonski, S. M. Kidwell, and K. Roy. 2006. Assessing the fidelity of the fossil record by using marine bivalves. *Proceedings of the National Academy of Sciences, USA* 103: 6599–6604.

Valentine, J. W., K. Roy, and D. Jablonski. 2002. Carnivore/noncarnivore ratios in northeastern Pacific marine gastropods. *Marine Ecology Progressive Series* 228:153–63.

Vamosi, S. M. 2005. On the role of enemies in divergence and diversification of prey: A review and synthesis. *Canadian Journal of Zoology* 83:894–910.

Van Valen, L. 1973. Festschrift. *Science* 180:488.

Vellend, M., and M. A. Geber. 2005. Connections between species diversity and genetic diversity. *Ecology Letters* 8:767–81.

Vermeij, G. J. 1987. *Escalation in evolution.* Princeton, NJ: Princeton University Press.

Vermeij, G. J. 2006. Historical contingency and the purported uniqueness of evolutionary innovations. *Proceedings of the National Academy of Sciences, USA* 103:1804–9.

Villier, L., and G. Eble. 2004. Assessing the robustness of disparity estimates: The impact of morphometric scheme, temporal scale, and taxonomic level in spatangoid echinoids. *Paleobiology* 30:652–65.

Wagner, P. J. 1995. Diversity patterns among early Paleozoic gastropods: Contrasting taxonomic and phylogenetic descriptions. *Paleobiology* 21:410–39.

———. 1996. Contrasting the underlying patterns of active trends in morphologic evolution. *Evolution* 50:990–1007.

———. 1997. Patterns of morphological diversification among the Rostroconchia. *Paleobiology* 23:115–50.

———. 2000. Exhaustion of morphologic character states among fossil taxa. *Evolution* 54:365–86.

Wagner, P. J., M. Ruta, and M. I. Coates, 2006. Evolutionary patterns in early tetrapods. II. Differing constraints on available character space among clades. *Proceedings of the Royal Society of London* B273:2113–18.

Wainwright, P. C. 2007. Functional versus morphological diversity in macroevolution. *Annual Review of Ecology Evolution and Systematics* 38:381–401.

Wainwright, P. C., M. E. Alfaro, D. I. Bolnick, and C. D. Hulsey. 2005. Many-to-one mapping of form to function: A general principle in organismal design? *Integrative and Comparative Biology* 45:256–62.

Wainwright, P. C. and S. M. Reilly, eds. 2004. *Ecological morphology*. Chicago: University of Chicago Press.

Walker, S. E., and C. E. Brett. 2002. Post-Paleozoic patterns in marine predaton: Was there a Mesozoic and Cenozoic marine predatory revolution? In *The fossil record of predation,* ed. M. Kowalewski and P. H. Kelley, 119–93, Paleontological Society Papers 8. Washington, DC: The Paleontological Society.

Wang, S. C., and P. J. Everson. 2007. Confidence intervals for pulsed mass extinction events. *Paleobiology* 33:324–36.

Wang, S. C., and C. R. Marshall. 2004. Improved confidence intervals for estimating the position of a mass extinction boundary. *Paleobiology* 30:5–18.

Webster, A. J., and A. Purvis. 2002. Ancestral character states and evolutionary rates of continuous characters. In *Morphology, shape and phylogeny*, ed. N. MacLeod and P. L. Forey, 247–68. London: Taylor & Francis.

Weinstock, J., E. Willerslev, A. Sher, W. Tong, S. Y. W. Ho, D. Rubenstein, J. Storer, et al. 2005. Evolution, systematics, and phylogeography of Pleistocene horses in the New World: A molecular perspective. *PLoS Biology* 3 (8): e241.

Wesley-Hunt, G. D. 2005. The morphological diversification of carnivores in North America. *Paleobiology* 31:35–55.

West, J. B., G. J. Bowen, T. E. Cerling, and J. R. Ehleringer. 2006. Stable isotopes as one of nature's ecological recorders. *Trends in Ecology and Evolution* 21:408–14.

Whibley, A. C., N. B. Langlade, C. Andalo, A. I. Hanna, A. Bangham, C. Thébaud, and E. Coen. 2006. Evolutionary paths underlying flower color variation in *Antirrhinum. Science* 313:963–66.

Williams, G. C. 1992. *Natural selection.* New York: Oxford University Press.

Whittaker, R. J., R. J. Ladle, M. B. Araujo, J. M. Fernandez-Palacios, J. D. Delgado, and J. R. Arevalo. 2007. The island immaturity-speciation pulse model of island evolution: An alternative to the "diversity begets diversity" model. *Ecography* 30:321–27.

Wiens, J. J., and M. J. Donoghue. 2004. Historical biogeography, ecology and species richness. *Trends in Ecology and Evolution* 19:639–44.

Wiens, J. J., C. Kuczynski, W. E. Duellman, and T. W. Reeder. 2007. Loss and re-evolution of complex life cycles in marsupial frogs: Can ancestral trait reconstruction mislead? *Evolution* 61:1886–99.

Willerslev, E., and A. Cooper. 2005. Ancient DNA. *Proceedings of the Royal Society of London* B272:3–16.

Willis, K. J., K. D. Bennett, and D. Walker, eds. 2004. The evolutionary legacy of the Ice Ages. *Philosophical Transactions of the Royal Society of London* B359:157–303.

Zahn, R., and D. A. T. Harper. 2006. Biotic diachroneity during the Ordovician Radiation: Evidence from South China. *Lethaia* 39:211–26.

Zuschin, M., and P. G. Oliver. 2005. Diversity patterns of bivalves in a coral dominated shallow-water bay in the northern Red Sea—high species richness on a broad scale. *Marine Biology Research* 1:396–410.

CHAPTER TWENTY-SIX

Punctuations and Paradigms: Has Paleobiology Been through a Paradigm Shift?

Michael Ruse

In their notorious paper introducing their theory of punctuated equilibrium, Niles Eldredge and Stephen Jay Gould wrote: "Science progresses more by the introduction of new world-views or 'pictures' than by the steady accumulation of information" (Eldredge and Gould 1972, 86). They qualified this with a footnote saying that they were not about to get into the "tedious debate" about whether they were talking about a "paradigm" in the sense of Thomas Kuhn (in his *Structure of Scientific Revolutions*). Those of us who knew Steve Gould will be fully aware that the only reason they were not about to get into the debate was because the issue in his mind was already settled. They were proposing a paradigm switch! And even if they hadn't been, the fact is that others were also thinking in the same terms. As the essays in this volume attest, some still do think in these terms, and agree that something of that kind did occur. There was a paradigm change or scientific revolution.

As it happens, philosophers and historians of science have shown that it is not always very helpful to try to force things into Kuhn's precise claims in his justly famous book (Ruse 1999). For a start, back then he was proposing an idealistic philosophy—the real world changes during revolutions—that few, including Kuhn himself in his more reflective moments, would want to accept. (And to be fair to Eldredge and Gould, that they never wanted to accept. They state explicitly that they are thinking of different ways of looking at the same facts.) Nevertheless, both because these are the terms in which the questions and conclusions have been framed, and because there is much value in using Kuhn as a guide if not a straightjacket, this is the question I want to ask now.

In some sense, are we looking at a paradigm change? In some sense, are we looking at a scientific revolution?

The preliminary question obviously is: What are we looking at?! I take it that few, probably not even Gould himself, would want to say that we are just looking at the theory of punctuated equilibrium. We are looking at something broader. We are looking at the creation in the past fifty years of the science, or subscience, of paleobiology. Just so the dice don't get loaded the wrong way at the beginning, let us use the description used by the journal of that name, on its Web site.

> *Paleobiology*, founded to provide a forum for the greater integration of paleontology and biology, began publication in 1975. *Paleobiology* publishes original articles that emphasize biological or paleobiological processes and patterns including: speciation, extinction, development of individuals and colonies, natural selection, evolution, and patterns of variation, abundance and distribution of organisms in space and time. Papers concerning recent organisms and systems are also included if they aid in understanding the fossil record and the history of life.

Integration of paleontology and biology. You can see already that it is not too useful to get caught up in a scholastic debate about paradigm changes, because you would probably never get beyond the question of whether, given the definition of paleobiology, we are looking at a paradigm switch or at the creation of a paradigm from or over or against a preparadigmatic science. My inclination is to think more in terms of the latter, although I suspect that both Charles Darwin (1859) and George Gaylord Simpson (1953), authors of major works in paleobiology that appeared more than fifty years ago, would disagree! So let us just say that the past fifty years have seen the creation of paleobiology as a functioning discipline, more than just a few works whistling in the wind— a functioning discipline with its own journal and practitioners and students and grants and everything else that we associate with a paradigm in the social sense: "an entire constellation of beliefs, values and techniques, and so on, shared by the members of a given community" (Kuhn 1962).

And let us also say that it has been very, very exciting creating this new discipline, and that no one associated with it regrets the time spent in the creation and building. That is precisely why we editors wanted to publish this book. There has been a terrifically worthwhile episode in the history of science and it is worth recording while most of the key figures are still living. What is it that has made this all so exciting and worthwhile? It was not just simply bringing two areas—paleontology and biology—together. It was making a subject with hypotheses and models and explanations—above all, making a subject that was not merely descriptive but law-bound or law-using, what several have called "nomothetic" (which is just a fancy way of talking about laws, since the Greek for "law" is "nomos"). You are not just describing one damn fossil after another, but trying to link them by showing how, when they were living animals or plants, they obeyed laws of nature like (perhaps identical to) the very laws of nature that govern animals and plants that are living today (Gould 1980b). Note that making something nomothetic does not simply mean finding laws and binding them together in models or theories. It also means applying them. In a historical area such as we are considering here, just as in astronomy, one might also be dealing with specific events—the Big Bang or the extinction of the dinosaurs—that were one-off phenomena. The point is that one is going to do more than just describe them. One is going to try to explain them, to understand them, and that in science means appealing to laws (Ruse 1973).

I am not even going to attempt to give a full catalogue of what this has all meant in the case of paleobiology—a kind of honor role of the best achievements—nor (and this would be more interesting) am I going to offer an exhaustive catalog of the kinds of achievements that have been scored. But a partial list and catalog is certainly going to include the following.

Theories from outside biology. Clearly the coming of plate tectonics to geology (something that surely is in major respects a paradigm switch) has made a big difference to paleobiology, as people have traced the movements of animals and plants around the globe, on and off continents (Ruse 1981). Another instance at a more-directed level are the Alvarez findings of the comet or meteor that hit the earth 65 million years ago, and how this has led to speculations about the demise of the dinosaurs (Alvarez et al. 1980).

Theories from inside organismic biology. One immediately thinks here of Eldredge and Gould's (1972) widely acknowledged appropriation of Mayr's (1954) ideas about the role of peripheral isolation in speciation. Similar examples abound, including the MacArthur-Wilson (1967) theory of island biogeography and of how this was used by Jack Sepkoski (1976, 1978, 1979, 1984) to model through time rather than space and to get his pictures of biodiversity through the ages.

Theories from inside molecular biology. The molecular clock is surely a big candidate here, and of how it has led to massive rethinking about the dates of crucial events (see the piece by Francisco Ayala in this volume). (It is worth noting how this has been a two-way process. Sometimes the paleontological side of things had to give way, as happened with the dating of the human-ape split. Sometimes, however, the paleontology has stood firm against the molecular approach, as happened when the molecules seemed to make the origins of the vertebrates, way, way before the Cambrian. Look at Michael Benton's comments in this volume).

Reexamination of old fossil material. The work done on the Burgess Shale qualifies here, as whole new worlds of soft-bodied animals are revealed and understood (Gould 1989; Conway Morris 1998).

Discovery of new fossil material. The human discoveries are significant here, as also are the pre-Cambrian discoveries (Johanson and Edey 1981; Knoll 2003). Note the point that was earlier made about one-off events. The revealing of the history of life up to the Cambrian has been anything but a matter of digging up fossils and labeling them. One has to make sense of them and to offer hypotheses and theories to explain why the organisms appeared when they did and why.

New methodologies. The coming of cladism has clearly been significant here. Even more significant, and really part of the same movement, has been the arrival of computers and the ways that they transform matters. Someone like Raup or Sepkoski could crunch up huge amounts of data and find patterns. Doing this would have been difficult or impossible before the computers arrived. The same is also true of the pictures or diagrams of simulated life histories and questions about randomness.

New theories or hypotheses. Obviously, punctuated equilibrium is the prime candidate here. To this you might want to add species selection and hierarchical thinking, although I am sure that a careful historical survey would show that this idea has come up before (Gould 2002). I am also sure that every paleobiologist has his or her own pet theory or hypothesis. Fill in the blanks yourselves, because, as I said, I am not pretending to be fully comprehensive. I am just trying to illustrate the sorts of things that have been entailed by or driven the coming of paleobiology.

I don't think that anything I have just listed suggests that there was something particularly mysterious or odd about the development of paleobiology. I am not sure that its coming was inevitable, but I am not sure what it would mean to say that its coming was inevitable. If computers had not developed as they did, then I doubt we would have had Jack Sepkoski's graphs of organic diversity. If Tom Schopf had not been born, then perhaps the development of paleobiology as a self-consciously separate field might not have occurred or perhaps not quite in the way that it did. If Steve Gould had never existed, I doubt that punctuated equilibrium would have been the phenomenon that it was. (That sounds a bit mean to Niles Eldredge and I don't mean to say that he would not have been an important scientist. But his original 1971 paper on punctuation events was nothing like the audacious, jointly written paper of 1972.)[1] I do want to emphasize that because I don't think that there was anything mysterious or odd about the arrival of paleobiology, this does not mean that I am now taking back anything I said about the excitement of the new field and the creative work in making it what it is. It does seem to me that there is a really good story waiting here to be told by a historian.

One thing that is implicit in my discussion is that, whatever the nature of the revolution that created paleobiology, it was not a revolution like that which we associate with Darwin or with plate tectonics. (Now don't immediately assume that I am saying it was not as important.) In the case of Darwin's theory of evolution through natural selection and of continental drift because of plate tectonics, you have a central unifying hypothesis that then goes on to explain in many different fields (Ruse 1979, 1981). Paleobiology, as I have been describing it, is not of this kind at all. Even if punctuated equilibrium was the greatest thing since Copernicus, it would only be part of what has been happening. As I see the paleobiology revolution, it has been a bit of this and a bit of that and a bit of the other. A bit of new theory, a bit of borrowing from here and another bit of borrowing from there, a bit of methodology, and more than a bit of new techniques. These have coalesced to make the new subject.

I think one can see the differences by the matter of disagreements. Of course, Darwinians can differ over whether natural selection applies in various cases or as much as others think. But to be a Darwinian you have got to be a selectionist. The same is true of plate tectonics. Start denying Gondwanaland and you are out of the paradigm. In the case of paleobiology, you can differ like mad and still be a paleobiologist. Look at this volume and the disagreements over punctuated equilibrium. Some clearly think that it was a major breakthrough. Others as clearly think it was an abomination unto the Lord. Supposedly it leads to no new discoveries and perverts the presentation of what we already know. There are also clearly differences over randomness and hierarchical theory and species selection and a lot more. Some think the history of life is sigmoidal. Others think it is exponential. Both sides are paleobiologists.

So, I do think it important to recognize exactly what has happened, and that is a major reason why I don't want to get too much hung up on paradigm talk. I am more than happy—as I have said already—to think of paradigms in the sociological sense of community building and the like. I think Kuhn is really insightful here. But conceptually I really don't think that paradigm talk is very helpful, although I myself am very happy to say that something revolutionary happened in the past fifty years. If you want to say that paleontology or part of paleontology or out of paleontology underwent or came a revolution, that seems to me to be both true and insightful.

Am I not being a bit dismissive here? I am, as I fully and happily admit, a hard-line Darwinian. As Richard Dawkins (1986) declares of himself, I stand somewhere to the right of Archdeacon Paley on the matter of adaptation (Ruse 2003). For me, it is natural selection all of the way down or up. Does this not mean I am rather belittling punctuated equilibrium for a start? I am refusing even to consider its anti- or beyond-Darwin paradigmatic status? At best, I am allowing that it is only a description of the phenomena—stasis and then instant change—rather than a full-blown theory? In reply, I would say that that is probably true, although I note (as I did just before) that there are established and respected paleobiologists who agree with me. But if you want to say that it has the potential (and, if you accept it, has the reality) to be a full-blown paradigm, I can live with that.[2] Kuhn noted explicitly that not all paradigms are monsters, like those of Copernicus and Darwin. Some paradigms apply only to small areas of experience. That would be my point about punk eek. Even if you accept it, there is more to paleobiology than just what the Eldredge-Gould paradigm talks about.

But, and I start to draw to an end of what I want to say, am I not missing the elephant in the room? What about Darwinism? Is this not the dominant paradigm in biology and (in the opinion of some) the paradigm in paleontology? At least, thanks to G. G. Simpson, was it not the dominant paradigm in paleontology around the time that paleobiology began to get off the ground? In which case, should the discussion not be phrased in terms of whether paleobiology still accepts the Darwinian paradigm or has rejected it? That way we really can tell whether we have had a Kuhnian-type revolution.

There are various kinds of possible answers you could have to this slew of questions. One would be that Simpson brought paleontology only so far into the Darwinian paradigm and that what paleobiologists have been doing for the past five decades is solidifying this movement. Note, however, that even if one granted this at the conceptual level, at the social level this seems not to be true at all. At least, as Todd Grantham (this volume) points out—and in fact, his findings are just what I found in a separate analysis some years ago (Ruse 1999)—the average neontologist simply could not care less about the fossils and the history of life. They just do not cross their intellectual horizon. So, Darwinian or not, paleobiology is not on the way to being fully integrated with the rest of Darwinian biological studies.

I suspect, incidentally, that there is a message here. A lot of paleobiological angst—and perhaps some actual science—has been devoted to getting attention and respect from the other biologists. The whole metaphor of being at the high table suggests this, and Gould particularly seemed often motivated by this drive. My suggestion is that you should forget it. It won't happen. You should just get on with your own work, which is worthwhile, and forget the others. I speak with some experience as a historian and philosopher of science—a trade that scientists think good only for the old and clapped out. As in: "Poor old Jones. He lost his grant and he is going through the philoso-pause." I have long since realized that scientists have their problems and I have my problems and that is all that matters. I certainly do not suffer from biology envy, as I am sure this paper is showing.

In any case, it is simply not true that paleobiology today is part and par-cel of the Darwinian paradigm. Some parts surely are. Remember that the journal statement of mission made explicit reference to natural selection. The work that has been done on dinosaur adaptation comes under the Darwinian heading. If you are spending time trying to work out if the fins on the back of the stegosaurus are for fighting or sexual display or for cooling the blood, then you are as Darwinian as someone looking at color morphs in butterflies. Other parts strike me as not particularly Darwinian and not particularly non-Darwinian. Using the molecular clock to work out dates, for example. The clock relies on the neutral theory of evolution, which is a theory about the changes in molecules at a level below that which selection can affect. No Dar-winian is bothered by this any more than any Darwinian is bothered by the fact that Newton's laws mean that planets go in ellipses rather than squares.

Some parts of paleobiology seem to me to be a bit ambiguous. Sepkoski's sigmoidal patterns have always struck me as a bit this way. There is no doubt that you can bring in some Darwinian explanations about how organisms enter new ecological niches and explode in number for a while and then cool off. But the overall pictures seem to me to owe more to Herbert Spencer's (1862) theory of dynamic equilibrium than to Darwin. (I should say that Jack would probably have disagreed on this one. A happy tale is that he got my book, *Mystery of Mysteries*, which has a chapter on him, shortly before his death. He was seen going around the corridors shouting with joy: "Ruse got me wrong! Ruse got me wrong!" My response would be that I may have got him wrong, but I got him interestingly wrong.)

Obviously some parts of paleobiology are pretty anti-Darwinian. Whatever he may have said otherwise, there were certainly versions of Gould's punctu-ated equilibrium picture that were not sympathetic to the theory of the *Ori-gin of Species*. I suspect that the same is also true of all of those computer-generated clade diagrams that supposedly showed that random factors were just as good at producing real-life patterns as were anything else. Probably some parts are not very friendly to Darwinism but don't want to make a big thing of it. Perhaps something like species selection is a case in point. It does not really strike me as very Darwinian—or if it is, can be reduced to a lower-level true Darwinian selection. On the other hand, it does not really strike me as something that goes out of its way to deny Darwinism, although perhaps it can if it is thought to override lower levels. I find it interesting that some paleobiologists find it worthwhile to call it "species sorting," as if to emphasize that they are not talking about a Darwinian process, but at the same time not being anti-Darwinian.

I suspect that by the very nature of the beast there is always going to be a certain tension between paleobiology and Darwinism. On the one hand, the paleobiologist does not see selection in action and never will. The paleobiologist is also talking about many other things that are not Darwinian, like plate tectonics and meteors from the sky and so on. There is also the huge time factor that seems to throw up phenomena that are not things that selection explains readily. Add to that the fact that, by the logical nature of evolutionary thinking, at one level the paleobiologist must take for granted what the neontologist tells them—not uncritically, of course, as the matter of the origin of vertebrates shows. On the other hand, what the paleobiologist has to say often (usually, always?) is not something of great concern to the neontologist.

This is not a matter of being better or worse. It is a matter of logic and metaphysics. No matter what the Marxists say (and I am not getting into whether Gould was a Marxist, although he certainly was sympathetic to some of the philosophical ideas of Engels), almost always you explain from the bottom up—small really is beautiful. I am not now endorsing or rejecting the emergence issue, whether there are large-scale (space or time) things that cannot be explained by the small things. I am saying that, by and large, if one level is explaining at another level, it is from small to large and not the other way around. This is true of physics and it is true of paleobiology. It is not like picking your nose in public. It is not something to be ashamed of. It is a fact of the way that scientific explanation works.

So let me conclude. Do we have a new paradigm in paleobiology? In some respects, yes. You certainly have a new group. In some respects, no. There is no new overarching theory, even if there are some new theories. But generally we are talking about a whole field and not just one area of conceptual breakthrough. And in some respects, maybe. There was some Darwinism before. There is some Darwinism after. Some are more Darwinian than others.

Things overall are probably not going to change dramatically, although of course they could. But ultimately, asking about paradigms is probably to ask the wrong questions. There has been a major advance in science, something rightly called revolutionary. That is reason enough to celebrate and to have put together this book.

NOTES

1. In fact, Eldredge has acknowledged as much himself: "Published in 1971 with a turgid title in the main journal devoted to evolutionary biology, my results sank pretty much without a trace. Repackaged a year later, with additions, in a jointly authored paper with a former fellow student with a knack for catchy phrases, the theory of 'punctuated equilibria' was born. That colleague, of course, was Stephen Jay Gould—who would never tolerate the subversion of paleontology to the interests of any other field. At last, a good choice!" Niles Eldredge, "Confessions of a Darwinist," *Virginia Quarterly Review*, Spring 2006, 34.

2. Actually, in an earlier paper looking at the paradigm claims of punctuated equilibrium, I did note that we seem to have various forms of punctuated equilibrium and that some—especially the earlier versions—strike me as being formulated explicitly in terms of already-existing theory. Here, I do not want to go back on that conclusion, but merely to say that if it is important to you to claim paradigmatic status for punctuated equilibrium, I am really not going to feel terribly uncomfortable. Part of this indifference stems from the fact that however well taken your claim may be, it is referring only to part of the paleobiological field.

REFERENCES

Alvarez, L. W., W. Alvarez, F. Asaro, and H. V. Michel. 1980. Extraterrestrial cause for the Cretaceous-Tertiary extinction. *Science* 208:1095–1108.

Conway Morris, S. 1998. *The crucible of creation: The Burgess Shale and the rise of animals.* Oxford: Oxford University Press.

Darwin, C. 1859. *On the origin of species.* London: John Murray.

Dawkins, R. 1986. *The Blind Watchmaker.* New York: Norton.

Eldredge, N. 1971. The allopatric model and phylogeny in paleozoic invertebrates. *Evolution* 25:156–67.

Eldredge, N., and S. J. Gould. 1972. Punctuated equilibria: An alternative to phyletic gradualism. In *Models in paleobiology,* ed. T. J. M. Schopf, 82–115. San Francisco: Freeman, Cooper.

Gould, S. J. 1980a. Is a new and general theory of evolution emerging? *Paleobiology* 6:119–30.

————. 1980b. The promise of paleobiology as a nomothetic, evolutionary discipline. *Paleobiology* 6:96–118.

————. 1989. *Wonderful life: The Burgess Shale and the nature of history.* New York: Norton.

————. 2002. *The structure of evolutionary theory.* Cambridge, MA: Harvard University Press.

Johanson, D., and M. Edey. 1981. *Lucy: The beginnings of humankind.* New York: Simon and Schuster.

Knoll, A. 2003. *Life on a young planet: The first three billion years of evolution on Earth.* Princeton, NJ: Princeton University Press.

Kuhn, T. 1962. *The structure of scientific revolutions.* Chicago: University of Chicago Press.

MacArthur, R. H., and E. O. Wilson. 1967. *The theory of island biogeography.* Princeton, NJ: Princeton University Press.

Mayr, E. 1954. Change of genetic environment and evolution. In *Evolution as a Process,* ed. J. Huxley, 157–80. London: Allen and Unwin.

Ruse, M. 1973. *The philosophy of biology.* London: Hutchinson.

————. 1979. *The Darwinian revolution: Science red in tooth and claw.* Chicago: University of Chicago Press.

————. 1981. What kind of revolution occurred in geology? *Philosophy of Science Association 1978* 2:240–73.

————. Is the theory of punctuated equilibria a new paradigm? *Journal of Social and Biological Structures* 12:195–212.

————. 1999. *Mystery of mysteries: Is evolution a social construction?* Cambridge, MA: Harvard University Press.

————. 2003. *Darwin and design: Science, philosophy, religion.* Cambridge, MA: Harvard University Press.

Sepkoski, J. J., Jr. 1976. Species diversity in the Phanerozoic: Species-area effects. *Paleobiology* 2 (4): 298–303.

————. 1978. A kinetic model of Phanerozoic taxonomic diversity. I. Analysis of marine orders. *Paleobiology* 4:223–51.

————. 1979. A kinetic model of Phanerozoic taxonomic diversity. II. Early Paleozoic families and multiple equilibria. *Paleobiology* 5:222–52.

————. 1984. A kinetic model of Phanerozoic taxonomic diversity. III. Post-Paleozoic families and mass extinctions. *Paleobiology* 10:246–67.

Simpson, G. G. 1953. *The major features of evolution.* New York: Columbia University Press.

Spencer, H. 1862. *First principles.* London: Williams and Norgate.

CONTRIBUTORS

RICHARD J. ALDRIDGE
Department of Geology
University of Leicester

Richard J. Aldridge is the F. W. Bennett Professor of Geology at the University of Leicester. His research interests include Paleozoic micropaleontology, especially conodont paleobiology and evolution; exceptionally preserved fossils, especially the late Ordovician Soom Shale of South Africa and the Early Cambrian Chengjiang biota of China; and evolutionary patterns in the Paleozoic phytoplankton.

FRANCISCO J. AYALA
Department of Ecology and Evolutionary Biology
University of California, Irvine

Francisco J. Ayala is University Professor and the Donald Bren Professor of Biological Sciences at the University of California, Irvine. He has published over 900 articles and is author or editor of 31 books. He is a member of the U.S. National Academy of Sciences and the American Philosophical Society, and received the 2001 U.S. National Medal of Science. *The New York Times* named Ayala the "Renaissance Man of Evolutionary Biology."

RICHARD BAMBACH
Department of Paleobiology
National Museum of Natural History
Smithsonian Institution

Richard Bambach is a research associate at the Smithsonian Institution's National Museum of Natural History, and professor emeritus of paleontology at the Virginia Polytechnic Institute and State University. His research focuses on paleoecology, especially paleocommunity ecology, change in ecosystem structure over time, and patterns of change in marine diversity over time. He has been awarded the Raymond C. Moore Medal, Society for Sedimentary Geology, and the Paleontological Society Medal.

MICHAEL BENTON
Department of Earth Sciences
University of Bristol

Michael Benton is professor of vertebrate paleontology at the University of Bristol. His research interests include the diversification of life, the quality of the fossil record, shapes of phylogenies, age-clade congruence, mass extinctions, Triassic ecosystem evolution, basal diapsid phylogeny, and basal archosaurs and the origin of the dinosaurs. He has been awarded the T. Neville George Medal of the Geological Society of Glasgow and the Lyell Medal of the Geological Society of London.

ARTHUR BOUCOT
Department of Zoology
Oregon State University

Arthur Boucot is professor emeritus of zoology and geology at Oregon State University. His research interests include taxonomy and morphology of Silurian-Devonian brachiopods and gastropods; community paleoecology; global historical biogeography; several aspects of the fossil record, such as rates of behavioral evolution, coevolution, community evolution, with evolutionary rate importance; Phanerozoic paleogeography, and climatic belts.

DEREK BRIGGS
Department of Geology and Geophysics
Yale University

Derek Briggs is the Frederick William Beinecke Professor of Geology and Geophysics at Yale University and director of the Yale Peabody Museum of Natural History. His research interests include the taphonomy and evolutionary significance of exceptionally preserved fossils; decay and mineralization, molecular preservation, the Cambrian radiation. He is a fellow of the Royal Society and has been awarded the Lyell Medal of the Geological Society of London and the Boyle Medal of the Royal Dublin Society.

JOE CAIN
Department of Science and Technology Studies
University College London

Joe Cain is senior lecturer in history and philosophy of biology at University College London. His current research interests include the history of evolutionary studies, history of American science, and history of natural history.

DAVID E. FASTOVSKY
Geosciences Department
University of Rhode Island

David E. Fastovsky is professor of geosciences at the University of Rhode Island. His research focuses on the evolution of Mesozoic terrestrial paleoenvironments, particularly those that contain dinosaurs and other terrestrial vertebrates. He is author (with D. B. Weishampel) of *Evolution and Extinction of the Dinosaurs.*

RICHARD FORTEY
Department of Palaeontology
The Natural History Museum

Richard Fortey has received numerous honors and awards throughout his career for his work in paleontology. He is a fellow of the Royal Society and has been awarded the Lyell medal of the Geological Society of London, the Frink medal of the Zoological Society of London, and the Royal Society's Michael Faraday Prize for Science Communication, among others. He is the author of several award-winning books, including *Life: An Unauthorized Biography* and *Trilobite! Eyewitness to Evolution.*

REBECCA Z. GERMAN
Department of Physical Medicine and Rehabilitation
Johns Hopkins University

Rebecca Z. German asks and occasionally answers questions about the interface between form and function. She has degrees in mathematics, geology, and biology, and is currently professor at the Johns Hopkins University School of Medicine. She has also been a Senior Fulbright Scholar at the University of Western Australia.

TODD GRANTHAM
Department of Philosophy
College of Charleston

Todd Grantham is associate professor of philosophy at the College of Charleston. His research is on philosophical issues within and about evolutionary biology, including reductionism and the unity of science, explanatory pluralism, and evolutionary epistemology. He has published widely on topics such as the units of selection controversy, the "species problem," phylogeny reconstructionism, and evolutionary psychology.

ANTHONY HALLAM
Earth Sciences ·
School of Geography, Earth and Environmental Sciences
University of Birmingham

Anthony Hallam is professor emeritus of geology at the University of Birmingham. His research interests include early Mesozoic paleoenvironments, the evolution and speciation of bivalves, and marine mass extinctions. He has also written widely on the history of geological ideas, including *Great Geological Controversies*. He has received a number of awards, including the Lyell Medal of the Geological Society of London, the Von Buch Medal of the German Geological Society, and the Lapworth Medal of the Palaeontological Association.

JOHN R. HORNER
Museum of the Rockies
Montana State University

John R. Horner is the Ameya Preseve Curator of Paleontology at the Museum of the Rockies and Regents Professor at Montana State University. He is one of the world's leading experts on dinosaur paleobiology, and has published eight books and numerous articles on the subject. He has been the recipient of a MacArthur Fellowship and was the technical advisor to Steven Spielberg for the movie *Jurassic Park* and its sequel, *The Lost World*.

JOHN HUSS
Department of Philosophy
The University of Akron

John Huss is assistant professor of philosophy at The University of Akron and a research associate of the Center for Genetic Research Ethics and Law at Case Western Reserve University. An alumnus of the University of Chicago's Committee on the Conceptual Foundations of Science and Department of Geophysical Sciences, he studied with paleobiological revolutionaries Sepkoski and Raup.

DAVID JABLONSKI
Department of Geophysical Sciences
University of Chicago

David Jablonski is the William R. Kenan, Jr., Professor in the Department of the Geophysical Sciences and the Committee on Evolutionary Biology at the University of Chicago. His research emphasizes the combining of data from living and fossil organisms to study the origins and the fates of lineages and adaptations, including the interplay or origination, extinction, and range shifts in shaping global diversity patterns. He also studies hierarchical approaches to large-scale evolutionary processes and the evolutionary significance of extinction events. He is a past winner of the Charles Schuchert award of the Paleontological Society of America.

MANFRED LAUBICHLER
School of Life Sciences
Arizona State University

Manfred D. Laubichler is professor of theoretical biology and history of biology and affiliated professor of philosophy at the School of Life Sciences and Centers for Biology and Society and Social Dynamics and Complexity at Arizona State University. He is the coeditor of *From Embryology to Evo-Devo, Modeling Biology, Der Hochsitz des Wissens,* and an associate editor of *Endothelial Biomedicine.* He is also an associate editor of the *Journal of Experimental Zoology, Part B Molecular and Developmental Evolution* and of *Biological Theory.*

ARNOLD MILLER
Department of Geology
University of Cincinnati

Arnold Miller is professor of geology at the University of Cincinnati, where his research focuses on the study of global biodiversity trends. He is author, with Michael Foote, of *Principles of Paleontology* (3rd edition), and he is currently writing a book about the significance of the Raup-Sepkoski-Bambach-Valentine 'Consensus Paper.'

KARL J. NIKLAS
Department of Plant Biology
Cornell University

Karl J. Niklas joined the Cornell faculty in 1978. He is the current Liberty Hyde Bailey Professor of Plant Biology at Cornell. His research focuses on biomechanics, allometry, and plant evolution, and he has published several books on those topics.

DAVID OLDROYD
School of History and Philosophy
University of New South Wales

David Oldroyd is an honorary visiting professor in the School of History and Philosophy at the University of New South Wales. He has written extensively on topics in the history of geology and is a recipient of awards from the Geological Society of London and the Geological Society of America for his historical work. He was Secretary-General of the International Commission on the History of Geological Sciences for eight years and is currently editor of *Earth Sciences History*.

PATRICIA PRINCEHOUSE
Department of Philosophy
Case Western Reserve University

Patricia Princehouse is lecturer in philosophy and evolutionary biology at Case Western Reserve University. Her research is on the history and philosophy of evolutionary biology, and she is active in promoting the teaching of evolution and combating Intelligent Design and other creationist theories.

DAVID M. RAUP
Department of Geophysical Sciences
University of Chicago

David M. Raup is professor emeritus of geophysical sciences at the University of Chicago. He is a member of the National Academy of Sciences and a past winner of the Paleontological Society of America Medal. He is author of *The Nemesis Affair* and *Extinction: Bad Genes or Bad Luck?*

MICHAEL RUSE
Department of Philosophy
Florida State University

Michael Ruse is professor of philosophy at Florida State University. He is the author of a number of books, including *Mystery of Mysteries: Is Evolution a Social Construction*, which discusses the science of some of the people who are the subjects of this collection.

J. WILLIAM SCHOPF
Department of Earth and Space Sciences
University of California, Los Angeles

J. William Schopf is professor of paleobiology and director of the IGPP Center for the Study of Evolution and the Origin of Life at UCLA. His interests focus on the origin and early evolution of living systems, spanning the Precambrian Eon of geological time.

DAVID SEPKOSKI
Department of History
University of North Carolina Wilmington

David Sepkoski is assistant professor of history at the University of North Carolina Wilmington. He has written extensively on the development of paleobiology and the history of evolutionary theory.

DEREK TURNER
Department of Philosophy
Connecticut College

Derek Turner is associate professor of philosophy at Connecticut College, where he is also affiliated with the Goodwin-Niering Center for Conservation Biology and Environmental Studies. He is the author of *Making Prehistory: Historical Science and the Scientific Realism Debate*, as well as several articles on philosophical issues in geology and paleobiology.

SUSAN TURNER
Monash University Geosciences

Susan Turner is an honorary research associate of Monash University School of Geosciences and former honorary research fellow at the Queensland Museum in Brisbane. She has worked in U.K. and Australian museums for forty years. Her historical work has particular emphasis on collectors and paleontologists, especially women, and also includes a recent history of the UNESCO-IUGS International Geosciences Programme (IGCP). She became a member of the International Commission on the History of Geological Sciences in 2004.

JAMES W. VALENTINE
Department of Integrative Biology
University of California

James W. Valentine is professor emeritus in the Department of Integrative Biology at the University of California, Berkeley. He is the recipient of numerous honors and awards in paleontology and is the author of several books, including *On the Origin of Phyla* (University of Chicago Press). He is a member of the National Academy of Sciences.

TIM WHITE
Department of Integrative Biology
University of California, Berkeley

Tim White is professor of integrative biology at the University of California at Berkeley, where he directs the Human Evolution Research Center. Specializing in hominid paleobiology, he has conducted fieldwork and laboratory studies throughout the world. He codirects the Middle Awash research project in the Afar rift of Ethiopia, compiling a six-million-year record of hominid origins and evolution.

INDEX

Page numbers in italics refer to photographs. Page numbers followed by f and t indicate figures and tables, respectively.

"Is a new and general theory of Evolution Emerging?" (Gould), 310

Ivany, L. C., 451

Jablonski, D.
  citations, 70, 218, 227, 232, 394, 411, 473, 476, 478, 479, 480, 483, 484, 485, 486, 487, 488, 489, 490, 491, 492, 495, 496
  Cope's rule, and studies of, 204
  "Paleontology in the twenty-first century," 471–517
  patterns of origination studies of, 70
  taxic paleobiology, and studies of, 218, 222

Jackson, J. B. C., 482, 488

Jackson, S. T., 483

Jaekel, O.
  biontology, and studies of, 281–82, 284–87
  citation, 285

James, N. P., 69

Janis, C. M., 231, 482

Jansson, R., 482

Janvier, P., 84

Janzen, D. H., 486

Jayasuriya, A. M. H., 227

Jeffery, J. E., 477

Jenkins, F. A., Jr., 477

Jenkins, R. J. F., 267–70, 273n6, 275n25

Jerison, H., 113

Jernvall, J., 478, 482, 493, 496

Ji Qiang, 116

Ji Shu'an, 116

Johanson, D., 136, 521

Johnson, A. L. A., 427

Johnson, D. W., 488

Johnson, R. G. J., 390, 401, 403, 406

Jolly, C. J., 130

Jones, C. G., 493

Jones, D. S., 488

Jones, K. E., 487

Jonsson, T., 493

Journal of Paleontology (journal), 406, 440

Junker, T., 280, 288

Jurassic Park (film), 117, 150, 204, 206, 208

Kahneman, D., 331

Kaiser, D., 340

Kaiser, H. E., 446, 447

Kappelman, J., 138–39

Katz, M. E., 482, 483

Kavanagh, K. D., 478, 482

Kear, A. J., 79

Kemp, A., 83, 84

Kemp, M., 479

Kidwell, S. M., 472, 473, 476, 484, 488

Kienast, F., 476

Kiessling, W., 485

Kiflawi, M., 494

Kimura, M., 158, 188, 330

King, A. A., 481

Kingsolver, J. G., 210, 487

Kinnison, M. T., 487

"Kiss and make up paper." *See* Consensus paper (Sepkoski, Bambach, Raup, and Valentine)

Klopfer, P. M., 387

Kluge, Arnold G., "Hierarchical Linear Modeling . . .", 308

Knappertsbusch, M. W., 488

Knight, C. R., 242, *242*

Knight, J. Brookes, 26–27

Knoll, A. H., 276, 276n27, 413, 483, 484, 521

Koh, L., 493

Kolm, N., 494

Konstruktionsmorphologie (Seilacher), 154, 158

Kowalewski, M., 228, 374, 473, 476

Krejsa, Dick, 83

Krijsa, D., 83

Krug, A. Z., 495

K/T (Cretaceous-Tertiary) boundary, 248–50, 410